Discovering Calculus

Jared M. Maruskin

Discovering Calculus

Cayenne Canyon Press

Jared M. Maruskin, Ph.D.

Published by Cayenne Canyon Press
San José, California

ISBN: 978-1-941043-14-1

10 9 8 7 6 5 4 3 2 1

Contents

Forward: To the Students

On Problem Solving

Calculus will challenge you. Most likely, your mathematical history to date has been a long string of similar problems, taken dozens at a time, each of which had a definitive, repeatable series of steps you could apply to render a solution. Calculus—the university experience in general—invites you to open a new door, through which you will find a multitude of complex problems, each requiring its own special care and ingenuity. Gone are the days of blindly memorizing a series of steps that allow you to solve only problems you have done one hundred times before. (There will be a few exceptions, such as learning the techniques of differentiation, which will require drill and practice.) The emphasis now is on the array of seemingly diverse problems that are unique in themselves and unlike any problem previously encountered. Calculus will require you to become an expert at diving into the unknown.

It is important to signal such a change in advance, lest you give up on every problem after five minutes with a shrug of the shoulders and an excuse, "I don't know how to solve this problem," or "Nobody showed me how to do this one." That's precisely the point: each problem will challenge you in a unique way, and you will seldom be able to solve them in under five minutes. Even though these problems may seem unrelated or unfamiliar at the surface level, there is a common skill set they all have in common—the ability to solve problems. This cabinet of critical-thinking skills is universal and transcends mathematics itself, being applicable to a broad spectrum of experiences in general. Here, it is almost always quixotic to search for that one formula that solves a given problem at hand. There is no such formula; you must come up with your own formula for each unique situation. Here are some tips that will help along the way:

1. **Real the Problem.** As obvious as it may seem, a common pitfall for novice university students is that they do not read the problem before going about trying to solve it. Read the problem closely. Read it twice. Write down in your own notation any information you are given. Write down what it is you are trying to solve.
2. **Draw a Picture.** If the problem loans itself to a visual rendering, draw a quick sketch of what is going on. This might be a literal picture (a boat being pulled toward a boat dock, with distances and rates of change properly la-

beled) or it might be an abstract one (an abstract box representing a changing quantity with input and output arrows symbolizing flows or change).

3. **Pay Attention to Units.** Many physical problems come with a set of units (m, kg, kW, V, etc.). Pay attention to the units. These are not peripheral distractions laced to trip you up; they are clues. On occasion, a physical principle can be understood or a problem can be solved by understanding the units alone.
4. **Identify the Equations.** What equations are in play? What is the problem asking for? What information were you given? What information are you trying to uncover? How can you form a logical bridge that takes you from what you are given to what you are trying to determine?
5. **Work the Equations—Carefully.** Do the algebra. Sometimes the logical bridge will have several steps: with the given information, you can use an equation to deduce an intermediary piece of information, which may be used to deduce new information, until the bridge takes you to the desired outcome.
6. **Perform a Reality Check.** A reality check is a quick mental computation done after you obtain your answer. Here, you are asking the question, "Does my answer make sense?" If you are trying to determine how much fuel a truck consumes while driving 30 minutes on the highway, 1,267 gallons is probably not the correct answer. (If it is, consider buying a new truck.) If you are computing how long it takes a rock to fall 30 feet, 89 seconds is probably too long. (When you drop a pebble from a 30 ft bridge, does it take a minute and a half before it hits the ground?) These are not flippant examples: real students actually came up with these numbers in formal homework write-ups as their final solution. A quick reality check ("Would a truck *really* consume 1,267 gallons of gasoline in half an hour?" "Would it *really* take a pebble over a minute to fall three stories?") could have tipped them off that something was awry.

Challenging problems such as the ones I describe here should be anticipated with excitement. They are the spice that makes this adventure exciting. Moreover, this process will cause you to grow and strengthen as a problem solver and as an individual. Even though the growth may *seem* slower—because you will spend your time constantly grappling with problems you *don't* know how to solve—in the end, the knowledge and the understanding will be deeper and you will have skills that will apply to a much broader range of disparate situations. The skills you will develop here are precisely the skills that will ultimately make you valuable to an employer. Nobody cares whether you can solve a canned problem that everybody already knows the solution to. When you graduate and get your first job, you will be faced with problems that *nobody* knows the answer to. (Imagine how well it would fly if you were to tell your boss that you can't solve a problem because he hasn't showed you how to solve that problem. If they already know how to solve all of their problems, what is the point of hiring you?) Employers want to hire people who are smart, accountable, and know how to get things done. Aside from learning things about calculus, which some of you may or may not ever use, it is the transcendent quality of being able to attack and dive into unknown and complex situations calmly and resolutely, and the ability to come out having made something out of the experience and have something worthwhile to say about it, that will make you valuable in your careers and in life.

For more advice on problem solving, please see my book *The Elements of Problem Solving*, available in hardcover and paperback.

On Practice and Perseverance

Mathematics is not a spectator sport. It will not suffice to read a number of examples and see examples worked out for you, expecting that you will be able to turn around and reproduce those skills on an exam. (Does a football team spend their practice time watching football on television, and then show up to the big game expecting that they will be able to perform? Or do they practice running the drills on the field over and over again?) Problem solving is a skill; as such, it must be acquired through repeated practice. Spend your time trying to solve as many problems as possible. Come up with your own problems and follow your intellectual curiosity. It is okay to spend hours or days working on a single problem: when problems capture your interest, follow them and try to learn everything you can from them.

Make math a habit. Create a daily ritual. If you reserve some time each day for mathematics—reading, skimming ahead, working problems, solving homework problems—it will be easier to get back into the zone each day. Stay on top of your game. Studying must occur throughout the course to allow for sufficient time to develop the skills you will need to be successful; these skills cannot be expected to be picked up during last-minute studying binges. You may start with a daily goal of one hour of math per day, but at some point you will realize that you have become completely absorbed in the material to the point where you loose track of time, completely enamored in a good problem. This should be celebrated! When you become so enraptured by a problem that you loose track of time or the outside world, that is when you know that you are an excellent problem solver: you are spending your efforts lost in the problem, and are not worrying about anything else. You are learning for the love of what you are learning.

No time spent thinking about the material is ever wasted. It will only be at the end of the course, when you look back, that you will be able to notice how far you really came. As you progress, change may seem slow, but it will happen. The best medicine here is to simply persevere. Everyone acquires skills at their own pace. Regardless your level, however, it is the countless hours you spend focusing on the material and on problems that will gradually make you a better problem solver. Have faith that this change will come, and spend your time getting into the material. Feeling stressed out or overwhelmed, on the other hand, does not bring you closer to becoming a better problem solver; when you feel these things, identify the emotions and then choose to focus on the material. Make the time you spend learning fun and enjoyable.

Keep these principles close at hand as you progress on your adventure through calculus. For additional advice and tips, please check out my book, *The Elements of Problem Solving*. It is a short, enjoyable read that can be consumed in one or two sittings.

San José, California
May 2014

Jared M. Maruskin

Chapter 1

Precalculus Review

1.1 Functions

We begin our story with a discussion of *functions*. The concept of a function plays an important role in calculus, as *the function* is the mathematical object that calculus seeks to understand and analyze. In this section, we define the meaning of a function, explore functional notation, and describe four fundamental ways in which functions may be represented: equations, graphs, data, and words.

1.1.1 Functions and Functional Notation

A function is essentially a rule that prescribes a *unique* output to every input. (Pay attention to this word "unique." We will explore precisely what we mean by it in short order.)

> **Definition 1.1.1** A *function* is a rule that assigns to each element in a specified set a unique output.
>
> The set of inputs (i.e., the specified set) is called the *domain* of the given function, and the set of all achieved outputs is the *range* of the function.

The concept of a function is illustrated in Figure 1.1. An input from the domain enters the function, the function operates on this input by applying a certain rule, and a unique output—unique to the input—is generated.

Example 1.1 Consider a classroom of 27 students. Is the birthdate a function of the student? Is the student a function of the birthdate?

First, let us consider whether the birthdate is a function of the student. In other words, given a student, is there a uniquely prescribed birthdate that is associated with him or her? The answer is yes: each student has a unique birthdate. (Assuming that each student was only born once.) Thus, the "birthdate function" can take any student as an input, and output his or her birthdate. Only one birthdate will be outputted for each student. Note that, for the first question asked, it is irrelevant

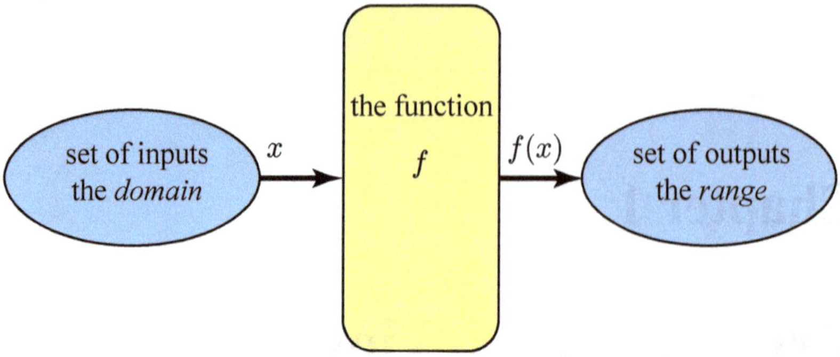

Figure 1.1: Schematic of a function.

whether or not two or more students share a given birthdate (maybe two of them were both born on March 1); what matters is that there is only one birthdate—only one output—assigned to each student.

Second, let us consider whether the student is a function of the birthdate. Now, the answer to this question depends on the particular group of 27 students, so we will have to consider two cases.

Case 1: Each student has a different birthdate, i.e., no two students share a common birthdate. In this case, the student is a function of the birthdate, because, given a birthdate, you may uniquely trace it back to the student. In other words, there is no ambiguity in determining which student has a given birthdate; each birthdate outputs a single, unique student. Thus, the student is a function of the birthdate.

Case 2: There is at least one shared birthdate, i.e., at least two students share a common birthdate. In this case, the student is not a function of the birthdate, because, for the particular, shared birthdate, you cannot assign a *unique* student output. Since two students correspond to that same birthdate, the output naming the student cannot be unique. Thus, the student is not a function of the birthdate.

■

The birthdate example is a favorite of mine in remembering which way functions work: the birthdate is a function of the person, i.e., each person is assigned a unique birthdate. Similarly, a function assigns a unique output to each of its inputs. So, given an input, there can only be a single, unique output that corresponds to that input.

(R) As the birthdate example further shows, sometimes functions can be reversed: sometimes there is a *one-to-one* correspondence between input and output. Not only does each input correspond to a unique output, but also, in some cases, the output corresponds to a unique input. In the birthdate example, this occurs when no two students share a common birthdate: there is a one-to-one correspondence between student and birthdate; the birthdate is a function of the student, *and* the student is a function of the birthdate. In such a situation, we say that the function is *invertible*, and we call the reverse function that sends each output back to its unique input the *inverse function*. We'll discuss this in greater detail in §1.5.

Functional Notation

The efficiency of mathematics relies on good notation. Notation is used to describe complex thoughts and relations in simple ways. It is therefore crucial to gain a thorough mastery of translating notation to its meaning and vice versa. No exception inheres here; a clean and precise use of functional notation will be our wings on this journey through calculus.

We use letters and symbols to represent the names of functions, in a similar way that letters and symbols may represent variables. (Sometimes they are used *both* ways: for example, x may represent the position of a particle, but this position may *also* be a function of time. Then we say that $x = x(t)$.) It is conventional to name functions f, g, and h, as they arise, in that order, when no other name is preferred. Thus, the most common name for a function is f.

Next, we use variables to refer to the inputs and the outputs. The most common instance of this is using the variable x to represent the input and using the variable y to represent the output. If the name of the function is f, we then would say: let f be a function of x. This tells the reader that f is the name of the function and that x is the name of the input variable. The *output* of the function is then referred to as $f(x)$, pronounced "f of x." Since we are using y to represent the output variable, we would then write

$$y = f(x).$$

This is read: "y is equal to f of x." Here, f is the name of the function, and $f(x)$ represents the *output* that is assigned to the input x. In terms of our birthdate example, suppose that f represents the birthdate as a function of the student, and suppose that Samuel was born on April 13[1]. Then, we may write

$$f(\text{Samuel}) = \text{April } 13.$$

Here, "Samuel" is the input, f is the name of the function, and f(Samuel) represents the output of the function, i.e., f(Samuel) represents the birthdate assigned to Samuel.

1.1.2 Four Ways of Expressing a Function

There are four main devices that people may use to describe a given function: an equation, a graph, data, and words. Many times, however, multiple devices may be used to represent the same function. For example, a function may at first be

[1]Bonus: What is Samuel's last name? He was a famous novelist.

described using words, and then one may interpret those words as an equation that represents the given function equally well.

An Equation

The most common way of describing a function, and perhaps the most familiar, is by using an equation. The rule that assigns a unique output to every input is, in this case, specified by an equation. For example, consider the function

$$f(x) = 3x^2 + 5x + 7. \tag{1.1}$$

In order to determine the output associated with a given input, we simply substitute that input in the place of the input variable x. Here are some examples:

$$\begin{aligned} f(0) &= 3 \cdot 0^2 + 5 \cdot 0 + 7 = 7, \\ f(1) &= 3 \cdot 1^2 + 5 \cdot 1 + 7 = 15, \\ f(2) &= 3 \cdot 2^2 + 5 \cdot 2 + 7 = 29, \\ f(s) &= 3s^2 + 5s + 7, \\ f(t) &= 3t^2 + 5t + 7, \\ f(x+h) &= 3(x+h)^2 + 5(x+h) + 7. \end{aligned}$$

If desired, this last equation can be expanded to read

$$f(x+h) = 3x^2 + 6xh + 3h^2 + 5x + 5h + 7.$$

Note that the function, as defined by equation (1.1), is simply the *rule* that assigns the output to the input, i.e., the function itself does not care what name is used as the input variable. Thus, the equations

$$f(x) = 3x^2 + 5x + 7 \qquad \text{and} \qquad f(u) = 3u^2 + 5u + 7$$

represent the same function f equally well.

Second, note that this rule works on the input in the same way, regardless of what the input is. That is, even in the case in which the input was the sum $x+h$, the variable x in equation (1.1) is simply replaced with $x+h$, the same way as one would replace the variable x with the number 2 if one wished to compute $f(2)$. Note, in particular, that $f(x+h)$ is *not* identical to

$$f(x) + h = 3x^2 + 5x + 7 + h.$$

These observations will be crucial to appreciate at several places in our subsequent discussions.

A Graph

The second most common way of specifying a given function is by using a graph. Graphs are convenient, visual representations of functions. The precise definition is as follows.

Definition 1.1.2 The *graph* of a function f is the complete set of points (x,y) that satisfy the relationship $y = f(x)$.

In other words, a point (x,y) on the plane is included in the graph if and only if $y = f(x)$. This set of points is then, typically, drawn on the plane to provide a visual representation of the given function.

(R) The phrase *if and only if* is used when each side implies the other. It may also be represented with the symbol "$\Longleftrightarrow$." For instance, if A and B are statements, each one either true or false, one may write

$$A \Longrightarrow B, \qquad B \Longrightarrow A, \qquad \text{or} \qquad A \Longleftrightarrow B.$$

1. $A \Longrightarrow B$ means "if A, then B," or "A implies B." Here, the truth of statement A implies the truth of statement B, but not necessarily the other way around. (E.g., if my car is out of gas, then it is not running. However, the *converse* statement, "if my car is not running, then it is out of gas," is not true.)
2. $B \Longrightarrow A$ means that if B is true, then A must be true as well. (But, as before, if B is not true, no information is given about A.)
3. $A \Longleftrightarrow B$ means that "A if and only if B," i.e., A and B are logically equivalent, i.e., the truth of A implies the truth of B *and vice versa*, i.e., A and B are either *both true* or *both false*.

An example of a graph is depicted in Figure 1.2. Notice that for each x-value, there is a unique y-value, which represents the output $y = f(x)$ of the given x value. To determine the output corresponding to a particular input x, one may do the following: first locate the given x-value, and then follow a vertical line to the point on the graph with the same x-value. Next, follow a horizontal line to the y-axis in order to see which y-value is assigned to the given x-value. For the function shown in Figure 1.2, for example, the output corresponding to $x = 3$ is given by $f(3) = 2$.

Since each x-value may only result in a single, unique output, we have the following result.

Proposition 1.1.1 — Vertical Line Test. The graph of a function may only intersect any vertical line once.

The reason for this is clear: if a graph intersects any one vertical line two (or more) times, then there would be two (or more) outputs corresponding to the particular x-value that defines that vertical line. (Recall that a vertical line is defined by an x-value, i.e., the equation $x = a$ represents the vertical line consisting of all the points whose x-value equals a.)

Also, note that the point $(3,2)$ is a point on the graph, which implies that $2 = f(3)$, from the definition of a graph.

Data

The third way in which a function may be represented is by using data. Data are typically tabulated, appearing in a table. The data points correspond to points on

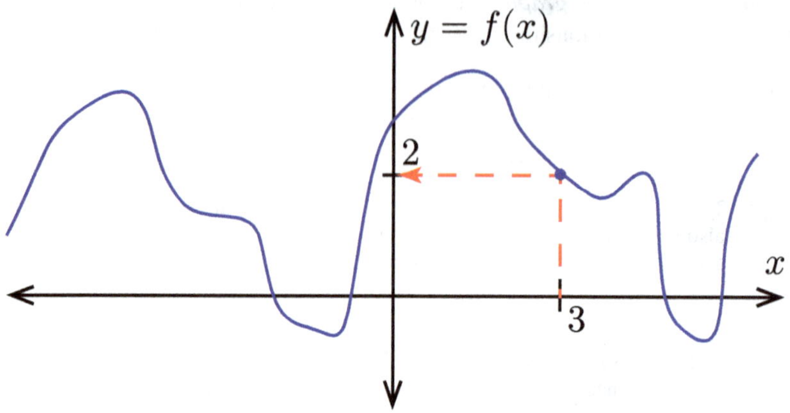

Figure 1.2: The graph of a given function.

the graph, i.e., a given data point (x_i, y_i) means that y_i is the output corresponding to x_i, that is $y_i = f(x_i)$. Here, the subscript refers to *which* data point we are referring to, the point (x_i, y_i) represents the ith data point.

■ **Example 1.2** The population of the United States is given as a function of time in Table 1.1. Each year corresponds to a unique population size, therefore the population is a function of the year.

year	1900	1910	1920	1930	1940	1950
population	76.0	92.0	105.7	122.8	131.7	150.7
year	1960	1970	1980	1990	2000	2010
population	179.3	203.2	226.5	248.7	281.4	308.7

Table 1.1: U.S. Population (in millions) by year. Source: U.S. Census Bureau, decennial census of population, 1900–2010.

Let $P(t)$ represent the U.S. population, measured in millions of people, t years after January 1, 1900. The data in the table do not give a complete picture of the function P, yet the general trends of population growth can be inferred.

The point $(60, 179.3)$ represents a point on the graph of P, therefore $P(60) = 179.3$. In practical terms, this means that the U.S. population was 179.3 million people in 1960.

We can further use the table to *interpolate* values of the population for years that are not listed. (To *interpolate* means to treat the function as *linear* between two given data points, as a means of inferring an intermediate value.) For instance, suppose we wish to approximate the population size in 1973. We begin by noting that

$$P(70) = 203.2 \qquad \text{and} \qquad P(80) = 226.5.$$

In order to interpolate the value of the population in 1973, we compute the following

$$P(73) \approx P(70) + 30\% \times (P(80) - P(70)) = 203.2 + 0.3(226.5 - 203.2) = 210.19.$$

We therefore infer that the population size of the U.S. was approximately 210.2 million people in 1973. This approximation assumes that the population growth may be approximated as being linear during the time period of the 1970's. (Of course, in general, the function might be changing wildly on any intermediate interval, which would result in an inaccurate approximation. Nevertheless, interpolation is a commonly used tool, especially in engineering, where the values of complicated functions are often tabulated for ease of use, such as in thermodynamic tables.) ■

Words

Finally, functions are sometimes specified using *words*. It is an important skill to be able to take the description of a function, in plain, ordinary English, and translate it into an equation, and vice versa.

■ **Example 1.3** A family's monthly electric bill consists of a fixed charge of \$4.42, including fees and taxes, and a charge of \$0.11 per kilowatt-hour (kWh) of electricity consumed during the billing cycle.

Here, the amount of the electric bill is a function of the amount of electricity used during the billing cycle. Let $f(x)$ represent the amount of the electric bill, measured in dollars, as a function of the amount of energy consumed, measured in kilowatt-hours. This is a true function, as each amount of electricity that can be used uniquely determines the total amount of the electric bill. In fact, this function can be translated into a simple equation:

$$f(x) = 4.42 + 0.11x.$$

The total cost of energy equals a \$4.42 fixed cost plus 11 cents times the number of kilowatt-hours consumed. ■

1.1.3 Properties of Functions

Finally, we explore some properties that functions may possess on their entire domain, or on different intervals of their domain. These properties are useful tools for describing the shape and behavior of certain functions.

Increasing and Decreasing Functions

Definition 1.1.3 A function f is

1. *increasing* if $f(x_1) \leq f(x_2)$, for every $x_1 < x_2$;
2. *decreasing* if $f(x_1) \geq f(x_2)$, for every $x_1 < x_2$;

3. *strictly increasing* if $f(x_1) < f(x_2)$, for every $x_1 < x_2$;
4. *strictly decreasing* if $f(x_1) > f(x_2)$, for every $x_1 < x_2$.

If any of these properties hold for every $x_1, x_2 \in I$, for some interval I, then we say that the function is [*insert name of property*] on the interval I.

If no reference to a particular interval is made, we assume that the properties hold on the entire domain.

If a function is *monotonic* if it is only increasing or only decreasing for all points on its domain.

Example function with these four different properties are shown in Figure 1.3. Notice that if a function is only increasing, it is allowed to level off for a while, whereas a strictly increasing function must continually maintain its climb. Similarly for decreasing and strictly decreasing functions.

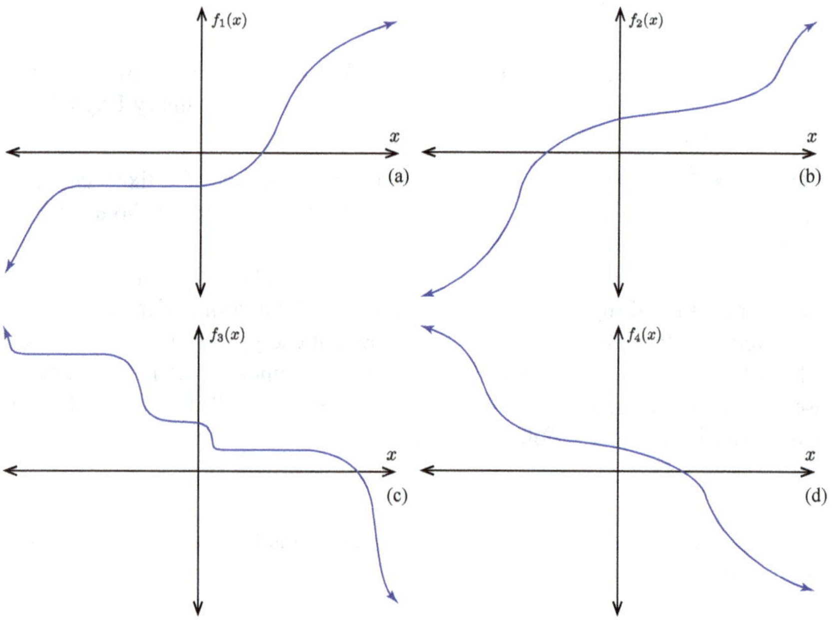

Figure 1.3: An example of an (a) increasing, (b) strictly increasing, (c) decreasing, and (d) strictly decreasing function.

Also, note that if a function is strictly increasing, then it is also increasing. Though, the converse is not necessarily true (that is, a function can be increasing without being strictly increasing). Thus, *strictly increasing* is a stronger condition than just plain *increasing*.

Tangent Lines and Concavity

The next property we will explore is called *concavity*. First, however, we introduce an important concept: the tangent line to a graph of a function at a point.

Definition 1.1.4 Consider a function f and a point $p = (a, f(a))$ on the graph of f. If the graph of f eventually looks like a straight line if you zoom in close enough to the point p, then this straight line, if continued outward, as a straight line, indefinitely, is the *tangent line* to the graph of f at the point p.

A function f is plotted concurrently with its tangent line T_a to a point $(a, f(a))$ in Figure 1.4. Notice that the tangent line to the graph at this point is simply the line that just touches the graph—tangentially—at that point. If one were to zoom in close enough to the point $(a, f(a))$, it would be nearly impossible to tell the difference between the tangent line and the actual graph.

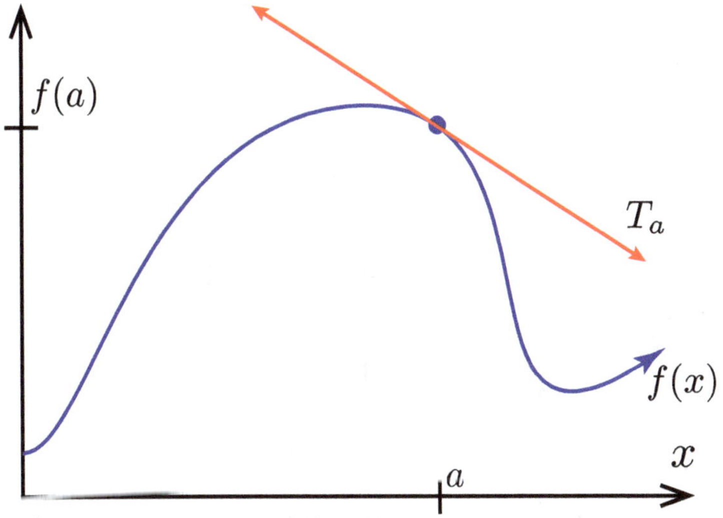

Figure 1.4: The graph of a function f, plotted concurrently with its tangent line T_a to the graph at the point $(a, f(a))$.

Finally, we introduce a property known as *concavity*. For now, we will base our description on the graph's relation to its tangent lines. In Section 2.3, however, we will come back and redefine concavity using the tools we develop concerning its instantaneous rate of change.

Definition 1.1.5 A function f is *concave up* on an interval I if the graph of f

lies above any of its tangent lines to any of the points on the interval I, i.e., if

$$T_a(x) \leq f(x),$$

for all $x \in I$ and for all $a \in I$, where T_a is the function that represents the tangent line to the graph of f at the point $x = a$.

Similarly, a function f is *concave down* on an interval I if the graph of f lies below any of its tangent lines to any of the points on the interval I, i.e., if

$$T_a(x) \geq f(x),$$

for all $x \in I$ and for all $a \in I$, where T_a is defined as above.

For example, consider the function shown in Figure 1.5. This function is concave up on the interval $[0, b]$ and concave down on the interval $[b, d]$.

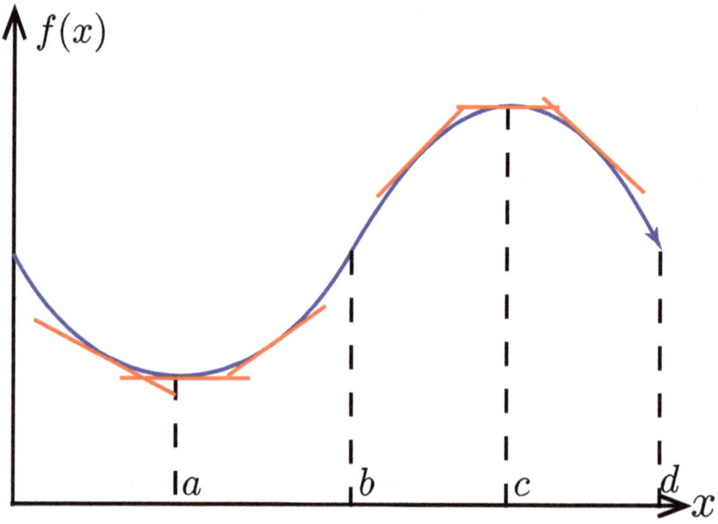

Figure 1.5: The graph of a function f, plotted concurrently with several of its tangent lines.

Also, we notice from Figure 1.5, that a function that is concave up may still be either increasing or decreasing. For instance, the function that is plotted is

- decreasing, concave up, on the interval $[0, a]$;
- increasing, concave up, on the interval $[a, b]$;
- increasing, concave down, on the interval $[b, c]$;
- decreasing, concave down, on the interval $[c, d]$.

Definition 1.1.6 A point at which a function changes concavity is called an *inflection point*. That is, the point $x = a$ is an inflection point for a function f if there exists some number $\varepsilon > 0$ such that f has a different concavity on the intervals $[a-\varepsilon, a]$ and $[a, a+\varepsilon]$.

For instance, the function shown in Figure 1.5 has an inflection point at $x = b$. For $x \in [0, b]$, the function is concave up, and for $x \in [b, d]$, the function is concave down. The function f therefore switches its concavity at the point $x = b$.

An equivalent way of qualifying whether or not a point is an inflection point is given in the following proposition.

Proposition 1.1.2 Consider a function f defined on an open interval containing the point $x = a$, and let T_a be the function that represents the tangent line to the graph of f at the point a. Then the point a is an inflection point for the function f if there exists a number $\varepsilon > 0$, such that either

- $f(x) < T_a(x)$, for $x \in (a-\varepsilon, a)$ and $f(x) > T_a(x)$, for $x \in (a, a+\varepsilon)$; or
- $f(x) > T_a(x)$, for $x \in (a-\varepsilon, a)$ and $f(x) < T_a(x)$, for $x \in (a, a+\varepsilon)$.

To illustrate this proposition, consider the function shown in Figure 1.6. This function has an inflection point at $x = a$. We observed that immediately to the left of the point $x = a$, the graph falls below the tangent line, whereas immediately to the right of the point $x = a$, the graph falls above the tangent line. This indicates a change of concavity at the point $x = a$.

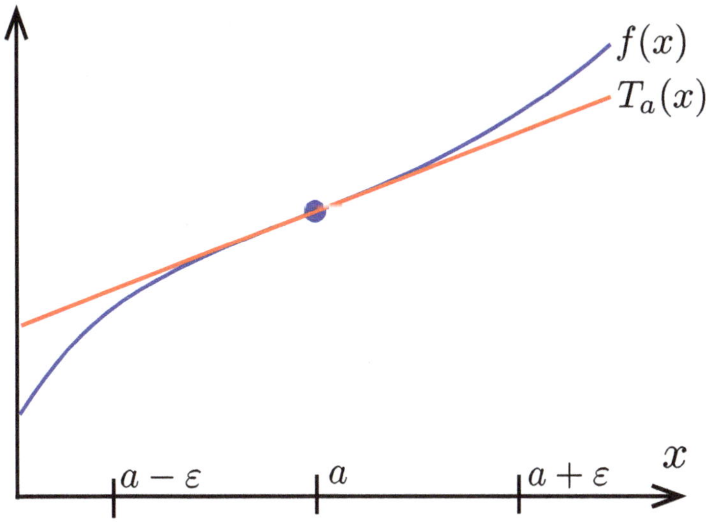

Figure 1.6: The graph of a function f, with an inflection point at $x = a$, plotted concurrently with the tangent line $T_a(x)$.

Exercises

For Exercises 1–5, identify whether a functional relationship exists between the pair of variables. For each, discuss whether the first is a function of the second, *and* whether the second is a function of the first. If the answer depends on any assumptions, state the conditions under which it is and is not a function. If you do any research, be sure to cite your sources. You may need to look up information and state the assumptions under which your conclusions are valid.

1. The name of a student in a classroom; the height of that student.
2. The depth below the surface; the salinity (salt concentration) of the sea water.
3. The air drag acting on an aircraft; the speed of the aircraft.
4. The temperature at which water boils; elevation above sea level.
5. The position of a car driving along a highway; the time.

For Exercises 6–10, approximate the population of the United States for the given year, using linear interpolation and the data from Table 1.1.

6. 1927.
7. 1985.
8. 1906.
9. 2008.
10. 1955.
11. Figure 1.7 shows four graphs of relationships between the x and y variables. For each graph, determine whether y is a function of x, and determine whether x is a function of y.
12. Consider the data given in Table 1.2. Determine whether x is a function of y, and whether y is a function of x.

x	0	1	2	3	4	5	6
y	3	5	7	9	11	13	15

Table 1.2: Data for Exercise 12.

13. Consider the data given in Table 1.3. Determine whether x is a function of y, and whether y is a function of x.

x	1	3	1	3	1	3	1
y	3	5	2	6	1	-4	13

Table 1.3: Data for Exercise 13.

14. Consider the locus[2] of points (x, y) such that $y = x^2$. Is y a function of x? Is x a function of y?

[2] a *locus* of points is simply the set of all points that satisfies the given requirements.

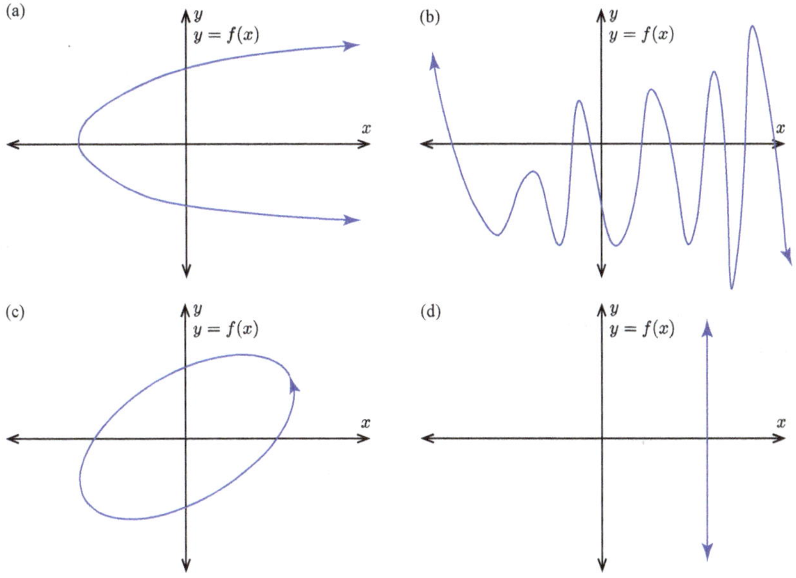

Figure 1.7: Four *x*-*y* relations for Exercise 11.

Problems

15. Consider the blood alcohol content chart for men as shown in Table 1.4[3].
    (a) For a fixed number of drinks, is the predicted blood alcohol content a function of the individual's bodyweight? Is the bodyweight a function of the blood alcohol content? (There is some discussion to be had here. State your assumptions. Think about whether the chart shows exact or approximate data.)
    (b) Is the blood alcohol content a function of the number of drinks *and* the bodyweight, i.e., is the blood alcohol content a function of the pair (no. drinks, bodyweight)? (This is an example of a *multivariable function*, i.e., a function for which the output is uniquely determined based on two or more input values.)
16. Mary randomly draws four cards from a standard deck of playing cards. Considering only the four cards that were drawn, is the suit (i.e., hearts, diamonds, clubs, and spades) a function of the face (i.e., 2, 3, 4, 5, 6, 7, 8, 9, Jack, Queen, King, Ace)? Is the face a function of the suit? The answer

[3] These data are intended for mathematical purposes only. Actual blood alcohol content may depend on various factors, including individual metabolism. Blood alcohol content levels for women are significantly higher than for men, for individuals with the same weight and same number of drinks. Always drink responsibly. Never drink and drive.

Drinks	Body weight (lbs)							
	100	120	140	160	180	200	220	240
0	.00	.00	.00	.00	.00	.00	.00	.00
1	.04	.03	.03	.02	.02	.02	.02	.02
2	.08	.06	.05	.05	.04	.04	.03	.03
3	.11	.09	.08	.07	.06	.06	.05	.05
4	.15	.12	.11	.09	.08	.08	.07	.06
5	.19	.16	.13	.12	.11	.09	.09	.08
6	.23	.19	.16	.14	.13	.11	.10	.09

Table 1.4: Blood alcohol content for men, for Problem 15. Please read problem footnote.

to these questions may depend on the particular four cards that were drawn. Discuss the conditions in each case under which there will or will not be a functional dependency.

1.2 Linear Functions

In this section, we explore the basic properties of linear functions. We also discuss piecewise-defined functions and, in particular, piecewise-linear functions.

1.2.1 Linear Functions and Their Properties

A linear function is a function that is always changing at the same rate. In other words, for a fixed change in the x-variable, the corresponding change in the y-variable will be automatically determined, regardless of where in the x-axis the change takes place. (In fact, the change in the y-variable will equal the *slope* times the change in the x-variable.)

Definition 1.2.1 A *linear function* is a function of the form

$$f(x) = mx + b. \tag{1.2}$$

The parameter m is called the *slope*, and the parameter b is called the *vertical intercept* (also, commonly, the *y-intercept*).

A linear function may be interpreted as follows: the graph of the function starts off, when $x = 0$, at the point $y = b$. Then, for each increment in the x-variable, the y-variable increments by m.

Proposition 1.2.1 Let (x_1, y_1) and (x_2, y_2) be two points on the graph of a linear function $f(x) = mx + b$, with $x_1 < x_2$. Then

$$m = \frac{y_2 - y_1}{x_2 - x_1}.$$

Moreover, this relation is true regardless of the points (x_1, y_1) and (x_2, y_2), as long as $x_1 < x_2$.

This proposition tells us that the slope of a linear function can be calculated by computing the ratio of the change in the y-variable divided by the change in the x-variable, over any interval on the function's domain. Moreover, this ratio is a constant, implying that the function is always changing at a constant rate. We prove the proposition as follows.

Proof. Let $f(x) = mx + b$ be a linear function, and consider two points (x_1, y_1) and (x_2, y_2) on its graph, with $x_1 < x_2$.

Since the points (x_1, y_1) and (x_2, y_2) are on the graph of f, it follows, from the definition of a graph, that $y_1 = f(x_1)$ and $y_2 = f(x_2)$. In particular, we have

$$\begin{aligned} y_1 &= mx_1 + b, \\ y_2 &= mx_2 + b. \end{aligned}$$

Next, let us use these relations to explicitly compute the right-hand side of equation (1.2). We obtain

$$\frac{y_2 - y_1}{x_2 - x_1} = \frac{(mx_2 + b) - (mx_1 + b)}{x_2 - x_1} = \frac{m(x_2 - x_1) + (b - b)}{x_2 - x_1} = m.$$

This proves the result. ■

The result of Proposition 1.2.1 is illustrated in Figure 1.8. The quantity $x_2 - x_1$ represents the total change in the x-variable, whereas the quantity $y_2 - y_1$ represents the total change in the y-variable. The ratio of the change in y to the change in x determines the slope of the linear function. Note that specifying the slope of a linear function is equivalent to specifying the angle of ascent, which, like the slope, is constant throughout.

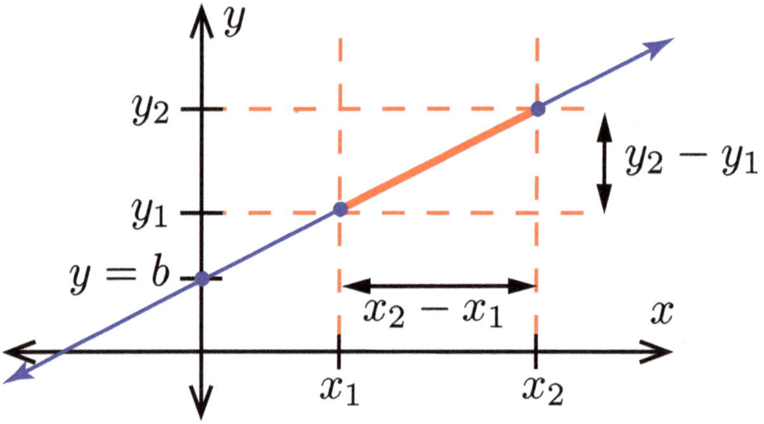

Figure 1.8: The slope of a linear function.

(R) If a function $f(x)$ is *not* linear, then one can, for a given interval $x_1 < x < x_2$, determine the quantity

$$\frac{f(x_2) - f(x_1)}{x_2 - x_1}.$$

This quantity, however, for a nonlinear function, will depend on the particular choice of x_1 and x_2. It represents the *average* rate of change of the function f over the given interval. A linear function, therefore, is a function that has a constant average rate of change, i.e., an average rate of change that is independent of the interval in question. We will explore this in greater depth in §2.1.

Next, we consider the following example.

x	0	1	2	3	4	5	6
$f(x)$	3	5	7	9	11	13	15

Table 1.5: Data for the function in Example 1.4.

■ **Example 1.4** Consider the data given in Table 1.5. These data describe a function $f(x)$. First, we suspect the function might be linear, because the change in y is consistent for each single change in x.

Assuming that these data accurately describe the function $f(x)$, i.e., assuming the function $f(x)$ is a linear function, we may therefore conclude that the vertical intercept is given by $b = f(0) = 3$ and that the slope is given by

$$m = \frac{f(1) - f(0)}{1 - 0} = \frac{5 - 3}{1 - 0} = 2.$$

The equation for this function is therefore $f(x) = 2x + 3$. ■

Example 1.4 illustrates a defining feature of linear functions: every time the independent variable x is incremented by 1, the dependent variable y is incremented by the same amount, an amount equal to the slope m. In other words, every time one adds 1 to the x-variable, one must *add* m to the previous y-value in order to obtain the new y-value. We will contrast this in §1.3 with exponential functions, for which one *multiplies* the previous y-value by a fixed constant in order to obtain the new y-value.

1.2.2 Piecewise-Defined Functions

As the name suggests, piecewise-defined functions are, quite literally, functions that are defined in pieces.

Definition 1.2.2 A *piecewise-defined function* is a function that uses a different rule to determine the output for different sections of the domain.

A *piecewise-linear function* is a piecewise-defined function such that the rule used for each section of the domain is of the form of equation (1.2).

A piecewise-defined function is typically written out using the format

$$f(x) = \begin{cases} f_1(x) & \text{for } x \in I_1, \\ f_2(x) & \text{for } x \in I_2, \\ \vdots & \vdots \\ f_n(x) & \text{for } x \in I_n \end{cases} . \tag{1.3}$$

Here, the functions $f_1, f_2, \cdots, f_n$ are the rules used to determine the output for the different pieces of the domain on which f is defined. Similarly, the intervals I_1, I_2, $\cdots$, I_n are the different intervals (or pieces of the domain) on which the function is defined. These intervals are typically back-to-back, forming a single, larger interval on which f is defined. The only requirement, however, is that the intervals $I_1, \cdots, I_n$ do not intersect.

In order to calculate the output of the function given by equation (1.3) for a given x-value input, one first identifies which piece of the domain the given x-value is located in, and then one uses the rule corresponding to that piece of the domain.

Example 1.5 Consider the piecewise-linear function defined by the equation

$$f(x) = \begin{cases} 1+2x & \text{for } x \in [0,2) \\ \dfrac{19}{3} - \dfrac{2}{3}x & \text{for } x \in [2,5) \\ \dfrac{-11}{3} + \dfrac{4}{3}x & \text{for } x \in [5,8] \end{cases}.$$

Compute $f(3)$ and plot the graph of the function f.

First, we identify $x = 3$ as being a part of the second section of the domain, i.e., $3 \in [2,5)$. Then, we use the local rule for the second section to compute the output:

$$f(3) = \frac{19}{3} - \frac{2}{3}(3) = \frac{13}{3}.$$

In order to graph the function f, we simply draw a graph of the given linear function on each of the three pieces of the domain. The result is shown in Figure 1.9.

■

1.2.3 Point-Slope Form of a Linear Function

A final useful tool for linear functions is as follows.

Proposition 1.2.2 — Point-Slope Form. Given a point (x_0, y_0) on the graph of a linear function f and the slope m of the linear function, the function may be written in the form

$$f(x) - y_0 = m(x - x_0). \tag{1.4}$$

Proof. This follows directly from Proposition 1.2.1 by identifying (x_1, y_1) with (x_0, y_0) (simply renaming the location of the given point) and by identifying (x_2, y_2) with $(x, f(x))$ (simply allowing the second point to be variable). The result follows by multiplying both sides of equation (1.2) by $(x - x_0)$. ■

Note that the point-slope form given by equation (1.4) is equivalent to writing

$$f(x) = y_0 + m(x - x_0). \tag{1.5}$$

This makes a great deal of sense: one starts at x_0, with a y-value of y_0. Then, one adds the slope times the change in the x-variable in order to obtain the change in the y-variable. Thus, $m(x - x_0)$ simply represents the change in the y-variable.

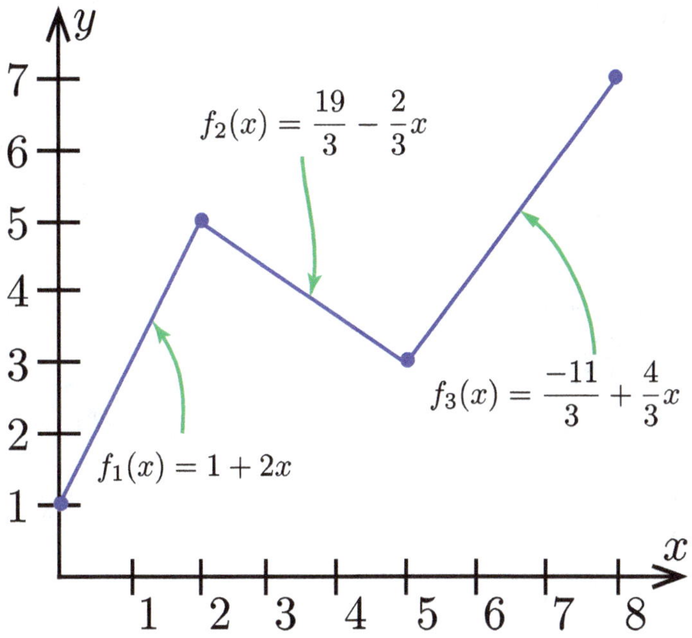

Figure 1.9: The piecewise-linear function of Example 1.5.

R Equation (1.5) is an important equation to understand. We will know it later, at the end of our discussion of integral calculus, as a *first-order Taylor polynomial* approximating f near the point x_0. The only difference, in that context, is that the slope m will be replaced with the *instantaneous* "slope" of a nonlinear function f at the point x_0. This instantaneous "slope" is called the derivative of f, as we will explore in Chapter 2.

1.2.4 Simple Interest

As an application to linear functions, we consider next the case of simple interest being applied to a loan or a savings account. First, some terminology.

Definition 1.2.3 The *principal* P_0 of a loan or a savings account is the initial amount of capital borrowed or invested, respectively. The *interest rate* r is the decimal expression of the annual interest rate.

Simple interest occurs when interest accrues on the principal balance, but not on any of the accumulated interest.

Proposition 1.2.3 The balance of a capital account (either a loan or savings account) $P(t)$, with an initial principal of P_0 and an annual interest rate of r, after t years, is given by

$$P(t) = P_0 + P_0 rt. \tag{1.6}$$

■ **Example 1.6** A small tech startup borrows \$500,000 at simple interest with an annual interest rate of 4.7% for a five-year period. Determine (a) the amount of interest that accrues each year, (b) the amount that the startup owes as a function of time, and (c) the total amount the startup will have to pay off when their loan becomes due in five years.

(a) Since the annual interest rate is 4.7%, and since it is simple interest (and is therefore only applied to the principal balance), the amount of interest that accumulates each year is given by

$$\text{annual interest charge} = 0.047 \times \$500,000 = \$23,500.$$

(b) As is evident by equation (1.6), the annual interest charge represents the slope of the amount due, thus

$$P(t) = \$500,000 + \$23,500t.$$

(c) After five years, the total amount due will be

$$P(t) = \$500,000 + \$23,500 \times 5 = \$617,500.$$

■

Exercises

For Exercises 1–5, determine the formula for a linear function that satisfies the indicated criteria. Draw a graph of the resulting function.

1. A slope of $-1/2$, and a vertical intercept of 4.
2. A slope of 3, passing through the point $(2,5)$.
3. A slope of $-1/3$, passing through the point $(2,5)$.
4. Passing through the points $(1,7)$ and $(7,1)$.
5. Horizontal intercept (x-intercept) at $x = 7$, and passing through the point $(11,13)$.

Problems

6. An airplane has an altitude of 26,000 ft at exactly 12:23pm, and an altitude of 20,000 ft at exactly 12:28pm. Assuming the airplane has a constant rate of descent, describe its altitude $h(t)$ as a linear function of the number of minutes t since 12:23pm. What is the airplane's rate of descent, in feet per second?

7. The temperature T, as measured in degrees Fahrenheit, is a linear function of the temperature x, as measured in degrees Celsius. Suppose that 0°C corresponds to 32°F, and that 100°C corresponds to 212°F. Determine $T(x)$.
8. Consider the 2013 single filer marginal tax rates for U.S. federal income tax, as shown in Table 1.6[4]. The total amount of income tax due is computed

Marginal Income	Marginal tax rate
\$0–\$8,925	10%
\$8,925–\$36,250	15%
\$36,250–\$87,850	25%
\$87,850–\$183,250	28%
\$183,250–\$398,350	33%
\$398,350–\$400,000	35%
\$400,000 and up	39.6%

Table 1.6: 2013 U.S. marginal tax rates for single filers, Problem 8.

as follows: the first \$8,925 an individual earns is taxed at 10%, the income earned after \$8,925 and below \$36,250 is taxed at 15%, and so forth. For example, if an individual has a taxable income of \$40,000 in 2013 (after deductions), he or she will pay

$$\begin{aligned} 10\% &\quad \text{of first } \$8,925, \text{ plus} \\ 15\% &\quad \text{of } (\$36,250 - \$8,925 = \$27,325), \text{ plus} \\ 25\% &\quad \text{of } (\$40,000 - \$36,250 = \$3,750). \end{aligned}$$

Thus, the total taxes due are

$$0.1(\$8,925) + 0.15(\$27,325) + 0.25(\$3,750) = \$5,928.75.$$

Let $T(x)$ represent the amount of federal income tax due as a function of an individual's total taxable income x.
   (a) Determine a formula for $T(x)$ as a piece-wise linear function of x. This can be achieved by constructing a linear function to represent the total tax liability for each of the seven intervals listed in Table 1.6.
   (b) Explain why the effective tax rate is given by $\frac{T(x)}{x}$. (In fact, an even more accurate representation of the true tax rate would be $\frac{T(x)}{x+d}$, where d is the amount (in dollars) of tax deductions claimed.)
9. In free fall, an object's velocity looses 32 feet per second each second, i.e., it accelerates at a rate of -32 ft/s^2. This is called acceleration due to gravity. (In metric, the acceleration is -9.8 m/s^2.) Here, positive velocity indicates upward motion, and negative velocity indicates downward motion. Suppose that a certain ball is shot upward, from ground level, at a speed of 80 ft/s.

[4]Taxes are computed based on one's adjusted gross income, which is computed by taking one's gross income and subtracting deductible expenses, such as home mortgage interest, state income tax, and property taxes.

(a) Determine a formula representing the ball's velocity as a linear function of time.
(b) When will the ball's velocity reach 0 ft/s? What part of the ball's trajectory does this correspond to physically?
(c) What is the total amount of time that the ball will be in the air, before impacting the ground again?

10. A certain company, *Backhanded Inventors, Corp.*, produces pingpong balls. There is a *fixed cost*[5] each year of $18,300. Each pingpong ball costs 7.2 cents to produce, and is sold at a wholesale rate of 56 cents per ball.
    (a) Represent the *cost* $C(x)$, measured in dollars, as a linear function of the number of pingpong balls x produced.
    (b) Represent the total *revenue* $R(x)$, measured in dollars, as a linear function of the number of pingpong balls x sold.
    (c) Assume that Backhanded Investors, Corp. has a contract in which a certain distributor, *Double Deuce Distributors, Inc.*, buys, at wholesale, all of the pingpong balls that Backhanded Investors, Corp. produce. (In other words, assume that the number of pingpong balls produced equals the number of pingpong balls that are sold.) Determine the *profit* $\Pi(x)$[6], measured in dollars, as a function of the number of balls produced/sold x.
    (d) How many pingpong balls must Backhanded Investors, Corp. produce in order to break even for the year?

11. A block with mass m is attached to a wall using a spring, with spring constant k. Let x represent the distance from the center of the block to the wall, as shown in Figure 1.10. Depending on the position of the block x, measured in centimeters, the spring will exert a certain force F on the block, measured in Newtons (N), known as the *restoring force*, since the spring wishes to be restored to its resting position. A positive force is directed in the positive x-direction (i.e., away from the wall), and a negative force is directed in the negative x-direction (i.e., toward the wall).
    (a) The restoring force (N) is given for two particular positions of the block (cm) in Table 11. Assuming that the restoring force depends linearly on the block's position, determine an equation for the restoring force $F(x)$ as a function of the block's position x.

x [cm]	3	11
$F(x)$ [N]	6.72	−2.88

Table 1.7: Restoring force of spring as a function of block's position for Problem 11.

[5]the *fixed cost* is the cost incurred before the first item can be produced, i.e., it represents the fixed cost of doing business, before any products can be produced. For example, a company may need to purchase or rent a factory, hire staff, pay business taxes, etc., before it begins manufacturing any goods. These costs are the fixed costs associated with managing or opening the business.

[6]Π is the capital Greek letter pi; the profit Π is given by revenue minus cost, i.e., $\Pi(x) = R(x) - C(x)$.

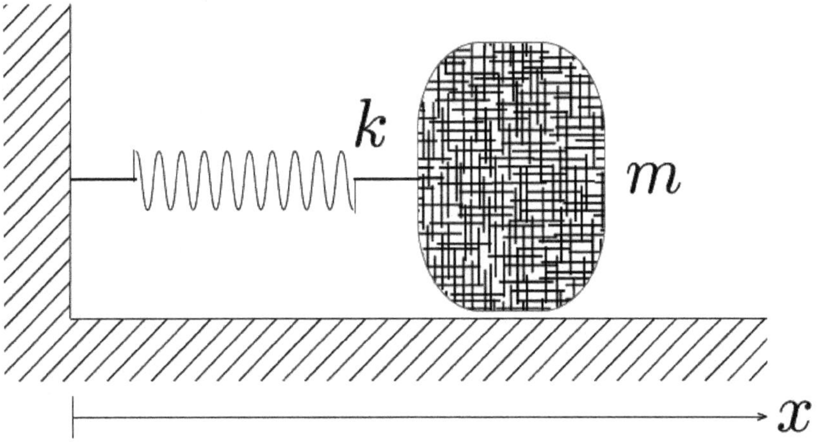

Figure 1.10: Block–spring system of Problem 11.

(b) Rewrite the formula for $F(x)$, which was obtained in part (a), in the form $F(x) = -k(x - x_0)$. What is k? What is x_0? Here, k represents the spring constant, and x_0 is the equilibrium position of the block, i.e., it is the resting positing at which the spring exerts no force on the block. Be sure to state the units of k and x_0.

(c) Explain what physical limitations may exist on our model of the restoring force. *Hint*: Is our function valid regardless of how far we pull the block from the wall, or is there a physical limitation on its accuracy?

1.3 Exponential Functions

In this section, we explore exponential growth and decay. We discuss two main forms for representing this class of functions, and we conclude with a derivation of Euler's number e.

.3.1 The Growth Factor

The growth factor (or decay factor, if the function is decreasing) of an exponential function is analogous to the slope of a linear function. Whereas the slope tells what number to *add* to the previous y-value in order to obtain the new y-value, as x is being incremented by 1, the growth factor tells what number to *multiply* to the previous y-value in order to obtain the new y-value, as x is being incremented by 1. In general, we can define the family of exponential functions as follows.

Definition 1.3.1 An *exponential function* is a function of the form

$$f(x) = ab^x, \tag{1.7}$$

for some constants $a \neq 0$ and $b > 0$, $b \neq 1$.

The parameter a is called the *initial condition* or the *vertical intercept.*

If $b > 1$, then the parameter b is called the *growth factor.* If $0 < b < 1$, then the parameter b is called the *decay factor.* (Though, sometimes, the term *growth factor* may be used in general for convenience.)

In our first example, we compare exponential and linear functions.

■ **Example 1.7** Consider the functions f and g, several of whose outputs are given in Table 1.8.

x	0	1	2	3	4	5	6
$f(x)$	3	5	7	9	11	13	15
$g(x)$	3	6	12	24	48	96	192

Table 1.8: Data for the functions in Example 1.7.

Assuming that the data accurately portray the functions, we can conclude that the function f is a linear function of the independent variable x, with slope $m = 2$ and vertical intercept $b = 3$. Also, we can conclude that the function g is an exponential function of the independent variable x, with growth factor $b = 2$ and initial condition $a = 3$.

The function f was already discussed in Example 1.4. For the function g, notice how the previous y-value is multiplied by a factor of 2 for each unit change in x. Further, since $g(0) = 3$, we conclude that the initial condition a has a value of $a = 3$. ■

In our next example, we will use a very important technique that will be used in many modeling problems to come.

■ **Example 1.8** Suppose that the points (2,3) and (5,21) are points on the graph of a certain exponential function f. Determine a formula for this function.

Since we know that the function f is an exponential function, we may write down the standard form for any exponential function, as given by equation (1.7): $f(x) = ab^x$.

Next, let us consider the data we are given. We know that (2,3) and (5,21) are points on the graph of f. Therefore, we know that

$$f(2) = 3 \qquad \text{and} \qquad f(5) = 21.$$

Given the general form of the exponential function, however, we also know that

$$f(2) = ab^2 \qquad \text{and} \qquad f(5) = ab^5.$$

Dividing, we therefore obtain

$$\frac{f(5)}{f(2)} = \frac{ab^5}{ab^2} = \frac{21}{3} = 7.$$

Thus, we have $b^3 = 7$, and, therefore,

$$b = 7^{1/3} \approx 1.91293118.$$

Now that we have identified the value for the parameter b, we can use the equation

$$f(2) = ab^2 = 3$$

to show that $a = 3b^{-2}$, or

$$a = \frac{3}{7^{2/3}} \approx 0.81982765.$$

Thus,

$$f(x) = \frac{3}{7^{2/3}} 7^{x/3} = 3 \cdot 7^{(x-2)/3}.$$

To check, we may compute $f(2) = 3 \cdot 7^0 = 3$ and $f(5) = 3 \cdot 7^{(5-2)/3} = 21$. ■

1.3.2 Half-life and Doubling Time

When a time-dependent quantity exhibits exponential growth or decay, one often assigns a certain charachterstic time to the quantity, as an alternative way of describing how quickly it is growing or decaying. In the case of exponential growth, this characteristic time is called the *doubling time*; in the case of exponential decay, it is called the *half-life*. These terms are defined precisely as follows.

Definition 1.3.2 Suppose the quantity Q is growing exponentially in time, i.e., $Q(t) = ab^t$, with $b > 1$. Then the *doubling time* is the length of time required for the quantity Q to double in size.

If the quantity Q is decaying exponentially in time, i.e., $Q(t) = ab^t$, with

$0 < b < 1$, then the *half-life* is the length of time required for the quantity Q to be reduced by a factor of one half.

(R) The doubling time (or half-life) of a quantity that is growing (or decaying) exponentially in time is independent of the initial time. That is, the quantity exhibiting exponential growth (or decay) will double (or halve) every time the doubling time (or half-life) is endured.

Proposition 1.3.1 Suppose that a quantity Q is growing exponentially in time, with initial value a and doubling time T. Then Q may be modeled by the equation

$$Q(t) = a2^{t/T}. \tag{1.8}$$

On the other hand, suppose that the quantity Q is decaying exponentially in time, with initial value a and half-life τ. Then Q may be modeled by the equation

$$Q(t) = a\left(\frac{1}{2}\right)^{t/\tau}. \tag{1.9}$$

Proof. First, let us suppose that Q is growing exponentially in time. Then Q may be written in the form of equation (1.7), where $b > 1$, that is, $Q(t) = ab^t$. We know that after a duration of T, the initial quantity $Q(0) = a$ will double to $Q(T) = 2a$. Thus,

$$Q(T) = ab^T = 2a,$$

or $b^T = 2$. Thus, $b = 2^{1/T}$, and

$$Q(t) = ab^t = a\left(2^{1/T}\right)^t = a2^{t/T},$$

as desired.

Similarly, if Q is decaying exponentially in time, we may still use equation (1.7) to write $Q(t) = ab^t$, except that now $0 < b < 1$. After a duration of τ, the initial quantity $Q(0) = a$ will be halved to $Q(\tau) = \frac{a}{2}$. Thus,

$$Q(\tau) = ab^\tau = \frac{a}{2},$$

or $b^\tau = 1/2$. Thus, $b = (1/2)^{1/\tau}$, and

$$Q(t) = ab^t = a\left((1/2)^{1/\tau}\right)^t = a\left(\frac{1}{2}\right)^{t/\tau},$$

as desired. ■

The concept of a doubling time is illustrated in Figure 1.11. A generic quantity $Q(t)$ that is growing exponentially in time is shown here; time is measured in multiples of the doubling time, and the output $Q(t)$ is measured in multiples of the initial value a.

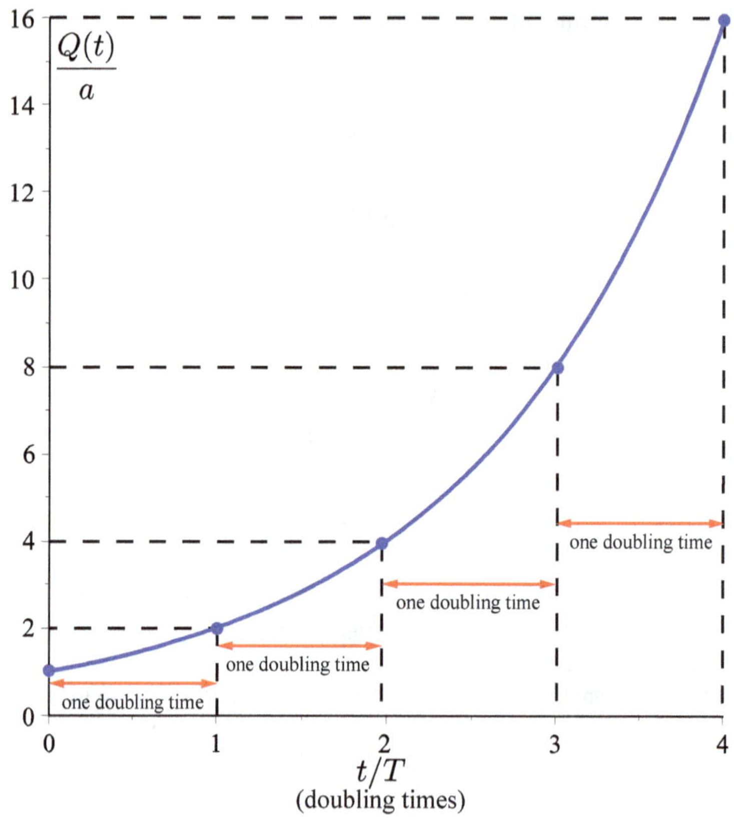

Figure 1.11: Graph of an exponentially growing quantity $Q(t)$, plotted against t (measured in doubling times).

One clearly sees that $Q(0) = a$, $Q(1) = 2$, $Q(2) = 4$, $Q(3) = 8$, and $Q(4) = 16$. Notice that for each doubling time (which has length 1), the quantity Q doubles. Even if one were to measure $Q(3.3)$ and $Q(4.3)$, one would find that $Q(4.3)$ is twice as large as $Q(3.3)$, and so forth.

A similar phenomenon is exhibited during exponential decay. That is, if we were to measure time in multiples of the half-life, we would find $Q(0) = a$, $Q(1) = \frac{1}{2}$, $Q(2) = \frac{1}{4}$, $Q(3) = \frac{1}{8}$, and $Q(4) = \frac{1}{16}$. For each half-life, the quantity Q is literally halved.

1.3.3 Continuous Growth Rate

Any exponential function may also be described in an alternate form, specifically using the base e. The number e is known as *Euler's number*[7], and has an approximate numerical value

$$e \approx 2.718281828459045.$$

Like the number π, Euler's number is an irrational number: the decimal expansion goes on forever without any pattern.

Proposition 1.3.2 An exponential function f can be written in the form

$$f(x) = ae^{rx}, \tag{1.10}$$

for some constants $a \neq 0$ and $r \neq 0$.

The parameter a is the same as used in Definition 1.3.1.

If $r > 0$, the parameter r is referred to as the *continuous growth rate*. If $r < 0$, the absolute value of the parameter r is referred to as the *continuous decay rate*. (Though, sometimes, the term *continuous growth rate* may be used in general for convenience.)

The proof of Proposition 1.3.2 is straightforward: one only need identify $b = e^r$. Consequently, $b > 1$ if and only if $r > 0$, and, similarly, $0 < b < 1$ if and only if $r < 0$.

The advantage of representing exponential functions in the form given by equation (1.10) will become fully apparent in §1.4 when we discuss the natural logarithm.

■ **Example 1.9** The most basic example of an exponential function is one with base e, that is, the exponential function $f(x) = e^x$. The graph of this function is shown in Figure 1.12. There are several key features of this function with which one should be familiar: first, the vertical intercept is one, i.e., the function $f(x) = e^x$ passes through the y-axis at the point $(0, 1)$. Second, for positive x-values, the function increases rapidly. (In fact, the rate at which the exponential function increases ultimately wins over a polynomial function of *any* degree.) Third, for larger and larger negative x-values, the exponential function rapidly approaches zero.

Another important feature to note about the function $f(x) = e^x$, is that its range is the open interval $(0, \infty)$. That is, the set of possible outputs include every positive number. ■

■ **Example 1.10** Another important function to understand graphically is $g(x) = e^{-x}$. First, let us compare this function with the function f given in Example 1.9. We note that a certain symmetry exists between the functions f and g. For instance,

[7] the "eu" in Euler is pronounced the same as the first syllable in "oil," or as in the "-oy" in "boy."

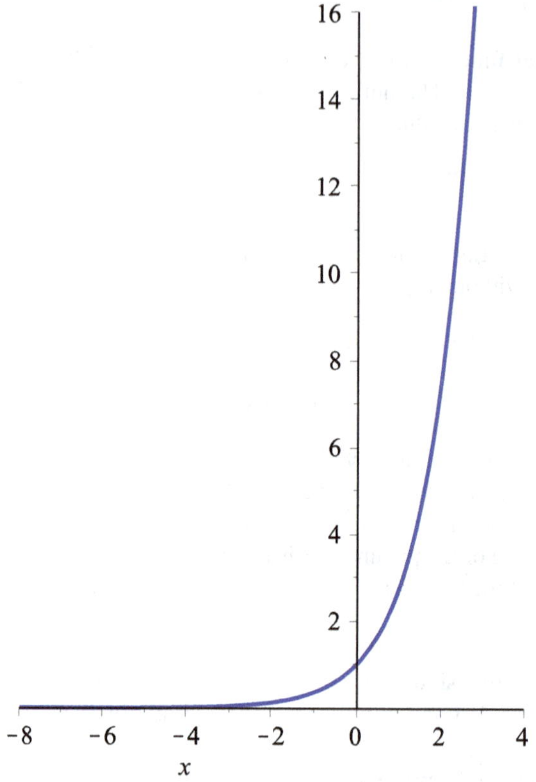

Figure 1.12: Graph of the exponential function $f(x) = e^x$.

consider the following:

$$f(1) = e^1, \quad \text{whereas} \quad g(-1) = e^{--1} = e^1;$$
$$f(2) = e^2, \quad \text{whereas} \quad g(-2) = e^{--2} = e^2.$$

This trend continues. In fact, $g(x) = e^{-x} = f(-x)$. We can understand the implication of this by plotting the functions f and g concurrently, as shown in Figure 1.13. We observe that the two functions are symmetric about the y-axis, i.e., the one is the mirror image of the other, and vice versa. The function $f(x) = e^x$ exhibits exponential growth as $x \to \infty$, whereas the function $g(x) = e^{-x}$ exhibits exponential decay as $x \to \infty$. ∎

1.3.4 Compound Interest

Compound interest occurs when interest is charged on *both* the principal *and* the accumulated interest.

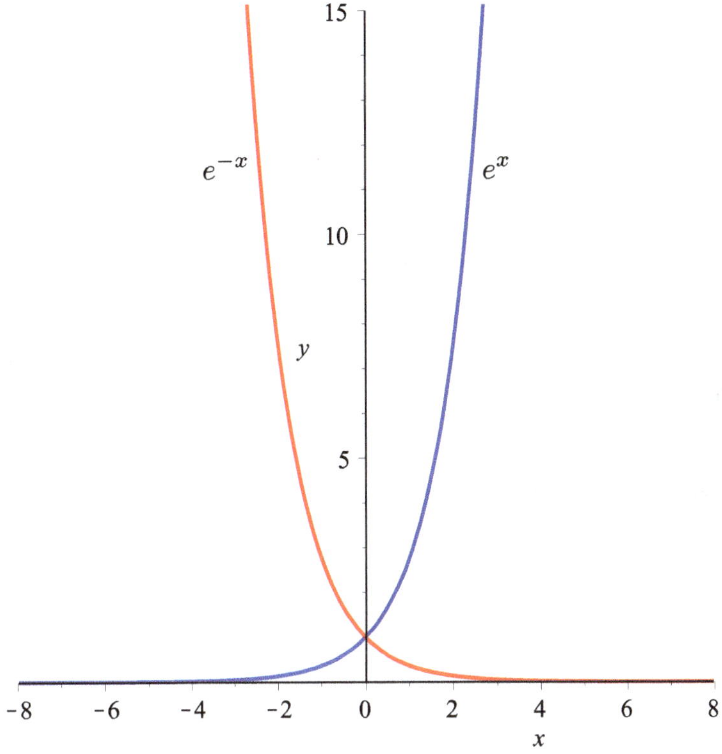

Figure 1.13: Concurrent graph of the exponential functions $f(x) = e^x$ (blue) and $g(x) = e^{-x}$ (red).

Definition 1.3.3 The *nominal interest rate* r is the annual rate at which interest charges are calculated, without taking into account any compounding of interest. The *compounding frequency* n is the number of times per year for which interest is compounded. The *effective interest rate* (or *annual percentage rate*) is the actual interest charge, as a percentage of the principal, that is accrued during the year, including the effects of compounding interest.

To investigate how to compute the balance of a capital account for which interest is compounded with a frequency n, let us consider the following. Suppose one deposits an amount P_0 into a savings account that advertises an annual interest rate r, to be compounded n times per year.

Let us consider how much capital is available in the savings account after the first several instances of compounding of the interest. That is, let P_0 represent the principal balance, let P_1 represent the amount available after the first time interest is compounded, let P_2 represent the amount available after the second time interest is compounded, and so forth.

Since interest is applied n times per year, the amount of interest applied each time that interest is compounded will be r/n. For instance, if the nominal interest rate is 4%, i.e., $r = 0.04$, and interest is compounded eight times per year, then each time interest is applied, an amount worth 4%/8, or 0.5%, of the current balance is added to the account. Thus, the new balance P_1 available once interest has been compounded for the first time is given by

$$P_1 = P_0\left(1+\frac{r}{n}\right).$$

The quantity $\left(1+\frac{r}{n}\right)$ represents 100% *plus* the interest. Using the numbers above, this quantity would equal 1.005, and would represent adding 0.5% of interest to the principal balance.

The second time interest is applied, it is not only applied to the original principal, but also to the interest that has accumulated to date. Therefore, interest is applied to the amount P_1. We obtain

$$P_2 = P_1\left(1+\frac{r}{n}\right) = P_0\left(1+\frac{r}{n}\right)^2.$$

Continuing on, we see that

$$\begin{aligned} P_3 &= P_0\left(1+\frac{r}{n}\right)^3, \\ P_4 &= P_0\left(1+\frac{r}{n}\right)^4, \\ P_5 &= P_0\left(1+\frac{r}{n}\right)^5, \end{aligned}$$

and so forth. At the end of the year, interest will have been compounded n times, bringing the balance to

$$P_n = P_0\left(1+\frac{r}{n}\right)^n.$$

After two years, we will have

$$P_{2n} = P_0\left(1+\frac{r}{n}\right)^{2n}.$$

We may therefore model compounded interest with frequency n as an exponential function with growth factor

$$\left(1+\frac{r}{n}\right)^n.$$

This amount (minus 1) also gives us the annual percentage rate. We thus arrive at the following proposition.

Proposition 1.3.3 If interest is compounded on a principal P_0, using a nominal annual interest rate r and compounding frequency n, the total balance after t years is given by

$$P(t) = P_0\left(1+\frac{r}{n}\right)^{nt}. \tag{1.11}$$

■ **Example 1.11** Suppose that $20,000 is deposited into an interest-bearing savings account on January 1, with a nominal annual interest rate of 6.0%, and interest is compounded monthly. Determine (a) a formula for the balance in the savings account as a function of time, (b) the effective interest rate, (c) the amount available after eight months, and (d) the amount available after eight months if interest were instead compounded daily.

(a) The quantity r/n works out as $r/n = 0.06/12 = 0.005$. Therefore, the balance in the account as a function of time can be approximated using equation (1.11) as

$$P(t) = \$20{,}000 \times 1.005^{12t}.$$

(b) The effective interest rate can be determined by first computing $1.005^{12} \approx 1.0616778$. Therefore, the effective interest rate is approximately 6.17%.

(c) Eight months is 2/3rds of a year, so the balance after eight months is given by

$$P(2/3) = \$20{,}000 \times 1.005^{12\times 2/3} = \$20{,}000 \times 1.005^{8} \approx \$20{,}814.14.$$

(d) If interest were instead compounded daily, the formula for $P(t)$ would instead be given by

$$P(t) = \$20{,}000 \times \left(1+\frac{0.06}{365}\right)^{365t}.$$

Since there are 243 days in the first eight months of the year, the balance at the end of August will be given by

$$P(243/365) = \$20{,}000 \times \left(1+\frac{0.06}{365}\right)^{243} \approx 20815.01.$$

We therefore see that compounding interest daily, as opposed to monthly, only results in an additional 87 cents of interest during the first eight months. In our next paragraph, we will ask: what happens in the limit as interest is compounded infinitely often, i.e., what happens when interest is *compounded continuously*? ■

1.3.5 Continuously Compounded Interest and The Number e

In this paragraph, we will examine what happens in the limit as interest is compounded infinitely often, i.e., continuously, and then use this result to explain where Euler's number e comes from.

The question now is, what happens as the compounding frequency n tends to infinity, i.e., as $n \to \infty$? In other words, what happens if interest, with a given nominal annual rate of r, is continuously compounded?

To answer this, let us rewrite our formula for compounded interest, as given by equation (1.11), by introducing the variable $x = \frac{r}{n}$. Thus, instead of the compounding frequency n, we will consider this ratio x. Now, as $n \to \infty$, it follows that $x \to 0$. We may express equation (1.11) in terms of this new variable x as

$$P(t) = a(1+x)^{rt/x} = a\left[(1+x)^{1/x}\right]^{rt}.$$

The key, therefore, is to determine what happens to the quantity $(1+x)^{1/x}$ as $x \to 0$. (This quantity is clearly undefined at $x = 0$, since 1/0 is undefined. We may, nevertheless, study the behavior of this quantity as x approaches 0.) Thus, we want to know the value of the limit

$$\lim_{x \to 0}(1+x)^{1/x}.$$

(We will study limits in §1.7. For now, it suffices to say that this limit is simply the number that the outputs of $(1+x)^{1/x}$ approach as x gets closer and closer to the value $x = 0$.)

In order to answer this, we may simply plug in a sequence of subsequently smaller values of x into the function $f(x) = (1+x)^{1/x}$. The values of this function for several input x values are given in Table 1.9.

x	$(1+x)^{1/x}$
10^{-1}	2.5937424601
10^{-2}	2.70481382942153
10^{-3}	2.71692393223559
10^{-4}	2.71814592682493
10^{-5}	2.7182682371923
10^{-6}	2.71828046909575

Table 1.9: The values of $(1+x)^{1/x}$ for several x values, as $x \to 0$.

We say that *in the limit* as $x \to 0$, this function approaches the number e. In fact, this is the definition of the number e.

Definition 1.3.4 — Euler's number. The number e is defined by

$$e = \lim_{x \to 0}(1+x)^{1/x}. \tag{1.12}$$

Given this definition for the number e, we may therefore write down a precise equation that describes the amount of capital in a savings account as a function of time, if interest is continuously compounded with an annual continuous growth rate of r.

Proposition 1.3.4 If a principal balance P_0 with nominal annual interest rate r is continuously compounded, i.e., compounded infinitely often, then the balance after t years is given by

$$P(t) = P_0 e^{rt}. \tag{1.13}$$

In this context, the parameter r is referred to as the *continuous growth rate.*

■ **Example 1.12** Continuing Example 1.11, let us compute the amount available in the account after eight months if interest is continuously compounded. This is given by

$$P(243/365) = \$20,000e^{0.06\times 243/365} = \$20,815.07.$$

Thus, compounding interest continuously only results in an additional 6 cents compared with compounding interest daily, after the first eight months. ■

Exercises

For Exercises 1–5, find the equation of the exponential function that satisfies the given criteria .

1. Passes through the points $(1, 0.5)$ and $(3, 1.125)$.
2. Passes through the points $(2, 8)$ and $(5, 1)$.
3. Has an initial condition of 3 and a half-life of 16.
4. Has a growth factor of 7 and passes through the point $(7, 7)$.
5. Has a doubling time of 7 and passes through the point $(7, 7)$.

For Exercises 6–10, determine whether the given data can be described with a linear or an exponential function. Find a formula for the function.

6.

x	1	2	3
$f(x)$	3	5	7

7.

x	1	5	13	29
$f(x)$	2	4	8	16

8.

x	3	6	9	12
$f(x)$	9/8	3/4	1/2	1/3

9.

x	1	2	3	4
$f(x)$	8	20	50	125

10.

x	1	5	15	40
$f(x)$	8	20	50	125

Problems

11. During the four years between 1999 and 2002, out-of-state tuition at the University of Michigan went from approximately \$20,000 per year to approximately \$28,000 per year. Let $P(t)$ be the price of tuition, t years after 1999.
    (a) Model $P(t)$ as a linear function. What is the prediction for tuition in 2008?
    (b) Model $P(t)$ as an exponential function. What is the prediction for tuition in 2008?
    (c) Out-of-state tuition in 2008 was approximately \$38,000. Which model most accurately predicts this value?
12. The atmospheric pressure can be modeled as an exponentially decreasing function of altitude by the relation

$$P(h) = P_0 e^{-h/h_0},$$

where P_0 is the atmospheric pressure at sea level and h_0 is a characteristic altitude. On Earth, $P_0 = 101.3$ kiloPascals (kPa) and $h_0 = 7{,}000$ meters (m). Suppose the peak of a certain mountaintop has an altitude of 3,350 m. What is the atmospheric pressure at the peak? Answer both as a fraction of sea level pressure (this is the number of "atmospheres" or atms) and in terms of kPa.
13. Consider the United States population data as shown in Table 1.1 on page 6. Let $P(t)$ represent the U.S. population as a function of the number of years t since 1900.
    (a) Model $P(t)$ as an exponential function using only the data points for 1900 and 1910. According to this model, what should the U.S. population be in 2010? What does this tell you about the validity of the model?
    (b) Model $P(t)$ as an exponential function using only the data points for 1900 and 2010. According to this model, what should the U.S. population be in 1950?
    (c) Considering only the data points for 1900 and 2010, approximately how many years should constitute the doubling time for the U.S. population?
14. A brand new sports car has a list price of \$45,000. When it is three years old, it will be worth \$30,000. Let $P(t)$ represent the value of the car when it is t years old.
    (a) Assuming the value of the car depreciates exponentially, determine a formula for $P(t)$.
    (b) How much will the car be worth when it is six years old?
15. A single bacterium founds a bacterial colony in a certain glass jar at 11:00. The population size doubles every minute, and the jar is completely filled to capacity with bacteria by 12:00.
    (a) At what time was the jar half filled?
    (b) At what time was the jar one quarter filled?

(c) Shortly before 12:00, three brand new jars were discovered. Assuming effortless jar-to-jar transit, at what time will the newly discovered jars also be filled, assuming the population continues to grow exponentially at the same rate?

1.4 Logarithmic Functions

Logarithms are useful when solving exponential growth and decay problems or when studying physical phenomena in which one or more variable commonly takes values ranging over multiple orders of magnitude, such as when studying sound intensity or earthquake intensity. In this section, we introduce logarithms, discuss their properties, study logarithmic functions, and take a look at how logarithmic functions are used in modeling various real-world phenomena.

1.4.1 Logarithms and Their Properties

We begin with the general definition of a logarithm operator.

Definition 1.4.1 The *logarithm* (or *log*), base b, of a number $a > 0$, written

$$\log_b a \qquad \text{or} \qquad \log_b(a),$$

is the power to which one can raise the base b in order to obtain a, i.e.,

$$\log_b a = c \qquad \text{if and only if} \qquad b^c = a. \tag{1.14}$$

If no base is given, the base is taken to be 10, i.e.,

$$\log(a) = c \qquad \text{if and only if} \qquad 10^c = a. \tag{1.15}$$

The *natural logarithm* of a, denoted $\ln(a)$, is the logarithm, base e, of a, i.e.,

$$\ln(a) = c \qquad \text{if and only if} \qquad e^c = a. \tag{1.16}$$

■ **Example 1.13** For example, consider the following:

$$\begin{aligned} \log(10) &= 1, \\ \log(100) &= 2, \\ \log(1{,}000) &= 3, \\ \log(10{,}000) &= 4, \\ \log(1{,}000{,}000) &= 6, \\ \log(1 \text{ googol}) &= 100. \end{aligned}$$

(One googol is equal to 10^{100}.) Written differently, these same equations appear as

$$\begin{aligned} \log(10^1) &= 1, \\ \log(10^2) &= 2, \\ \log(10^3) &= 3, \\ \log(10^4) &= 4, \\ \log(10^6) &= 6, \\ \log(10^{100}) &= 100. \end{aligned}$$

The reason is straightforward: $\log(10^4)$, for example, asks us what number you must raise the base 10 to in order to obtain 10^4. The answer is clearly 4. ■

Considering, now, the logarithm as a function, i.e., $f(x) = \log(x)$. The previous example reveals two interesting features: first, the logarithmic function increases without bound, i.e., for any positive y-value, there will always be some x, large enough, so that $\log(x) = y$. (In fact, the x-value that solves this equation is just 10^y.) Second, the logarithmic function increases incredibly slowly; it takes, for instance, an x-value of one googol, which is an astronomically large number[8], just for the y-value to reach a mere 100. Keep these properties in mind. We will return to our consideration of the logarithm as a function momentarily, but first we introduce several important algebraic properties of logarithms.

Theorem 1.4.1 — Properties of Logarithms. The following relations hold for the natural logarithm:

$$e^{\ln(a)} = a, \qquad \text{for } a > 0; \tag{1.17}$$

$$\ln(a^b) = b\ln(a), \qquad \text{for } a > 0; \tag{1.18}$$

$$\ln(ab) = \ln(a) + \ln(b), \qquad \text{for } a, b > 0; \tag{1.19}$$

$$\ln\left(\frac{a}{b}\right) = \ln(a) - \ln(b), \qquad \text{for } a, b > 0. \tag{1.20}$$

(R) Equations 1.17–1.20 also hold for logarithms in general, as long as one replaces e with the base of the logarithm in equation (1.17).

Proof. Equation 1.17. This identities follows directly from the definition of the natural logarithm. The quantity $\ln(a)$ asks: to what power must one raise e in order to obtain a? That is, whatever $\ln(a)$ equals, if you raise e to that power, you obtain a. But this is the content of equation (1.17).

Equation 1.18. To show this identity, let us introduce a variable x to represent the value of $\ln(a^b)$, so that

$$\ln(a^b) = x.$$

Following the definition of the natural logarithm, as given in equation (1.16), we notice that $\ln(a^b) = x$ if and only if $e^x = a^b$. Raising both sides of this equation to the power of $(1/b)$, we then obtain

$$e^{x/b} = a.$$

Again following equation (1.16), we conclude that this statement is true if and only if the statement

$$\ln(a) = \frac{x}{b}$$

[8]Consider that there are approximately 10^{66} atoms in the Milky Way Galaxy, and approximately 10^{81} atoms in the observable Universe. So one googol is over a billion billion times, that is, over one quintillion times, the number of atoms in the observable Universe.

is also true. Therefore, $x = b\ln(a)$. The definition of x, however, was $x = \ln(a^b)$. Therefore $\ln(a^b)$ and $b\ln(a)$ must be equal to each other.

Equation 1.19. For this identity, we start with the right-hand side. Let us define

$$x = \ln(a) \qquad \text{and} \qquad y = \ln(b).$$

Once again, following the definition of the natural logarithm, we then find that

$$e^x = a \qquad \text{and} \qquad e^y = b.$$

Multiplying, we therefore have

$$ab = e^x e^y = e^{x+y}.$$

Using the definition of the natural logarithm yet again, we obtain

$$\ln(ab) = x + y.$$

The variables x and y, however, represent $\ln(a)$ and $\ln(b)$, respectively, by definition. The result follows.

Equation 1.20. This follows from the previous identities. In particular,

$$\begin{aligned}\ln\left(\frac{a}{b}\right) &= \ln(ab^{-1}) \\ &= \ln(a) + \ln(b^{-1}) \\ &= \ln(a) + (-1)\ln(b) \\ &= \ln(a) - \ln(b).\end{aligned}$$

The second equality, here, follows from equation (1.19), whereas the third equality follows from equation (1.18). This completes the proof. ■

Corollary 1.4.2 For any $x > 0$, we have

$$\ln(x^{-1}) = -\ln(x). \tag{1.21}$$

Proof. This follows directly from equation (1.18) by identifying $b = -1$ (and $a = x$, which is merely a cosmetic change). ■

Corollary 1.4.2 is useful when simplifying equations involving logarithms. It turns out that $\ln(x) < 0$ for $0 < x < 1$, and $\ln(x) > 0$ for $x > 1$. (Why?) Thus, as a matter of style, it is typically preferred to replace a number less than one (in a logarithm) with its greater-than-one reciprocal. For instance, instead of leaving $\ln(1/2)$ in a formula, it would be preferred to replace it with $-\ln(2)$. In the latter case, it is more readily apparent that this quantity is negative.

Our next corollary provides a quick method for computing logarithms with arbitrary base by using a calculator, since most modern scientific calculators have a natural logarithm button.

Corollary 1.4.3 For any $a, b > 0$, we have

$$\log_b(a) = \frac{\ln(a)}{\ln(b)}. \tag{1.22}$$

Proof. Define $c = \log_b(a)$. By the definition of the logarithm, we therefore have

$$b^c = a.$$

Next, let us compute the natural logarithm of both sides of this equation, thereby obtaining

$$\ln(b^c) = \ln(a).$$

Using equation (1.18), however, this reduces to

$$c\ln(b) = \ln(a).$$

Therefore

$$c = \frac{\ln(a)}{\ln(b)}.$$

But, by definition, c was equal to $\log_b(a)$. This proves the result. ■

▪ **Example 1.14** Consider $\log_2(8)$. Clearly, this should equal 3, as $2^3 = 8$. But let us check the validity of equation (1.22):

$$\log_2(8) = \frac{\ln(8)}{\ln(2)} \approx \frac{2.079441541679836}{0.6931471805599453} \approx 3.$$

(Note that the quantity $\ln(8)/\ln(2)$ is exactly equal to 3, but only approximately equal to the first fifteen decimal places shown in numerator and denominator of the approximation.) Similary,

$$\log(1000) = \frac{\ln(1000)}{\ln(10)} \approx \frac{6.907755278982137}{2.302585092994046} \approx 3.$$

▪

The logarithm operator, along with its various properties, turns out to be quite useful in solving exponential growth and decay problems, as is evident in our next example.

▪ **Example 1.15** A 100g sample of the californium-253 (^{253}Cf) is prepared in a laboratory. Exactly 3 days later, only 88.98g of the sample remain. Determine the continuous decay rate of this radioactive isotope.

To solve this, we may assume that the amount of californium-253 $a(t)$, as measured in grams, that is remains after t days is of the form

$$a(t) = a_0 e^{-rt},$$

where a_0 is the initial value and r is the continuous decay rate. Substituting the initial value of $a_0 = 100$ into the formula, we obtain $a(t) = 100e^{-rt}$. Here, we used the first piece of information that was given: the initial amount of californium-253 is 100g.

Next, let us use the second piece of information that was given: after three days, only 88.98g of the sample remain, i.e., $a(3) = 88.98$. On the other hand, the formula for $a(t)$ tells us

$$a(3) = 100e^{-3r}.$$

Comparing this with the given information, we conclude that

$$88.98 = 100e^{-3r}.$$

We can use this equation to solve for the value of the parameter r as follows:

$$\begin{aligned}
\frac{88.98}{100} &= e^{-3r}, \\
\ln(.8898) &= -3r, \\
r &= \frac{-1}{3}\ln(0.8898) \\
r &= \frac{1}{3}\ln\left(\frac{5000}{4449}\right) \approx 0.03891952020339856.
\end{aligned}$$

Note that the last line follows from equation (1.21), since $0.8898^{-1} = \frac{5000}{4449}$. We did this since $\ln(0.8898) < 0$.

Substituting this value for r back into our formula for $a(t)$, we find

$$\begin{aligned}
a(t) &= 100e^{-rt} = 100e^{-\frac{t}{3}\ln\left(\frac{5000}{4449}\right)} \\
&= 100e^{\ln\left[\left(\frac{5000}{4449}\right)^{-t/3}\right]} = 100\left(\frac{5000}{4449}\right)^{-t/3} = 100\left(\frac{4449}{5000}\right)^{t/3}.
\end{aligned}$$

This makes sense, since when $t = 3$, we obtain

$$a(3) = 100\left(\frac{4449}{5000}\right) = 88.98,$$

verifying our result. ■

1.4.2 The Logarithmic Function

The logarithmic function is simply the function

$$f(x) = \ln(x),$$

on the domain $x > 0$. We next discuss several features of the logarithmic function, which is graphed in Figure 1.14. First,

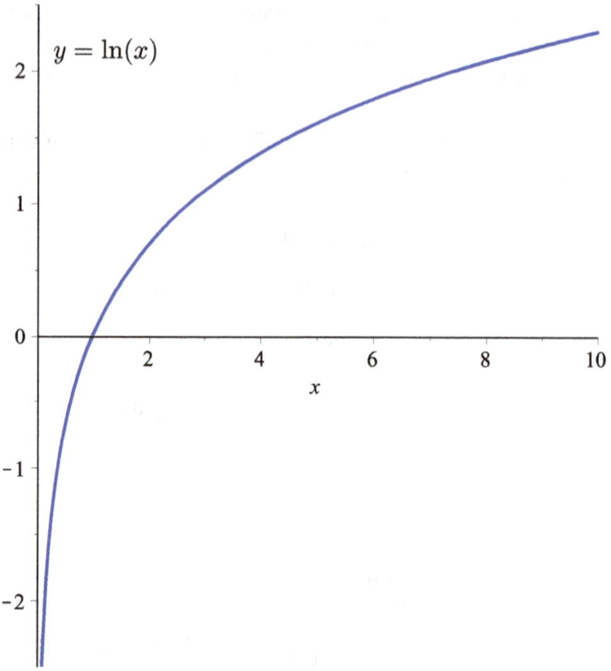

Figure 1.14: Graph of the logarithmic function $f(x) = \ln(x)$.

$$f(1) = 0 \qquad \text{and} \qquad f(e) = 1.$$

Thus, the horizontal intercept of the function is located at $x = 1$. This follows since $\ln(1) = 0$, due to the fact that $e^0 = 1$.

Next, we note that $f(x)$ continuous increasing without bound, i.e., it never levels off at a "ceiling," but grows forever. In fact, given any real number y, one can solve $f(x) = y$ for x, by simply unraveling the definition of the natural logarithm. To wit, $\ln(x) = y$ if and only if $e^y = x$. For any y-value, therefore, there is some x-value that produces it. As y gets larger and larger, so, too, must x. However, the function increases incredibly slowly. To see this, consider what happens when we evaluate f at one googol:

$$f(10^{100}) = \ln(10^{100}) = 100\ln(10) \approx 230.2585092994046.$$

Thus, as x moves along the interval from $x = 1$ to $x = 10^{100}$, the output of this function only increases by about 230. So the function grows incredibly slowly. Finally, as previously discussed, the outputs of $f(x)$ are negative for values of x less than 1.

It is also instructive to examine the graph of the logarithmic function $\ln(x)$ concurrently with the graph of the exponential function e^x, as shown in Figure 1.15.

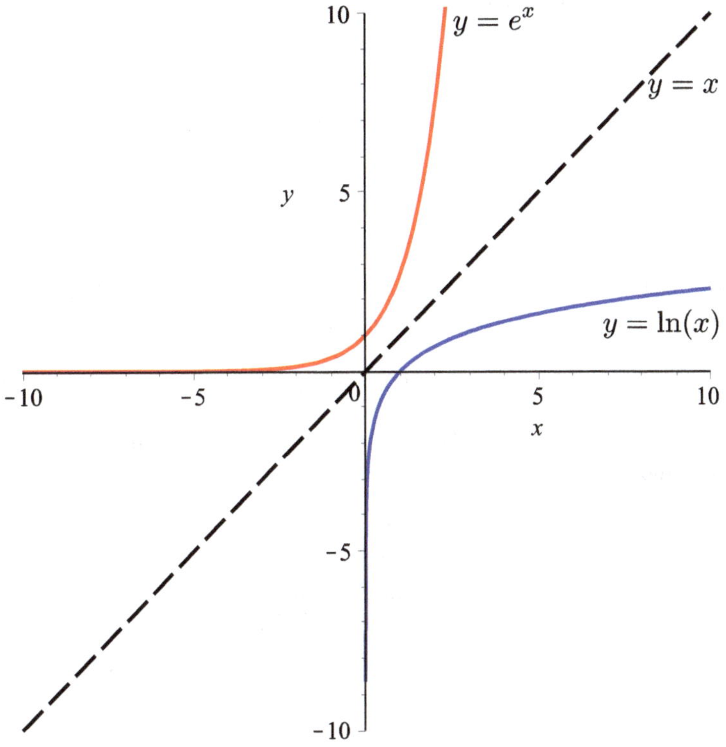

Figure 1.15: Graph of the logarithmic function $f(x) = \ln(x)$ (blue) shown concurrently with the graph of the exponential function $y = e^x$ (red).

It turns out that these two functions are simply a *reflection* of each other about the line $y = x$. This has to do with the fact that

$$e^{\ln x} = x \qquad \text{and} \qquad \ln(e^x) = x,$$

that is, the exponential function an the logarithmic function "undo" each other. We will explore this in greater depth in §1.5.

1.4.3 Application to Half-life and Doubling Time Problems

Natural logarithms are also important in relating the continuous growth or decay rate of exponential functions to the doubling time or half-life, as we explore in our next theorem.

Theorem 1.4.4 Suppose the function $f(t)$ is growing exponentially in time and may therefore be described by the equation

$$f(t) = ae^{rt}, \tag{1.23}$$

where the parameter $r > 0$ is the continuous growth rate. Then the doubling time T is given by the formula

$$T = \frac{\ln 2}{r}. \tag{1.24}$$

Similarly, suppose the function $f(t)$ is decaying exponentially in time and may therefore be described by the equation

$$f(t) = ae^{-rt}, \tag{1.25}$$

where the parameter $r > 0$ is the continuous decay rate. Then the half-life τ is given by the formula

$$\tau = \frac{\ln 2}{r}. \tag{1.26}$$

Even though equations (1.24) and (1.26) appear the same, they are actually different, as the positive parameter r is defined differently in equations (1.23) and (1.25). In equation (1.25), the minus sign is explicitly written out, so that the parameter r takes a positive value.

Proof. First, let us consider a quantity that exhibits exponential growth, so that its size is described by equation (1.23). The initial value is $f(0) = a$, and we are told that the doubling time is T. Therefore, $f(T)$ must equal $2a$, and we can write

$$f(T) = 2a = ae^{rT}.$$

The second equality is obtained by evaluating the formula for $f(t)$ at time T. From this equation, we obtain

$$\begin{aligned} 2 &= e^{rT}, \\ \ln(2) &= rT, \\ T &= \frac{1}{r}\ln(2), \end{aligned}$$

which verifies equation (1.24).

Next, let us suppose that we have a quantity that exhibits exponential decay, so that its size is described by equation (1.25). The initial value is $f(0) = a$, and we are told that the half-life is τ. Therefore, $f(\tau)$ must equal $a/2$, and we can write

$$f(\tau) = \frac{a}{2} = ae^{-r\tau}.$$

From this, we obtain

$$\begin{aligned} \frac{1}{2} &= e^{-r\tau}, \\ \ln\left(\frac{1}{2}\right) &= -r\tau, \\ \tau &= -\frac{1}{r}\ln\left(\frac{1}{2}\right), \\ \tau &= \frac{1}{r}\ln(2). \end{aligned}$$

This verifies equation (1.26). ■

It is preferred to remember the steps of the proof over the actual formulas. Remembering the logic of the derivation is simpler, and one can apply the same steps to answering similar questions in which the formula may not be directly applicable. (For example, what if you are given the *tripling time*?)

Exercises

For Exercises 1–5, use logarithms to solve the algebraic equation for the variable x.

1. $10^x = 325$.
2. $10^x = 0.012$.
3. $e^{2x/3} = 9$.
4. $20(1/2)^{x/57} = 15$.
5. $8(1/3)^{2x/9} = 3$.
6. Compute the logarithm (base 10) of the following numbers: 0.0023, 13, 146, 10,119, and 5,000,346. Describe any observations.

For Exercises 7–11, use logarithms to compute the half-life or doubling time (as appropriate) of the indicated exponential function.

7. $f(x) = 8(1.12)^x$.
8. $f(x) = 16e^{0.5x}$.
9. $f(x) = 10^x$.
10. The exponential function passing through the points $(1,1)$ and $(3,3)$.
11. The exponential function with initial condition 11 and tripling time 9.

Problems

12. A certain bacterial population doubles in size every three hours. How long does it take for the population to triple in size?
13. Use the years 1900 and 1910 from Table 1.1 to model the population of the United States using an exponential function. According to this model, what is the predicted U.S. population in 2000? Repeat, instead fitting an exponential function to the data points at 1900 and 1950. What is the prediction for the U.S. population in 2000 this time?
14. Atmospheric pressure (kPa) as a function of altitude (m) can be modeled using an exponential function as

$$P(h) = P_0 e^{-kh},$$

for some positive constants P_0 and k. Assuming that sea level pressure is 101.3 kPa and that air pressure at an altitude of 10,000 m is 24 kPa, determine the atmospheric pressure at an altitude of 5,000 m.
15. A frozen turkey, with an initial temperature 32°, is placed in a 350° oven. After 2 hours, its temperature is 150°. Newton's law of heating and cooling states that the turkey's temperature can be described by the function

$$u(t) = R + ae^{-kt},$$

for an appropriate choice of constants R, a, and k.
    (a) Using the formula, what is the turkey's temperature in the long run? (You may answer in terms of the parameters R, a, and k.) Use this and the given information to determine the value of one of the parameters.
    (b) Use the initial temperature of the turkey to determine another one of the parameters.
    (c) Use the fact that $u(2) = 150$ to determine the value for the third parameter.
    (d) Next, use the formula for the function u and the known value for the three parameters to determine how long it will take for the turkey's temperature to reach 300°.
16. The half-life of carbon-14 is 5,370 years. A certain fossil contains 2.7% of its original amount of carbon-14, i.e., 2.7% of its original carbon-14 has not yet decayed. How old is the fossil?
17. Uranium-238 has a half-life of 4.468 billion years. A piece of moon rock is discovered that still has 49.5% of its original uranium-238. How old is this moon rock? (Answer to the nearest million years.)
18. A certain stock is forecasted to grow at a rate of 7% per year. Assuming this forecast is accurate, approximately how long will it take for the value of the stock to double?

1.5 Transformations of Functions

In this section, we analyze various transformations of functions. A *function transformation* is essentially a systematic way of constructing a new function from a given function using a prescribed formula. Function transformations afford us a way of visualizing the relationship between similar functions, decomposing certain functions using a standard vocabulary, and understanding certain families of functions.

1.5.1 Horizontal and Vertical Translations

We begin with a look at horizontal and vertical translations. A translation is essentially a way of creating a copy of a given function with a different placement, either shifted left, right, up, or down.

Definition 1.5.1 Given a function f, the *horizontal translation* (or *horizontal shift*) of f by an amount a is given by the new function T_a^{hor}, defined by

$$T_a^{\text{hor}}(x) = f(x-a). \tag{1.27}$$

If $a > 0$, this is called a horizontal translation to the right; if $a < 0$, this is called a horizontal translation to the left.

Similarly, the *vertical translation* (or *vertical shift*) of f by an amount a is given by the new function T_a^{ver}, defined by

$$T_a^{\text{ver}}(x) = f(x)+a. \tag{1.28}$$

If $a > 0$, this is called a vertical translation upward; if $a < 0$, this is called a vertical translation downward.

(R) The names T_a^{hor} and T_a^{ver} are simply the names of the new functions defined by Definition 1.5.1, treated no differently than if the new functions were named g and h. We only use this notation for clarity, to avoid a proliferation of new function names—g, h, k, l—that would otherwise undoubtedly ensue.

The graph of a horizontal translation to the right of a certain function is shown in Figure 1.16. The new function is literally a copy of the function f, only shifted by a units to the right.

The reason that a translation to the *right* is defined, for $a > 0$, by equation (1.27) is as follows. In order to evaluate the new function T_a^{hor} at a point x, one must evaluate the function f at the point $(x-a)$, that is, one must *subtract* a units from the current x position, and therefore evaluate the function f exactly a units to the *left*. For example, suppose that we have a translation of 3 units to the right, so that $T_3^{\text{hor}}(x) = f(x-3)$. To evaluate the new function T_3^{hor} at a given point x, we must evaluate the function f at the point $(x-3)$. For example, $T_3^{\text{hor}}(10) = f(10-3) = f(7)$.

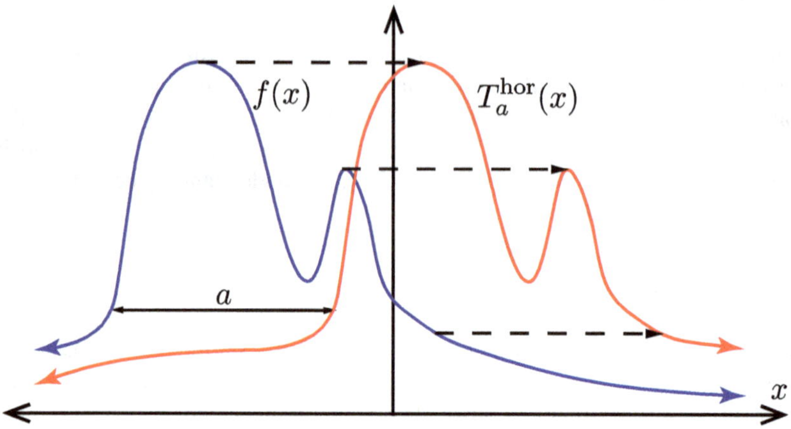

Figure 1.16: Graph of a horizontal translation to the right $T_a^{\text{hor}} = f(x-a)$ of a function f.

Therefore, the output of the function T_3^{hor} at the point $x = 10$ is identical to the output of the function f at the point $x = 7$. Similarly, if $a < 0$, equation (1.27) would represent a translation to the left.

Whereas horizontal translations are obtained by subtracting the constant a from the *inside* of the function f, the vertical translations are obtained by adding the constant a to the *output* of the function f. For example, if $a = 3$ is added to the output of the function f at a given point x, the new output, i.e., the output of T_a^{ver}, will simply be the old output plus 3. Thus, the graph of f is shifted *upward* by $a = 3$ units. For example, if $f(10) = 20$, then $T_3^{\text{ver}}(10) = f(10) + 3 = 23$.

The graph of a vertical translation upward of a certain function is shown in Figure 1.17. The new function is literally a copy of the function f, only shifted upward by a units.

▪ **Example 1.16** Suppose that the salary structure for a given position at a certain company is determined solely based on the number of years that an employee has worked on the job. Let $f(x)$ represent an employee's salary after working for x years. The company decides to give every employee a fixed \$960 per year salary increase. The new salary structure g will be given by $g(x) = f(x) + 960$. Thus, the graph of the pay scale has been shifted upward by \$960. ▪

1.5.2 Horizontal and Vertical Scalings and Reflections

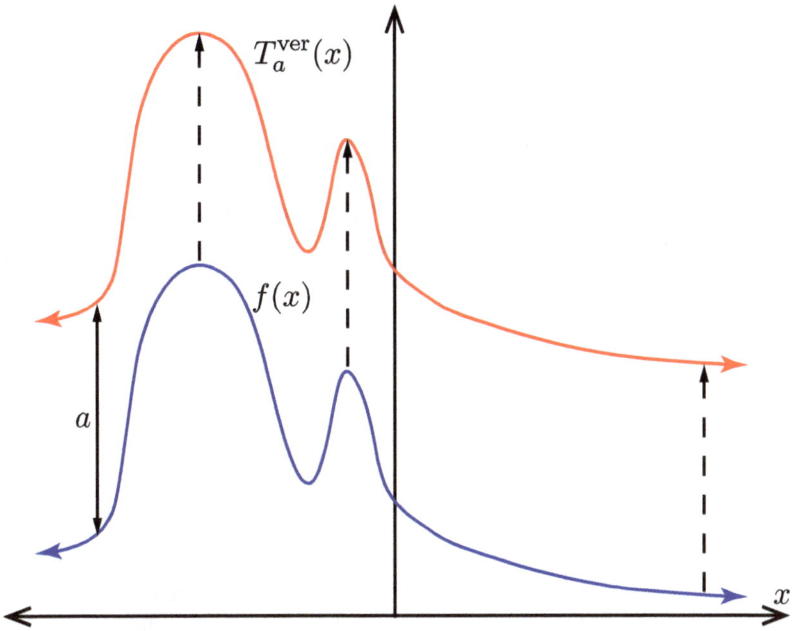

Figure 1.17: Graph of a vertical translation upward $T_a^{\mathrm{ver}} = f(x) + a$ of a function f.

Horizontal and Vertical Scalings

Next, we consider horizontal and vertical scalings. A scaling is simply a stretch or a compression of the graph in a certain direction, like the shape of a spring may be stretched by pulling the ends apart or compressed by pushing them together.

Definition 1.5.2 Given a function f, the *horizontal scaling* of f by an amount a is given by the new function S_a^{hor}, defined by

$$S_a^{\mathrm{hor}}(x) = f\left(\frac{x}{a}\right). \tag{1.29}$$

If $a > 1$, this is called a *horizontal stretch*; if $0 < a < 1$, this is called a *horizontal compression*.

Similarly, the *vertical scaling* of f by an amount a is given by the new function S_a^{ver}, defined by

$$S_a^{\mathrm{ver}}(x) = af(x). \tag{1.30}$$

If $a > 1$, this is called a *vertical scaling*; if $0 < a < 1$, this is called a *vertical compression.*

Similar to horizontal and vertical translations, changing the *inside* of the function f affects the horizontal direction, whereas changing the *outside* of the function f affects the vertical direction. Here, we can achieve either a stretch or a compression of the graph of the function f depending on the size of the parameter a.

Let us begin by looking at horizontal scalings. Consider, in particular, the horizontal stretch $S_3^{\text{hor}}(x) = f\left(\frac{x}{3}\right)$. Using this formula to compare several example outputs of the functions f and S_3^{hor}, we find

$$
\begin{aligned}
S_a^{\text{hor}}(1) &= f\left(\frac{1}{3}\right), \\
S_a^{\text{hor}}(3) &= f(1), \\
S_a^{\text{hor}}(6) &= f(2), \\
S_a^{\text{hor}}(9) &= f(3),
\end{aligned}
$$

and so forth. We can see what is happening: the outputs of the graph of the function f are being pulled outward, away form the vertical axis (i.e., y-axis). An example of a horizontal stretch is shown in Figure 1.18. Observe that it is as though someone is pulling the function f, stretching it out horizontally.

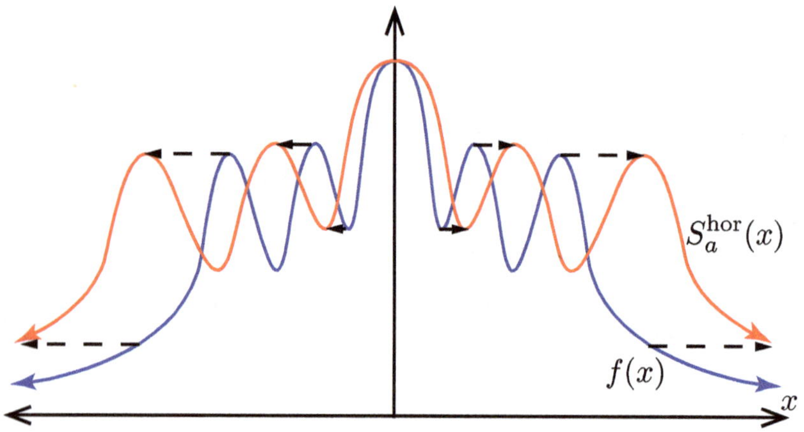

Figure 1.18: Graph of a horizontal stretch $S_a^{\text{hor}}(x)$ of a function f.

The opposite of a stretch is a compression: the function is pushed in toward the vertical axis instead of being stretch out from the vertical axis.

An example of a vertical compression is shown in Figure 1.19. The output of the function f is reduced by the same factor at each point in the domain. Notice that the graph is being squeezed in toward the horizontal axis (or, the x-axis). The farther the point on the graph from the horizontal axis, the farther that point moves toward the horizontal axis. The distance to the horizontal axis, however, is reduced by an equal percentage at every point. For example, every output may be reduced by a factor of 2, i.e., reduced by 50%. This function transformation would be given by $S_{1/2}^{\text{ver}}(x) = \frac{f(x)}{2}$. The output at every point is simply halved.

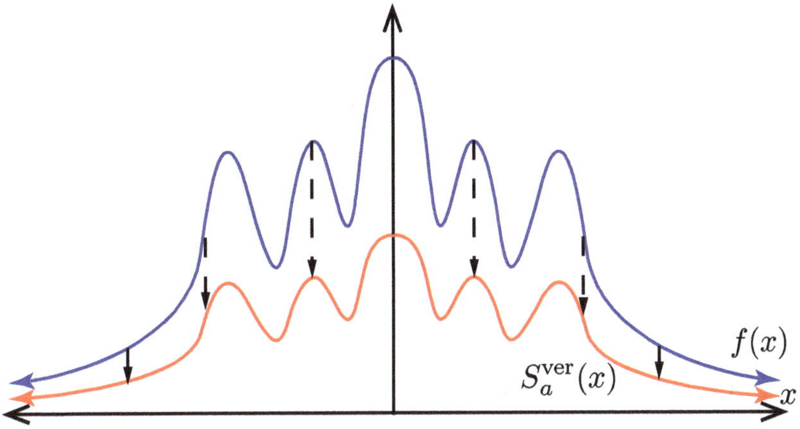

Figure 1.19: Graph of a vertical compression $S_a^{\text{ver}}(x)$ of a function f $(0 < a < 1)$.

Horizontal and Vertical Reflections

We next consider horizontal and vertical reflections; this operation is tantamount to creating a mirror image of the function reflected about one of the axes.

Definition 1.5.3 Given a function f, the *horizontal reflection* of f is given by the new function R^{hor}, defined by

$$R^{\text{hor}}(x) = f(-x). \tag{1.31}$$

Similarly, the *vertical reflection* of f is given by the new function R^{ver}, defined by

$$R^{\text{ver}}(x) = -f(x). \tag{1.32}$$

The graph of a horizontal and vertical reflection of a function f is shown in Figure 1.20. As evident from the graph, the horizontal reflection is the mirror

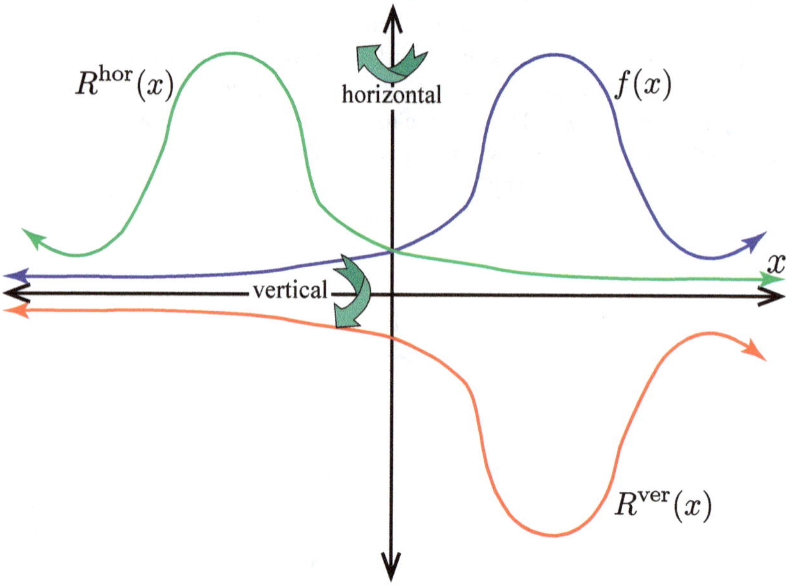

Figure 1.20: Graph of a horizontal and vertical reflection, R^{hor} and R^{ver}, of a function f.

image of the original function, as reflected about the vertical axis, so that the image creates a horizontal parallel with the original. Similarly, the vertical reflection is the mirror image of the original function, as reflected about the horizontal axis, so that the image creates a vertical parallel with the original. To create the vertical reflection, one simply multiplies the *output* of the function by -1. Similarly, to create the horizontal reflection, one multiplies the *input* of the function by -1.

Summary of Function Transformations

We summarize the various function transformations in Table 1.10. Notice that for each type of function transformation, changing the inside of the function f, i.e., changing the input, affects a horizontal change, whereas changing the outside of the function f affects a vertical change.

Also, notice that the notation is slightly changed in Table 1.10, though still consistent with the definitions. That is, in Table 1.10, only positive values of the parameter a are used for translations, and only values of the parameter a that are greater than 1 are used for scalings. The separate cases listed (e.g., horizontal, vertical, left, right, up, down) therefore more closely mirror the notation that you would use in practice: $f(x-3)$ is a translation to the right, $f(x+2)$ is a translation

horizontal translation	right	$T_a^{\text{hor}}(x) = f(x-a)$	$a > 0$
	left	$T_{-a}^{\text{hor}}(x) = f(x+a)$	$a > 0$
vertical translation	up	$T_a^{\text{ver}}(x) = f(x)+a$	$a > 0$
	down	$T_{-a}^{\text{ver}}(x) = f(x)-a$	$a > 0$
horizontal scaling	stretch	$S_a^{\text{hor}}(x) = f\left(\frac{x}{a}\right)$	$a > 1$
	compression	$S_{1/a}^{\text{hor}}(x) = f(ax)$	$a > 1$
vertical scaling	stretch	$S_a^{\text{ver}}(x) = af(x)$	$a > 1$
	compression	$S_{1/a}^{\text{ver}}(x) = \frac{f(x)}{a}$	$a > 1$
reflection	horizontal	$R^{\text{hor}}(x) = f(-x)$	
	vertical	$R^{\text{ver}}(x) = -f(x)$	

Table 1.10: Summary of basic function transformations.

to the left, etc.

1.5.3 Function Composition

In this paragraph, we examine how to *compose* one function with another. By function composition, we mean creating a new function that works by sequentially applying one function to an input, and then applying the second function to the output of the first. To be precise, we have the following definition.

Definition 1.5.4 Given two functions f and g, with range(g) $\subset$ domain(f) (that is, every output of the function g is a possible input for the function f), the *composition* of f with g, denoted $f \circ g$, is the new function defined by

$$f \circ g(x) = f(g(x)). \tag{1.33}$$

(R) Supposing that these function compositions are both well defined, it is not true, in general, that $f \circ g$ will be identical to $g \circ f$. In general, the order in which one composes the two functions will yield a different result.

(R) The function composition $f \circ g$ means that one should *apply function g first*. That is, we read function compositions from *right to left*. The rationale for this is as follows: the function closest to the input variable acts on the input first. That is, in order to calculate $f \circ g(x)$, we first let g act on x; we then take the output from function g and input it into function f. Thus, g acts first, then f.

Example 1.17 Consider the functions

$$\begin{aligned} f(x) &= x^2 + 1, \\ g(x) &= \ln(x). \end{aligned}$$

We first observe that

$$\begin{aligned} \text{domain}(f) = \mathbb{R}, \qquad & \text{range}(f) = [1, \infty), \\ \text{domain}(g) = (0, \infty), \qquad & \text{range}(g) = \mathbb{R}. \end{aligned}$$

This confirms that range$(g) \subset$ domain(f) and range$(f) \subset$ domain(g), so that both compositions $f \circ g$ and $g \circ f$ are well defined. We may therefore apply equation (1.33) as follows:

$$\begin{aligned} f \circ g(x) &= f(g(x)) = f(\ln(x)) = [\ln(x)]^2 + 1, \\ g \circ f(x) &= g(f(x)) = g(x^2 + 1) = \ln\left(x^2 + 1\right). \end{aligned}$$

It is evident that the result of forming the composition $f \circ g$ is different than the result obtained from the composition $g \circ f$. In fact, one can check that domain$(f \circ g) = (0, \infty)$, whereas domain$(g \circ f) = \mathbb{R}$. The composite functions $f \circ g$ and $g \circ f$ are shown in Figure 1.21. ∎

1.5.4 Inverse Functions

In our final paragraph of the section, we introduce the concept of the inverse of a function. This is analogous to taking the inverse of a nonzero number x, which is the unique number x^{-1} such that $xx^{-1} = 1$. Here, 1 is the number with the property that $1x = x$ for all other numbers x, i.e., the number 1 is the *multiplicative identity*. To introduce invertible functions, we therefore need a concept of an identity element, which is given in our next definition.

Definition 1.5.5 The *identity function* I is the unique linear function that passes through the origin with unit slope, i.e.,

$$I(x) = x. \tag{1.34}$$

The identity function has the property that $f \circ I = I \circ f = f$. To see this, consider, for example, the function $f(x) = x^2 + 1$. We see that

$$f \circ I(x) = f(I(x)) = f(x), \qquad \text{and} \qquad I \circ f(x) = I(f(x)) = f(x),$$

which verifies our claim (at least for this example function).

The inverse of a function is defined as follows.

Definition 1.5.6 A function f is said to be *invertible* if there exists a unique function f^{-1}, called the *inverse* of the function f, such that

$$f \circ f^{-1} = f^{-1} \circ f = I, \tag{1.35}$$

for all $x \in \text{domain}(f)$. Here, I is the identify function, as defined in equation (1.34).

Example 1.18 The inverse of the exponential function $f(x) = e^x$ is the natural logarithm function $f^{-1}(x) = \ln(x)$. We check by performing the computations

$$\begin{aligned} f \circ f^{-1}(x) &= f(f^{-1}(x)) = f(\ln(x)) = e^{\ln(x)} = x, \\ f^{-1} \circ f(x) &= f^{-1}(f(x)) = f^{-1}(e^x) = \ln(e^x) = x. \end{aligned}$$

Thus, as we previously explored, the exponential function and the natural logarithm functions "undo" each other. ■

The inverse of a function f essentially takes a y-value in the range of f, and returns the x-value that produces this output. For example, if $f(3) = 11$, then $f^{-1}(11) = 3$. This statement relies on the assumption that 3 is the *only* number that the function f sends to 11.

Proposition 1.5.1 A function is invertible if and only if it is *one-to-one*, i.e., if there is a one-to-one correspondence between points in the domain and points in the range.

The proof is straightforward: if there were two distinct x-values, say x_1 and x_2, that both resulted in the same output, say y_0, then that output can not be traced back to its unique origin. In other words, suppose that $f(x_1) = y_0$ and $f(x_2) = y_0$. The inverse f^{-1} then takes y_0 as an input, but cannot return a unique output (and, therefore, cannot be a function).

Functions that are either strictly increasing or strictly decreasing must also be one-to-one, and therefore invertible.

Example 1.19 Determine the inverse of the linear function

$$f(x) = 2x + 7.$$

To achieve this, let us define a variable y to represent the function's output, so that $y = 2x + 7$. Next, let us solve for the input variable x, thereby obtaining

$$x = \frac{y-7}{2}.$$

Thus, the function that traces a point in the range back to its starting point in the domain is the function

$$f^{-1}(x) = \frac{x-7}{2}.$$

To check, consider, for example, the point $x = 3$. We compute $f(3) = 13$. Also, $f^{-1}(13) = (13-7)/2 = 3$. ■

1.5.5 Even and Odd Functions

We conclude this section with a brief discussion of two particular classifications of certain functions: *even* and *odd*.

Definition 1.5.7 A function f is an *even function* if

$$f(-x) = f(x), \tag{1.36}$$

for all x.

A function f is an *odd function* if

$$f(-x) = -f(x), \tag{1.37}$$

for all x.

(R) A given function does not have to be either even or odd; i.e., there are plenty of functions that are neither even nor odd. The properties (1.36) and (1.37) are properties that some functions may have; they are a type of *symmetry*. Thus, in general, a function can be even, odd, or neither.

An even function is a function that is equal to its own horizontal reflection. An example of an even function and an odd function is shown in Figure 1.22.

Notice that the even function is symmetric about the y-axis: the function evaluated at −3, for example, yields the same result as the function evaluated at +3. The odd function, on the other hand, has the opposite property: the function evaluated at −3 yields the opposite number, in sign, as the function evaluated at +3.

Example 1.20 Show that the function

$$f(x) = 4x^3 + 7x$$

is an odd function.

To achieve this, let us compute $f(-x)$:

$$f(-x) = 4(-x)^3 + 7(-x) = -4x^3 - 7x = -f(x).$$

Thus, f is an odd function as it satisfies equation (1.37). ∎

Example 1.21 Show that the function

$$f(x) = 3x^4 + 7x^2 + 9$$

is an even function.

Proceeding in a similar fashion, we obtain

$$f(-x) = 3(-x)^4 + 7(-x)^2 + 9 = 3x^4 + 7x^2 + 9 = f(x).$$

Therefore, the function f is an even function, as it satisfies equation (1.36). ∎

Exercises

For Exercises 1–5, use the given functions f and g to determine (a) $f(g(x))$, and (b) $g(f(x))$.

1. $f(x) = \sqrt{x}$, $g(x) = x^2 + 3$.
2. $f(x) = \sin(x)$, $g(x) = x^2$.
3. $f(x) = e^x$, $g(x) = \sin(x)$.
4. $f(x) = x^2 + x + 7$, $g(x) = \cos(x)$.
5. $f(x) = 8x + 2$, $g(x) = \frac{x}{8} - \frac{1}{4}$.

For Exercises 6–10, compute the inverse of the given function, or show that the function is not invertible.

6. $f(x) = 7x + 11$.
7. $f(x) = \ln(3x + 9)$, for $x \geq 0$.
8. $f(x) = x^3 + 1$.
9. $f(x) = e^{x^3}$.
10. $f(x) = \ln|x|$.

For Exercises 28–15, simplify the given expression, using the function $f(x) = 3x^2 + 2x + 7$.

11. $f(x+1)$.
12. $f(x+h)$.
13. $f(x) + h$.
14. $f(x+h) - f(x)$.
15. $\frac{f(x+h)-f(x)}{h}$.

For Exercises 16–20, explain which function transformations (in which order) can be used to transform the given function f into the given function g.

16. $f(x) = e^x$, $g(x) = e^{2x+6}$.
17. $f(x) = 3x^2 + 2x + 1$, $g(x) = 9x^2 - 6x + 1$.
18. $f(x) = x^3 + x^2 + x$, $g(x) = x^3 - x^2 + x$.
19. $f(x) = \sin(x)$, $g(x) = 8 + 2\sin(x)$.
20. $f(x) = \cos(x)$, $g(x) = 8 + 2\cos(3x + 7)$.

For Exercises 21–25, determine whether or not the stated function is invertible. Explain your reasoning and state any assumptions you make.

21. The temperature of a turkey as a function of the amount of time it's been in a hot oven.
22. The mileage of an automobile as a function of the number of years since it was first sold.
23. The fuel economy of an automobile as a function of its cruising speed.
24. The average cost of a movie ticket as a function of the year.
25. The size of a harvest as a function of the number of seeds that were planted.

Problems

26. A certain audio signal can be described by the equation
$$A(t) = 120\cos(120\pi t).$$
A circuit filter that the signal is processed through creates a delay of 4 ms (milliseconds) in the delay. Determine the new function that describes the signal. What type of function transformation has occurred.
27. The atmospheric pressure P, measured in kPa, at an altitude of h km is given by the relation
$$P(h) = 101e^{-0.14h}.$$
If the altitude of a hot air balloon can be described by the relation
$$h(t) = 1 + 0.05t,$$
t minutes into its flight, write a function that represents the atmospheric pressure as a function of the time into the flight.

28. A certain self-publishing company that allows authors to publish their own works charges a printing fee of \$11.00 for binding plus \$0.19 per page in order to print a large sized, paperback, full-color book.

(a) Express the total printing cost $C(p)$ as a linear function of the number of pages p contained in the book. Evaluate $C(300)$, the cost to print a 300-page full-color book.

(b) Let x represent the retail price of the book. The bookstore/distribution share of this amount is typically 55%, i.e., 45% of the list price is paid to the publisher. After this discount is taken away from the retail price, the cost C to print the book is subtracted. From this amount, the publisher pays the author a certain percentage of what is left over as a royalty for each sale. For mainstream publishers, the royalty percentage is typically between 5%–15%, and for self-publishing companies, the royalty can be as high as 80%. (Self-publishing companies, however, do not invest any capital in terms of editing, formatting, design, advertising, etc.) Suppose that Mr. Joyce wishes to self-publish a 300-page, full-color book with a self-publishing company that gives an author royalty of 80%. Express Mr. Joyce's royalty amount $R(x)$ as a function of the list price of the book.

(c) What must the list price be in order for Mr. Joyce to earn a royalty of \$20 for each book sold?

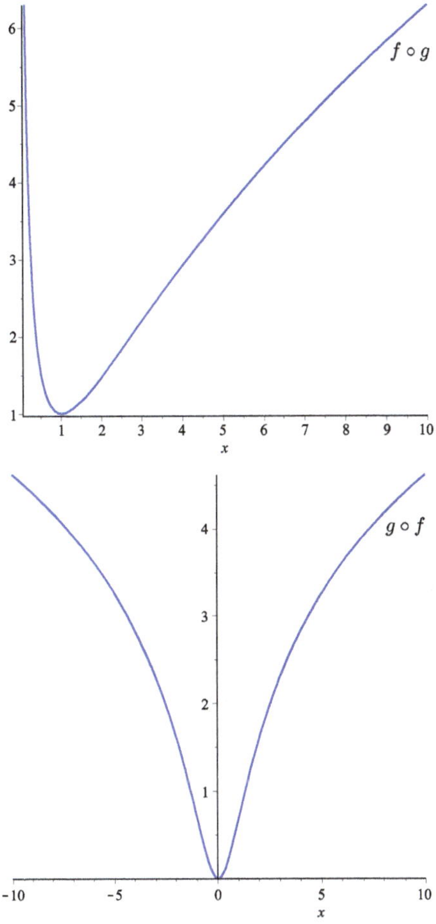

Figure 1.21: Function composition $f \circ g$ (left) and $g \circ f$ (right).

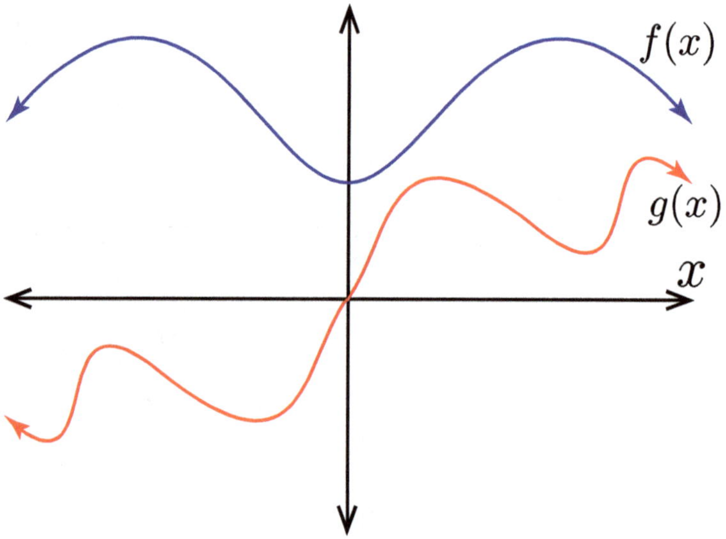

Figure 1.22: An even function f and an odd function g.

1.6 Trigonometric Functions

1.6.1 Basic Trig Functions and the Unit Circle

The two most basic trigonometric functions are the cosine and sine functions, which are denoted by $\cos(\theta)$ and $\sin(\theta)$, respectively. These functions represent the coordinates of a point on the unit circle, as we see in the following definition.

Definition 1.6.1 The *cosine* and *sine* functions, denoted by $\cos(\theta)$ and $\sin(\theta)$, respectively, return the x- and y-coordinates for a point on the unit circle, as a function of the given point's angle θ, as measured in radians, counterclockwise from the positive x-axis.

The *tangent* function, denoted $\tan(\theta)$, is defined as the ratio

$$\tan(\theta) = \frac{\sin(\theta)}{\cos(\theta)}. \tag{1.38}$$

A picture of the unit circle showing the cosine and sine functions is given in Figure 1.23. The angle θ is measured *from* the positive x-axis, so as the point on

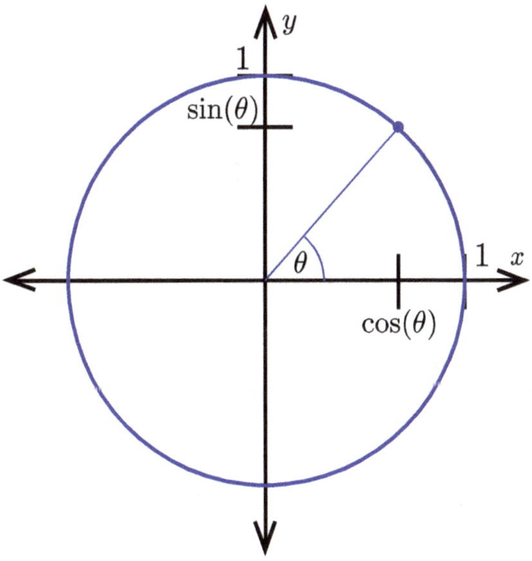

Figure 1.23: The unit circle and the cosine and sine functions.

the unit circle moves counterclockwise, the angle θ increases.

(R) For the purpose of these trig functions, the angle θ is **always** measured in radians.

Radians are a natural way of measuring angles, as the angle, when measured in radians, is literally equal to the corresponding arc length along the circumference of the unit circle. This is why one full circle has an angular measurement of 2π radians—the circumference of a unit circle is 2π.

Given these geometric definitions of the cosine and sine functions, we immediately have the following result.

Proposition 1.6.1 For any angle θ, the trigonometric identity

$$\sin^2(\theta) + \cos^2(\theta) = 1 \tag{1.39}$$

holds.

To see why the preceding proposition is true, we must only recognize that the x- and y-coordinates form two legs of a right triangle, as shown in Figure 1.24. Since the point (x, y) is located on the unit circle, the hypotenuse has length 1. The result therefore follows from the Pythagorean theorem.

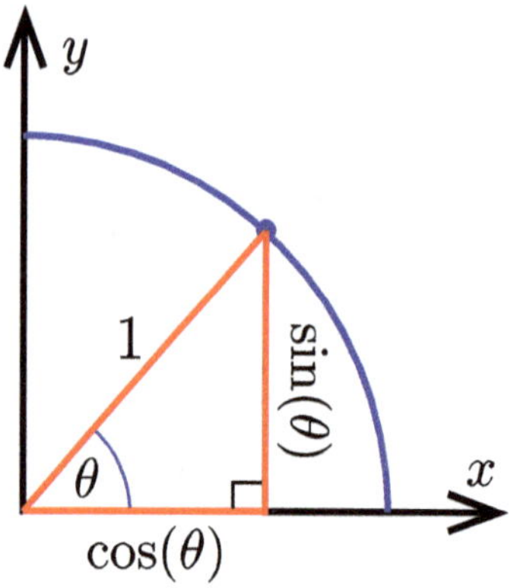

Figure 1.24: Graphical representation of the identity $\sin^2(\theta) + \cos^2(\theta) = 1$.

We may further infer from Definition 1.6.1 the following fact: a point (x, y) on the unit circle can be expressed as a function of the angle θ via the relations

$$x(\theta) = \cos(\theta), \tag{1.40}$$
$$y(\theta) = \sin(\theta). \tag{1.41}$$

Similarly, a point (x, y) on a circle of radius r can be described as a function of the angle θ via the relations

$$x(\theta) = r\cos(\theta), \tag{1.42}$$
$$y(\theta) = r\sin(\theta). \tag{1.43}$$

(Both the x- and y-coordinates of the unit circle are multiplied by a factor of r, therefore promoting the point to lie on a circle of radius r instead of a circle of radius 1.)

The functions $x(\theta) = \cos(\theta)$ and $y(\theta) = \sin(\theta)$ are plotted in Figure 1.25. To

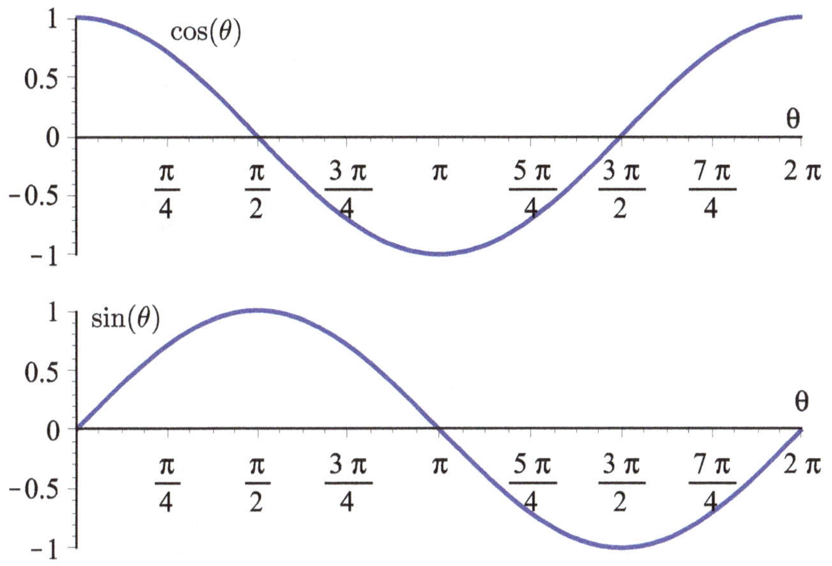

Figure 1.25: The functions $x(\theta) = \cos(\theta)$ and $y(\theta) = \sin(\theta)$.

see why these functions take this shape, one can literally trace the unit circle, in the counterclockwise sense, with one's finger, starting at the point (1,0), where the unit circle intersects the positive x-axis. If one tracks just the x-coordinate as a function of the angle θ, one will produce the cosine curve shown in the top of Figure 1.25. (When $\theta = 0$, the x-coordinate is 1. When $\theta = \pi/2$, i.e., 90 degrees, the x-coordinate is now zero, and so on.) Similarly, by tracing the unit circle and tracking the y-coordinate, as it varies with θ, one can produce the sine curve.

The sine and cosine functions are plotted concurrently in Figure 1.26. Notice that these functions are nearly identical, except that the cosine function is a copy of the sine function that is shifted $\pi/2$ units to the left, i.e., it is an example of a horizontal translation. Alternatively, we could say that the sine function is the cosine function shifted $\pi/2$ units to the right. This implies the following fact.

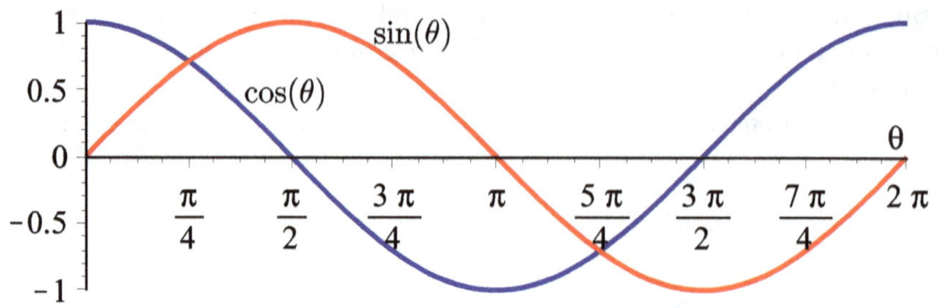

Figure 1.26: The function $x(\theta) = \cos(\theta)$, plotted concurrently.

Proposition 1.6.2 The following identities holds:

$$\sin(\theta) = \cos\left(\theta - \frac{\pi}{2}\right), \tag{1.44}$$

$$\cos(\theta) = \sin\left(\theta + \frac{\pi}{2}\right). \tag{1.45}$$

Some notable points along the unit circle are given in Table 1.11. The corresponding values of $\tan(\theta)$ are also listed. Also, recall that $\pi/6$ radians is equivalent to 30°; $\pi/4$ radians is equivalent to 45°; $\pi/3$ radians is equivalent to 60°; and $\pi/2$ radians is equivalent to 90°. Note that $\tan(\pi/2)$ is actually undefined; we

θ	$\sin(\theta)$	$\cos(\theta)$	$\tan(\theta)$
0	0	1	0
$\frac{\pi}{6}$	0.5	$\frac{\sqrt{3}}{2}$	$\frac{1}{\sqrt{3}}$
$\frac{\pi}{4}$	$\frac{1}{\sqrt{2}}$	$\frac{1}{\sqrt{2}}$	1
$\frac{\pi}{3}$	$\frac{\sqrt{3}}{2}$	0.5	$\sqrt{3}$
$\frac{\pi}{2}$	1	0	"∞"

Table 1.11: Notable points on the unit circle; $\tan(\pi/2)$ is undefined.

listed it as "∞" only to remind the reader that the tangent function has a vertical asymptote at this point.

The tangent function, which is defined as the ratio of sine divided by cosine, is plotted in Figure 1.27. Whereas sine and cosine repeat every 2π units, the tangent function is defined once on the interval $\left(\frac{-\pi}{2}, \frac{\pi}{2}\right)$ and repeated every π units. The tangent function is undefined for the odd multiples (positive and negative) of $\pi/2$—the same places where the cosine function returns an output of zero.

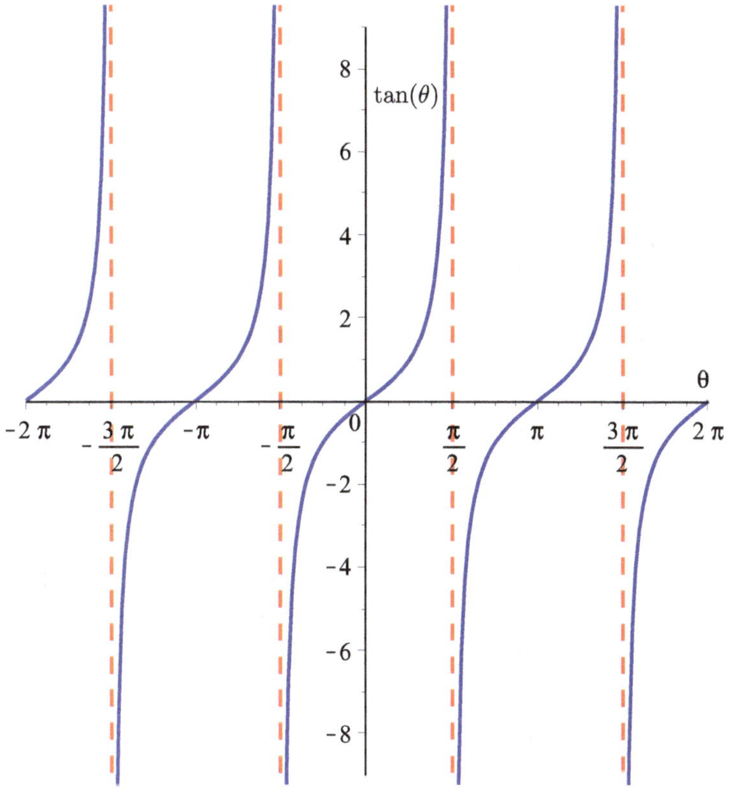

Figure 1.27: The function $\tan(\theta)$; vertical asymptotes are represented by the red dashed lines.

The sign of the tangent function is determined by the quadrant. If the point $(\cos(\theta), \sin(\theta))$ on the unit circle is in the first quadrant, then $\tan(\theta) > 0$. This is apparent in the graph of the tangent function, since $\tan(\theta) > 0$ over the interval $\theta \in \left(0, \frac{\pi}{2}\right)$, which corresponds to the angular range of the first quadrant. Similarly, if the point (x, y) is in the second or fourth quadrants, the ratio x/y will be negative, and, therefore $\tan(\theta)$ will be negative. The second quadrant corresponds to the angles $\frac{\pi}{2} < \theta < \pi$, and the fourth quadrant corresponds to the angles $\frac{-\pi}{2} < \theta < 0$. Finally, one can check that the tangent function returns positive values if the angle θ lies in the fourth quadrant, i.e., if $-\pi < \theta < \frac{-\pi}{2}$.

In the preceding discussion, we treated the unit circle as it ranges from $-\pi < \theta < \pi$. Since the domain is circular, the values of θ are not unique. For example, the angles $\theta = -\pi$ and $\theta = \pi$ both refer to the negative x-axis, and so forth. In fact, whenever two angles differ by an integer multiple of 2π, they refer to the same point.

1.6.2 Anatomy of Sinusoidal Functions

When one first learns trigonometry, one learns about triangles. Then, at some point, one eventually learns that these trigonometric functions are related to the unit circle. However, their efficacy at curing problems and modeling real-world situations does not stop at triangles and circles—trigonometric functions are powerful devices that are commonly used to model a variety of practical situations. They are used extensively in signal processing, the mathematics of waves, electrical circuit design, mechanics, quantum mechanics, and so forth. In fact, an advanced study of applied mathematics will ultimately reveal that all periodic phenomenon can be decomposed into an infinite collection of sines and cosines, and this fact has been instrumental in much of the development of modern theoretical physics and engineering. It would have been impossible to design cell phones, telecommunications networks, computers, and the shock absorbers in automobiles if it were not for trigonometric functions. (And calculus!)

In this paragraph, we will examine the simplest sort of periodic functions—those composed of sines and cosines. Such functions are often called *sinusoidal*. To begin, we lay out some basic nomenclature for general periodic functions.

Definition 1.6.2 A function f is *periodic* if it repeats every p units, for some positive number p, i.e., if $f(x+p) = f(x)$, for every $x \in \mathbb{R}$. The *period* of a periodic function f is the smallest positive number p such that $f(x+p) = f(x)$ for every $x \in \mathbb{R}$. A periodic function with period p is sometimes referred to as being *p-periodic*.

Given a periodic function f, its amplitude A is half of the difference between its maximum value and its minimum value, if this number exists.

■ Example 1.22 The functions $\sin(\theta)$ and $\cos(\theta)$ are periodic functions with period $p = 2\pi$, i.e., they are 2π-periodic. Their amplitude is 1, since the maximum and minimum output of the sine and cosine functions are 1 and -1, respectively.

The function $\tan(\theta)$ is a periodic function with period $p = \pi$, i.e., the tangent function is π-periodic. It does not have an amplitude. ■

As general sinusoidal functions are commonly used to model quantities that vary sinusoidally with time, we will use the independent variable t to describe the general sinusoidal function, as we now define.

Definition 1.6.3 A general *sinusoidal function* can be expressed as

$$f(t) = M + A\cos(\omega t - \varphi), \tag{1.46}$$

where M is the *midline*, A is the *amplitude*, ω is the *angular frequency*, and φ is the *phase shift*.

To understand equation (1.46), let us break it down into its constituents. First, let us focus on what is happening *outside* the cosine function. The midline is the horizontal line $y = M$ about which the function is centered, and the amplitude tells us how far from the midline that the outputs can vary. This is shown for the function $f(\theta) = M + A\cos(\theta)$ in Figure 1.28.

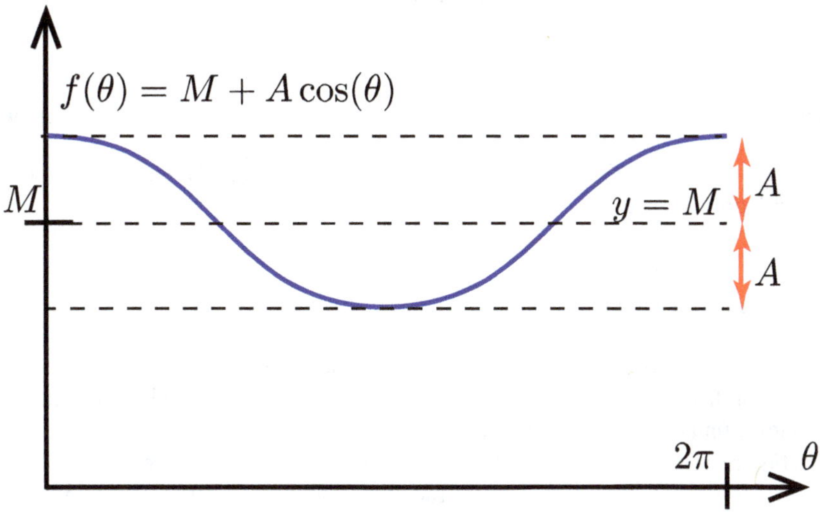

Figure 1.28: Midline and amplitude for the function $f(\theta) = M + A\cos(\theta)$.

The function $f(\theta) = M + A\cos(\theta)$ can be obtained from the cosine function by applying two vertical transformations: starting with $\cos(\theta)$, we first perform a vertical scaling (stretch or compression) by a factor of A, to obtain the function $A\cos(\theta)$. Instead of oscillating between $y = -1$ and $y = +1$, the function $A\cos(\theta)$ now oscillates between $y = -A$ and $y = +A$. Next, we apply a vertical translation (up or down) in the amount M, to obtain the function $f(\theta) = M + A\cos(\theta)$. This new function now oscillates between $y = M - A$ and $y = M + A$, as shown in Figure 1.28.

Similarly, the horizontal transformations can be regarded as follows: first, we do a horizontal shift to the right by φ radians (or *to the left*, if $\varphi < 0$). Then we perform a horizontal compression by a factor of ω (or a horizontal stretch, if $\omega < 1$). While this describes the horizontal transformation, which happen on the *inside* of the cosine function, there is also a more practical way to think about these transformations.

First, the phase shift φ represents the angle (radians) by which the cosine function is shifted *to the right*. For instance, if $\varphi = \pi/2$, the cosine function is shifted $\pi/2$ radians to the right, and the resulting graph is the sine wave, and so forth. It is important to recognize that the units of φ are radians. The angle is sometimes referred to as the *phase*, which is why φ is known as the *phase shift*—it literally represents a shift in the phase, for functions that do not start at the crest of the wave like the cosine function does.

Second, the angular frequency ω can be related to the period of the wave. To see this, we notice that the inside function $\theta = \omega t - \varphi$ causes the graph to repeat every time θ is incremented by 2π radians. Thus, ωt must be incremented by 2π. Setting $t = p$, and $\theta(p) - \theta(0) = 2\pi$, we see that

$$\omega p = 2\pi.$$

Therefore, the period and the angular frequency are related to each other as follows.

Proposition 1.6.3 Given a general sinusoidal function (1.46), the period of the oscillation is equal to

$$p = \frac{2\pi}{\omega}, \tag{1.47}$$

where ω is the angular frequency.

If time is measured in seconds, the units of the angular frequency ω are radians per second. In general, the units of the angular frequency are radians divided by whatever units the variable t is measured in.

Figure 1.29 shows how the function $f(\theta) = M + A\cos(\theta - \varphi)$ represents a horizontal translation to the right by φ radians. The new function is drawn in red. Finally, to capture the effect of the angular frequency, we can replace the θ-axis with the t-axis, and relabel the point $\theta = 2\pi$ as the point $t = 2\pi/\omega$. Thus, the angular frequency simply rescales the horizontal axis, so that each cycle occurs over a period of $p = 2\pi/\omega$ (units of time), as opposed to over an angular period of 2π radians. Thus, it factors in the notion of *dynamics*—that the sinusoidal wave occurs for a duration in *time*, not just abstractly as a range of angles.

The preceding discussion is intended to instill in the reader a graphical understanding of the equation (1.46). Next, we turn to some particular problems and examine how we can *apply* this knowledge to model various real-world situations.

▪ **Example 1.23** On a certain day at Monterey Bay, California, low tide was at 6:30 a.m., and high tide was at 1:10 p.m. If the water level at low tide was 0.4 ft, and the water level at high tide was 4.8 ft, model the water level in Monterey Bay as a sinusoidal function of time t, measured in hours since noon. (Time before noon will be negative.)

First, let us convert the time at low tide and high tide into the variable t, which represents the number of hours after noon. High tide, which occurred at 1:10 p.m.,

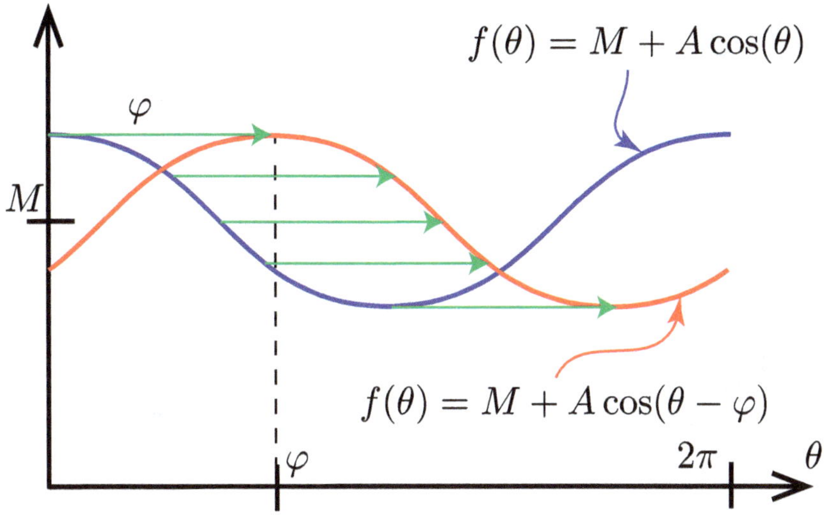

Figure 1.29: The function $f(\theta) = M + A\cos(\theta - \varphi)$.

is therefore represented by $t = 7/6$, since 1:10 p.m. is 70 minutes, or 7/6ths of an hour, after noon. Low tide, on the other hand, occurred at 6:30 a.m., which is five and a half hours before noon, so that $t = -5.5$ at low tide.

Next, we can make a sketch of the water level $h(t)$ as a function of time t, as shown in Figure 1.30. The midline and the amplitude are easy to find, so let's start with those. The amplitude is going to be half the distance between high tide and low tide. Therefore, we compute

$$A = \frac{4.8 - 0.4}{2} = 2.2 \text{ ft.}$$

Since the low tide is one amplitude below the midline, we may compute the midline as

$$M = 0.4 + 2.2 = 2.6 \text{ ft.}$$

To check, we verify that $M + A$ is the water level at high tide, and $M - A$ is the water level at low tide, which can easily be confirmed.

Next, let us compute the angular frequency ω. The period is going to be twice the time between low tide and high tide, as the length of time between low and high tide represents one half of a complete cycle. Thus, half the period is given by $p/2 = 7/6 - (-5.5) = 20/3$. The period is therefore $p = 40/3$. Using the relation (1.47), the angular frequency works out to be

$$\omega = \frac{3\pi}{20} \text{ rad/hr.}$$

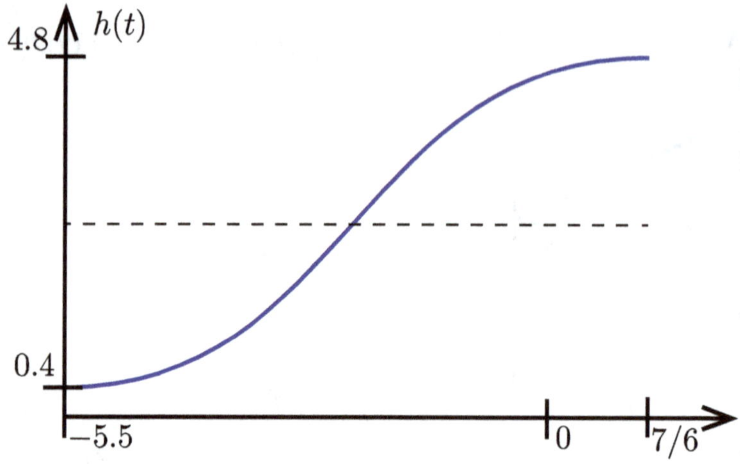

Figure 1.30: Graph of the water level at Monterey Bay.

The final step is to figure out the phase of the tide at 12:00 noon. To achieve this, we must basically convert time into radians, and think about where in the cycle the tides are at time $t = 0$. The water level has its peak at time $t = 7/6$; we can therefore view it as a cosine function shifted 7/6 hours to the right. To convert from hours to radians, we simply identify the period of the oscillation with a complete cycle, so that

$$40/3 \text{ hours} = 2\pi \text{ radians.}$$

Therefore, the phase shift will be

$$\varphi = -\frac{7}{6} \text{ hrs} \times \frac{2\pi \text{ radians}}{40/3 \text{ hours}} = \frac{-7\pi}{40} \text{ radians.}$$

The phase shift is negative because we are shifting the cosine function to the right. The final equation is therefore given by

$$h(t) = 2.6 + 2.2\cos\left(\frac{3\pi t}{20} - \frac{7\pi}{40}\right) \text{ feet.}$$

To verify our work, we can plot this function using computer software or a graphing calculator, with which we obtain an image like the one in Figure 1.31. (Note that, in this Figure, the t-axis does not pass through the vertical axis at the origin.) This function matches the physical description of the tide data, so we may conclude that our solution is correct. ■

Finally, we can relate the general form of a sinusoidal function, as given by equation (1.46), to a simple sum of sines and cosines, using the following proposition.

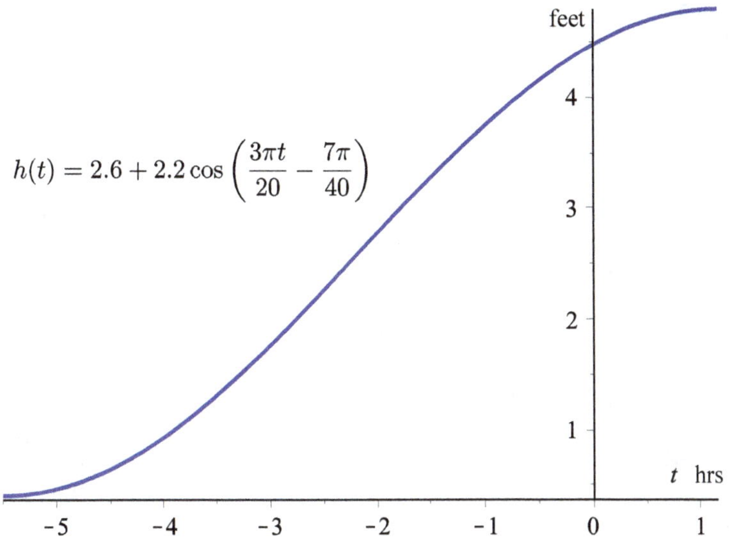

Figure 1.31: Graph of the water level at Monterey Bay.

Proposition 1.6.4 A sinusoidal function may alternatively be described using

$$f(t) = M + A\cos(\omega t - \varphi) \tag{1.48}$$

or

$$f(t) = M + c_1\cos(\omega t) + c_2\sin(\omega t). \tag{1.49}$$

In doing so, the parameters (A, φ) are related to the parameters (c_1, c_2) by the following formulas

$$c_1 = A\cos(\varphi), \tag{1.50}$$

$$c_2 = A\sin(\varphi), \tag{1.51}$$

$$A^2 = c_1^2 + c_2^2, \tag{1.52}$$

$$\tan(\varphi) = \frac{c_2}{c_1}. \tag{1.53}$$

Thus, if one knows A and φ, one can use equations (1.50) and (1.51) to solve for c_1 and c_2. Similarly, if one knows c_1 and c_2, one can use equations (1.52) and (1.53) to solve for the amplitude and phase shift.

Proof. To start, let us evaluate $f(0) - M$ using both equations (1.48) and (1.49). Setting these two evaluations equal to each other, we obtain

$$c_1 = A\cos(-\varphi).$$

But since cosine is an even function, it follows that $\cos(-\varphi) = \cos(\varphi)$, and equation (1.50) follows.

Next, let us evaluate $f\left(\frac{\pi}{2\omega}\right) - M$ using the two different forms. We obtain

$$A\cos\left(\frac{\pi}{2} - \varphi\right) = c_2,$$

since $\cos(\pi/2) = 0$ and $\sin(\pi/2) = 1$. Now, $\cos(\theta + \pi/2) = -\sin(\theta)$, as adding $\pi/2$ to the argument represents a shift by $\pi/2$ radians to the left, which therefore produces the negative of the sine wave. Thus, the preceding equation is equivalent to

$$-A\sin(-\varphi) = c_2.$$

Since sine is an odd function, it follows that $\sin(-\varphi) = -\sin(\varphi)$, and equation (1.51) follows.

Equation (1.52) follows by summing the squares of equations (1.50) and (1.51). Equation (1.53) follows by computing the ratio of equation (1.51) to equation (1.50). This completes the proof. ■

1.6.3 Inverse Trigonometric Functions

The three basic trigonometric functions—sine, cosine, and tangent—are, by themselves, not invertible functions, when considered on their full domain. (None of these passes the horizontal line test.) An inverse function for each of these can, however, be identified, by appropriately restricting the section of the domain on which these functions are defined.

First, let us consider the sine function. The function $\sin(\theta)$ *is* invertible if we restrict it to the domain $\left[-\frac{\pi}{2}, \frac{\pi}{2}\right]$, since each value of $\theta \in \left[-\frac{\pi}{2}, \frac{\pi}{2}\right]$ matches with exactly one output value on the range $[-1, 1]$, as shown in Figure 1.32.

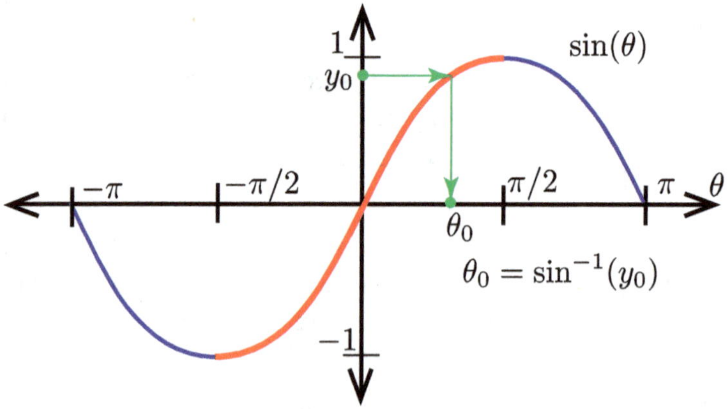

Figure 1.32: The sine function and computing $\sin^{-1}(y_0)$.

The inverse sine function, also called the arcsine function, is usually denoted using the notation $\sin^{-1}$. Evaluated at a point on the range $y_0 \in [-1,1]$, the arcsine function returns the unique point $\theta_0 = \sin^{-1}(y_0)$ in the interval $\left[-\frac{\pi}{2}, \frac{\pi}{2}\right]$ such that $\sin(\theta_0) = y_0$.

The inverse of the cosine function can similarly be constructed. We observe that the cosine function is one-to-one on the interval $\theta \in [0,\pi]$, i.e., each input $\theta \in [0,\pi]$ uniquely corresponds to a single output value in the range $[-1,1]$. By restricting the cosine function to this domain, we can therefore construct the inverse cosine function, also known as the arccosine function, as shown in Figure 1.33

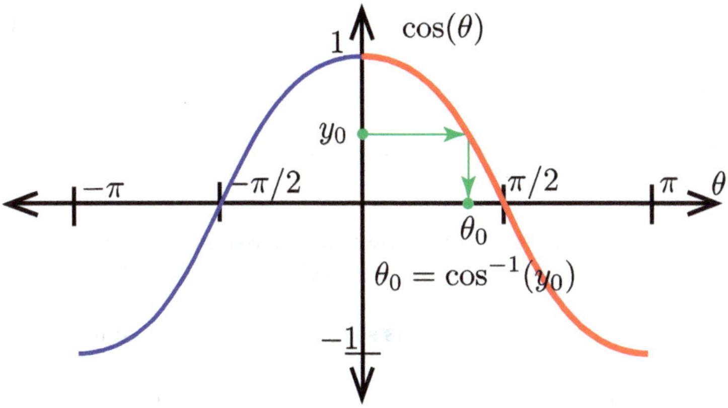

Figure 1.33: The cosine function and computing $\cos^{-1}(y_0)$.

We may formalize the preceding discussion with the following definition.

Definition 1.6.4 — Inverse Trig Functions. The arcsine function, $\sin^{-1}$, assigns to any input $y_0 \in [-1,1]$ the unique number $\theta_0 = \sin^{-1}(y_0) \in \left[-\frac{\pi}{2}, \frac{\pi}{2}\right]$ such that $y_0 = \sin(\theta_0)$.

The arccosine function, $\cos^{-1}$, assigns to any input $y_0 \in [-1,1]$ the unique number $\theta_0 = \cos^{-1}(y_0) \in [0,\pi]$ such that $y_0 = \cos(\theta_0)$.

R The notation $\sin^{-1}(x)$ and $\cos^{-1}(x)$ does *not* indicate the reciprocal of $\sin(x)$ and $\cos(x)$, respectively, i.e.,

$$\sin^{-1}(x) \neq \frac{1}{\sin(x)} \quad \text{and} \quad \cos^{-1}(x) \neq \frac{1}{\cos(x)}.$$

Only for values of n not equal to -1 does the shorthand

$$\sin^n(x) = [\sin(x)]^n, \qquad \cos^n(x) = [\cos(x)]^n, \qquad \text{for } n \neq -1$$

apply. The notation $\sin^{-1}(x)$ and $\cos^{-1}(x)$ is reserved exclusively for the arcsine and arccosine functions, respectively, i.e., it is reserved for the *function inverse* of sine and cosine.

In order to denote the reciprocal of sine or cosine, we instead utilize the following notation.

Definition 1.6.5 The secant function, $\sec(\theta)$, is the reciprocal of the cosine function, i.e.,

$$\sec(\theta) = \frac{1}{\cos(\theta)}, \tag{1.54}$$

defined whenever $\cos(\theta) \neq 0$.

The cosecant function, $\csc(\theta)$, is the reciprocal of the sine function, i.e.,

$$\csc(\theta) = \frac{1}{\sin(\theta)}, \tag{1.55}$$

defined whenever $\sin(\theta) \neq 0$.

Next, let us define the inverse tangent function (i.e., the arctangent function) and the cotangent function.

Definition 1.6.6 The arctangent function, $\tan^{-1}$, assigns to any input $y_0 \in \mathbb{R}$ the unique number $\theta_0 = \tan^{-1}(y_0) \in \left(-\frac{\pi}{2}, \frac{\pi}{2}\right)$ such that $y_0 = \tan(\theta_0)$.

An illustration of the arctangent is shown in Figure 1.34. Notice that, unlike sine and cosine, the range of the tangent function is all real numbers. Thus, for any real number y_0, we can trace it back to a unique value θ_0 on the interval $\left(\frac{-\pi}{2}, \frac{\pi}{2}\right)$. Recall that the tangent function is undefined at $\pi/2$ and $-\pi/2$, so that the arctangent function returns a value on the *open interval* $\frac{-\pi}{2} < \theta < \frac{\pi}{2}$.

Again, the arctangent function is the function inverse—not the reciprocal. Thus,

$$\tan^{-1}(x) \neq \frac{1}{\tan(x)},$$

even though that for values of n other than -1 we have

$$\tan^n(x) = [\tan(x)]^n, \qquad \text{for } n \neq -1.$$

The particular notation $\tan^{-1}(x)$ is again reserved for the arctangent function, which is the *function inverse*. To describe the reciprocal, we use the cotangent function.

Definition 1.6.7 The *cotangent function* is the reciprocal of the tangent function, i.e.,

$$\cot(x) = \frac{1}{\tan(x)}, \tag{1.56}$$

defined whenever $\tan(x) \neq 0$.

Exercises

For Exercises 1–5, write a formula for the sinusoidal function with the given

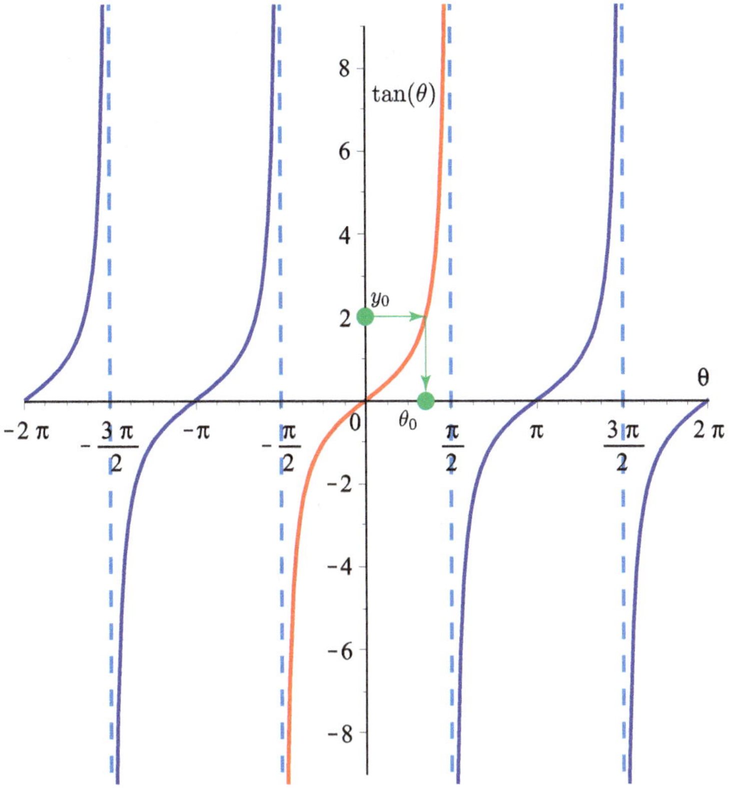

Figure 1.34: The tangent function and computing $\theta_0 = \tan^{-1}(y_0)$.

characteristics.

1. midline $y = 3$, amplitude 2, angular frequency 4, phase shift 0.
2. midline $y = 7$, amplitude 9, angular frequency 2π, phase shift π.
3. midline $y = -3$, amplitude 3, period 5, phase shift 6.
4. midline $y = 1$, amplitude 2, period 3, phase shift $-\pi/2$.
5. midline $y = 7$, amplitude 8, period 8, phase shift 9.

For Exercises 6–15, identify the midline, amplitude, frequency, period, and phase shift for the given sinusoidal function.

6. $f(t) = 8 + 9\cos(3t)$.
7. $f(t) = \cos(\pi t) - 1$.
8. $f(t) = 2 + 2\cos(t - \pi)$.
9. $f(t) = 3 - 2\cos(2\pi t)$.
10. $f(t) = 2 + 7\sin(8t)$.
11. $f(t) = 3\cos(4t) + 4\sin(4t)$.
12. $f(t) = 5 + 5\cos(t/\pi) + 12\sin(t/\pi)$.
13. $f(t) = \cos(8t) + 3\sin(8t) - 1$.
14. $f(t) = 3 + 3\cos(4t) - 4\sin(4t)$.
15. $f(t) = 2 + 8\cos(\pi t) + 9\sin(\pi t)$.

For Exercises 16–20, use the inverse-trigonometric functions to find the solution(s) of the indicated equation on the specified domain.

16. $\sin(\theta) = \frac{\sqrt{3}}{2}$, $\theta \in \left[\frac{-\pi}{2}, \frac{\pi}{2}\right]$.
17. $\sin(\theta) = \frac{-1}{2}$, $\theta \in [\pi, 2\pi]$.
18. $\cos(\pi t) = \frac{-1}{\sqrt{2}}$, $t \in [0,1]$.
19. $\cos(\pi t) = \frac{-\sqrt{3}}{2}$, $t \in [1,2]$.
20. $\tan(\theta) = 1$, $\theta \in \left[\frac{\pi}{2}, \frac{3\pi}{2}\right]$.

Problems

21. The voltage in an electrical outlet in the U.S. is given by

$$V(t) = 170\sin(120\pi t).$$

What is the amplitude and period of the oscillation?

22. The voltage in an electrical outline in Europe is given by

$$V(t) = 325\sin(100\pi t).$$

What is the amplitude and period of the oscillation?

23. Suppose that average temperature varies sinusoidally as a function of the time of the year. Suppose that the highest temperature occurs on the summer solstice, which is 2/3rds through the month of June, and that the lowest temperature occurs on the winter solstice, which is six months later, 2/3rds through the month of December. If the high temperature is 100° F and the low temperature is −10° F, express the average temperature $T(t)$ as a function of the number of months t since the beginning of the year. Identify the midline, amplitude, angular frequency, and phase shift.
24. On a given day, the high temperature is at 2:00 p.m., when it is 72° F. The low temperature occurs at 2:00 a.m., when it is 45° F.
    (a) Model the temperature as a sinusoidal function of time, where time t is measured in hours since midnight.
    (b) Predict what the temperature will be at 7:00 p.m. that evening.
25. The London Eye ferris wheel stands 135 m high and has a diameter of 120 m. It takes 30 minutes to make a complete revolution around the wheel. Determine the height of a passenger above the ground as a function of time t, measured in minutes since the ride began.
26. In 2014, Anchorage, Alaska will have 19 hours and 21 minutes of daylight on the day of June 21, the summer solstice, but only 5 hours and 28 minutes of daylight on December 21, the winter solstice. Approximating the summer and winter solstices as 17/3 and 35/3 months into the year, respectively, determine a sinusoidal function that models $D(t)$, the number of hours of daylight t months after the start of the year.

1.7 Limits and Continuity

In this section, we explore the concept of a limit. At first the concept may seem simple or obvious, but it turns out to be foundational in the establishment of calculus.

1.7.1 Limits Defined

We begin with the definition of a limit. Limits allow us to say things about the *behavior* of a function near a point at which the function may not even be defined.

Definition 1.7.1 Suppose that a function f is defined on some open interval containing the point $x = a$, except, perhaps, at a itself. We say that *the limit as x approaches to a of $f(x)$* is equal to L, written

$$\lim_{x \to a} f(x) = L, \tag{1.57}$$

if the outputs $f(x)$ become closer and closer to the value L as the inputs x become closer and closer to the point $x = a$. The value of the function f at the point $x = a$, if it is even defined at that point, does not affect the value of the limit.

Limits are especially useful when a function is not defined at a point, but when we nevertheless wish to know what value the function *would* take, if we were to fill in that hole using the function's behavior close to the missing point. As we shall see in our next chapter, this becomes a crucial operation when trying to understand the *instantaneous* rate of change of a function, for example, when trying to understand the speed of a particle at a given instant in time.

▪ **Example 1.24** Consider the function f as shown in Figure 1.35. For this function, observe that $f(4) = 3.5$, even though the output values of the function f are not close to the output 3.5 when x is close to, but not equal to, the input point 4. This is an example of a function "with a hole in it."

The behavior of the function f near the point $x = 4$, but not at $x = 4$, suggests that the output of the function at $x = 4$ *should* equal 2. In other words, as the x-values become closer and closer to the point $x = 4$, the y-values are becoming closer and closer to an output of $y = 2$. We therefore capture this information by writing

$$\lim_{x \to 4} f(x) = 2.$$

This tells us about what is happening as x *nears* the point $x = 4$, ignoring what the output at the point $x = 4$ actually is, if it is even defined. ▪

Sometimes the behavior of a function near a point depends on whether you are approaching that point from the left or from the right, i.e., the function may be nearing two different y-values depending on which side of the point you are on. In

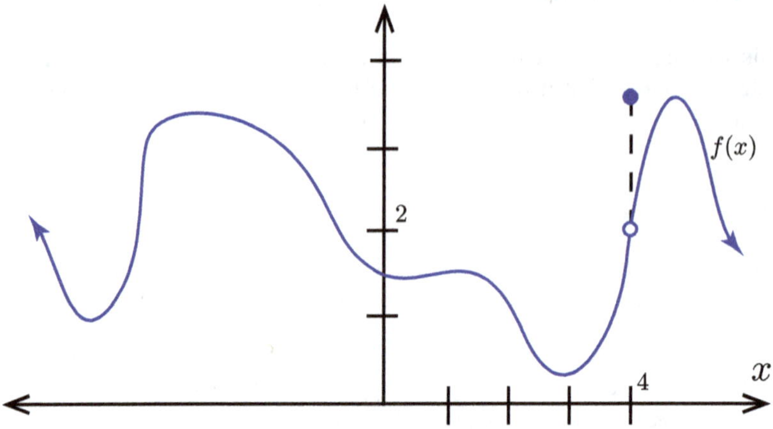

Figure 1.35: Function from Example 1.24.

this case, the limit does not exist; one can, however, define two creatures known as left-hand and right-hand limits to nevertheless capture this behavior.

Definition 1.7.2 — Left-hand and Right-hand Limits. We say that the *limit as x approaches a from the left* of $f(x)$ is equal to L, written as

$$\lim_{x\to a^-} f(x) = L,$$

if the values of $f(x)$ become closer and closer to L as the inputs x become closer and closer to the point a, only considering inputs that satisfy $x < a$.

Similarly, we say that the *limit as x approaches a from the right* of $f(x)$ is equal to L, written as

$$\lim_{x\to a^+} f(x) = L,$$

if the values of $f(x)$ become closer and closer to L as the inputs x become closer and closer to the point a, only considering inputs that satisfy $x > a$.

Example 1.25 Consider the function f shown in Figure 1.36. Here, the limit as x approaches 4 of $f(x)$ does not exist, because the y-values do not approach a consistent output for all values of x near enough to $x = 4$.

The one-sided limits, however, do exist. In fact, we see that

$$\lim_{x\to 4^-} f(x) = 1 \qquad \text{and} \qquad \lim_{x\to 4^+} f(x) = 3.$$

That is, as x approaches 4 *from the left*, the outputs are approaching an output of 1, and as x approaches 4 *from the right*, the outputs are approaching an output of 3.

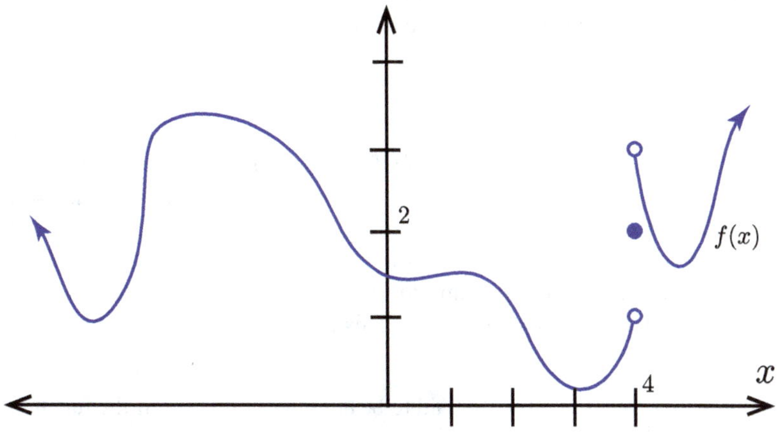

Figure 1.36: Function from Example 1.25.

Furthermore, we observe that neither the left-hand or right-hand limits equal the actual function output $f(4) = 2$. ■

1.7.2 Properties of Limits

There are certain properties of limits that often come to aid when attempting to evaluate limits using algebraic techniques. These properties are as follows.

Proposition 1.7.1 — Properties of Limits. Let c be a constant, let n be a positive integer, and suppose that the functions f and g have the property that

$$\lim_{x\to a} f(x) = L \qquad \text{and} \qquad \lim_{x\to a} g(x) = K.$$

Then each of the following properties hold:

$$\lim_{x\to a} cf(x) = cL, \tag{1.58}$$

$$\lim_{x\to a} \left[f(x)+g(x)\right] = L+K, \tag{1.59}$$

$$\lim_{x\to a} f(x)g(x) = LK, \tag{1.60}$$

$$\lim_{x\to a} \frac{f(x)}{g(x)} = \frac{L}{K}, \quad \text{if } K \neq 0, \tag{1.61}$$

$$\lim_{x\to a} \left[f(x)\right]^n = L^n, \tag{1.62}$$

$$\lim_{x\to a} \left[f(x)\right]^{1/n} = L^{1/n}, \quad \text{if } L^{1/n} \text{ is defined.} \tag{1.63}$$

Notice that these relations only hold if the limits of f and g, as x approaches a, exist, i.e., only if the separate limits are defined.

1.7.3 Continuity

As we previously discussed, the limit of a function f as the independent variable x approaches a point a is the single output that would best fit the graph at $x = a$. In other words, if the actual output value $f(a)$ is different from the limit $\lim_{x\to a} f(x)$, then there is some sort of hole or break in the graph of the function at that single point. The y-values are approaching the value of the limit, though the function itself takes a different value at this point. This observation motivates the following definition.

Definition 1.7.3 A function f is said to be *continuous at $x = a$* if the function f is defined at the point $x = a$ and if

$$\lim_{x\to a} f(x) = f(a). \tag{1.64}$$

If a function is not continuous at a point $x = a$, then it is said to be *discontinuous* at a.

If a function f is continuous for each point on some interval I, we say that f is continuous on I.

If a function f is continuous for each point on its domain, we say that f is continuous.

We can think of continuity as follows: a function is continuous if you can draw its graph without breaking contact between the pencil and the page, i.e., the graph is a single curve with no "holes" or "jumps" in it. The functions shown in Figure 1.35 and 1.36 are both discontinuous at the point $x = 4$.

An interesting result of continuity is given by the following theorem.

Theorem 1.7.2 — Intermediate Value Theorem. Suppose that a function f is continuous on a closed interval $I = [a,b]$, and suppose that L is a number between $f(a)$ and $f(b)$ (i.e., either $f(a) < L < f(b)$ or $f(b) < L < f(a)$), then there exists a point $c \in (a,b)$, such that $f(c) = L$.

The intermediate value theorem is represented visually in Figure 1.37. We start off with the endpoints $x = a$ and $x = b$, with $a < b$, and suppose that the function f is continuous on the closed interval $a \le x \le b$. All the intermediate value theorem says, is that every y-value between $f(a)$ and $f(b)$ *must* be achieved. This only stands to reason: if one is to draw a path connecting the points $(a, f(a))$ and $(b, f(b))$ without breaking or loosing contact with the page, each y-value must be actually realized by that path.

In Figure 1.38, we randomly selected a y-value, which we named L, that lies between the y-values $f(a)$ and $f(b)$. As is evident from the graph, there indeed exists an x-value c, that lies between a and b, such that $f(c) = L$.

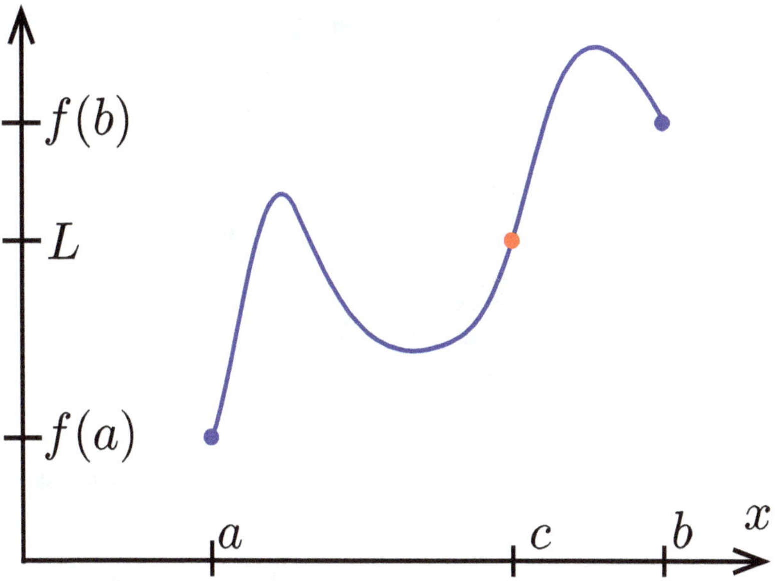

Figure 1.37: An illustration of the intermediate value theorem.

The intermediate value theorem, though simple in its statement, is sometimes profound in its implication.

■ **Example 1.26** Consider a continuous function f defined at all points on the surface of a sphere. For example, the function f may represent the temperature, atmospheric pressure, density, or elevation for different points on Earth. Show that along any great circle, there must be a pair of antipodal points for which the function f returns the same value. (A *great circle* is any circle along the surface of the sphere that lies on a plane that passes through the center of the sphere, i.e., it is a circle that divides the sphere into halves. *Antipodal points* are two points that are opposite each other on the sphere, such as the north and south pole. A great circle is pictured along with two antipodal points in Figure 1.38.)

Before proceeding, let us bask, for a moment, in the wonder of what we are saying. Along Earth's equator (or any other great circle), there will always be two antipodal points, i.e., two points along the equator that are exactly opposite of one another, for which the temperature will be the same. (And there must also be two antipodal points along the equator that have the same elevation, two that have the same atmospheric pressure, etc.) This is quite a marvelous result.

To prove this result, we must first set some notation. Consider a particular great circle on the sphere, though its selection is arbitrary. For concreteness, we can think

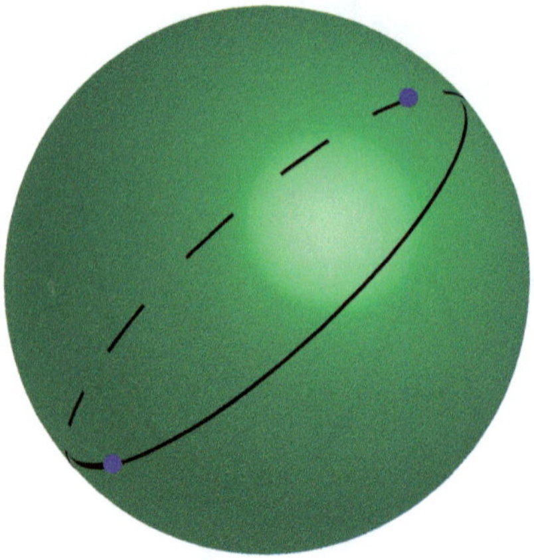

Figure 1.38: A sphere, one of its great circles, and two antipodal points.

of the equator. Next, let $f(x)$ represent the output of the function at longitude x, for $x \in [-180, 180]$, such that $f(-180) = f(180)$ (since these two x-values represent the same point along the circle). The idea behind the result is to next define a new function that measures the difference between a point and its antipodal point. Thus, let us define the function

$$d(x) = f(x) - f(x - 180),$$

on the domain $x \in [0, 180]$. Next, observe that

$$\begin{aligned} d(0) &= f(0) - f(-180) = f(0) - f(180), \\ d(180) &= f(180) - f(0) = -(f(0) - f(180)). \end{aligned}$$

Thus, the outputs $d(0)$ and $d(180)$ are equal in magnitude but opposite in sign. Therefore, the value 0 lies in the interval between $d(0)$ and $d(180)$. By the intermediate value theorem, there must exist a point $c \in (0, 180)$ such that $d(c) = 0$. Thus, for that point, we have $d(c) = f(c) - f(c - 180) = 0$ and, therefore, $f(c) = f(c - 180)$. This proves that there must exist a pair of antipodal points along any great circle for which the continuous function f takes the same value. ∎

1.7.4 Squeeze Theorem

In our final paragraph, we discuss a theorem that offers us a way of showing that certain limits exist, and what they are equal to, when other methods fail us. Essentially, if you can bound a function between a lower function and an upper function, and the lower and upper functions have the same limit at a point, then the intermediate function must be "squeezed" between its two bounds to have the same limit as well. To be precise, we have the following.

Theorem 1.7.3 — The Squeeze Theorem. Suppose that

$$l(x) \leq f(x) \leq u(x)$$

for all x on an open interval containing the point $x = a$, except, perhaps, at the point a itself. If

$$\lim_{x \to a} l(x) = \lim_{x \to a} u(x) = L,$$

then the limit as x goes to a of $f(x)$ also exists and is equal to L, i.e., then

$$\lim_{x \to a} f(x) = L.$$

The rationale for the squeeze theorem is straightforward: if the outputs of the function f are trapped between the outputs of the lower function l and the upper function u, and if the limits of l and u both approach the same value L, then the limit of the intermediate function f must also approach the same value L as x approaches the point $x = a$.

■ **Example 1.27** Use the squeeze theorem to determine the limit

$$\lim_{x \to 0} f(x),$$

for the function

$$f(x) = x^2 \sin\left(\frac{1}{x}\right).$$

First, we remark that the function $\sin(x)$ oscillates between $+1$ and -1 infinitely many times as $x \to \infty$. Moreover, the transformation that sends $x \mapsto 1/x$ sends the infinite interval $[1, \infty)$ into the finite interval $(0, 1]$. Thus, the infinitely many oscillations that $\sin(x)$ achieves between one and infinity are achieved by $\sin(1/x)$ between one and zero: the function $\sin(1/x)$ oscillates more and more rapidly as $x \to 0$. The function $\sin(1/x)$ is shown in Figure 1.39. As is evident by the graph, the limit

$$\lim_{x \to 0} \sin(1/x)$$

does not exist, as the function does not "settle" near any particular y-value as $x \to 0$. Rather, the outputs pass through *every* y-value between $-1 \leq y \leq 1$ as x becomes closer and closer to zero, again and again.

The function $f(x)$ is the same as $\sin(1/x)$, only multiplied by an additional factor of x^2. We can view this coefficient function as an amplitude that varies with x. The graph of the function $f(x)$, plotted concurrently with the functions $y = \pm x^2$, is shown in Figure 1.40. As can be seen by this figure, the function $f(x) = x^2 \sin(1/x)$ is "squeezed" between the upper and lower bounds $u(x) = x^2$ and $l(x) = -x^2$ as x approaches zero. Since,

$$-x^2 \leq x^2 \sin\left(\frac{1}{x}\right) \leq x^2,$$

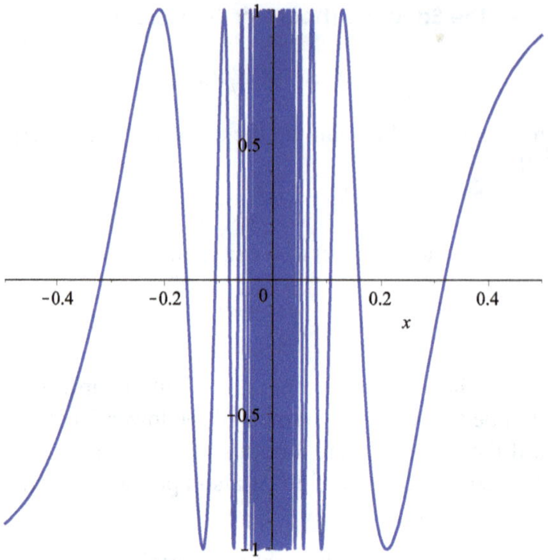

Figure 1.39: Graph of the function $\sin(1/x)$.

and since

$$\lim_{x\to 0}(-x^2) = \lim_{x\to 0}(+x^2) = 0,$$

we conclude, by the squeeze theorem, that

$$\lim_{x\to 0} x^2 \sin\left(\frac{1}{x}\right) = 0.$$

■

Exercises

For Exercises 1–10, approximate the value of the following limits.

1. $\lim_{x\to 0} \dfrac{\sin x}{x}$.
2. $\lim_{x\to 0} \dfrac{\sin x - x}{x^3}$.
3. $\lim_{x\to 3} \dfrac{x^2 - x - 6}{x - 3}$.
4. $\lim_{x\to 0} \dfrac{\cos(x) - 1}{x^2}$.
5. $\lim_{x\to 0} \dfrac{\cos x - 1 + x^2/2}{x^4}$.
6. $\lim_{x\to 0} \dfrac{e^x - 1}{x}$.
7. $\lim_{x\to 1} \dfrac{\ln|x|}{x - 1}$.
8. $\lim_{x\to 0^+} \dfrac{|x|}{\sin(x)}$.
9. $\lim_{x\to\pi/2} \left(x - \dfrac{\pi}{2}\right)\tan(x)$.
10. $\lim_{x\to 0} x^x$.

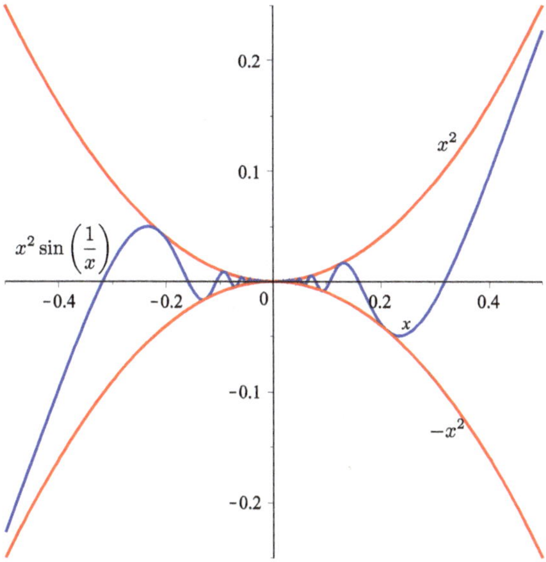

Figure 1.40: Graph of the function $f(x) = x^2 \sin(1/x)$, plotted concurrently with the graphs of $y = \pm x^2$.

For Exercises 11–15, determine the value of the parameter k that will make the given function continuous.

11. $f(x) = \begin{cases} x^2 & \text{for } x < 0 \\ 3x + k & \text{for } x \geq 0 \end{cases}$.

12. $f(x) = \begin{cases} x^2 & \text{for } x < 3 \\ 3x + k & \text{for } x \geq 3 \end{cases}$.

13. $f(x) = \begin{cases} e^{2x} & \text{for } x \leq 1 \\ 2 + kx & \text{for } x \geq 1 \end{cases}$.

14. $f(x) = \begin{cases} e^{kx} & \text{for } x \leq 1 \\ 3 + 4x & \text{for } x > 1 \end{cases}$.

15. $f(x) = \begin{cases} x^3 & \text{for } x < 2 \\ k\sqrt{x} & \text{for } x \geq 2 \end{cases}$.

For Exercises 16–19, use the intermediate-value theorem to prove that the given function has a root on the indicated interval.

16. $f(x) = \arctan(x) - x^3$; on $[0.5, 1]$.

17. $f(x) = \ln(x) + x$; on $[0.5, 1]$.

18. $f(x) = x - 0.5 \sin x - 1$; on $[1, 2]$.

19. $f(x) = \sin(x^3 - 1)$; on $[-1.5, 0.5]$.

For Exercises 20–29, use the graph of the function f, as shown in Figure 1.41, to determine the indicated limits. If a given limit does not exist, write "DNE."

20. $\lim_{x \to 1^-} f(x)$

21. $\lim_{x \to 1^+} f(x)$

22. $\lim_{x \to 3^-} f(x)$

23. $\lim_{x \to 3^+} f(x)$

24. $\lim_{x \to 5^-} f(x)$

25. $\lim_{x \to 5^+} f(x)$

26. $\lim_{x \to 6^-} f(x)$

27. $\lim_{x \to 6^+} f(x)$

28. $\lim_{x \to 8^-} f(x)$

29. $\lim_{x \to 8^+} f(x)$

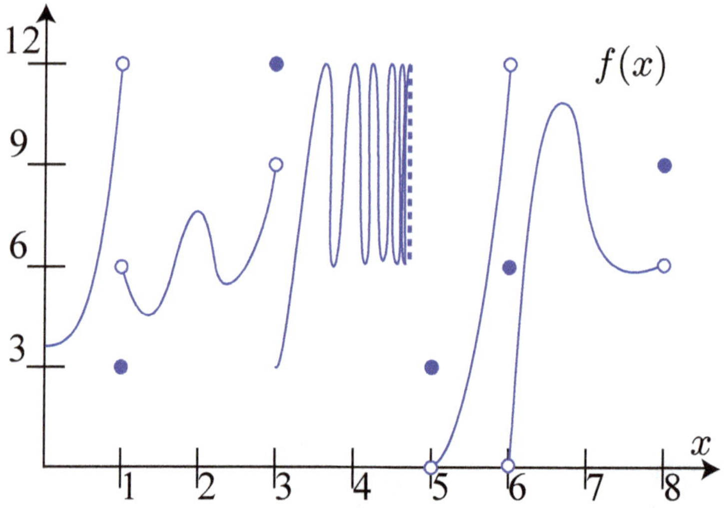

Figure 1.41: Graph of the function f; Exercises 20–29.

30. The *floor function* is defined by $\lfloor x \rfloor$ = *greatest integer that is less than or equal to x*. A graph of $\lfloor x \rfloor$ for $0 \le x \le 6$ is shown in Figure 1.42.
    (a) Determine $\lim_{x \to 1^-} \lfloor x \rfloor$.
    (b) Determine $\lim_{x \to 1^+} \lfloor x \rfloor$.
    (c) Find $\lim_{x \to 3} \lfloor x \rfloor$, if it exists. If it does not exist, explain why.
31. The *ceiling function* is defined by $\lceil x \rceil$ = *smallest integer that is greater than or equal to x*.
    (a) Graph the ceiling function on the domain $-5 \le x \le 5$.
    (b) Determine $\lim_{x \to 1^-} \lceil x \rceil$.
    (c) Determine $\lim_{x \to 1^+} \lceil x \rceil$.
    (d) Find $\lim_{x \to 3} \lceil x \rceil$, if it exists. If it does not exist, explain why.
32. The *signum* function, sgn(x), is defined for nonzero values of x by the relation

$$\text{sgn}(x) = \frac{|x|}{x} \qquad \text{for } x \neq 0,$$

and by $\text{sgn}(x) = 0$ for $x = 0$.
    (a) Draw a graph of the signum function.
    (b) Compute $\lim_{x \to 0^+} \text{sgn}(x)$.

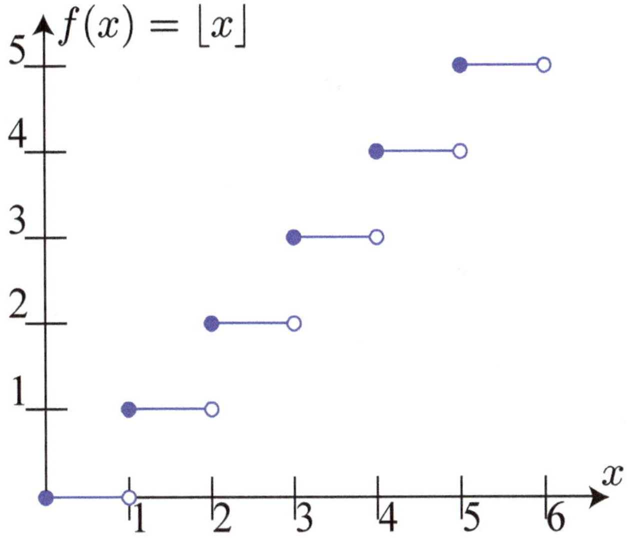

Figure 1.42: Graph of the floor function; Exercises 30.

(c) Compute $\lim_{x\to 0^-} \operatorname{sgn}(x)$.

(d) Is the signum function continuous at $x = 0$? Explain your answer.

33. For what values of a and b will the following piecewise-defined function be continuous?

$$f(x) = \begin{cases} \cos x & \text{for } x \le 0 \\ ax + b & \text{for } 0 < x < \pi \\ \cos x & \text{for } \pi \le x \end{cases}$$

34. For what values of a and b will the following piecewise-defined function be continuous?

$$f(x) = \begin{cases} \ln x & \text{for } 0 < x < 1 \\ ax + b & \text{for } 1 \le x \le e \\ \ln x & \text{for } e < x \end{cases}$$

Problems

35. Use the squeeze theorem to determine the limit, as x approaches 0, of the piecewise-defined function

$$f(x) = \begin{cases} x^2 & \text{if } x \in \mathbb{Q} \\ 0 & \text{if } x \notin \mathbb{Q} \end{cases},$$

i.e., the function defined to output x^2 if x is a rational number, but to output zero if x is irrational. Is $f(x)$ continuous at $x = 0$?

36. At noon on a certain day, Jody runs the 26-mile course of the Boston marathon, completing the course exactly five hours later. Her running is erratic: she speeds up, slows down, stops, and takes various breaks, in a seemingly random and chaotic fashion. At noon the following day, Tim runs the same 26-mile course in the opposite direction. His run is just as erratic and unpredictable as Jody's was. Must there be a time, between noon and four o'clock, at which both Jody and Tim were at the same point of the course at the same time (except on different days)? Use the intermediate value theorem to justify your answer.
37. A patient at St. Elsewhere's Hospital is required to take a 150 mg pill every 6 hours. Suppose that the half-life of the medicine is also 6 hours. A graph of the amount of the drug $f(t)$, measured in mg, in the patient's body, as a function of the number of hours t since the patient took the first pill is shown in Figure 1.43.

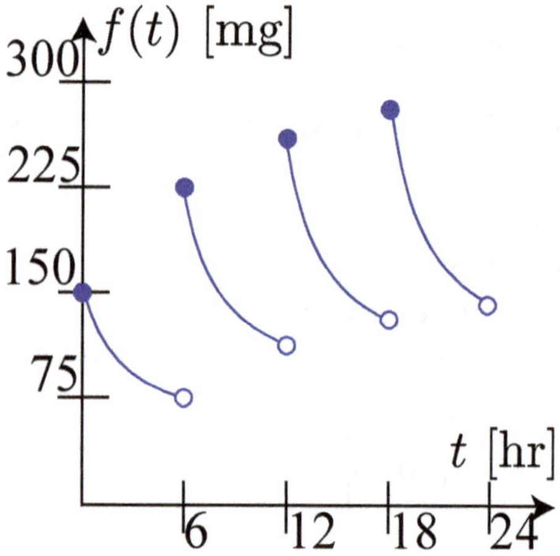

Figure 1.43: Amount of drug in a patient's bloodstream; Exercises 37.

(a) Use the graph and your knowledge of exponential functions to estimate
   i. $\lim_{t\to 6^-} f(t)$,
   ii. $\lim_{t\to 6^+} f(t)$,
   iii. $\lim_{t\to 24^-} f(t)$, and
   iv. $\lim_{t\to 24^+} f(t)$.

(b) Explain the physical meaning of the quantities in part (a).

38. The water level $h(x)$, measured in feet, in a certain, shallow river, x feet downstream of a measuring sensor, is shown in Figure 1.44. This particular

river demonstrates an instance of a *hydraulic jump*, which is a sudden discontinuity in the water level.

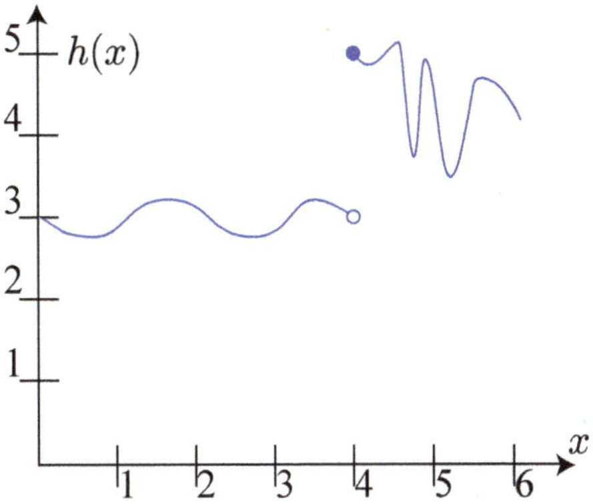

Figure 1.44: Profile of a hydraulic jump; Exercises 38.

(a) Compute $\lim_{x \to 4^-} h(x)$ and $\lim_{x \to 4^+} h(x)$.

(b) For which points is $h(x)$ discontinuous? Explain.

39. Let $A(t)$ be the amount of money in dollars that Ingrid has in her Lucky Charm Savings Account, t days after the start of the month. Lucky Charm continuously compounds interest at a *continuous* growth rate of 3.65% per year. (Assume that 1 year = 365 days).

(a) If Ingrid has P_0 dollars in her account at time t, write a formula describing how much she will have in her account d days later, assuming that she makes no withdrawals or deposits during those d days.

(b) Assume that Ingrid has \$8,800.00 in her account at the start of the month, and makes the withdrawals and deposits indicated in the table below.

t [days]	Action	Amount
1.35	Deposit	\$2,900.00
14.52	Withdrawal	\$650.00
18.69	Deposit	\$1,347.52
25.92	Withdrawal	\$120.00

Determine the following limits. Explain your reasoning.

i. $\lim_{t \to 1.35^-} A(t)$

ii. $\lim_{t \to 1.35^+} A(t)$

iii. $\lim_{t \to 14.52^-} A(t)$

iv. $\lim_{t \to 14.52^+} A(t)$

v. $\lim_{t \to 18.69^-} A(t)$

vi. $\lim_{t \to 18.69^+} A(t)$

vii. $\lim_{t \to 25.92^-} A(t)$

viii. $\lim_{t \to 25.92^+} A(t)$

(c) Ingrid's last withdrawal, on the 26th day of the month, was from an ATM. What time did she withdrawal her cash from this ATM?

(d) Compute the total amount of interest Ingrid earned during the month, assuming that the month had 30 days.

40. A box weighing $mg = 98$ N (Newtons) has a static and kinetic *coefficient of friction* equal to $\mu_s = 0.5$ and $\mu_k = 0.2$, respectively. If the box is stationary, the frictional force acting on the box will equal the applied force with which you are pushing the box, as long as it is not greater than $\mu_s mg$, the minimum amount of applied force which causes the box to start sliding. When the box begins to slide, the frictional force will equal $\mu_k mg$. Suppose you apply a time-dependent force $F_a(t)$ to this box for 10 seconds, given by the equation $F_a(t) = 10t$ N. Then the frictional force $F_f(t)$ acting on the box will be given by

$$F_f(t) = \begin{cases} 10t & \text{for } 0 \le t \le 4.9 \\ 19.6 & \text{for } 4.9 < t \le 10 \end{cases}$$

(a) Draw a graph of $F_f(t)$.

(b) Determine

$$\lim_{t \to 4.9^-} F_f(t) \quad \text{and} \quad \lim_{t \to 4.9^+} F_f(t).$$

(c) At what time did the box start moving?

(d) The *net* force acting on the box is equal to $F(t) = F_a(t) - F_f(t)$. Determine a piece-wise defined expression for $F(t)$. Draw a graph of $F(t)$.

(e) Determine

$$\lim_{t \to 4.9^-} F(t) \quad \text{and} \quad \lim_{t \to 4.9^+} F(t).$$

(f) The box's acceleration $a(t)$ is proportional to the net force acting on the box. At what point during the 10 second interval its acceleration the greatest?

41. The gravitational force at a distance r from the center of the Earth is given by

$$F(r) = \begin{cases} \dfrac{GMr}{R^3} & \text{for } 0 \le r \le R \\ \dfrac{GM}{r^2} & \text{for } R < r \end{cases},$$

where R is the radius of the Earth, M is the mass of the Earth, and G is the gravitational constant. Is $F(r)$ continuous at $r = R$?

1.8 Limits Involving Infinity

In this section, we take a look at limits that involve infinity. This may occur in one of two different ways: either the independent variable can approach positive or negative infinity, or the dependent variables can approach positive or negative infinity (as the independent variable approaches a particular constant). These two phenomena are closely related to horizontal and vertical asymptotes, which we will also examine in the scope of our discussion.

1.8.1 Limits to Infinity

We begin with our main definition.

Definition 1.8.1 We say that

$$\lim_{x\to\infty} f(x) = L \tag{1.65}$$

if the output values of $f(x)$ eventually approach value L as the independent variable x become larger and larger.

Similarly, we say that

$$\lim_{x\to-\infty} f(x) = L \tag{1.66}$$

if the output values of $f(x)$ eventually approach value L as the independent variable x become more and more negative.

Thus, "the limit as x goes to infinity of $f(x)$" is the unique output value that the function approaches *in the long run*, i.e., as x becomes arbitrarily large.

Limits at infinity have a variety of applications. First, when modeling time-dependent phenomena, they can be used to determine the eventual state of the system in the long run, i.e., as $t \to \infty$. This is useful for systems with a *steady-state* or *equilibrium* condition, as the limit usually approaches such a state. For example, if $u(t)$ models the temperature of a hot cup of coffee, then $\lim_{t\to\infty} u(t)$ will typically equal the constant room temperature, as the temperature of the coffee will ultimately come into thermodynamic equilibrium with its surroundings. If $x(t)$ models the vertical displacement of your automobile after hitting a pothole, then $\lim_{t\to\infty} x(t) = 0$, i.e., the car's vertical displacement will eventually relax back to its equilibrium value, perhaps after a series of shortening "bounces" if the vehicle has underdamped shock absorbers.

Another use for limits at infinity is when one wants to sum a thing over an infinite domain. We will examine this, in different contexts, later, in Chapters 5 and 7. In each instance, however, we will need to first do a finite sum, and then look at the limit as we expand the domain to infinity. Additionally, the theory of integration, which we will discuss in Chapter 6, has its roots in limits at infinity, as such limits are used in the construction of the definite integral.

Often, determining the limit of a function as its independent variable x tends to infinity relies on an intuitive understanding of the function's behavior for large values of x. To begin, let us offer a few observations concerning such behaviors.

Proposition 1.8.1 Let $a > b > 0$. Then, as $x \to \infty$:

1. a polynomial of a higher degree wins against a polynomial of a lesser degree;
2. the exponential function e^{ax} wins against any polynomial, regardless its degree;
3. the exponential function e^{-ax} goes to zero;
4. the exponential function e^{ax} wins against the exponential function e^{bx} (when $a > b > 0$);
5. a polynomial with degree $n > 0$ wins against a logarithmic function.

Similarly, as $x \to -\infty$:

1. a polynomial of a higher degree wins against a polynomial of a lesser degree;
2. the exponential function e^{-ax} wins against any polynomial, regardless its degree;
3. the exponential function e^{ax} goes to zero;
4. the exponential function e^{-ax} wins against the exponential function e^{-bx} (when $a > b > 0$).

If a function f *wins* against a function g, as $x \to \pm\infty$, we mean that

$$\lim_{x \to \pm\infty} \frac{f(x)}{g(x)} = \infty,$$

i.e., that the function$f$ goes to infinity faster than the function g.

The preceding proposition can be summarized as follows: larger polynomials win against smaller polynomials. An exponential function e^{ax} (positive growth rate) can be regarded as an infinite-degree polynomial—it wins out against polynomials of any other order. Note, however, that e^{ax} goes to infinity as $x \to \infty$, but it goes to zero as $x \to -\infty$. On the other hand, the exponential function e^{-ax} goes to zero as $x \to \infty$, but it goes to infinity as $x \to -\infty$. (This is because e^{ax} and e^{-ax} are reciprocals: as one goes to infinity, the other goes to zero.) Finally, logarithmic functions go to infinity, but at a rate slower than a polynomial of any degree.

Proposition 1.8.2 Consider two functions f and g. If f wins against g as $x \to \infty$, then

$$\lim_{x \to \infty} \frac{f(x)}{g(x)} = \pm\infty.$$

If g wins against f as $x \to \infty$, then

$$\lim_{x \to \infty} \frac{f(x)}{g(x)} = 0.$$

If f and g tie as $x \to \infty$, then the limit of their ratio will equal the ratio of the coefficients of the dominant terms.

To explore the consequences of this theory, let us apply it to compute the limits of several *rational functions*.

Definition 1.8.2 A function f is a *rational function* if it can be written as a ratio of two polynomials, i.e., if

$$f(x) = \frac{P(x)}{Q(x)},$$

where P and Q are polynomials.

Example 1.28 Determine the limit

$$\lim_{x\to\infty} \frac{x^7 + 62x^5 + 99x^3 + 17}{33x^6 + 28x^4 + 5x}.$$

Here, the top function wins over the bottom function, because it is a polynomial of a greater degree. We conclude that the limit is equal to ∞. ■

Example 1.29 Determine the limit

$$\lim_{x\to\infty} \frac{x^7 + 62x^5 + 99x^3 + 17}{33x^6 + 28x^4 + 5x + 88e^{0.01x}}.$$

In this example, the dominate term in the denominator is $88e^{0.01x}$, since exponentials win over any polynomial. The bottom function, this time, will win over the top, and therefore the limit is 0. ■

Example 1.30 Determine the limit

$$\lim_{x\to\infty} \frac{9x^7 + 55x^5 + 99x^3 + 17}{8x^7 + 88x^4 + 13x}.$$

Here, we have a balance of powers. The $9x^7$ dominates in the numerator, whereas $8x^7$ dominates in the denominator. The limit is therefore the ratio of the dominate coefficients, and so

$$\lim_{x\to\infty} \frac{9x^7 + 55x^5 + 99x^3 + 17}{8x^7 + 88x^4 + 13x} = \frac{9}{8}.$$

In fact, we would get the same answer even in the limit as $x \to -\infty$, since the ratio of the dominate terms, as $x \to -\infty$, would still be $9x^7/(8x^7)$. ■

Example 1.31 Determine the limit

$$\lim_{x\to\infty} \frac{22x^{77} + 101x^3 + 99e^{-0.003x}}{7x^{77} + 13x^6 + 22e^{-0.003x}}.$$

Here, the dominate terms of the numerator and denominator, as $x \to \infty$, are $22x^{77}$ and $7x^{77}$, respectively. (The exponentials with a negative exponent will decay as $x \to \infty$.) Therefore

$$\lim_{x\to\infty} \frac{22x^{77} + 101x^3 + 99e^{-0.003x}}{7x^{77} + 13x^6 + 22e^{-0.003x}} = \frac{22}{7}.$$

On the other hand, as $x \to -\infty$, the dominate terms in the numerator and denominator will be $99e^{-0.003x}$ and $22e^{-0.003x}$, as these terms grow exponentially with the same rate as $x \to -\infty$. Therefore

$$\lim_{x\to-\infty} \frac{22x^{77}+101x^3+99e^{-0.003x}}{7x^{77}+13x^6+22e^{-0.003x}} = \frac{99}{22} = 4.5.$$

■

Next, we use limits at infinity to define the concept of a horizontal asymptote.

Definition 1.8.3 The horizontal line $y = H$ is a *horizontal asymptote* of the function f if either

$$\lim_{x\to\infty} f(x) = H \qquad \text{or} \qquad \lim_{x\to-\infty} f(x) = H.$$

Given this definition, and the uniqueness of limits, a function can have *at most* two horizontal asymptotes.

■ **Example 1.32** Determine the horizontal asymptotes of the function

$$f(x) = \frac{22x^{77}+101x^3+99e^{-0.003x}}{7x^{77}+13x^6+22e^{-0.003x}}.$$

In Example 1.31, we saw that

$$\lim_{x\to\infty} f(x) = \frac{22}{7} \qquad \text{and} \qquad \lim_{x\to-\infty} f(x) = 4.5.$$

Therefore, the horizontal asymptotes are $y = 4.5$ and $y = 22/7$.

Despite the relative ease with which we computed these asymptotes algebraically, a graphical examination would not have been so immediate. By plotting the function f on the interval $[-5,5]$, the only apparent asymptote would be $y = 22/7$, on *both* sides of the y-axis. Even enlarging this interval, the function appears to have the same horizontal asymptote for both large positive and negative x-values. However, around the point $x = -325,400$, the function f has a vertical asymptote. Only for values of $x < -325,401$ will the horizontal asymptote of $y = 4.5$ be apparent. This can be seen in Figure 1.45. For the graph on the left, the function f is plotted from $[-5,5]$. Here, the horizontal asymptote $y = 22/7$ is already apparent. It is not until the values of x drop below $-325,400$, however, that the other horizontal asymptote of $y = 4.5$ becomes apparent in the graph. Thus, analytical tools, such as the algebraic approach we used to compute these limits, can often reveal phenomena that can be missed if one only relies on graph of a function alone. ■

■ **Example 1.33** Determine the horizontal asymptotes of the function

$$f(x) = \frac{5x^2+3e^{-x}}{5x^2+e^{-x}}.$$

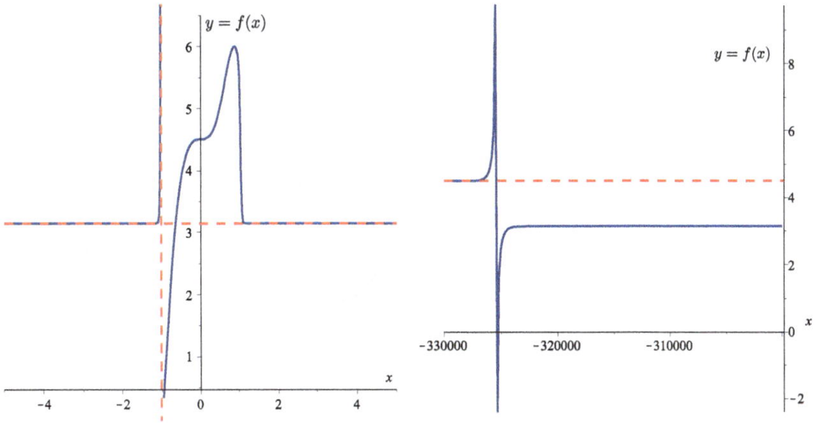

Figure 1.45: Graph of the function f from Example 1.32 on the domain $[-5,5]$ (left) and $[-330000,-300000]$ (right).

As $x \to \infty$, the quadratic term in both the numerator and denominator will dominate over the term with exponential decay, which will tend towards zero. Therefore, we have

$$\lim_{x\to\infty} \frac{5x^2+3e^{-x}}{5x^2+e^{-x}} = \frac{5}{5} = 1.$$

On the other hand, as $x \to -\infty$, the exponential term will grow, winning out over the quadratic terms. Thus,

$$\lim_{x\to-\infty} \frac{5x^2+3e^{-x}}{5x^2+e^{-x}} = \frac{3}{1} = 3.$$

Therefore, the two horizontal asymptotes of $f(x)$ are at $y = 1$ and $y = 3$, as can be seen in Figure 1.46.

■

1.8.2 Infinite Limits

Infinities can also become entangled with limits when the output of a function approaches infinity as its input approaches a definitive value.

Definition 1.8.4 We say that

$$\lim_{x\to a} f(x) = \infty$$

if the outputs of the function f become arbitrarily large as the inputs x become closer and closer to the point $x = a$.

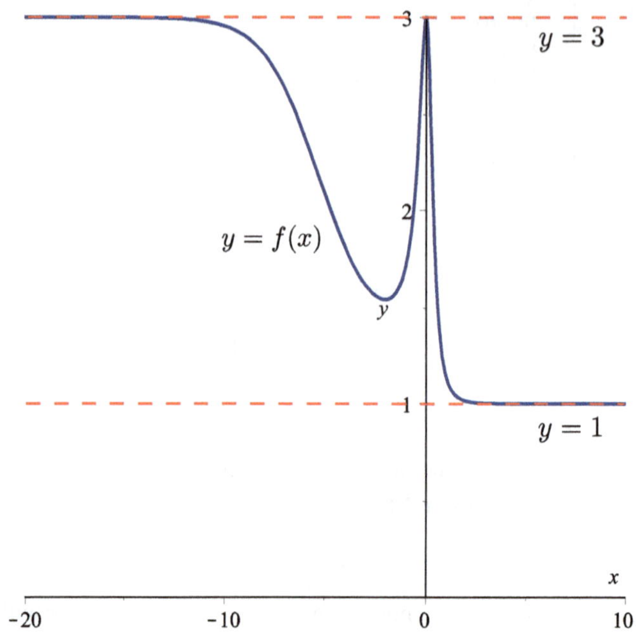

Figure 1.46: Graph of the function f from Example 1.33 on the domain $[-20, 10]$.

Similarly, we say that

$$\lim_{x \to a} f(x) = -\infty$$

if the outputs of the function f become arbitrarily large and negative as the inputs x become closer and closer to the point $x = a$.

(R) If the output values of a function f approach $+\infty$ as $x \to a$ on one side and approach $-\infty$ as $x \to a$ on the other side, we should simply say that the limit does not exist. We can, however, say that the right-hand and left-hand limits are $\pm\infty$, as appropriate.

Definition 1.8.5 The vertical line $x = a$ is a *vertical asymptote* of the function f if

$$\lim_{x \to a^{\pm}} f(x) = \pm\infty,$$

for any choice of the + or − signs.

Thus, a line $x = a$ is a vertical asymptote if the function is unbounded, i.e., the outputs approach *either* plus or minus infinity, as x approaches a from at least one side, if not both.

■ **Example 1.34** Consider the function $f(x) = 1/x^2$. This function has a vertical asymptote at $x = 0$, since

$$\lim_{x\to 0} \frac{1}{x^2} = \infty,$$

as shown in Figure 1.47. ■

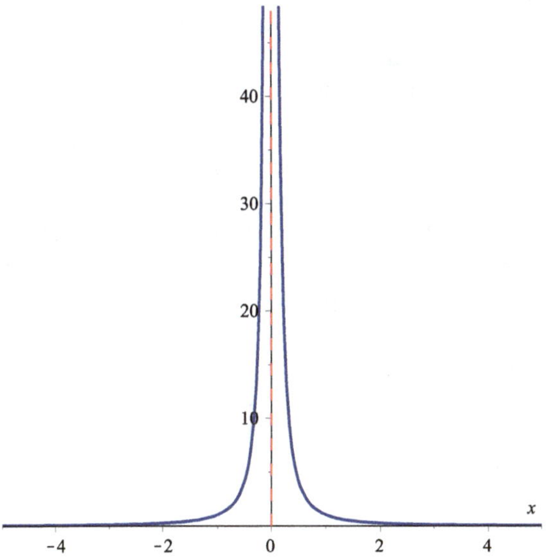

Figure 1.47: Vertical asymptote of the function $f(x) = 1/x^2$.

■ **Example 1.35** Determine the limit

$$\lim_{x\to 3} \frac{x^3 - 3x^2 + 9x - 27}{x - 3}.$$

At first, it looks as though there might be a vertical asymptote at $x = 3$, as there is a zero in the denominator for the value $x = 3$. Upon factoring the numerator, however, we find that this function is equivalent to

$$\lim_{x\to 3} \frac{(x^2 + 9)(x - 3)}{x - 3} = 18.$$

This follows since the factor of $(x - 3)$ in both the numerator and denominator cancel for all values of x except for $x = 3$, at which the function is undefined. Since the value of the function at the point $x = 3$ is irrelevant for the purpose of computing the limit (what matters is the *behavior* of the function as x approaches 3), we can simply evaluate $x^2 + 9$ at $x = 3$ to obtain our answer. ■

■ **Example 1.36** Next, let us consider the behavior of the function

$$f(x) = \frac{x^2+9}{x-3}$$

as x approaches 3. In this case, the numerator will approach a value of $3^2+9 = 18$. Since we have a finite number divided by zero, we can safely assume that there will be a vertical asymptote at $x = 3$. Next, we must consider the left-hand and right-hand limits separately. If $x < 3$, then $(x-3)$ will be negative, whereas if $x > 3$, then $(x-3)$ will be positive. We conclude that

$$\lim_{x\to 3^-} \frac{x^2+9}{x-3} = -\infty \qquad \text{and} \qquad lim_{x\to 3^+} \frac{x^2+9}{x-3} = +\infty,$$

that is, if we approach 3 from the left, the outputs approach negative infinity, whereas if we approach 3 from the right, the outputs approach positive infinity. This behavior is verified in the graph shown in Figure 1.48. ■

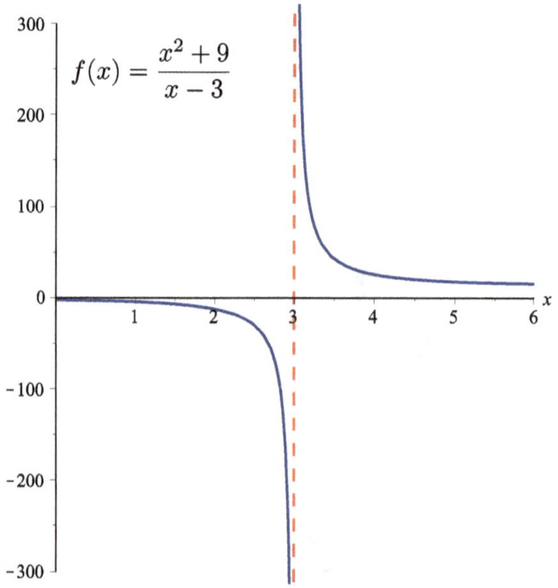

Figure 1.48: Vertical asymptote of the function $f(x)$ from Example 1.36.

Exercises

For Exercises 1–10, determine the value of the given limit, if it exists. If it does not exist, state whether it diverges to plus or minus infinity, or neither.

1. $\lim_{x\to\infty} \dfrac{\sin(x)}{x}$.
2. $\lim_{x\to\infty} \dfrac{x^3+8x^2+99x+101}{99x^3+100x^2+101x+107}$.
3. $\lim_{x\to\infty} \dfrac{\sqrt{7x}+\pi}{\sqrt[3]{x}}$.
4. $\lim_{x\to\infty} \dfrac{x^7+99x^5+556x^3+999x}{x^3+x+e^{0.001x}}$.
5. $\lim_{x\to\infty} \dfrac{x^3+3x+e^{-0.001x}}{x^2+2x+7}$.
6. $\lim_{x\to\infty} x^{100}e^{-0.01x}$.
7. $\lim_{x\to-\infty} \dfrac{x^5+3x^2+e^{-0.1x}}{9x^5+12x^3+5e^{-0.1x}}$.
8. $\lim_{x\to-\infty} \dfrac{x^2+e^{5x}}{x^3+\pi}$.
9. $\lim_{x\to-\infty} x^9e^{0.01x}$.
10. $\lim_{x\to-\infty} \dfrac{x^5+3x^2+11}{8x^4+13x+e^{3x}}$.

For Exercises 11–20, determine the value of the given limit, if it exists. If it does not exist, determine the left-hand and right-hand limits, if they exist.

11. $\lim_{x\to 3} \dfrac{x^3+5x^2-17x-21}{x-3}$.
12. $\lim_{x\to 0} \dfrac{\sin(x)}{x}$.
13. $\lim_{x\to 2} \dfrac{\cos(8x)}{x-2}$.
14. $\lim_{x\to 5} \dfrac{x^3-10x^2+25x}{x-5}$.
15. $\lim_{x\to 5} \dfrac{x^3-10x^2+25x}{(x-5)^2}$.
16. $\lim_{x\to -1} \dfrac{x^2+1}{x+1}$.
17. $\lim_{x\to -1} \dfrac{x^2+1}{x^2+2x+1}$.
18. $\lim_{x\to 0} \dfrac{4e^{3x}+7x^3}{x}$.
19. $\lim_{x\to 0} \dfrac{5e^{-3x}+2x^2}{\sqrt{x}}$.
20. $\lim_{x\to 4} \dfrac{x-4}{x^2+2}$.

Problems

21. A hot cup of coffee has a temperature $u(t)$ degrees Fahrenheit, t minutes after it is fresh, given by the equation

$$u(t) = 72 + 108e^{-0.05t}.$$

   (a) What is the initial temperature of the coffee?
   (b) How long will it take to cool to a temperature of 120° F?
   (c) Determine $\lim_{t\to\infty} u(t)$. Interpret your answer in practical terms.
   (d) What is the temperature of the surrounding environment, where the cup of coffee is located?

22. In Einstein's theory of relativity, *time dilation* can be described by the formula

$$\Delta t' = \gamma \Delta t,$$

which shows how a duration of time Δt is magnified, due to the effect of its relative velocity, as observed by another observer who records a time $\Delta t'$. The factor *gamma* is a function of the velocity and is given by the formula

$$\gamma(v) = \frac{1}{\sqrt{1 - \frac{v^2}{c^2}}},$$

where v is the velocity and c is the speed of light. Compute the limit

$$\lim_{v \to c^-} \gamma(v).$$

Explain what this means in practical terms.

1.9 Precise Definition of the Limit

In this section, we examine the precise ways in which mathematicians define limits. A rigorous undertaking of the theory of limits and their applications to calculus is the subject of a branch of mathematics known as *analysis*. Here, we will offer a glimpse into this field by introducing several of the actual definitions of limits, in order to foster a greater appreciation for and understanding of what limits actually mean.

1.9.1 Limits to Infinity

We begin with limits at infinity, as these turn out to be the simplest to define and understand.

Definition 1.9.1 The statement

$$\lim_{x\to\infty} f(x) = L$$

means the following: for every $\varepsilon > 0$, there exists a number T, such that

$$x > T \qquad \text{implies that} \qquad |f(x) - L| < \varepsilon. \tag{1.67}$$

(R) The statement

$$|f(x) - L| < \varepsilon$$

is equivalent to the statement

$$(L - \varepsilon) < f(x) < (L + \varepsilon).$$

In other words, the statement $|f(x) - L| < \varepsilon$ *requires* the output value of $f(x)$ to be within a distance of ε to the number L.

(R) When reading these precise mathematical definitions, keen attention must be paid to phrases such as *for every* and *there exists*.

What makes Definition 1.9.1 work is that the implication (1.67) holds true *for every* epsilon greater than zero, regardless of how small. That is, regardless of how small you make ε, the outputs of the function f will *eventually* end up no farther away from $y = L$ than a distance of ε. That is, *for every* ε, there is always some point of no return $x = T$ such that, whenever x lies beyond T, the outputs are guaranteed to be within an ε-distance of L.

Definition 1.9.1 is illustrated in Figure 1.49. For a given, positive value of ε (usually thought of as being a small number), the condition $|f(x) - L| < \varepsilon$ creates an *ε-tube* around the line $y = L$, as shaded in the figure.

At some point—if you wait long enough—the outputs of the function f will all fall within this ε-tube. That is, there exists some point $x = T$, such that whenever x is greater than T, the outputs of f fall within this tube, as shown in the figure.

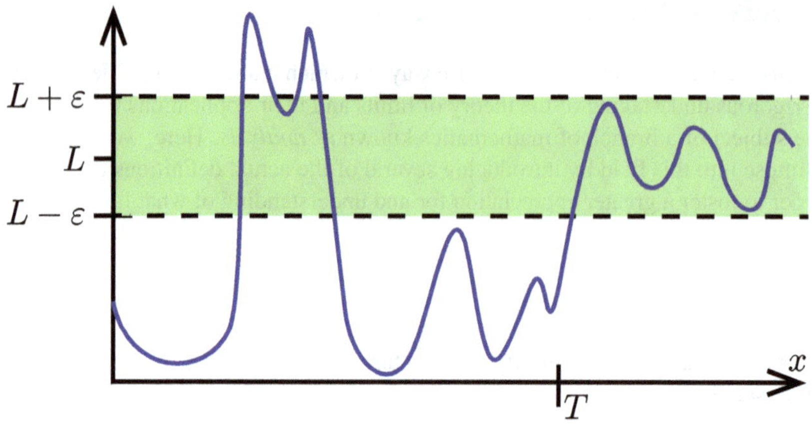

Figure 1.49: Illustration of $\lim_{x\to\infty} f(x) = L$.

What makes the definition work, is that this must be true *for every positive value of* ε! Even if we choose a smaller value of ε, *eventually* the outputs will be within that ε-tube.

To illustrate this, suppose that for one value of epsilon, $\varepsilon = \varepsilon_1$, there exists a point $x = T_1$, such that equation (1.67) holds, as shown in the previous Figure. If we choose a smaller value of epsilon, say $\varepsilon = \varepsilon_2$, then there will also be some point $x = T_2$ (typically occurring later than T_1) at which all of the outputs of the function f land within the ε_2-tube about $y = L$, as shown in Figure 1.50.

This process continues, since *for any* value of $\varepsilon > 0$, there will always be some point of no return $x = T$, after which the output values of the function f will be trapped within that ε-tube about the line $y = L$. Only then do we say that the limit of $f(x)$, as x approaches infinity, is equal to L.

■ **Example 1.37** Use Definition 1.9.1 to prove that

$$\lim_{x\to\infty} \frac{3x^2+5}{x^2} = 3.$$

Our proof must reproduce the elements of the definition: for every $\varepsilon > 0$, there must exist some point T (that can depend on ε), such that

$$x > T \qquad \text{implies} \qquad |f(x) - 3| < \varepsilon.$$

Our task is to determine the value of T, as it depends on ε, that will make this logical implication true. We start with the conclusion, i.e., we start with writing

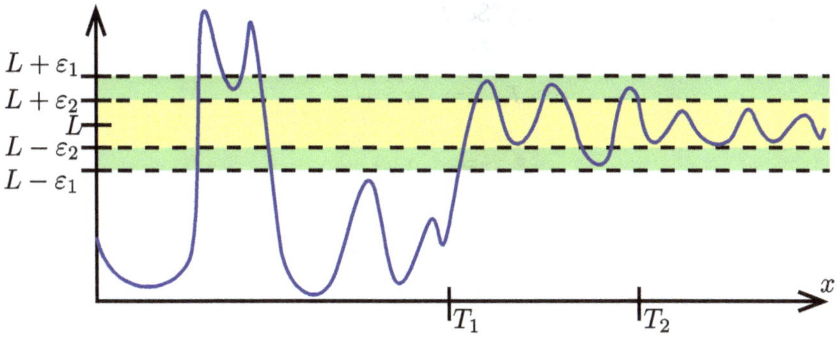

Figure 1.50: Illustration of $\lim_{x\to\infty} f(x) = L$ with two values of epsilon, $\varepsilon_1 > \varepsilon_2 > 0$. For $x > T_1$, the graph lies within the green tube. For $x > T_2$, the graph lies within the yellow tube.

out $|f(x)-3| < \varepsilon$, and then working backwards. Thus:

$$\begin{aligned}
|f(x)-3| &< \varepsilon,\\
\left|\frac{3x^2+5}{x^2}-3\right| &< \varepsilon,\\
\left|3+\frac{5}{x^2}-3\right| &< \varepsilon,\\
\frac{5}{x^2} &< \varepsilon.
\end{aligned}$$

We dispensed with the absolute-value signs as $5/x^2$ must be positive. Solving this inequality for x, we can then show that

$$\frac{5}{\varepsilon} < x^2, \qquad \text{or} \qquad \sqrt{\frac{5}{\varepsilon}} < x.$$

Our proof can now be stated as follows:

Proof. Let $\varepsilon > 0$ be arbitrary. Set $T = \sqrt{5/\varepsilon}$. Now,

$$\begin{aligned} x > T \quad &\Longrightarrow \quad x > \sqrt{\frac{5}{\varepsilon}} \\ &\Longrightarrow \quad x^2 > \frac{5}{\varepsilon} \\ &\Longrightarrow \quad \frac{1}{x^2} < \frac{\varepsilon}{5} \\ &\Longrightarrow \quad \frac{5}{x^2} < \varepsilon \\ &\Longrightarrow \quad 3 + \frac{5}{x^2} - 3 < \varepsilon \\ &\Longrightarrow \quad |f(x) - 3| < \varepsilon. \end{aligned}$$

We conclude that

$$\lim_{x \to \infty} \left(3 + \frac{5}{x^2} \right) = 3,$$

by Definition 1.9.1. ■

1.9.2 Infinite Limits

Next, let us consider infinite limits.

Definition 1.9.2 The statement

$$\lim_{x \to a} f(x) = \infty$$

means the following: for every $M > 0$, there exists a $\delta > 0$, such that

$$0 < |x - a| < \delta \qquad \text{implies that} \qquad f(x) > M. \tag{1.68}$$

(R) The condition $0 < |x - a| < \delta$ has a twofold meaning: first, the requirement that $0 < |x - a|$ simply means that the point $x = a$ itself is excluded from the requirement. This makes sense, as we are concerned with the behavior of f *as x approaches a.* The value of $f(a)$ need not even be defined!

The second condition, that $|x - a| < \delta$, is equivalent to requiring that

$$(a - \delta) < x < (a + \delta),$$

that is, x must be within a δ-distance of the point $x = a$.

Taken together, the condition $0 < |x - a| < \delta$ means that x must be closer to a than δ (i.e., x must be within a δ-distance to $x = a$), but not at the point $x = a$ itself.

Thus, Definition 1.9.2 essentially states the following: *for every* number $M > 0$, regardless of how large, there is always some (small) number δ, such that if x is

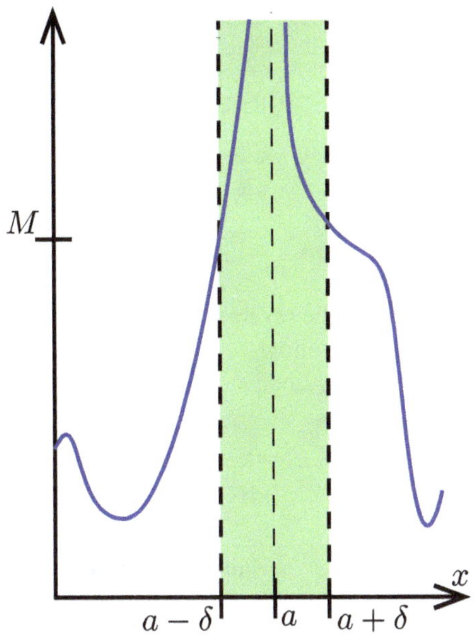

Figure 1.51: Illustration of $\lim_{x\to a} f(x) = \infty$.

close enough to a (i.e., within a distance of δ to $x = a$), then the output will be larger than M.

Definition 1.9.2 is illustrated in Figure 1.51: it shows how the outputs of the function f will become arbitrarily large as x gets closer and closer to the point $x = a$. Of course, if we wish to make M larger, the definition guarantees a smaller value of $\delta > 0$, such that if x is within the corresponding δ-tube about $x = a$, the outputs will be larger than the new, higher value of M.

Example 1.38 Prove that

$$\lim_{x\to 0} \frac{1}{x^2} = \infty.$$

The proof must mirror the definition of this type of limit; it must go something like this: let $M > 0$ be arbitrary. Then we will define a $\delta > 0$, which will depend on the value of M, so that

$$0 < |x| < \delta \qquad \Longrightarrow \qquad \frac{1}{x^2} > M.$$

To achieve this, we start at the end and work backwards. That is, we start with the inequality we would like to show:

$$\frac{1}{x^2} > M.$$

Then, we work backwards to find what must be true of x in order to make this work. We begin by reciprocating:

$$x^2 < \frac{1}{M}.$$

Taking the square root of both sides, we find

$$x < \frac{1}{\sqrt{M}}.$$

Each of the preceding steps can be reversed, so that we may identify $\delta = 1/\sqrt{M}$. The proof can then be written as follows.

Proof. Let $M > 0$ be arbitrary. Let $\delta < 1/\sqrt{M}$. For $x \neq 0$, we have

$$x < \delta \implies x < \frac{1}{\sqrt{M}} \implies x^2 < \frac{1}{M} \implies \frac{1}{x^2} > M.$$

Therefore,

$$0 < |x| < \delta \qquad \text{implies} f(x) > M,$$

and thus

$$\lim_{x\to 0} \frac{1}{x^2} = \infty,$$

by Definition 1.9.2. ∎

1.9.3 Ordinary Limits

Next, we take a look at regular limits, which do not involve infinity. For these sorts of ordinary limits, we need to combine elements of limits at infinity and infinite limits, as the y-values need to get close to a particular horizontal line as the x values get close to a particular vertical line. The precise definition is as follows.

Definition 1.9.3 The statement

$$\lim_{x\to a} f(x) = L$$

means the following: for every $\varepsilon > 0$, there exists a $\delta > 0$, such that

$$0 < |x-a| < \delta \qquad \text{implies that} \qquad |f(x)-L| < \varepsilon. \tag{1.69}$$

To understand Definition 1.9.3, we start with the horizontal ε-tube about the line $y = L$. Consider any positive ε, and the corresponding ε-tube created by the condition $|f(x)-L| < \varepsilon$, as shown in Figure 1.52. The definition says that for *any* of these ε-tubes, if x is "close enough" to $x = a$, the outputs of the function f must be contained within the ε-tube.

Here, "close enough" also means something very precise. For our fixed ε-tube, there will be a vertical δ-tube about the line $x = a$, such that as long as x is within the range $(a-\delta, a+\delta)$, the outputs will be within the corresponding ε-tube. This

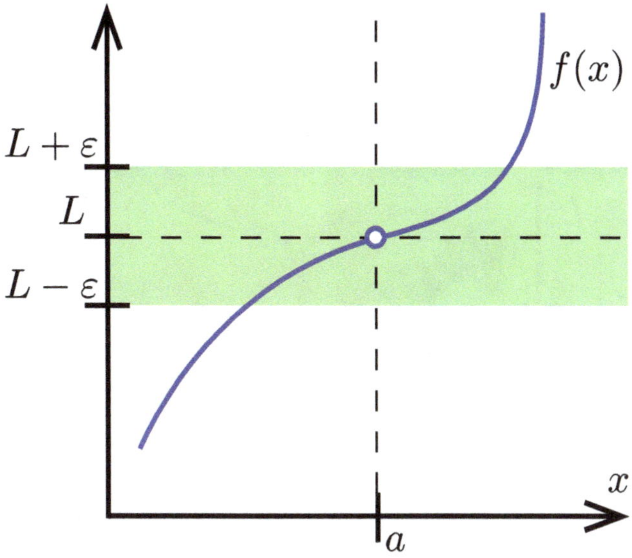

Figure 1.52: Schematic of the ε–δ definition of the limit.

is shown in Figure 1.53: For the given ε-tube, we have constructed a range of x-values near $x = a$, such that the output of the function always ends up within our ε-tube target space.

Notice that the value of δ must depend on the value of ε: for a smaller value of ε, we will need to choose a smaller δ in order to guarantee that the outputs fall within a distance of ε from the target output of $y = L$.

Also, just like we saw with infinite limits, the condition that $0 < |x-a|$ excludes the point $x = a$ from consideration: it does not matter what the function f is equal to at $x = a$, or if $f(a)$ is even defined. What matters is the behavior of f as x approaches a. Thus, the outputs must fall within the ε-tube for all x values within the interval $(a-\delta, a+\delta)$, except at the point $x = a$ itself, where the function need not even be defined.

Example 1.39 Prove that

$$\lim_{x\to 2} x^3 = 8.$$

Our proof must go something like this: let $\varepsilon > 0$ be arbitrary, and for a given $\delta > 0$ (which will depend on ε), we will have the implication

$$0 < |x-2| < \delta \qquad \Longrightarrow \qquad |x^3-8| < \varepsilon.$$

To figure out how to define δ, which depends on the fixed, arbitrary value of ε, we start at the end and work backwards, i.e., we start with the inequality

$$|x^3-8| < \varepsilon,$$

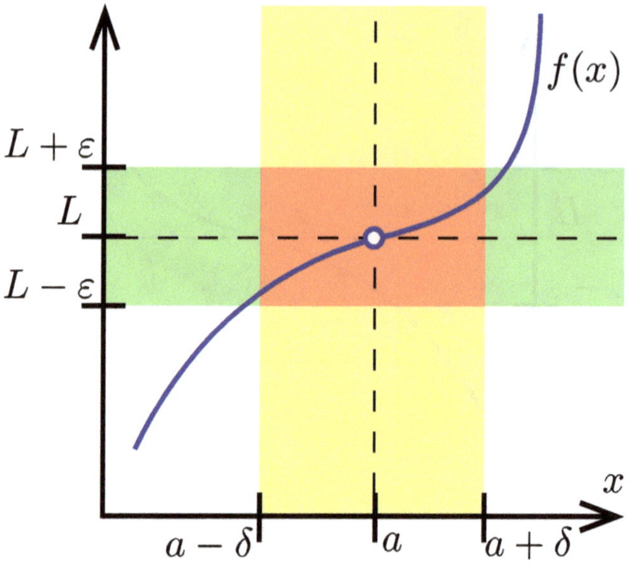

Figure 1.53: Schematic of the ε–δ definition of the limit.

and figure out how close to the point $x = 2$ that x needs to be in order for this to be true.

The inequality $|x^3 - 8| < \varepsilon$ is equivalent to

$$(8-\varepsilon) < x^3 < (8+\varepsilon),$$

which, of course, is equivalent to

$$(8-\varepsilon)^{1/3} < x < (8+\varepsilon)^{1/3}.$$

Subtracting 2 from each part, we obtain the inequality

$$(8-\varepsilon)^{1/3} - 2 < x - 2 < (8+\varepsilon)^{1/3} - 2.$$

What sense can we make of this? Well, first note that $(8+\varepsilon)^{1/3} - 2$ is a small *positive* number, whereas $(8-\varepsilon)^{1/3} - 2$ is a small *negative* number. Thus, the absolute value of $x-2$ must be less than the magnitude of both of these numbers. We therefore proceed as follows.

Proof. Let $\varepsilon > 0$ be arbitrary, and define

$$\delta = \min\left\{2-(8-\varepsilon)^{1/3}, (8+\varepsilon)^{1/3} - 2\right\}.$$

Then, the condition $|x-2| < \delta$ implies that

$$-\delta < x - 2 < \delta.$$

Furthermore, from the definition of δ, we have

$$-\left(2-(8-\varepsilon)^{1/3}\right) \leq -\delta < x-2 < \delta \leq (8+\varepsilon)^{1/3}-2.$$

Equivalently, we have

$$(8-\varepsilon)^{1/3} < x < (8+\varepsilon)^{1/3}.$$

Therefore, $(8-\varepsilon) < x^3 < (8+\varepsilon)$, from which we conclude that $|x^3-8| < \varepsilon$. This completes the proof. ∎

Exercises

For Exercises 1–5, determine a value for T, such that $|f(x)-L| < \varepsilon$ whenever $x > T$. What is the smallest value of T that works?

1. $\lim_{x\to\infty} \left(5+7e^{-0.02x}\right) = 5;\ \varepsilon = 0.1.$
2. $\lim_{x\to\infty} \dfrac{x^2+7}{3x^2+9x} = \dfrac{1}{3};\ \varepsilon = 0.1.$
3. $\lim_{x\to\infty} e^{-3x}\sin(x) = 0;\ \varepsilon = 0.1.$
4. $\lim_{x\to\infty} \dfrac{3x^2+5}{x^2} = 3;\ \varepsilon = 0.01.$
5. $\lim_{x\to\infty} \dfrac{x^4+5x^2}{4x^4+4} = \dfrac{1}{4};\ \varepsilon = 0.1.$

For Exercises 6–10, prove that the limit has the given value. In particular, for an arbitrary $\varepsilon > 0$, determine $T > 0$ such that $|f(x)-L| < \varepsilon$ whenever $x > T$.

6. $\lim_{x\to\infty} \left(5+7e^{-0.02x}\right) = 5.$
7. $\lim_{x\to\infty} \dfrac{x^2+7}{3x^2+9x} = \dfrac{1}{3}.$
8. $\lim_{x\to\infty} e^{-3x}\sin(x) = 0.$
9. $\lim_{x\to\infty} \dfrac{3x^2+5}{x^2} = 3.$
10. $\lim_{x\to\infty} \dfrac{x^4+5x^2}{4x^4+4} = \dfrac{1}{4}.$

For Exercises 11–20, prove the following limits.

11. $\lim_{x\to 4} 3x+2 = 14.$
12. $\lim_{x\to 2} x^2 = 4.$
13. $\lim_{x\to a} c = c.$
14. $\lim_{x\to 0} x^2 = 0.$
15. $\lim_{x\to 0^+} \dfrac{1}{x} = \infty.$
16. $\lim_{x\to 1} \dfrac{1}{(x-1)^2} = \infty.$
17. $\lim_{x\to 0^+} \ln|x| = -\infty.$
18. $\lim_{x\to\infty} \dfrac{6x^2+7}{3x^2} = 2.$
19. $\lim_{x\to\infty} e^{-x} = 0.$
20. $\lim_{x\to\infty} \dfrac{1}{x} = 0.$

Problems

21. Consider the function

$$f(x) = \begin{cases} 0 & \text{if } x \text{ is rational,} \\ 1 & \text{if } x \text{ is irrational} \end{cases}.$$

Show that

$$\lim_{x\to 0} f(x)$$

does not exist.

Chapter 2

Theory of Differentiation

2.1 Average and Instantaneous Rates of Change

Calculus is the study of change: it is the study of the rates at which certain quantities change in time as well as the study of the cumulative change that is accrued over periods of time. The first part of this story, in which we study the rates of change, is called *differential calculus*, whereas the second part of this story, in which we study cumulative change over long periods of time, is called *integral calculus*. Of course, we will take a step toward abstraction, and study not only how quantities change in time, but also how quantities change with respect to any other independent variable. We begin by building the bridge from the average rate of change (for instance, the average speed of a car during a long road trip), to the instantaneous rate of change (for instance, the actual speedometer reading of the vehicle at one moment in time).

2.1.1 Average Rate of Change

Without further ado, we define the average rate of change as follows.

Definition 2.1.1 The *average rate of change* of a function f over an interval $[a,b]$ is defined as the quantity

$$AROC = \frac{f(b)-f(a)}{b-a}. \tag{2.1}$$

R If we let $y = f(x)$, and define *the change in* y as $\Delta y = f(b) - f(a)$ and *the change in* x as $\Delta x = b - a$, then the average rate of change, given by equation (2.1), is equivalent to

$$AROC = \frac{\Delta y}{\Delta x}. \tag{2.2}$$

Thus, the average rate of change is simply "the rise over the run," i.e., the net change in the output y divided by the net change in the input x.

The average rate of change of a function f over a specified interval $[a,b]$ has a particularly simple understanding: the average rate of change is the slope of the line that passes through the points $(a, f(a))$ and $(b, f(b))$. The quantity measured by equation (2.1) is the net change in the y-variable divided by the net change in the x-variable, as measured over the specified interval from $a \le x \le b$.

A graphic representation of the average rate of change of a function f over an interval $[a,b]$ is shown in Figure 2.1. We observe that the quantity $f(b) - f(a)$

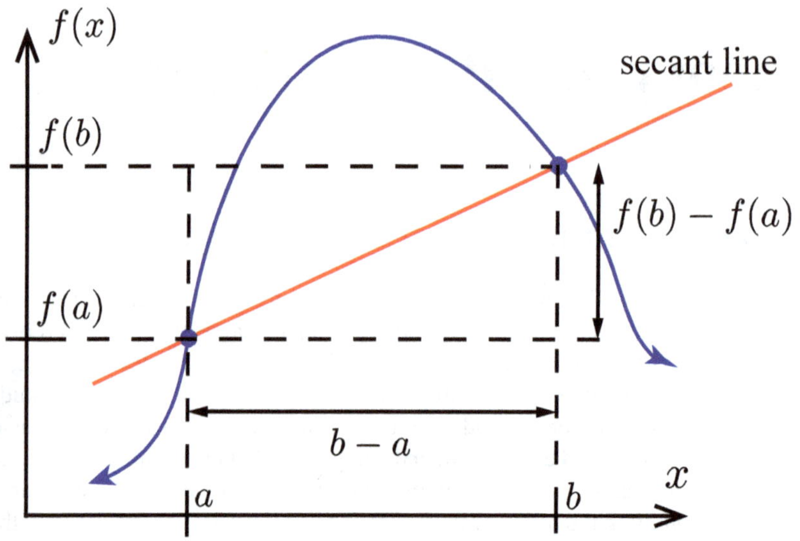

Figure 2.1: The average rate of change of a function.

represents the total change of the output variable (let's call it y), and the quantity $b-a$ represents the total change of the input variable x. Thus, the ratio

$$m = \frac{f(b) - f(a)}{b - a}$$

simply represents the slope of the line that passes through the two endpoints $(a, f(a))$ and $(b, f(b))$. This line is called a *secant line* through the function f.

The preceding interpretation is important enough that we will now state it as a proposition.

Proposition 2.1.1 The average rate of change of a function f on the interval $[a,b]$ is equal to the slope of the secant line of f passing through the points $(a, f(a))$ and $(b, f(b))$.

As an immediate consequence of this proposition, we have the following.

Corollary 2.1.2 If the function f is a linear function, then its average rate of change over *any* interval is equal to its slope.

Proof. Corollary 2.1.2 follows directly from the preceding proposition, though it can also be proved algebraically as follows. Assuming that f is a linear function, we can write it in the form

$$f(x) = mx + c,$$

where m represents the slope of the function (or, more precisely, the slope of its graph). (Notice that I used the parameter c for the y-intercept so as to reserve b for the second endpoint of the interval.) Next, let us compute the average rate of change of f over the interval $[a,b]$. We find

$$AROC = \frac{f(b) - f(a)}{b - a} = \frac{(mb + c) - (ma + c)}{b - a} = \frac{m(b - a)}{b - a} = m.$$

Since the interval $[a,b]$ was arbitrary, this completes the proof. ■

▪ **Example 2.1** On a certain day, low tide at Monterey Bay was at 6:30 a.m., when the water level was 0.4 ft. High tide was at 1:10 p.m., when the water level was 4.8 ft. Determine the average rate of change of the water level with respect to time between low tide and high tide.

One can match a sinusoidal function to these data, thereby obtaining

$$h(t) = 2.6 + 2.2 \sin\left(\frac{3\pi t}{20} + \frac{13\pi}{40}\right), \tag{2.3}$$

where $h(t)$ represents the water level in Monterey Bay t hours after noon on that given day. (So that 6:30 a.m. represents $t = -5.5$, and 1:10 p.m. represents $t = 7/6$.) The actual water level, as modeled by this sinusoidal function, is shown concurrently with the secant line in Figure 2.2.

Modeling the water level as a sinusoidal function, however, is not necessary to answer the given question. We are only asked to compute the average rate of change of the water level during the period of time between low tide and high tide. We are given the water levels at these moments in time:

$$h(-5.5) = 0.4 \qquad \text{and} \qquad h(7/6) = 4.8.$$

Thus, the average rate of change of the water level is given by

$$AROC = \frac{h(7/6) - h(-5.5)}{7/6 - (-5.5)} = \frac{4.8 - 0.4}{7/6 + 5.5} = \frac{22/5}{20/3} = \frac{33}{50}.$$

Thus, the average rate at which the tides were rising between low tide and high tide was 0.66 feet per hour. ■

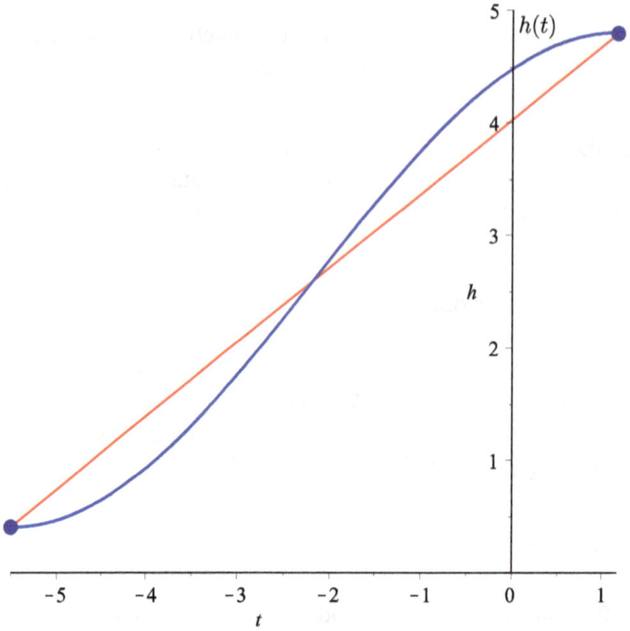

Figure 2.2: Water level at Monterey Bay, between low tide (6:30 a.m.) and high tide (1:10 p.m.), shown with the secant line (red).

■ **Example 2.2** At 3:00 p.m., a certain car passes mile marker 116 on a certain highway. At 5:30 p.m., they pass mile marker 296. What was the car's average velocity between 3:00 p.m. and 5:30 p.m.?

First, let $x(t)$ represent the car's position along the highway as a function of time t, measured in hours since 3:00 p.m. We are thus given the data $x(0) = 116$ and $x(2.5) = 296$. Using these data, we obtain

$$AROC = \frac{x(2.5) - x(0)}{2.5 - 0} = \frac{296 - 116}{2.5} = \frac{180}{2.5} = 72.$$

The car's average velocity was therefore 72 miles per hour. ■

The preceding examples reveal yet another interpretation of the average rate of change, and that is this. The average rate of change represents the rate of change that the function must have if it were to change at a constant rate over the specified interval. In Example 2.1, if the water level were to rise at a *steady* rate of 0.66 feet per hour, starting at a low-tide level of 0.4 feet, the water level would have reached 4.8 feet by high tide. We know from Figure 2.2, however, that the water level did not rise at a steady rate; it began rising at a slow rate, then the water level began rising more rapidly, and it finally, gently, "leveled off" at 4.8 feet at high tide. Similarly, if the car in Example 2.2 were to drive at a constant speed, it would have to have driven at 72 mph in order to traverse that distance in the same amount of time. In reality, however, the car might have sped up, slowed down, or even stopped at some point during those two and a half hours.

2.1.2 Instantaneous Rate of Change

Next, we define the concept of an *instantaneous* rate of change, that is, the rate of change of a function at one *instant*. Though we all have an intuitive understanding of an instantaneous rate of change—e.g., the velocity of your automobile at a particular instant in time—defining such a concept mathematically is, at first, elusive. If we are trying to measure the rate of change of a function *at an instant*, the Δx (or Δt, as the case may be) in the denominator will equal zero. In fact, even the function itself cannot change by a finite amount in an infinitesimal amount of time. (How much distance does your car traverse in one instant of time?)

The idea behind remedying this situation is to, instead of *directly* computing the instantaneous rate of change, compute the average rate of change over a sequence of successively smaller intervals, each containing our given instant in time.

> **Definition 2.1.2** The *instantaneous rate of change* of a function f at a point a is the single value that the average rate of change of f approaches if computed over a sequence of successively smaller intervals, each containing the point a, and if such a value exists.

Thus, we think of the instantaneous rate of change as a *process*. We compute the average rate of change over a small interval containing the point a, and then we do it again for a smaller interval containing the point a, and then we do it again over a smaller interval containing the point a, and so forth. If the average rate of change *levels off* to some fixed value, we say that this value is the instantaneous rate of change of the function at the point a.

The process we just described is exactly the process behind the concept of the *limit*. We may therefore, alternatively, express the instantaneous rate of change as

$$IROC = \lim_{\Delta x \to 0} \frac{\Delta y}{\Delta x}, \tag{2.4}$$

where it is understood that each interval for which the Δy and Δx are computed must contain the point a.

The instantaneous rate of change of a function f at a point a can be understood graphically, as shown in Figure 2.3. We begin by computing the the average rate of change over an interval, say $[a, b_1]$. This average rate of change represents the slope of the secant line labeled S_1. Next, we compute the average rate of change over a smaller interval, say $[a, b_2]$. This average rate of change represents the slope of the secant line S_2. Similarly, the slopes of secant lines S_3 and S_4 represent the average rates of change over the intervals $[a, b_3]$ and $[a, b_4]$.

We notice that as the interval becomes smaller and smaller, the secant lines are limited, in their slope, by the *tangent line* T to the graph of f at the point a. That is to say, in the limit as $n \to \infty$, and as $b_n \to a$, the secant lines S_n are

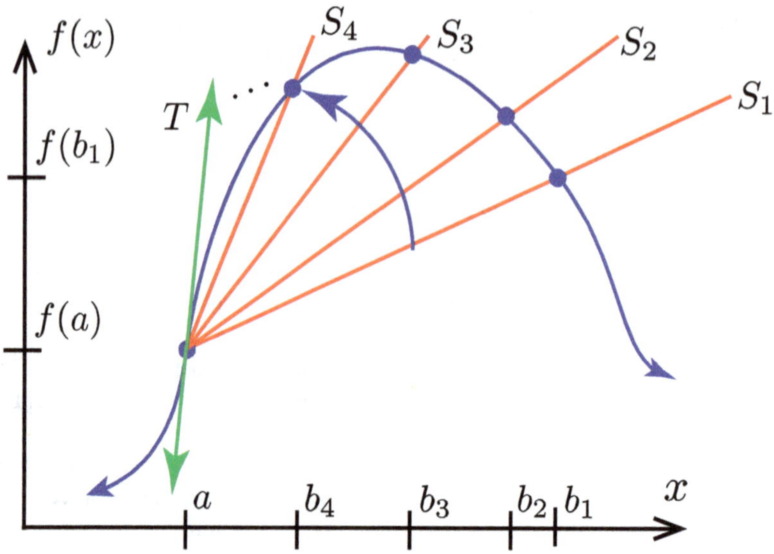

Figure 2.3: The instantaneous rate of change of a function at a point a.

approaching the tangent line T. Not only does this give us a visual understanding the of the instantaneous rate of change of a function, but it also yields the following proposition.

Proposition 2.1.3 The instantaneous rate of change of a function f at a point a is equal to the slope of the tangent line to f at the point a.

In other words, the tangent line, i.e., the unique line that touches the graph of f tangentially at the point $(a, f(a))$, has a slope that equals the *instantaneous* rate of change of the function f at the point a.

(R) Proposition 2.1.3 does not give us a method for computing the instantaneous rate of change of a function, but, rather, a graphical interpretation for the instantaneous rate of change. We will develop our own host of techniques for computing instantaneous rates of change in the remainder of the chapter and into Chapter 3. These techniques will then enable us, pursuant to the preceding proposition, to compute a formula for any tangent line to a graph.

An alternate way of thinking about the instantaneous rate of change and the tangent line is as follows.

(R) Consider a function f. If one were to zoom in on the graph of f, while keeping the point $(a, f(a))$ in the window, until the graph *looks* like a straight line (to the human eye), this line, if continued outward as a line, would be the tangent line to f at the point a, and its slope would be the instantaneous rate of change of f at the point a.

The preceding remark is illustrated in Figure 2.4. If we continually zoom in on a point $(a, f(a))$ on the graph of a function, eventually it will look like a straight line. (If it does not, then the instantaneous rate of change is not defined at that point.) This line, if continued outward as a line, represents the tangent line to the

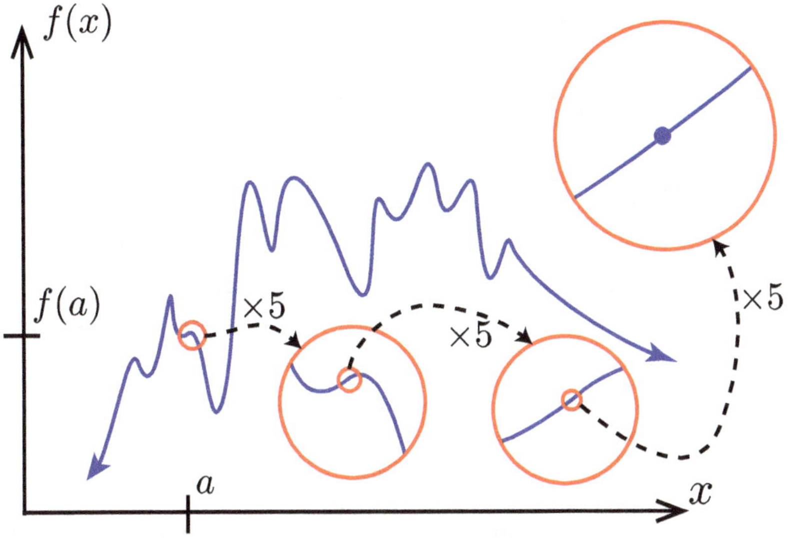

Figure 2.4: Zooming in on a point $(a, f(a))$ of the graph of a function f.

graph at the point $(a, f(a))$, and its slope represents the instantaneous rate of change of f at the point a.

■ **Example 2.3** Consider again the tide levels from Example 2.1. This time we will use the function (2.3) to model the water level, and we will determine the instantaneous rate of change of the water level half way between low tide and high tide, i.e., at 9:50 a.m., or when $t = -13/6$.

To do this, we compute the average rate of change over a sequence of successively smaller intervals, each of which contain the point $t = -13/6$. For instance, our first interval might be $[-13/6, -2]$. Computing the average rate of change over this interval, we obtain

$$AROC(-13/6, -2) = \frac{h(-2) - h(-13/6)}{-2 - (-13/6)} = \frac{(2.77261\cdots) - 2.6}{1/6} \approx 1.03566 \text{ ft/hr}.$$

This is our first approximation for the instantaneous rate of change at 9:50 a.m. To further facilitate this process, we may construct a table of values to more efficiently proceed with these computations. In this table, we will list a Δx, which will define the interval, on which we compute the average rate of change over, as the interval $[-13/6, -13/6+\Delta x]$. (In the preceding calculation, $\Delta x = 1/6$.) In the next column, we enumerate the output $h(-13/6+\Delta x)$, and in the final column, we enumerate the average rate of change, as given by $AROC = [h(-13/6+\Delta x) - h(-13/6)]/\Delta x$. These outputs are shown in Table 2.1 using 10 significant figures for the computations, and again in Table 2.2 using 20 significant figures.

Δx	$h(-13/6+\Delta x)$	$AROC(-13/6, -13/6+\Delta x)$	error
10^{-1}	2.703634192	1.036341920	3.8×10^{-4}
10^{-2}	2.610367217	1.036721700	3.9×10^{-6}
10^{-3}	2.601036726	1.036726000	4.2×10^{-7}
10^{-4}	2.600103673	1.036730000	4.4×10^{-6}
10^{-5}	2.600010367	1.036700000	2.6×10^{-5}
10^{-6}	2.600001037	1.037000000	2.7×10^{-4}
10^{-7}	2.600000104	1.040000000	3.3×10^{-3}
10^{-8}	2.600000010	1.000000000	3.7×10^{-2}
exact		1.036725576	

Table 2.1: Some tide data, as generated by equation (2.3), along with approximate AROC's. Computed using 10 significant figures.

Tables 2.1 and 2.2 show us what tricky business estimating the instantaneous rate of change can be, and why it is advisable to actually compute the average rate of change over a sequence of intervals, as opposed to just choosing one really small interval and going with that. The "error" is simply the absolute value of the difference between the predicted and exact value of the instantaneous rate of change of h at the point $t = -13/6$, i.e.,

$$\text{error} = |IROC - AROC(-13/6, -13/6+\Delta x)|.$$

The exact value for the instantaneous rate of change turns out to be $IROC = 0.33\pi$; this was computed, however, using techniques that we will discuss later in the text. We observe the error to decrease, as we use smaller and smaller values of Δx, but then it starts to increase again, as we loose accuracy of our computation. What is at play here is numerical error—specifically, *roundoff error* (or, *truncation error*)—and it is an unavoidable pitfall of technology. Roundoff error inevitably results as one's calculator or computer only tracks so many decimal places when it performs a computation. Here, we are attempting to compute a ratio $\Delta y/\Delta x$ of two quantities that are both going to zero. The Δy's become smaller and smaller as do the Δx's. Computing this ratio numerically, therefore, yields results that are more and more accurate, but only up to a certain point. After this point, the results become less and less accurate. This transition happens when the roundoff error of the computer hides the significant changes in y, resulting in a poorer estimation of the ratio $\Delta y/\Delta x$.

Δx	$h(-13/6+\Delta x)$	$AROC(-13/6,-13/6+\Delta x)$	error
10^{-1}	2.7036341915612138540	1.0363419156121385400	3.8×10^{-4}
10^{-2}	2.6103672173866215295	1.0367217386621529500	3.8×10^{-6}
10^{-3}	2.6010367255373143648	1.0367255373143648000	3.9×10^{-8}
10^{-4}	2.6001036725575300929	1.0367255753009290000	3.8×10^{-10}
10^{-5}	2.6000103672557568079	1.0367255756807900000	3.8×10^{-12}
10^{-6}	2.6000010367255756846	1.0367255756846000000	3.2×10^{-14}
10^{-7}	2.6000001036725575685	1.0367255756850000000	3.7×10^{-13}
10^{-8}	2.6000000103672557568	1.0367255756800000000	4.6×10^{-12}
exact		1.0367255756846317687	

Table 2.2: Some tide data as generated by equation (2.3), along with approximate AROC's. Computed using 20 significant figures.

Table 2.1 was generated using 10 decimal places. We observe that the optimal accuracy occurs when $\Delta x = 10^{-3} = 0.001$. For smaller Δx's, ten decimal places is not sufficient to accurately measure the ratio of $\Delta y/\Delta x$, and poorer accuracy results. Table 2.2 was generated using 20 decimal places. Here, the optimal accuracy is achieved by using $\Delta x = 10^{-6} = 0.000001$. ■

(R) In Example 2.3, we benefitted from *knowing* the exact value of the instantaneous rate of change. If we know this value, however, there is typically no need to approximate it. This begs a question: how then will we know when we've reached the optimal accuracy for our particular calculator or computer? The answer is as follows.

Before the optimal accuracy is achieved, some of the decimal places of the *AROC* will be changing. E.g., the fifth decimal place changes from a four to a two as we change Δx from 10^{-1} to 10^{-2} in Table 2.1.

After the optimal accuracy is achieved, all of the decimal places will remain the same (except, perhaps a ±1 change in the last nonzero decimal due to rounding), only there will be one fewer nonzero decimal place with every order of magnitude change in Δx. E.g., the *AROC* changes from 1.03673 to 1.0367 as we change Δx from 10^{-4} to 10^{-5} in Table 2.1. The *AROC* then changes to 1.037 as we change Δx to 10^{-6}, etc. (Notice the $+1$ change in the third decimal place due to rounding.)

2.1.3 One-dimensional Particle Kinematics

Calculus was first developed in conjunction with physics (Newtonian mechanics, in particular) as an analytical tool for solving and understanding complex problems of motion. Applications to physics therefore remain fundamental to any progression through calculus, though the applications and use of calculus have blossomed to a much grander palette. To facilitate simple applications to physics, however, we next present some background on one-dimensional particle motion.

In classical physics, a *particle* is any body or object whose size, shape, and internal motion is negligible. Particles are commonly visualized and act as though they are single points with definite mass, though they are an approximation of some macroscopic body like a ball, a stone, a sky-diver, or an automobile. In this paragraph, we consider particle motion in one dimension.

In studying motion, we always want to define a coordinate system and remain consistent. This is particularly important when studying *vertical* motion, e.g., free fall, in which the positive x-axis can either be defined as pointing upward or pointing downward.

Suppose now that our particle is free to move along the x-axis, and we wish to describe its motion as a function of time, i.e., we wish to express the relation $x = x(t)$. For example, consider the various snapshots of a particle moving along the x-axis, as shown in Figure 2.5. Here, the particle starts with an initial position at the origin.

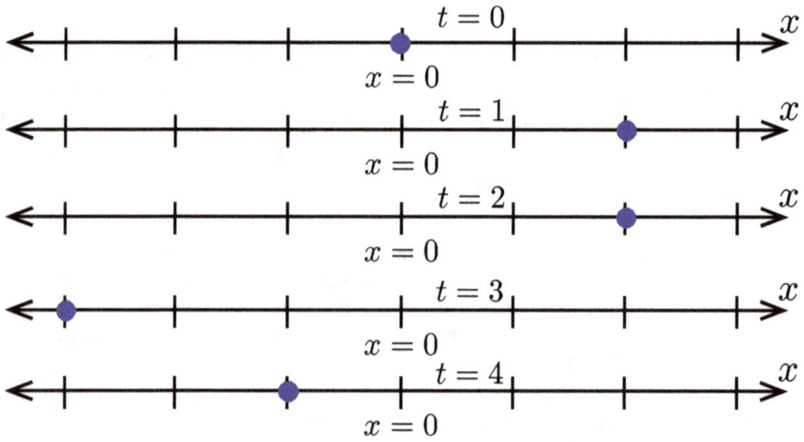

Figure 2.5: Various snapshots of a particle moving along the x-axis.

When $t = 1$, the particle is located at $x = 2$; when $t = 2$, the particle is still located at $x = 2$; when $t = 3$, the particle is located at $x = -3$; and when $t = 4$, the particle is located at $x = -1$. Assuming linear motion in between the unit-time intervals, the graph of the function $x(t)$ will appear as shown in Figure 2.6(a). Given only the information from Figure 2.5, however, any continuous function passing through the points $(0,0)$, $(1,2)$, $(2,2)$, $(3,-3)$, and $(4,-1)$ may be a possible motion of the particle, such as the graph shown in Figure 2.6(b).

What is important here is simply that we have a particle moving back and forth along the x-axis, and that this motion may be described as a function of time, i.e., $x = x(t)$. When the graph of the function $x(t)$ is increasing, this corresponds to the particle moving to the right (or toward the positive x-direction), and when the

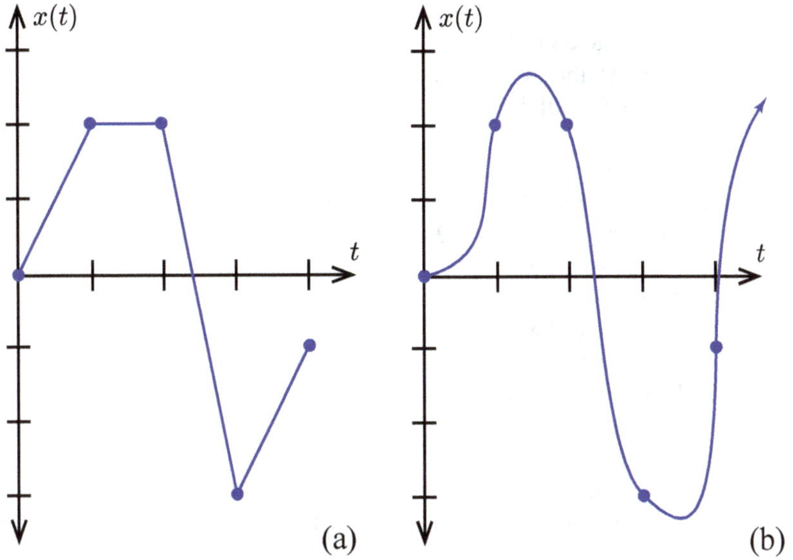

Figure 2.6: The particle's position $x(t)$ as a function of time t.

graph of the function $x(t)$ is decreasing, this corresponds to the particle moving to the left (or toward the negative x-direction).

Exercises

For Exercises 1–5, compute the average rate of change of the given function over the indicated interval(s).

1. $f(x) = \sin x$; (a) $[0,\pi/6]$, (b) $[0,\pi/3]$, (c) $[0,\pi/2]$, (d) $[0,\pi]$.
2. $f(x) = \cos x$; (a) $[0,\pi/6]$, (b) $[0,\pi/3]$, (c) $[0,\pi/2]$, (d) $[0,\pi]$.
3. $f(x) = 3x+7$; (a) $[0,1]$, (b) $[0,3]$, (c) $[6,7]$.
4. $f(x) = e^x$; (a) $[0,1]$ (b) $[0,2]$, (c) $[0,3]$, (d) $[4,9]$.
5. $f(x) = \ln|x|$; (a) $[1,2]$, (b) $[1,4]$, (c) $[2,4]$.

For Exercises 6–10, approximate the instantaneous rate of change of the given function at the given point(s). For each, determine an equation for the tangent line to the graph at the given point(s) and plot the function with its tangent line.

6. $f(x) = \sin x$; (a) $x = 0$, (b) $x = \pi/4$, (c) $x = \pi/2$, (d) $x = \pi$.
7. $f(x) = e^x$; (a) $x = 0$, (b) $x = 1$, (c) $x = 2$.
8. $f(x) = \sqrt{x}$; (a) $x = 1$, (b) $x = 2$, (c) $x = 4$.
9. $f(x) = \ln|x|$; (a) $x = 1$, (b) $x = 2$, (c) $x = 4$.
10. $f(x) = x^2$; (a) $x = 0$, (b) $x = 1$, (c) $x = 2$, (d) $x = 3$.
11. Report any observations you made from Exercises 6–10.

Problems

12. Consider the function $f(x) = |x|$.
    (a) Compute the average rate of change of f over the intervals $[0,1]$, $[0,0.1]$, $[-1,0]$, $[-0.1,0]$, $[-0.1,0.1]$, and $[-0.1,0.5]$.
    (b) Explain why the instantaneous rate of change of f is undefined at $x = 0$.
13. The Wright brothers performed the first powered airplane flights on December 17, 1903 at Kitty Hawk, North Carolina.
    (a) Their first flight lasted 12 seconds and covered 120 feet. What was the average speed of the plane during this flight?
    (b) Their fourth flight that day covered a distance of 852 feet, and the average speed was 14.44 ft/s. How long did this flight last?
14. The distance s that a car has traveled t hours into a road trip is given in Table 2.3 for various values of t. Find the average velocity between the first and fifth hour of the trip.

s (miles)	0	37	102	175	212	269
t (hours)	0	1	2	3	4	5

Table 2.3: Data for Problem 14

15. The distance s that Jill drove to see her aunt, who lives 50 miles away, as a function of the number of minutes t since she left her home, is shown in Table 2.4. During which 10 mile segment of her trip did she have the greatest average velocity? What was the average velocity for those 10 miles? During which 10 minute segment did she have the least average velocity? What was the average velocity for those 10 miles?

s (miles)	0	10	20	30	40	50
t (mins)	0	17	26	34	49	64

Table 2.4: Data for Problem 15.

16. The position x of a termite that is walking along a ruler is given at various points in time, t minutes after he began his trek, in Table 2.5. What is the termite's average velocity during the first four minutes of his trek? What is his average velocity between $t = 2$ and $t = 3$? What does the sign of your answer tell you about how the termite was walking?

x (inches)	5.25	7.0	4.25	6.5	7.25	5.0
t (mins)	0	1	2	3	4	5

Table 2.5: Data for Problem 16.

17. Suppose that a car has driven $x(t) = 80\ln(t+1)$ miles during the first t hours of a certain trip.
    (a) Graph $x(t)$. What is the domain of this function? Explain why this is so using practical terms.
    (b) Compute the car's average speed for the first two hours of the trip.
    (c) Estimate the car's speed at the moment it started the trip.
    (d) Estimate the car's speed at exactly one hour into the trip.
    (e) Challenge! Estimate the car's instantaneous acceleration one hour into the trip. *Hint: To estimate the instantaneous acceleration 1 hour into the trip, you need to compute the average rate of change of* $v(t)$ *over a small interval containing* $t = 1$, *where* $v(t)$ *represents the car's speed.* Is the acceleration positive or negative? Why?
18. Suppose that $h(t)$ represents the height of the water level in a harbor above some reference height, t hours after the high tide on a particular day, as shown in Figure 2.7.
    (a) Find the average rate of change of the water level over the intervals $6 \le t \le 12$ and $0 \le t \le 24$.
    (b) Carefully reproduce the graph in Figure 18. Draw two secant lines whose slopes represent the two average rates of change you computed in part (a).
    (c) When was the water level at its minimum? What was the instantaneous rate of change of $h(t)$ at that point?

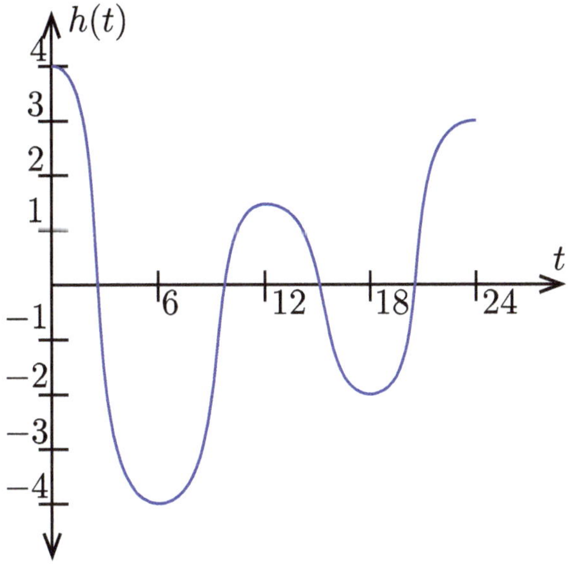

Figure 2.7: Water level (ft) in a harbor as a function of time (hr); Problem 18.

19. Let $d(s)$ represent the density (kg/m^3) of salt water at 4°C as a function of its salinity s (g of salt / L of water). Various data points are included in Table 2.6.

$d(s)$ (kg/m^3)	1000	1004.000	1007.972	1011.938	1015.902
s (g/L)	0	5	10	15	20

Table 2.6: Density of salt water as a function of its salinity; Problem 19.

(a) Use dimensional analysis to show that 1 g / L = 1 kg / m^3. (*Note*: 1 m^3 = 1000L.)
(b) What are the units of the average rate of change of the density of salt water with respect to its salinity?
(c) Compute the average rate of change of the density of salt water over the intervals $0 \le s \le 5$, $5 \le s \le 15$, and $15 \le s \le 20$.

20. Consider $f(x)$ as depicted in Fig. 2.8. Place the following in increasing order:
(a) The instantaneous rate of change of $f(x)$ at $x = a$.
(b) The instantaneous rate of change of $f(x)$ at $x = c$.
(c) The average rate of change of $f(x)$ over the interval $a \le x \le b$
(d) The average rate of change of $f(x)$ over the interval $a \le x \le c$
(e) The average rate of change of $f(x)$ over the interval $c \le x \le d$
(f) The slope of the tangent line of $f(x)$ at $x = b$.
(g) The slope of the tangent line of $f(x)$ at $x = d$.

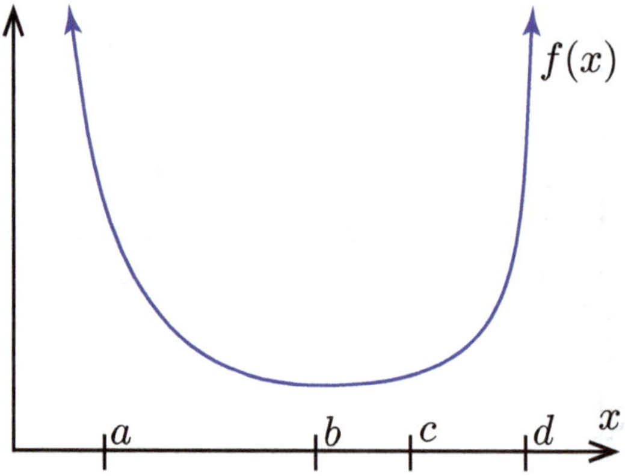

Figure 2.8: Function for Problem 20.

21. Suppose $f(x)$ is concave down on the interval $a \le x \le b$. Place the following in increasing order. Be sure to explain your reasoning.
(a) The instantaneous rate of change of $f(x)$ at $x = a$.

(b) The instantaneous rate of change of $f(x)$ at $x = b$.
(c) The average rate of change of $f(x)$ over the interval $a \le x \le b$.

22. Suppose $f(x)$ is concave up on the interval $a \le x \le b$. Place the following in increasing order. Be sure to explain your reasoning.
    (a) The instantaneous rate of change of $f(x)$ at $x = a$.
    (b) The instantaneous rate of change of $f(x)$ at $x = b$.
    (c) The average rate of change of $f(x)$ over the interval $a \le x \le b$.
23. Suppose $f(x)$ is an increasing function on the interval $a \le x \le b$. Why is it impossible, based on this information alone, to determine the order of:
    (a) The instantaneous rate of change of $f(x)$ at $x = a$.
    (b) The instantaneous rate of change of $f(x)$ at $x = b$.

    Draw two different functions that are both increasing. Draw one function with the property $(a) > (b)$. Draw the other function with the property $(b) > (a)$.
24. A car starts from at rest, and starts driving with an increasing speed. Draw a possible graph of its distance as a function of time.
25. A speeding car starts reducing its speed. Draw a possible graph of its distance as a function of time.
26. A ball is dropped from a tall building. Its height from the ground, measured in feet, t seconds after it is dropped, is given by $h(t) = 100 - 16t^2$.
    (a) How tall is the building?
    (b) At what time will the ball hit the ground?
    (c) Determine the average velocity of the ball over the time intervals $[2, 2.5]$, $[2, 2.1]$, $[2, 2.05]$, and $[2, 2.01]$.
    (d) Estimate the instantaneous velocity of the ball at $t = 2$s.
27. Suppose the number of members of a certain social club who show up to the annual party is a function $M(b)$ of the number of wine barrels, b, initially present at the party, as shown in Figure 2.9. The more barrels of wine ordered for the party, the more popular the party is believed to be, and the more members show up. There is a limit, however, to how fast the word can be spread about the party, so eventually the total number of members levels off. If too many barrels are present, this may even turn off some members, as some may feel that they will be lost in the crowd or that the club is trying too hard to be popular.
    (a) Estimate the average rate of change of $M(b)$ on the interval $0 \le b \le 30$.
    (b) Estimate the average rate of change of $M(b)$ on the interval $0 \le b \le 60$.
    (c) Estimate the instantaneous rate of change of $M(b)$ at $b = 10$. Interpret the meaning of this number in words. *Hint: If ten barrels are purchased, purchasing one extra barrel will result in approximately how many extra members attending the party?*
    (d) Estimate the instantaneous rate of change of $M(b)$ at $b = 40$. What is the sign of your answer? What is the real-world significance of the sign of your answer? Interpret your answer in words.
    (e) How many barrels should be purchased so that a maximum number of members will show up at the party?

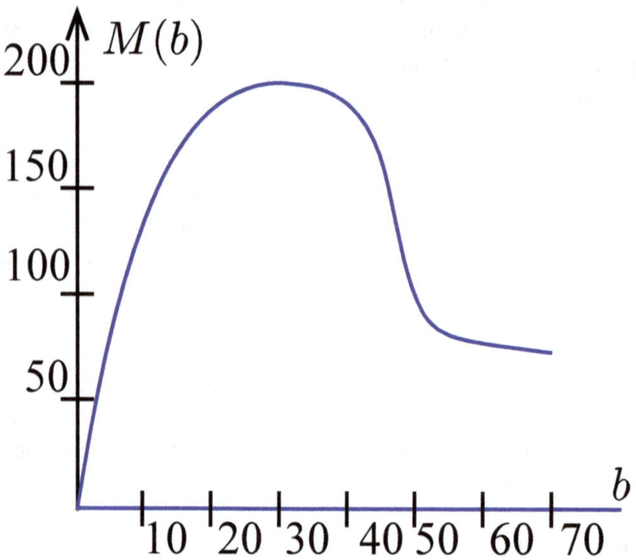

Figure 2.9: Function for Problem 27.

(f) What is the instantaneous rate of change of $M(b)$ if b is equal to the answer to part (e)?

2.2 The Derivative of a Function

In this section, we formally introduce the concept of a *derivative* of a function. Simply put, given one function, its derivative is a new function that outputs the instantaneous rate of change of the original function at any x-value in its domain. We next link this notion of a derivative with the tangent line to the functions graph, hint at some properties of derivatives, and then define velocity and acceleration in a return to our discussion on one-dimensional particle motion.

2.2.1 Derivatives Defined

In this paragraph, we present our first definition for a derivative, discuss various notations used for expressing this concept, and discuss the units of the derivative function.

Definition

We present an intuitive, working definition of a derivative here; we will formalize this definition later in Section 2.4.

> **Definition 2.2.1** Given a function f, the *derivative* of f, denoted f', is the new function that outputs, at any point in the domain of f, the instantaneous rate of change of the function f at that point, whenever it is defined.
>
> If the derivative of a function f is defined at a point $x = a$ on the domain, we say that f is *differentiable* at a.

■ **Example 2.4** In Examples 2.2 and 2.3, we studied the water level $h(t)$, measured in feet, in Monterey Bay, as a function of the number of hours t since noon on a particular day. We modeled $h(t)$ as a sinusoidal function, as given by equation (2.3), and went on to examine how the instantaneous rate of change of this function at 9:50 a.m., i.e., when $t = -13/6$, was equal to 0.33π. Since the derivative function h' yields an output, at any point, equal to the instantaneous rate of change of h at that point, we conclude that

$$h'(-13/6) = 0.33\pi \text{ ft/hr}.$$

(Of course, this exact value was only approximated in Example 2.3, but the point remains nevertheless.) ■

Notation

As mentioned previously, the derivative of a function f is denoted by adding a "prime" to the name of the function, i.e., f' (read "f prime"). This is *Lagrange's notation*. An alternate notation is sometimes also used; in particular, it is used when the dependent variable also has its own name. For instance, let $y = f(x)$. Then the derivative may also be denoted by the symbol

$$\frac{dy}{dx}.$$

Note that this collection of variables represents one symbol—dy/dx—that is used to refer to the derivative in the same way that y is used to refer to the original function. This latter notation is referred to as *Leibniz' notation.*

Often, the two notations work in conjunction with each other. For instance, we may state that $y = f(x)$, and then, therefore,

$$\frac{dy}{dx} = f'(x).$$

A final variation: we also use the notation

$$\frac{d}{dx}[\cdots]$$

to represent "the derivative of" $\cdots$ with respect to x. (Stating "with respect to x" is a statement as to which independent variable we are taking the derivative with respect to. This is an important distinction later on, when we take derivatives of functions of functions, e.g., suppose the temperature u is a function of a particle's position x, but this position is also a function of time t, thus, we have $u(x(t))$. More on this later.)

The following example illustrates the numerous ways in which we may say one thing, that the derivative of one function is another function.

▪ **Example 2.5** Given the function $f(x) = x^3$, we assert that its derivative is given by the function $f'(x) = 3x^2$. (We will see why this is true later, in Chapter 3. For now, we will treat it as a given.) The following two statements are equivalent ways of expressing this fact.

1. The derivative of the function $f(x) = x^3$ is $f'(x) = 3x^2$.
2. $\frac{d}{dx}\left[x^3\right] = 3x^2$.

Alternatively, the following three statements are equivalent.

1. The original function is $f(x) = x^3$; its derivative is $f'(x) = 3x^2$.
2. $f(x) = x^3$; $f'(x) = 3x^2$.
3. $y = x^3$; $\frac{dy}{dx} = 3x^2$.

Each of the above five statements is a different expression of the same information.
▪

The rationale behind Leibniz' notation is as follows. The instantaneous rate of change of a function at a point expresses the ratio of an infinitesimal change in the dependent (output) variable y to the corresponding infinitesimal change in the independent (input) variable x. Whereas the *average* rate of change is symbolized as

$$AROC = \frac{\Delta y}{\Delta x},$$

the *instantaneous* rate of change can be symbolized as

$$IROC = \frac{dy}{dx},$$

where it is understood that dy and dx are both infinitesimally small quantities, and that their ratio is formed in the limit as the length of the interval shrinks to zero. Allowing the instantaneous rate of change to be evaluated at any point on the domain, the symbol dy/dx represents a *function* of x—the derivative of the original function f.

Units

In real-world applications, one often studies problems in which the variables represent physical quantities. Such quantities typically carry along with them certain *units of measurement*, or *units*, for short. It is important, then, to keep these units straight. In particular, given a physical quantity y that is a function of another physical quantity x, so that $y = f(x)$, we will need to understand how to correctly report the units of the derivative function f'. To aid in our discussion, we use the notation units$(\cdots)$ to represent "the units of $\cdots$." For example, suppose $x = 3.7$ft. Then we would say that units$(x) =$ ft.

The units of a derivative of a function are given pursuant to the following proposition.

Proposition 2.2.1 Suppose a certain physical quantity y is a function of the physical quantity x according to the relation $y = f(x)$. Then the units of the derivative of f are given by

$$\text{units}\left(f'(x)\right) = \frac{\text{units}(y)}{\text{units}(x)}.$$

This follows since the units of the average rate of change of the function f are similarly given by the ratio of the units of y to the units of x, and since the instantaneous rate of change is only a limit of average rates of change.

▪ **Example 2.6** Consider again the water level in Monterey Bay, as discussed in Examples 2.2, 2.3, and 2.4. The output of the function h has units of feet, whereas the input variable t has units of hours. Therefore, the units of the derivative are given by

$$\text{units}\left(h'(t)\right) = \frac{\text{units}(h)}{\text{units}(t)} = \frac{\text{ft}}{\text{hr}}.$$

The units of the derivative function h' are therefore given by feet per hour. ▪

2.2.2 The Derivative and the Tangent Line

In Proposition 2.1.3, we saw that the instantaneous rate of change of a function at a point is equal to the slope of the tangent line to the graph of that function at that same point. Since the derivative is a function that outputs the rate of change of the original function f, when evaluated at a point, we arrive at the following proposition.

Proposition 2.2.2 Given a function f, its derivative f', and a point a on the domain, the value of the derivative $f'(a)$, if defined at $x = a$, is equal to the slope of the tangent line to the graph of f at the point $(a, f(a))$.

The preceding proposition tells us that the derivative reveals the slope of the tangent line at each point on the domain. Consider, for example, the graph of the function f, as shown in Figure 2.10. Short tangent lines are plotted at various points

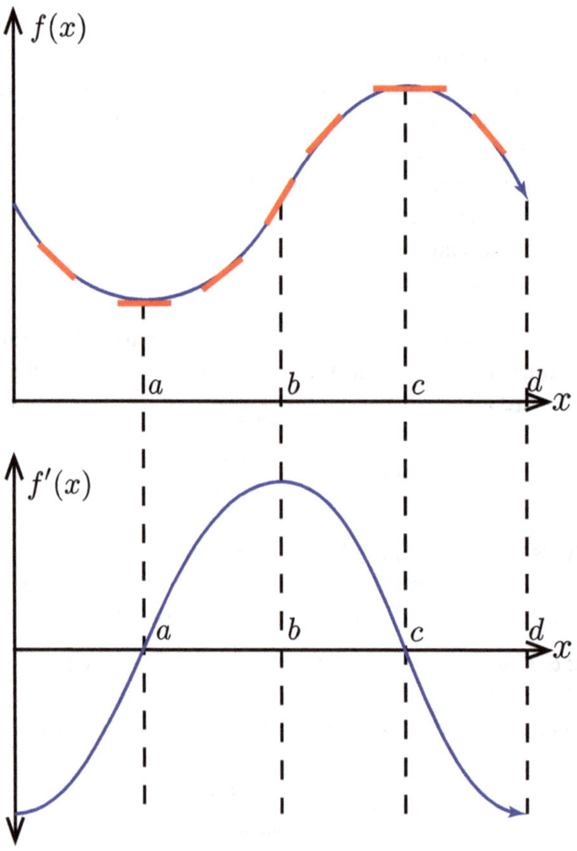

Figure 2.10: A function f plotted along with its derivative.

along the graph of f. Below this graph is a graph of the derivative function f'. Notice that the outputs of the derivative function f' change according to the slopes of the tangent lines.

Notice that when the function f is increasing (i.e., for $x \in [a, c]$), its tangent line is sloped upward, and, therefore, the derivative takes on a positive value. Similarly, when the function is decreasing (i.e., for $x \in [0,a]$ and $x \in [c,d]$), the derivative takes on a negative value.

Also, note that when the function f is concave up (i.e., for $x \in [0,b]$), the derivative function f' is increasing, whereas when the function f is concave down (i.e., for $x \in [b,d]$), the derivative function f' is decreasing.

The relation between a function and its derivatives, in terms of when the function is increasing, decreasing, concave up, or concave down, and when the derivative is positive, negative, increasing, or decreasing, is an important one, which we will explore in greater depth in Section 2.3. For now, however, we will be content with making these preliminary observations.

2.2.3 Differentiability Implies Continuity

In this paragraph, we discuss the relation between differentiability and continuity: in particular, we will see that a differentiable function must be continuous, but the converse of this statement—the assertion that a continuous function be also differentiable—need not be true.

Theorem 2.2.3 If a function is differentiable at a given point, it must also be continuous at that given point.

Proof. We present a heuristic argument to explain why this must be true. Differentiability at the point $x = a$ means that one can determine the instantaneous rate of change of the given function f at the given point, which, in turn, entails that when one computes the average rate of change over a sequence of successively smaller intervals, the average rates of change must settle on one particular value. In other words, in order for a function to be differentiable, the limit

$$\lim_{\Delta x \to 0} \frac{\Delta y}{\Delta x}$$

must exist, i.e., the average rates of change, given by $\Delta y/\Delta x$, must settle down on a particular value as we shrink the width of the interval, Δx, down to zero. If the denominator Δx shrinks to zero, as the ratio $\Delta y/\Delta x$ approaches a constant, it must be the case that the numerator Δy must shrink to zero as well, commensurately with Δx. (If Δy were to remain finite, such as would be the case if the function were discontinuous at the point in question, then the ratio $\Delta y/\Delta x$ would be undefined in the limit, as it would approach $1/0$.)

Now, if Δy goes to zero as Δx goes to zero, it must, finally, be the case that the function f is continuous. This conclusions follows since Δy is defined as $\Delta y = f(x) - f(a)$, where $x = a + \Delta x$. Since

$$\lim_{\Delta x \to 0} \Delta y = \lim_{x \to a} \left[f(x) - f(a) \right] = 0,$$

it follows that

$$\lim_{x \to a} f(x) = f(a)$$

as well. ∎

One can alternatively understand Theorem 2.2.3 as follows: if a function is differentiable at a point, then one must be able to sufficiently zoom in on that point and see the tangent line, i.e., the graph must approximately look linear in a neighborhood about the given point. This is impossible if the function fails to even be continuous at the point in question, because, at the very least, discontinuity would imply a gap or a jump in the function at that point.

The converse of Theorem 2.2.3 is not true, i.e., if a function is continuous, it does not necessarily imply that it is differentiable.

■ **Example 2.7** Consider the *absolute value function*

$$f(x) = |x|.$$

We will discuss the continuity and differentiability of this function at the point $x = 0$.

The graph of the absolute value function is shown in Figure 2.11. Intuitively, on

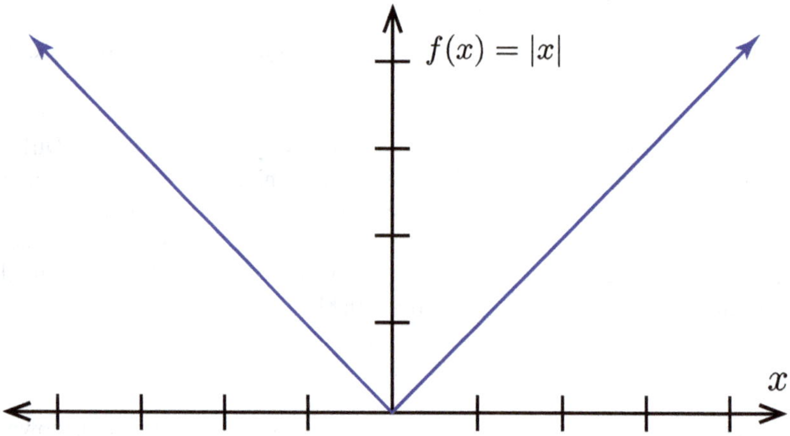

Figure 2.11: Graph of the absolute value function $f(x) = |x|$.

the one hand, we observe that the absolute value function is continuous everywhere; in particular, it is continuous at the point $x = 0$. This is because the graph can be drawn without lifting the pencil from the page. On the other hand, however, the absolute value function fails to be differentiable at the point $x = 0$, as the graph has a "corner" in it; i.e., no matter how closely one zooms into the point $(0,0)$, the absolute value function will never look like a single line. The corner will be always be there, no matter how far we zoom in on that point. ■

■ **Example 2.8** Consider the function f that is pictured in Figure 2.12. First, we observe that the function f is discontinuous at the points $x = -4, -3, -1, 0, 1$, and

3. Since continuity is prerequisite to differentiability, the function f also fails to be differentiable at those points. In addition, the function f is nondifferentiable at the points $x = -2$, 2, and 4, though f *is* continuous at those points.

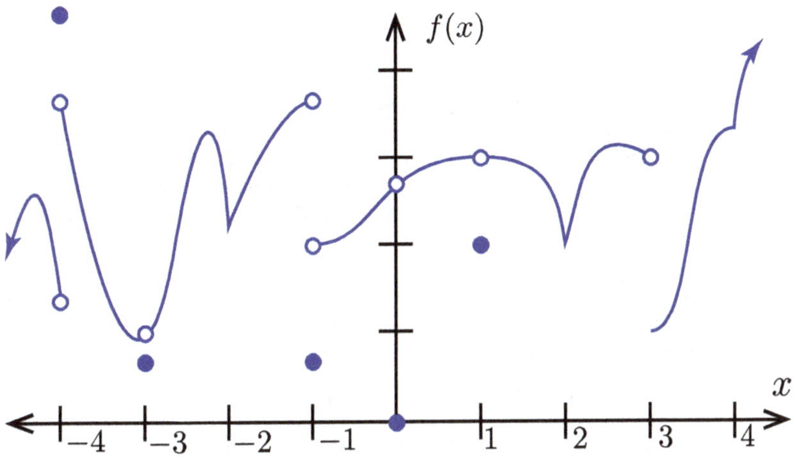

Figure 2.12: A function with various discontinuities and points of nondifferentiability.

■

The preceding examples show how a function can fail to be differentiable at a given point, and yet still be continuous. In short, the function can have some sort of corner in it, where its slope, if you were to only look on one side of the point, is different than its slope if you were to look on the other side of that same point. On the other hand, whenever a function is differentiable, it must also be continuous.

2.2.4 Velocity and Acceleration

Toward the end of Section 2.1, we discussed how one-dimensional particle motion can be described by representing the position x as a function of time t. We continue that discussion here, as we delve into the relation between derivative, position, velocity, and acceleration.

Definition 2.2.2 Suppose a particle moves along a one-dimensional axis with coordinates x, so that its position may be represented as a function $x = x(t)$. The *velocity* $v(t)$ of the particle as a function of time t is given by

$$v(t) = x'(t). \tag{2.5}$$

Similarly, the *acceleration* $a(t)$ of the particle as a function of time t is given by

$$a(t) = v'(t). \tag{2.6}$$

The *speed* of the particle is the magnitude of its velocity, i.e., the particle's speed is given by $|v(t)|$.

(R) The speed of a particle measures *how fast* the particle is traveling. The velocity of a particle tells us *how fast and in what direction* the particle is traveling. For one-dimensional motion, the direction of the particle may only take two values: toward the positive x-axis or toward the negative x-axis. Hence, the direction is indicated by the *sign* of the velocity function $v(t)$.

Example 2.9 A ball is thrown vertically upward with an initial upward velocity of 24.5 m/s. Let $x(t)$ represent the balls height above the ground, t seconds after the ball is released. Its position may be represented by the function

$$x(t) = 24.5t - 4.9t^2, \qquad \text{for } t \in [0,5]. \tag{2.7}$$

The ball's height, as a function of time, is illustrated in Figure 2.13.

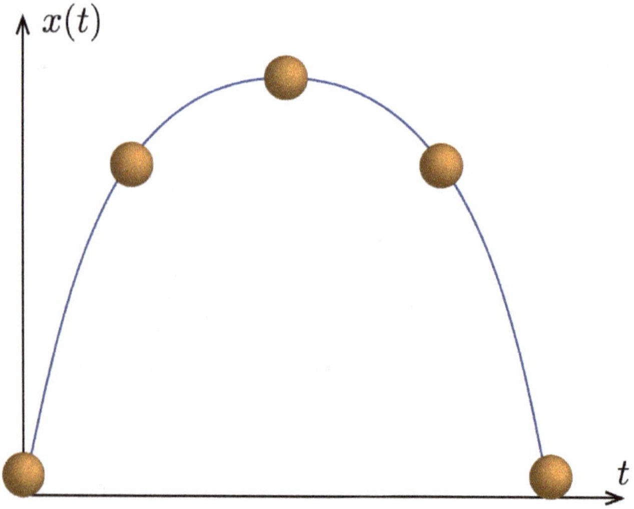

Figure 2.13: A ball thrown vertically upward, shown at several points along its trajectory.

Once we learn techniques for quickly computing the derivative of a function, we will be able to immediately see that the derivative of the function given by equation (2.7) is the function

$$v(t) = x'(t) = 24.5 - 9.8t, \tag{2.8}$$

which represents the ball's vertical velocity as a function of time.

We can verify the validity of equation (2.8) directly, by computing the instantaneous rate of change of the function x at various moments in time. For instance, we can use equation (2.7) to approximate the velocity of the ball after 1 second, thereby obtaining

$$v(1) \approx \frac{x(1.001) - x(1)}{0.001} = \frac{19.6146951 - 19.6}{0.001} = 14.6951 \text{ m/s}.$$

On the other hand, we can use the derivative (2.8) to compute the *exact* value of the ball's velocity at this single moment in time, as $v(1) = 24.5 - 9.8 = 14.7$ m/s. If we were to compute the average rate of change of the function x over a sequence of successively smaller intervals, each containing the point $t = 1$, we would obtain a sequence of numbers that would become closer and closer to 14.7. ∎

Exercises

For Exercises 1–8, a function is given along with its derivative. Choose a random point $x = a$ and experimentally determine the instantaneous rate of change of the function at that point. Now compute the value of the given derivative at that point. How do these two answers compare?

1. $f(x) = e^{\cos x}$, $f'(x) = -\sin x e^{\cos x}$.
2. $f(x) = e^{\sin^2 x}$, $f'(x) = 2\sin x \cos x e^{\sin^2 x}$.
3. $f(x) = xe^{\sin x}$, $f'(x) = xe^{\sin x}\cos x + e^{\sin x}$.
4. $f(x) = \sin^2 x$, $f'(x) = 2\sin x \cos x$.
5. $f(x) = \cos^2 x$, $f'(x) = -2\sin x \cos x$.
6. $f(x) = \tan x$, $f'(x) = \sec^2 x$.
7. $f(x) = x^3 + 9x^2 + 3x + 17$, $f'(x) = 3x^2 + 18x + 3$.
8. $f(x) = x^{3/2}$, $f'(x) = 3\sqrt{x}/2$.

For Exercises 9–14, a table is given that contains certain values of a function f or its derivative f'. Assume that the values you see in the table are a very good representation of the function (or its derivative). Based on the information available in the table, state everything you can about the function f and its derivative f'; e.g., the intervals on which each function is positive/negative, increasing/decreasing, concave up/down. State when there is not enough information to determine any of these things. Explain your reasoning.

9.

x	0	1	2	3	4
$f(x)$	0	1	3	6	10

10.

x	0	1	2	3	4
$f(x)$	0	-1	-3	-6	-10

11.

x	0	4	7	9	10
$f(x)$	0	1	2	3	4

12.

x	0	1	2	3	4
$f'(x)$	0	1	2	3	4

13.

x	0	1	2	3	4
$f'(x)$	2	6	8	6	2

14.

x	0	1	2	3	4
$f'(x)$	-2	0	4	0	-2

For Exercises 15–18, (a) approximate the instantaneous rate of the given function at the given points, (b) use your estimates to plot the values of the derivative f' at those points, and (c) state which basic function the derivative most closely resembles.

15. $f(x) = \sin x$, at $x = 0, \pi/4, \pi/2, 3\pi/4, \pi, 3\pi/2$, and 2π.
16. $f(x) = \cos x$, at $x = 0, \pi/4, \pi/2, 3\pi/4\ \pi, 3\pi/2$, and 2π.
17. $f(x) = e^x$, at $x = 0, 1, 2$, and 3.
18. $f(x) = \ln x$, at $x = 1, 2, 3$, and 4.
19. Consider the statements listed in Table 2.7. Each entry in Column I is true exactly when one corresponding entry from Column II is true. Match the corresponding entries.

Column I	Column II
$f(x)$ is increasing	$f'(x)$ is increasing
$f(x)$ is decreasing	$f'(x)$ is decreasing
$f(x)$ is concave up	$f'(x)$ is positive
$f(x)$ is concave down	$f'(x)$ is negative

Table 2.7: Matching choices for Problem 19

20. Consider the function $f(x) = \ln|\cos x|$.
    (a) By experimenting (i.e. by computing the average rate of change of $f(x)$ over subsequently smaller intervals containing the appropriate point), approximate the instantaneous rate of change of $f(x)$ at $x = 0$.
    (b) The derivative of this function is $f'(x) = \tan x$. Evaluate $f'(0)$.
21. Consider the function $f(x) = e^{x^2}$.
    (a) By experimenting, approximate the instantaneous rate of change of $f(x)$ at $x = 0$.
    (b) The derivative of this function is $f'(x) = 2xe^{x^2}$. Evaluate $f'(0)$.
    (c) Explain why this answer makes sense. *Hint:* You may want to graph $f(x)$.

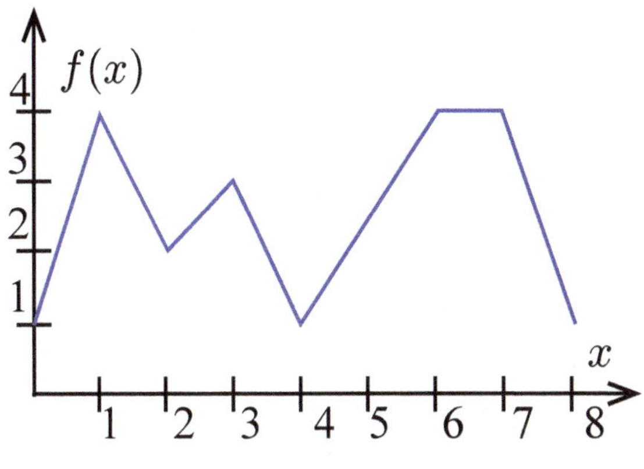

Figure 2.14: The function f from Exercise 22.

22. Draw a graph of the derivative f' of the function f shown in Figure 2.14. Is the function f continuous? Is the derivative f' continuous?
23. Let $f(x)$ represent the loudness of a rocket, measured in deciBels (dB), at a distance x miles from the rocket. What are the units of the derivative $f'(x)$?
24. Let $f(x)$ represents the sound intensity of a rocket, measured in Watts per meter squared (W/m^2), at a distance of x meters from the rocket. What are the units of the derivative $f'(x)$?

Problems

25. Let $h(t)$ represent the height of a rock that was thrown off of a tall sand dune on Mars as a function of the number of seconds t since the throw. Various data are given in Table 2.8.

$h(t)$ [ft]	20	20.0494	19	18.9294	6	5.8094
t [s]	0	0.01	1	1.01	2	2.01

Table 2.8: Height h of a rock that is in free fall on Mars; Problem 25.

(a) Approximate the instantaneous vertical velocity $h'(t)$ of the rock at $t = 0, 1$, and 2.
(b) What does the sign of the vertical velocities approximated in part (a) tell you about the path of the rock?
(c) Does $h'(t)$ appear to be increasing or decreasing? What does this tell us about the shape of the graph of $h(t)$?
(d) Using your approximations from (a), approximate the gravitational acceleration on the surface of Mars.

26. Let $s(t)$ represent the distance (miles) that a Boeing 747 has travelled down

a runway, during takeoff, t seconds after it began down the runway. Various data are given in the Table 2.9. Clearly state any assumptions you make and why they might be justified.

$s(t)$ [mi]	0.0625	0.0689	0.25	0.2627	0.5625	0.5814
t [s]	10	10.5	20	20.5	30	30.5

Table 2.9: Runway distance as a function of time; Problem 26.

(a) Approximate the airplane's velocity in mph at $t = 10, 20$, and 30.
(b) Using the three values of the airplane's velocity you approximated in part (a), approximate the airplane's acceleration.
(c) If the airplane requires a velocity of 180 mph before liftoff, how long will it take for the plane to lift off the ground?
(d) Approximately how far down the runway will the plane be when it takes off?

27. Let $P(h)$ represent the total hydrostatic pressure, measured in kiloPascals (kPa), at a depth of h meters beneath the ocean's surface. Various data are given in Table 2.10.

$P(h)$ [kPa]	101.3	103.2992	10,128.91	10,179.20	40,518.56	40,549.13
h [m]	0	0.2	1,000	1,005	4,000	4,003

Table 2.10: Hydrostatic pressure (kPa) as a function of depth (m); Problem 27.

(a) Approximate the values of the derivative $P'(h)$ at $h = 0, 1000$, and 4000.
(b) Approximate the depth at which the water pressure is twice as much as the surface pressure.
(c) Is the pressure increasing at a greater rate (per meter of depth) near the surface of the ocean, or deep down beneath the surface?
(d) Based on your approximations in (a), do you suspect that the graph of $P(h)$ would be concave up or concave down? Explain your reasoning.
(e) The formula for the total hydrostatic pressure, $f(h)$ kPa, at a depth of h meters beneath the surface, is often given in elementary physics texts by the formula

$$f(h) = P_0 + \frac{\rho g h}{1000},$$

where P_0 is the surface pressure (kPa), ρ is the water density (kg/m^3), $g = 9.8$ m/s^2 is acceleration due to gravity, and h is the depth (m) beneath the surface. The density of pure water is 1,000 kg/m^3. The function $f(h)$ is a linear function of the depth. Compute $f'(h)$. How closely does $P'(h)$, as estimated from the data above, agree with $f'(h)$. How might you explain this difference?

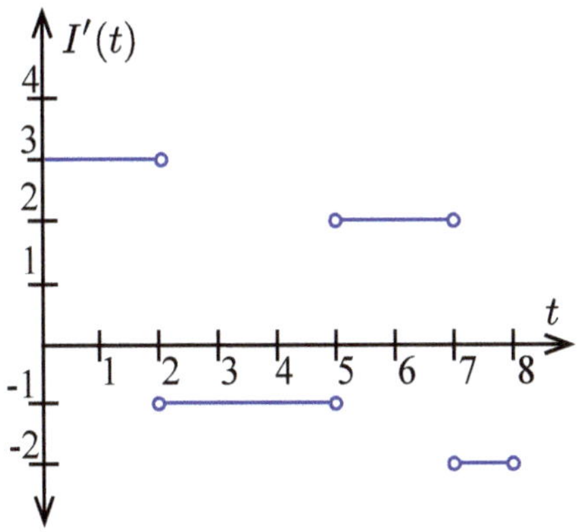

Figure 2.15: The derivative I' of a stock's value as a function of time t; Problem 30.

28. Let $T(t)$ be the temperature of a hot cup of coffee, t minutes after the coffee is poured. Assume that the ambient, room temperature is a constant, R. You are told that *the rate of change of the coffee's temperature is proportional to the difference between its temperature and the room temperature*, with constant of proportionality $k > 0$. Write down an equation that relates the temperature of the cup to its rate of change. Make sure you have the correct sign in front of the constant of proportionality.
29. The velocity $v(t)$ of an F-16 Falcon jet aircraft at various points in time t during its 7.5 s takeoff is given in the table below.

$v(t)$ [m/s]	0	24.2	24.8	25.5	26.6	27.6
t [s]	0	3.8	3.9	4.0	4.1	4.2

   (a) An afterburner in the jet engine of the aircraft is ignited at $t = 4$, causing a discontinuous jump in the vehicle's acceleration. Approximate

$$\lim_{t\to 4^-} v'(t) \qquad \text{and} \qquad \lim_{t\to 4^+} v'(t).$$

   (b) Is the acceleration of the aircraft continuous at $t = 4$?
   (c) Estimate the speed of the aircraft at takeoff. State any assumptions you might make.
30. Let $I(t)$ represent the value (dollars) of one share of a certain stock t milliseconds (ms) after the start of a discontinuous growth surge in the stock's value. The rate at which the stock's value is changing $I'(t)$ is plotted in Figure 2.15. How much more is a single share of this stock worth at the end of the 8 millisecond surge than it was at the beginning? If each share of the stock was initially worth \$52.00 at $t = 0$, what is each share of the stock worth 8 ms later?

2.3 Properties of Functions and Their Derivatives

In this section, we explore some of the implications of the derivative; in particular, we explore the relation between properties of the graph of a function and the function's derivative. Finally, we introduce the second derivative of a function.

2.3.1 Linear Functions and the Derivative

We begin with a brief discussion of linear functions. Recall from Corollary 2.1.2, the average rate of change of a linear function *over any interval* is simply equal to its slope. It follows, therefore, that the instantaneous rate of change of a linear function at any point must also equal its slope.

Proposition 2.3.1 The derivative of a linear function f is the constant function whose single output value is the slope of the function f, i.e., if we let $f(x) = mx + b$, then $f'(x) = m$.

This proposition is illustrated in Figure 2.16. A linear function is shown, along with two of its tangent lines. Notice that the tangent line for a linear function is coincident with the graph of the function itself. The slopes of the tangent lines therefore each equal the slope to the function f.

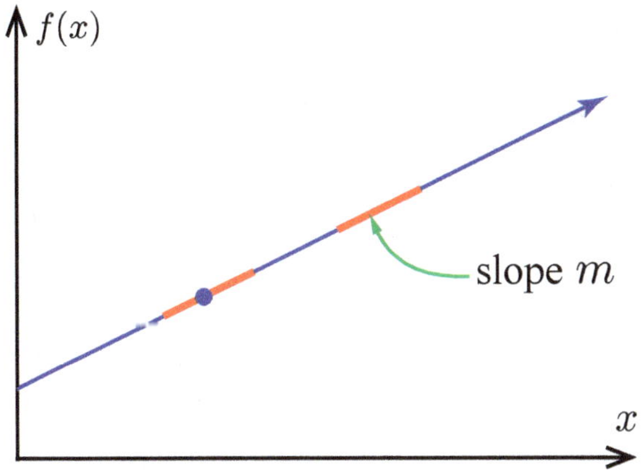

Figure 2.16: A linear function plotted with two of its tangent lines.

2.3.2 Increasing and Decreasing Functions

In Section 1.1, we first defined what is meant by increasing and decreasing functions, in particular, whether a function is increasing or decreasing on a particular interval of its domain.

Proposition 2.3.2 A differentiable function f is strictly increasing on an open interval I if and only if its derivative is strictly positive on the interval I, i.e., if and only if $f'(x) > 0$, for all $x \in I$.

A differentiable function f is strictly decreasing on an open interval I if and only if its derivative is strictly negative on the interval I, i.e., if and only if $f'(x) < 0$, for all $x \in I$.

To illustrate this, we refer again to Figure 2.10 on page 132. The set of points on which the function f is increasing corresponds precisely to the set of points at which the tangent line to the graph has a positive slope, which, in turn, corresponds to the set of points for which the derivative is positive.

Similarly, the function is decreasing precisely where the tangent line has a downward, or negative, slope, which, in turn, coincides with the points for which the derivative is negative.

Intuitively, the derivative measures the instantaneous rates of change of the function at any of the points on its domain. Since the function should be increasing when its rate of change is positive and decreasing when its rate of change is negative, it follows that the function should also be increasing when its derivative is positive and decreasing when its derivative is negative.

2.3.3 Concavity

In Section 1.1, we defined a function's concavity on an interval by the relation of the graph of the function to its tangent lines. In particular, when the graph of a function lies consistently above its tangent lines, then the function is concave up. Conversely, when the graph of a function lies consistently below its tangent lines, then the function is concave down. In this paragraph, we explore concavity more closely, redefining the concavity of a function in relation to its derivative.

Definition 2.3.1 We say that a function f is *concave up* on an open interval I if it is differentiable on I and its derivative f' is an increasing function on I.

Similarly, we say that a function f is *concave down* on an open interval I if it is differentiable on I and its derivative f' is a decreasing function on I.

This new definition is equivalent to Definition 1.1.5, but slightly more sophisticated. Their equivalency can again be seen by examining Figure 2.10 on page 132. The original function f, as shown in this figure, is clearly concave up on the interval $(0,b)$. As is immediately evident, the derivative f' is increasing on precisely this same interval. Similarly, the function f is concave down on the interval (b,d), the same interval on which the derivative f' is decreasing.

To see why this is true, let us consider Figure 2.17, in which is graphed a function that is concave up. By drawing several of its tangent lines, we immediately observe that these slopes must increase as we move from left to right. In particular, perhaps the slopes of the shown tangent lines are -4, -2, -1, 0, 1, 2, and 4. Regardless of whether the slope is positive or negative, the slope is always

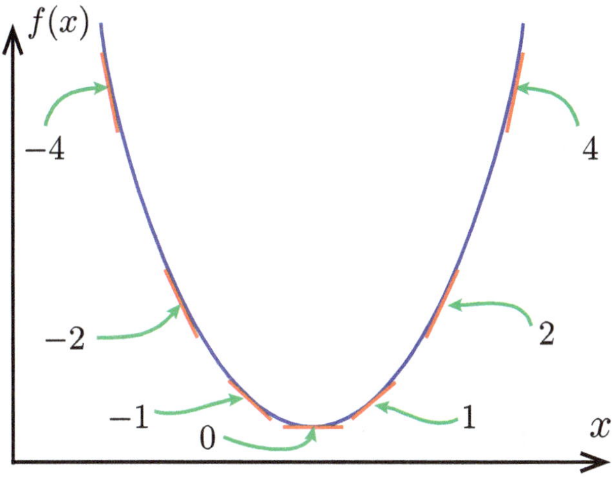

Figure 2.17: A concave-up function, plotted with various of its tangent lines.

increasing. Alternatively, imagine we place a pencil, tangent to the curve, on the far left end. This pencil is to represent the tangent line. Next, imagine moving the pencil along the curve, left to right, so that the center of the pencil is always touching the curve, and so that the length of the pencil is always tangent to the curve. As you visualize this, notice that the pencil must move counterclockwise, thus, its slope always has an upward trend.

The analogue picture for a concave-down function is shown in Figure 2.18. In this case, notice how the tangent line rotates clockwise, and that the slopes of the tangent lines decrease as we move from left to right along the graph.

Finally, notice that in Figures 2.17 and 2.18, when the function is decreasing, the tangent lines have a negative slope. Similarly, when the function is increasing, the tangent lines have a positive slope.

2.3.4 The Mean-Value Theorem

In this paragraph, we discuss an important result known as the mean-value theorem. Of course, we are not trying to anthropomorphize values in this name; rather, "mean" is used in its mathematical sense, i.e., as a synonym for the "average" value.

The mean-value theorem states that a function's instantaneous rate of change must, at least once, equal its average rate of change over an interval. Intuitively, this assertion makes a good deal of sense. Imagine a man is on a lengthy road trip, and his average velocity was 65 mph. Well, it stands to reason that he couldn't have been going faster than 65 mph the whole time. If he only drove faster than 65 mph, his average velocity could not be only 65 mph, it would have to be higher. Similarly, he couldn't have been driving less than 65 mph the entire time. Therefore, if he

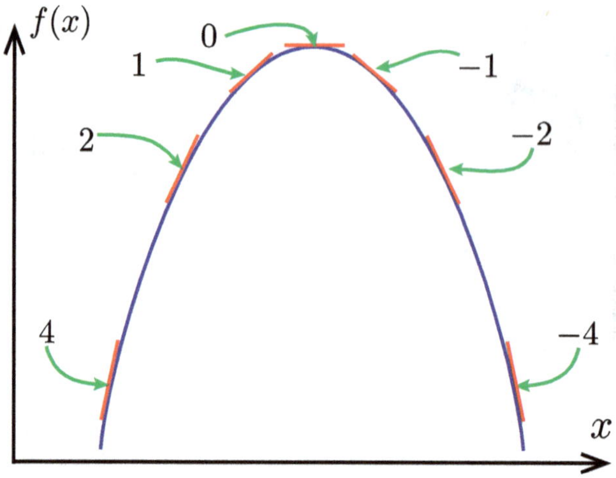

Figure 2.18: A concave-down function, plotted with various of its tangent lines.

was traveling faster than 65 mph for part of the time, then, also, he must have been traveling slower than 65 mph for part of the time as well. Thus, at some particular moment in time, his velocity must have actually passed through 65 mph. Maybe this occurred only for an instant, but his instantaneous velocity must have been 65 mph at some point in time nevertheless. This concept is codified precisely in the following theorem.

Theorem 2.3.3 — Mean-Value Theorem. If a function f is differentiable on a closed interval $I = [a,b]$, then there must exist some point $c \in (a,b)$ at which the derivative of f equals the average rate of change of f over the interval I, i.e., there exists a point $c \in (a,b)$, such that

$$f'(c) = \frac{f(b) - f(a)}{b - a}. \tag{2.9}$$

The mean-value theorem is illustrated in Figure 2.19. We start with the graph of a differentiable function f on a closed interval $I = [a,b]$. The points $(a, f(a))$ and $(b, f(b))$ are shown. Since the function is differentiable, it must also be continuous. Thus, the graph cannot have any jumps, holes, or sharp corners. Next, we draw the secant line that connects the points $(a, f(a))$ and $(b, f(b))$. The slope of this line is equal to the average rate of change of the function f over the interval I, i.e., the slope of this secant line is given by the formula

$$AROC = \frac{f(b) - f(a)}{b - a}.$$

As can be observed from the figure, there must be a point whose tangent line

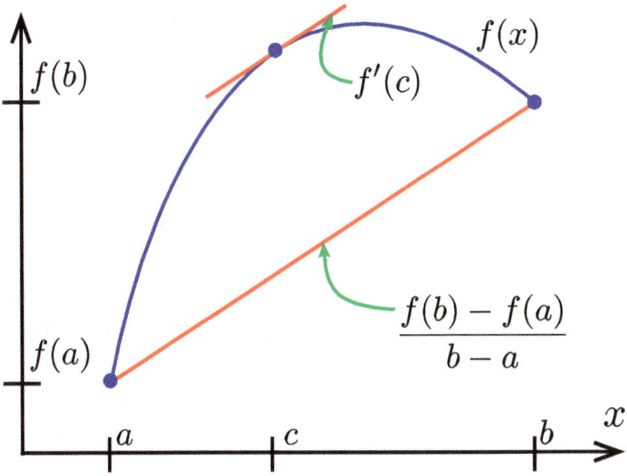

Figure 2.19: An illustration of the mean-value theorem.

is parallel to this secant line, which is an alternate statement of the mean-value theorem. We identified this point as the point c, and we observe that the tangent line to the graph of f at the point c is parallel to the secant line connecting the points $(a, f(a))$ and $(b, f(b))$. We conclude that the value of the derivative at the point c, i.e., $f'(c)$, is equal to the average rate of change of f over the interval I, as guaranteed by our theorem.

The mean-value theorem does not prohibit the case in which there are more than one points that satisfy the equation (2.9). Rather, it only guarantees that there is *at least one* such point.

2.3.5 The Second Derivative

If the derivative of a function is itself a function, then it is natural to ask whether or not a derivative has its own derivative. This is often the case, and the derivative of the derivative is known as the second derivative of the original function, as laid out in the following definition.

> **Definition 2.3.2** Suppose that a function f is differentiable on an open neighborhood containing the point $x = a$, i.e., the derivative f' is defined for all points nearby the point $x = a$. If the derivative f' is, itself, differentiable at the point $x = a$, then we say that the function f is *twice differentiable* at $x = a$. The derivative of the function f' is called the *second derivative* of the function f,

and is denoted by f'', or by

$$\frac{d^2y}{dx^2} = f''(x).$$

(R) When ambiguity threatens, one often refers to the derivative f' as the *first derivative* of the function f.

Definition of the second derivative in hand, we may now describe the relation between a function, its monotonicity or trend (i.e., whether it is increasing or decreasing), its concavity, and the properties of its various derivatives. This information is summarized in Table 2.11.

function	trend		concavity	
f	$\nearrow$	$\searrow$	UP	DN
f'	$+$	$-$	$\nearrow$	$\searrow$
f''	N/A	N/A	$+$	$-$

Table 2.11: Corresponding properties of a function f and its derivatives. Here, $\nearrow$ represents increasing; $\searrow$ represents decreasing; + represents positive; – represents negative; UP represents concave up; DN represents concave down.

First, we observe the basic information that whether or not a function is increasing or decreasing corresponds precisely to whether or not its derivative is positive or negative. If we only know the trend of a function, i.e., increasing or decreasing, then we cannot say anything for certain about the second derivative without additional information.

Next, we observe that the table further captures the relationship between the concavity of a function and the trend of its derivative. When a function is concave up, its derivative is increasing. (Therefore, its second derivative must be positive.) Similarly, when a function is concave down, its derivative is decreasing. (Therefore, its second derivative must be negative.)

Table 2.11 is a useful friend in a variety of contexts. Understanding these simple relationships—amongst trend, concavity, and sign—is a primary objective of any decent calculus training.

Exercises

1. If the functions f and g are related by the equation $f(x) = g(x) + ax + b$, for some nonzero constants a and b, is it necessarily true that f and g are both increasing and decreasing over the same intervals? Are f and g both concave up and concave down over the same intervals? Explain your reasoning.
2. Consider the linear functions

$$f(x) = ax + b \qquad \text{and} \qquad g(x) = cx + d.$$

Determine the derivative of the functions f, g, and $f+g$, in terms of the parameters a, b, c, and d. What do you observe? Do you expect that this might be true in general? Explain your reasoning.

3. Consider the function f shown in Figure 2.20. For each point $x = a, b, c, d$, and e, decide whether the derivative should look closest to the choice (i), (ii), (iii), or (iv) near that point. Some choices will be used more than once. Explain your reasoning.

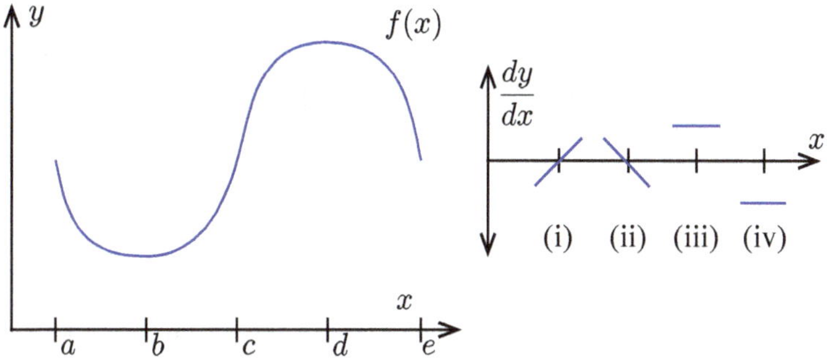

Figure 2.20: The function for Exercise 3.

4. Consider the data from a function f, as given in Table 2.12. Use these data

x	0	0.1	0.2	0.3	0.4	0.5
$f(x)$	17	17.1	17.25	17.5	17.8	18.05
x	0.6	0.7	0.8	0.9	1.0	—
$f(x)$	18.2	18.3	18.3	18.0	17.6	—

Table 2.12: Points on the graph of a function f for Problem 4.

to answer the following questions.

(a) On which intervals is the function f (i) increasing; (ii) decreasing?

(b) Estimate the value of the derivative $f'(0.25)$.

(c) Approximate the point(s) at which the derivative f' is (i) equal to zero; (ii) the greatest; (iii) the least.

(d) On which intervals is the function f (i) concave up; (ii) concave down?

5. Consider the function f that is graphed in Figure 2.21.

(a) On which intervals is the function f (i) concave up; (ii) concave down; (iii) increasing; (iv) decreasing?

(b) On which intervals is the derivative f' (i) positive; (ii) negative; (iii) increasing; (iv) decreasing?

(c) Draw a graph of the derivative f'.

6. Suppose there is a function g whose derivative is given by $g'(x) = f(x)$, where f is the function that is graphed in Figure 2.21.

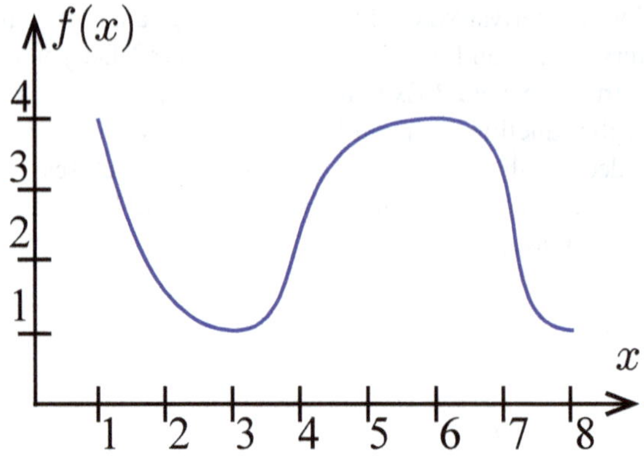

Figure 2.21: The function for Exercise 5.

(a) On which intervals is the function g (i) concave up; (ii) concave down; (iii) increasing; (iv) decreasing?
(b) On which intervals is the derivative g' (i) positive; (ii) negative; (iii) increasing; (iv) decreasing?
(c) Draw a graph of the function g.

Problems

7. Let $V(t)$ represent the total volume, in gallons (gal), of water in a tank, t minutes after noon on a given day. Suppose that, exactly at noon, water begins to be simultaneously pumped into the tank and drained from the tank, possibly at different rates.
   (a) Draw a diagrammatical sketch of the tank. Make sure to indicate water inlets and outlets.
   (b) Suppose the tank contained exactly 35 gals of water at noon. Let $V_{\text{in}}(t)$ be the total amount of water that has been pumped *into* the tank within the first t minutes after noon and let $V_{\text{out}}(t)$ be the total amount of water drained *from* the tank within the first t minutes after noon. Explain why
   $$V(t) = 35 + V_{\text{in}}(t) - V_{\text{out}}(t).$$
   (c) Briefly explain the physical meaning of the functions $V'_{\text{in}}(t)$ and $V'_{\text{out}}(t)$ in practical terms.
   (d) Following the assumptions of (b), write an expression for $V'(t)$.
   (e) Suppose that the tank contained exactly V_0 gals of water at noon, instead of 35 gals. Write an expression for $V(t)$ and $V'(t)$.
8. Let $M(t)$ be the total mass of saltwater, in kilograms (kg), in a tank, t minutes after salt and water begin being added to the tank. Let $S(t)$ be the total mass (kg) of salt in the tank at time t, and let $W(t)$ be the total mass of water (kg)

in the tank at time t. Various values of $S(t)$ and $W(t)$ are given in Table 2.13.

$S(t)$ - [kg]	5	14	14.7	20	20.8
$W(t)$ - [kg]	95	130	134.4	175	184.4
t - [min]	0	1	1.1	2	2.2

Table 2.13: Data for Problem 8.

(a) How much saltwater was initially in the tank?
(b) Approximate $S'(1)$, $W'(1)$, $S'(2)$, and $W'(2)$. Explain what these numbers mean physically.
(c) Based on your results to (b), predict the concavity of the functions $S(t)$ and $W(t)$. Explain your reasoning.
(d) At what rate was the total mass of saltwater increasing at $t = 1$ and 2?
(e) Let $f(t) = 100S(t)/M(t)$ be the percentage of salt in the saltwater at time t. Approximate $f'(1)$ and $f'(2)$. What are the units of these numbers? Do you predict the graph of $f(t)$ is concave up or concave down?
(f) Based on your estimates for $f'(1)$ and $f'(2)$, when would you predict that the percentage of salt in the saltwater reached its maximum value? Before 1 min, between 1 and 2 mins, or after 2 mins?

9. A patient in a hospital begins receiving a certain drug intravenously at 3:00pm. Let $I(t)$ be the total amount of the drug, in mL, injected into the patient's bloodstream t hours after 3:00pm, and let $A(t)$ be the total amount (mL) of the drug absorbed from the bloodstream t hours after 3:00pm. The derivatives $I'(t)$ and $A'(t)$ are graphed in Figure 2.22.
   (a) Briefly explain why the total amount of drug (mL) in the patient's bloodstream is given by $T(t) = I(t) - A(t)$.
   (b) What is the physical meaning of the functions $I'(t)$ and $A'(t)$?
   (c) At what rate was the total amount of drug in the patient's bloodstream changing at 5:00pm and 9:00pm? State whether the amount of drug in the patient's bloodstream was increasing or decreasing at each time.
   (d) At what time did the patient have the *most* amount of the drug in his bloodstream?
   (e) Over which time intervals is the graph of $I(t)$ concave up? concave down?

10. In economics, the *profit* of a given venture is defined as the *revenue* minus the *cost*. Suppose that the cost for a company to produce q quantity of goods is $C(q)$ dollars, and that the revenue they would bring in is $R(q)$ dollars. Then the profit, as a function of the quantity q of goods produced, is given by the relation $\Pi(q) = R(q) - C(q)$. Consider for a moment the following definitions.
    - The *marginal revenue*, if q quantity of goods are produced, is the amount of *additional* revenue the company would bring in if one

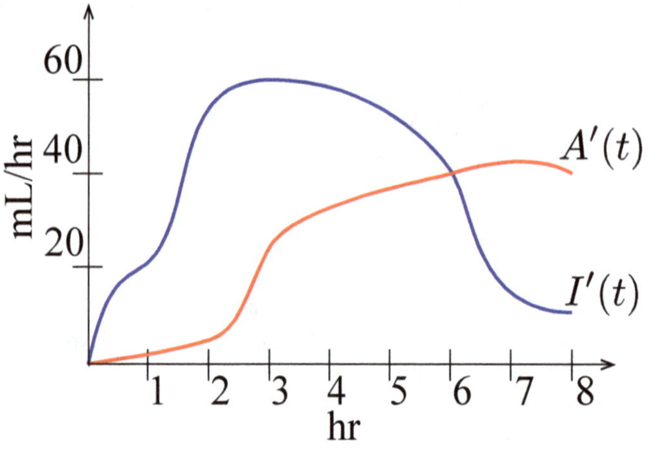

Figure 2.22: The injection and absorption rates, I' and A', of a drug; Problem 9.

additional unit of goods were to be produced.

- The *marginal cost*, if q quantity of goods are produced, is the amount of *additional* cost that would be required in order to produce one additional unit of goods.
- The *marginal profit*, if q quantity of goods are produced, is the amount of *additional* profit that the company would earn if one additional unit of goods were to be produced.

(a) Express the marginal revenue and marginal cost in terms of the functions $\Pi(q)$, $R(q)$, $C(q)$, and their derivatives.

(b) How is the derivative $\Pi'(q)$ related to the marginal revenue and marginal cost?

(c) If the marginal revenue is greater than the marginal cost, should the company produce at least one more quantity of goods?

(d) If the marginal profit is negative, would you suggest that the company produce a greater or lesser quantity of goods?

11. During U.S. President Jimmy Carter's 1977 energy speech, he stated:

> *During the 1950s, people used twice as much oil as during the 1940s. During the 1960s, we used twice as much as during the 1950s. And in each of those decades, more oil was consumed than in all of mankind's previous history.*

(a) According to Carter's description, is the rate at which oil was being consumed increasing or decreasing from 1940 - 1970?

(b) Would the graph of the total, cumulative oil consumption in all of preceding history, as a function of time, be concave up or concave down?

(c) What type of function describes Carter's statement about the way in which the rate of oil consumption was increasing?

12. While steaming milk for a latte on an espresso machine, Kate recorded the time, t, on her stopwatch whenever the temperature of the milk, u, increased by 10°F in Table 2.14.

t - [s]	0	15	36	50	63	73
u - [°F]	57	60	70	80	90	100
t - [s]	81	87	92	96	100	—
u - [°F]	110	120	130	140	150	—

Table 2.14: Temperature of milk u as a function of time t; Problem 12.

   (a) Estimate the rate at which the temperature of the milk was increasing at the beginning and end of the period for which Kate recorded her data.
   (b) Will the graph of the temperature as a function of time be concave up or concave down?
   (c) Now suppose that $t = f(u)$ represents the time on Kate's stopwatch as a function of the temperature of the milk. What are the units of $f'(u)$? What does the quantity $f'(100)$ represent physically? Estimate the numerical value of $f'(100)$.

13. Let $\rho(T)$ be the density (kg/m^3) of Nitrogen gas when stored at temperature T degrees Kelvin (K) and at a pressure of 100 kPa. Various data are given in Table 2.15.
   (a) What are the units of $\rho'(T)$?
   (b) Estimate $\rho'(100)$ and $\rho'(180)$.
   (c) Is the graph of $\rho(T)$ concave up or concave down?

14. In physics, a *blackbody* is an object that does not reflect light, but yet can emit thermal radiation. Thermal radiation can be carried by photons at different wavelengths. Let $I(\lambda)$ be the total intensity (W/m^2) of all electromagnetic radiation emitted by a blackbody carried by photons with wavelength less than λ micrometers (μm). A graph of $I'(\lambda)$ for a certain blackbody is shown in Figure 2.23.
   (a) On which intervals is the intensity function $I(\lambda)$ (i) increasing; (ii) decreasing; (iii) concave up; (iv) concave down? Draw a sketch of $I(\lambda)$.
   (b) The function $I'(\lambda)$ actually has its own name; it is called the *radiancy*,

$\rho(T)$ - [kg/m^3]	3.43607	2.84026	2.42407
T - [K]	100	120	140
$\rho(T)$ - [kg/m^3]	2.11582	1.87779	1.68831
T - [K]	160	180	200

Table 2.15: Density ρ of nitrogen gas at 100 kPa as a function of temperature T; Problem 13.

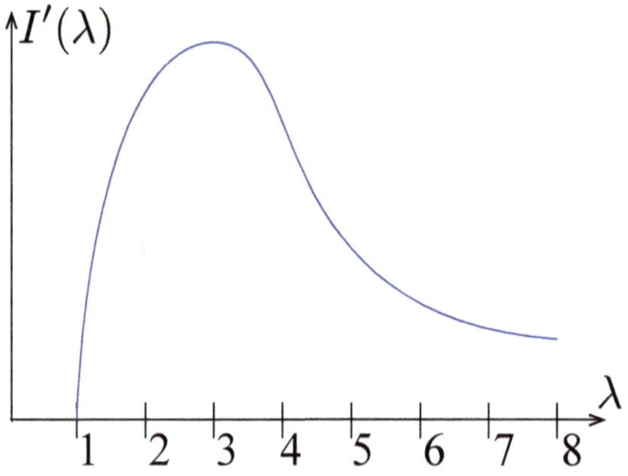

Figure 2.23: The radiancy $I'(\lambda)$ of a blackbody; Problem 14.

$R(\lambda) = I'(\lambda)$. On which intervals is $R'(\lambda)$ (i) positive; (ii) negative; (iii) increasing; (iv) decreasing? Draw a sketch of $R'(\lambda)$.

15. Let $P(t)$ represent the population of the Snowy Owl on the Great Plains of North America, t years after January 1, 2000. The graph of $P'(t)$ is given in Figure 2.24. For what years was the Snowy Owl population (a) increasing; (b) decreasing; (c) concave up; and (d) concave down? Draw a sketch of the Snowy Owl population as a function of time.

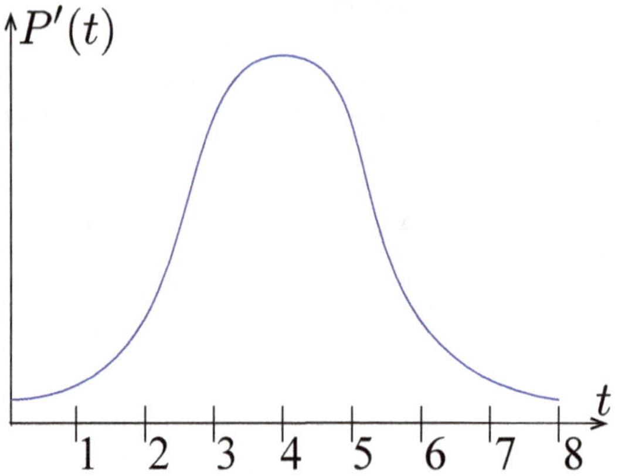

Figure 2.24: The rate of change of the Snowy Owl population; Problem 15.

2.4 Limit Definition of the Derivative

In this section, we present the precise definition of the derivative of a function. Our new definition will be wholly consistent with the intuitive definition we previously discussed in Section 2.2, but will fully utilize our understanding of limits.

2.4.1 The Derivative at a Point

We begin by simply considering the definition of the derivative at a point.

Definition 2.4.1 Given a function f, its derivative at a point a on its domain is defined by

$$f'(a) = \lim_{h \to 0} \frac{f(a+h) - f(a)}{h}. \tag{2.10}$$

If the preceding limit exists, we say that the function f is *differentiable* at the point a. Otherwise, f is *nondifferentiable* at a.

We observe that this is a precise statement of the instantaneous rate of change of the function f at the point a. It is, in fact, equivalent to equation 2.4. In order to fully understand the definition of the derivative at a point, as given by equation (2.10), we must appreciate a few facts: first, the point a is held *fixed*. That is, it is pinned down to a particular value and is not allowed to move around at all. Second, the parameter that is varied in computing the limit given by equation (2.10) is the parameter h. Here, h replaces what we previously referred to as Δx and it represents the width of our sample interval, on which we compute the average rate of change. Allowing the parameter h to shrink to zero, our limit finally returns what can only be understood as the instantaneous rate of change of the function f at the point a.

The mechanism expressed by equation (2.10) is illustrated in Figure 2.25. This figure depicts how we can understand the quantity

$$\frac{f(a+h) - f(a)}{h}$$

as the average rate of change of the function f over the interval $[a, a+h]$.

Next, as we allow the parameter h to shrink to zero, we note that a similar picture is formed as was previously given in Figure 2.3 on page 118. The secant line formed by connecting the points $(a, f(a))$ and $(a+h, f(a+h))$ eventually creeps up on the tangent line to the graph of f at the point a, which, in turn, represents the instantaneous rate of change of the function f at the point a.

■ **Example 2.10** Compute the derivative of the function $f(x) = 3x^2 + 2x + 7$ at the

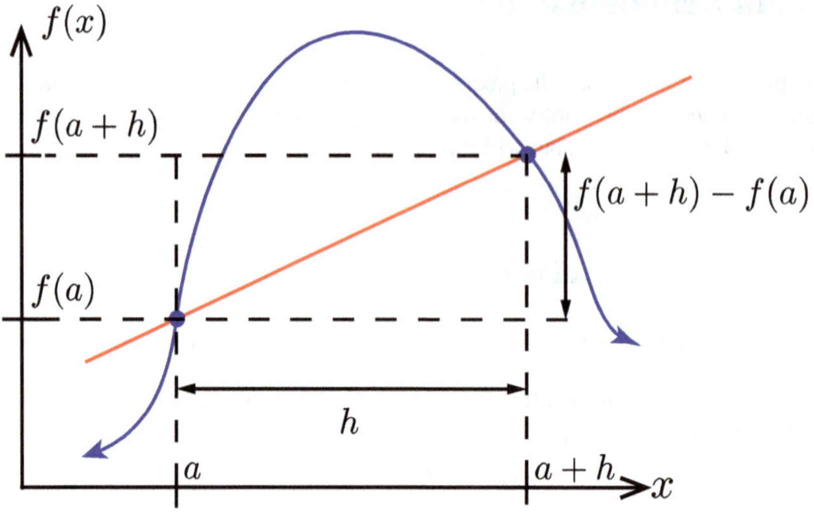

Figure 2.25: The average rate of change of f over the interval $[a, a+h]$, as used in the definition of the derivative of the function f at the point a.

point $x = 2$. To do this, we implement equation (2.10) as follows:

$$
\begin{aligned}
f'(2) &= \lim_{h\to 0}\left(\frac{f(2+h)-f(2)}{h}\right) \\
&= \lim_{h\to 0}\left(\frac{\left(3(2+h)^2+2(2+h)+7\right)-\left(3(2)^2+2(2)+7\right)}{h}\right).
\end{aligned}
$$

Notice that we simply evaluated the function f at the points $2+h$ and 2, as prescribed by the definition. Continuing, we find:

$$
\begin{aligned}
f'(2) &= \lim_{h\to 0}\left(\frac{3(h^2+4h+4)+(4+2h)+7-(12+4+7)}{h}\right) \\
&= \lim_{h\to 0}\left(\frac{3h^2+12h+12+4+2h+7-12-4-7}{h}\right) \\
&= \lim_{h\to 0}\left(\frac{3h^2+14h}{h}\right) \\
&= \lim_{h\to 0}(3h+14) = 14.
\end{aligned}
$$

Also, notice that we did nothing with the limit until the very end: first, we simplified the expression $(f(a+h)-f(a))/h$ as much as possible. Once this is achieved, we then compute the limit of our result in order to obtain the value of the derivative.

We conclude that $f'(2) = 14$. One can verify this result by the much more tedious procedure of computing the average rate of change of f over a sequence of successively smaller intervals containing the point $x = 2$, for instance, computing the average rate of change over $[2, 2.001]$, $[2, 2.00001]$, and so forth. The advantage of using limits and the precise definition of the derivative should be clear. ∎

2.4.2 The Derivative as a Function

The leap from the derivative at a point to the derivative as a function is as simple as replacing the point a with the independent variable x.

Definition 2.4.2 Given a function f, its derivative is the function f' that is defined by the relation

$$f'(x) = \lim_{h \to 0} \frac{f(x+h) - f(x)}{h}. \tag{2.11}$$

The function f is *differentiable* on the set of points for which the preceding limit exists.

The fact that there are now *two* variables—x and h—slightly muddies the picture, though this ailment may easily be cured. The key to understanding equation (2.11) is that the variable x is treated as though it were a fixed constant, for the purpose of computing the limit. That is, the limit only regards the parameter h as a variable, leaving the independent variable x, temporarily, as a fixed, but unspecified, constant. Only after the limit is computed do we then regard x as a variable—in fact, as the independent variable of this newly constructed function f'.

The logic behind the concept of the derivative as a function may further be rendered as follows. The derivative at a point, which represents the instantaneous rate of change of the original function f at that point, may, itself, be computed at any point on the domain (except for those points at which the function is discontinuous or possesses a corner). If we allow this point to vary, we therefore obtain a function. This completes the circle, returning us back to Definition 2.2.1: the derivative is a new function that, when evaluated at a given point, yields the instantaneous rate of change of the original function at that point. Notice, also, that if we evaluate the derivative function, as defined by equation (2.11), at the point $x = a$, we obtain precisely the definition of the derivative of the point a, as given by equation (2.10).

■ **Example 2.11** Compute the derivative of the function $f(x) = 3x^2 + 2x + 7$.

To achieve this, we implement equation (2.10) as follows:

$$\begin{aligned}
f'(x) &= \lim_{h\to 0}\left(\frac{f(x+h)-f(x)}{h}\right)\\
&= \lim_{h\to 0}\left(\frac{[3(x+h)^2+2(x+h)+7]-(3x^2+2x+7)}{h}\right)\\
&= \lim_{h\to 0}\left(\frac{3x^2+6xh+3h^2+2x+2h+7-3x^2-2x-7}{h}\right)\\
&= \lim_{h\to 0}\left(\frac{6xh+3h^2+2h}{h}\right)\\
&= \lim_{h\to 0}(6x+3h+2) = 6x+2.
\end{aligned}$$

We conclude that $f'(x) = 6x+2$. We can, in particular, confirm our result, as obtained in Example 2.10; evaluating $f'(x) = 6x+2$ at $x = 2$, we obtain $f'(2) = 14$. This new formula, however, is naturally much more potent than the simple computation of the derivative at a single point: with $f'(x) = 6x+2$, we now have knowledge of the derivative at *any* point! Thus, with a single computation, we now know what the rate of change of the function f is at any of the points on its domain.

■

2.4.3 The Tangent Line

We previously saw, in Proposition 2.2.2 on page 132, how the derivative of a function at a point is equal to the slope of the tangent line to the graph of the function at that point. Now that we may precisely compute the value of the derivative at a point, using limits, we next introduce a formula for the tangent line.

Proposition 2.4.1 Suppose the function f is differentiable at the point $x = a$. Then the tangent line to the graph of f at the point a is the graph of the function

$$T(x) = f(a) + f'(a)(x-a). \tag{2.12}$$

This function is also known as the *linearization* of the function f near the point a.

Proof. Let $y = T(x)$ represent the tangent line to the graph of f near the point a. The function T is therefore the unique linear function that passes through the point $(a, f(a))$ with a slope equal to $f'(a)$. Using the point-slope formula for a line, we obtain

$$\begin{aligned}
y - y_0 &= m(x-x_0),\\
y - f(a) &= f'(a)(x-a),\\
y &= f(a) + f'(a)(x-a).
\end{aligned}$$

This verifies the result. ■

Proposition 2.4.1 is illustrated in Figure 2.26. The tangent line is an approximation of the actual function for points near the point a, the point at which the tangent line is attached to the graph of the function. The quantity $f(a)$ represents the starting value of the function at the point $x = a$. The quantity $f'(a)(x-a)$ is equivalent to $m\Delta x$, i.e., it represents the slope of the tangent line multiplied by the change in x. Of course, we immediately recall that $\Delta y = m\Delta x$, for linear functions. Thus, the quantity $f'(a)(x-a)$ yields the change in the y-coordinate for the tangent line, corresponding to a change in the x-coordinate from a to x.

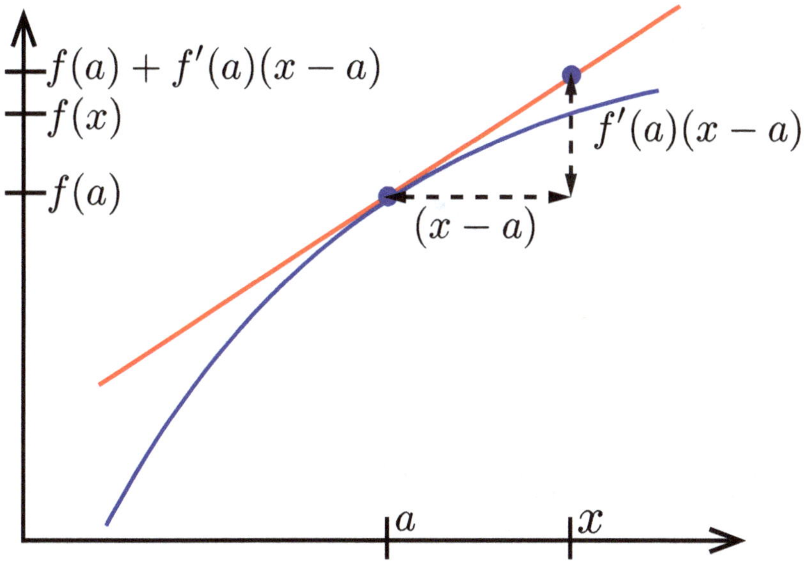

Figure 2.26: The graph of a function and its tangent line.

The reason equation (2.12) is referred to as the linearization of the function f is as follows. Since the function closely resembles its own tangent line for x-values "close enough" to the point $x = a$, the tangent line should give us a good approximation for the function f near the point for which its derivative is computed. This gives us a tool for approximating complicated, wily functions locally, using simply the instantaneous rate of change at a single point.

■ **Example 2.12** We previously saw in Example 2.10 that the derivative of the function $f(x) = 3x^2 + 2x + 7$ at the point $x = 2$ is given by $f'(2) = 14$. Moreover, we can easily compute $f(2) = 23$. Therefore, according to equation (2.12), the tangent line to the function f at the point 2 is given by the equation

$$T(x) = 23 + 14(x-2).$$

We plot the function f and its tangent line T concurrently in Figure 2.27. This

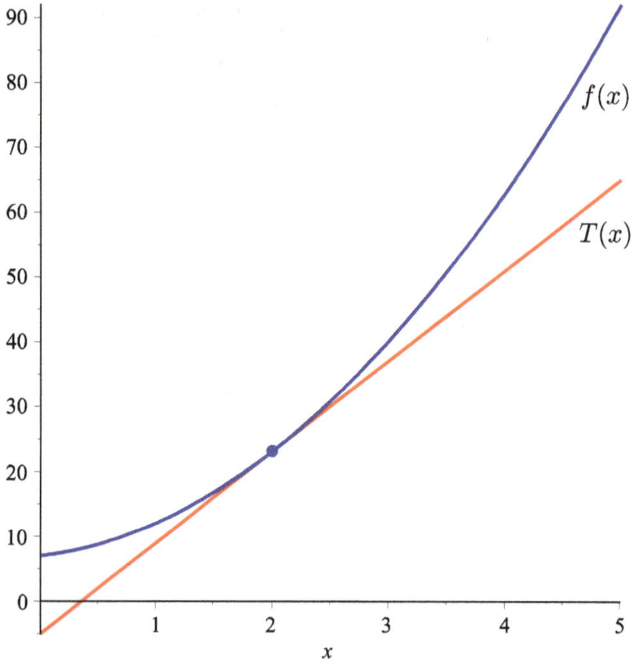

Figure 2.27: The graph of the function $f(x) = 3x^2 + 2x + 7$ and its tangent line $T(x) = 23 + 14(x-2) = 14x - 5$.

confirms the validity of equation (2.12).

We may further demonstrate the concept of linearization by approximating the output of the function f near the point 2. For example, suppose we wished to approximate $f(2.1)$. Using the linearization, i.e., the tangent-line approximation, we obtain

$$T(2.1) = 23 + 14(0.1) = 24.4.$$

The actual output of the function f at the point $x = 2.1$, on the other hand, is given by

$$f(2.1) = 3(2.1)^2 + 2(2.1) + 7 = 24.43.$$

We observe how the linearization of the function f, as given by its tangent line, fairly closely approximates the actual function, for input points nearby the centering point $x = 2$. The usefulness of this exercise may be presently obscured by the ease at which the actual value of $f(2.1)$ may be obtained. This simple technique, however, was a hallmark in the development of mathematics, especially in the age when computers were not so readily available. Even today, a more advanced form of this approximation is behind the programmed functions of the modern scientific calculator. (How does your calculator know the sine and logarithm of every imaginable number?) ∎

2.4.4 Linearity of the Derivative

Much of this section, so far, has been a formalizing of concepts previously digested on an intuitive level in our preceding discourse. Now for something new.

Theorem 2.4.2 — Linearity of the Derivative. The derivative is a *linear operator*; i.e., given two differentiable functions f and g, and constants a and b, the following relation hold:

$$\frac{d}{dx}\left[af(x)+bg(x)\right] = af'(x)+bg'(x), \tag{2.13}$$

where f' and g' are the derivatives of f and g, respectively.

The preceding theorem is powerful, for it allows us to compute derivatives term-by-term. For instance, if we know the derivative of x^5 and the derivative of x^2, then we immediately know the derivative of $3x^5+7x^2$. It decomposes immediately as

$$\frac{d}{dx}\left[3x^5+7x^2\right] = 3\frac{d}{dx}\left[x^5\right]+7\frac{d}{dx}\left[x^2\right].$$

In particular, it turns out that

$$\frac{d}{dx}\left[x^5\right] = 5x^4 \qquad \text{and} \qquad \frac{d}{dx}\left[x^2\right] = 2x.$$

Therefore, it follows that

$$\frac{d}{dx}\left[3x^5+7x^2\right] = 15x^4+14x.$$

Proof. The reason for Theorem 2.4.2 is straightforward: to begin, consider the function $y = af(x)$. Since a is a constant, we have

$$\Delta(af) = a\Delta f.$$

Meanwhile, Δx is unchanged. Therefore

$$\Delta y/\Delta x = \Delta(af)/\Delta x = a\Delta f/\Delta x.$$

In the limit as $\Delta x \to 0$, we therefore obtain

$$\frac{d}{dx}\left[af(x)\right] = a\frac{d}{dx}\left[f(x)\right].$$

Similarly, consider the sum $y = f(x)+g(x)$. Here, $\Delta y = \Delta(f+g) = \Delta f+\Delta g$. Therefore,

$$\Delta y/\Delta x = \Delta(f+g)/\Delta x = \Delta f/\Delta x+\Delta g/\Delta x.$$

We conclude that

$$\frac{d}{dx}\left[f(x)+g(x)\right] = \frac{d}{dx}\left[f(x)\right] + \frac{d}{dx}\left[g(x)\right].$$

Combining the preceding results yields the theorem. ■

Exercises

For Exercises 1–13, use the following given derivatives

$$\begin{aligned}
\frac{d}{dx}\left[x^3\right] &= 3x^2, \\
\frac{d}{dx}\left[x^7\right] &= 7x^6, \\
\frac{d}{dx}[\sin(3x)] &= 3\cos(3x), \\
\frac{d}{dx}[\cos(2x)] &= -2\sin(2x), \\
\frac{d}{dx}[\ln|x|] &= \frac{1}{x}, \\
\frac{d}{dx}[\tan x] &= \sec^2 x,
\end{aligned}$$

in order to determine the derivative of the given function.

1. $f(x) = x^3 + x^7$.
2. $f(x) = 4x^3$.
3. $f(x) = 6x^7$.
4. $f(x) = 4x^3 + 6x^7$.
5. $f(x) = 2x^3 + 8x^7$.
6. $f(x) = 2x^3 + 8x^7 + \sin(3x)$.
7. $f(x) = 3x^3 + 16x + 17$.
8. $f(x) = \sin(3x) + 8\cos(2x)$.
9. $f(x) = 12\tan(x) + \ln x$.
10. $f(x) = \tan(x) + 2\ln x + 3\cos(2x) + 4\sin(3x) + 5x^7 + 6x^3 + 7x + 8$.
11. $f(x) = \ln(x^2)$.
12. $f(x) = \ln(22x)$.
13. $f(x) = \ln\left(\dfrac{x^3}{e^{\sin(3x)}}\right)$.

For Exercises 14–23, use the limit definition of the derivative to compute the derivative of the given function.

14. $f(x) = 5$.
15. $f(x) = x$.
16. $f(x) = 3x + 7$.
17. $f(x) = x^2 + 8x + 6$.
18. $f(x) = 4x^2 + 2x + 2$.
19. $f(x) = x^3$.
20. $f(x) = 4x^2$.
21. $f(x) = x^3 + 4x^2$.
22. $f(x) = x^4$.
23. $f(x) = x^{-1}$.

For Exercises 24–29, write out the difference quotient

$$\frac{f(x+h) - f(x)}{h}$$

for the given function.

24. $f(x) = e^x$.
25. $f(x) = \sin x$.
26. $f(x) = \cos x$.
27. $f(x) = \tan x$.
28. $f(x) = \sqrt{x}$.
29. $f(x) = \ln x$.

For Exercises 30–33, identify each limit as the derivative of a certain function f at a given point a.

30. $\displaystyle\lim_{h\to 0} \frac{\cos(h^2+\pi)+1}{h}$.
31. $\displaystyle\lim_{h\to 0} \frac{\sqrt{(1+h)^2+3}-2}{h}$.
32. $\displaystyle\lim_{h\to 0} \frac{\tan(\pi/4+h)-1}{h}$.
33. $\displaystyle\lim_{h\to 0} \frac{e^2(e^h-1)}{h}$.

Problems

34. Let $u(r)$ be the velocity of a fluid in a cylindrical pipe at a distance r from the central axis of the pipe as shown in Figure 2.28. The *shear stress* τ between

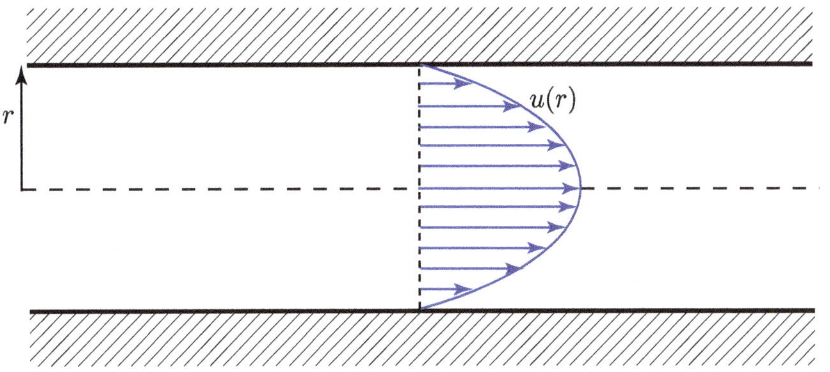

Figure 2.28: The velocity profile of a fluid in a cylindrical pipe; Problem 34.

the fluid and the wall of the pipe is given by the relation

$$\tau = -\mu u'(R),$$

where R is the radius of the pipe and μ is a constant equal to the *dynamic viscosity* of the fluid. Use the limit definition to calculate the shear stress for a flow if its velocity profile is

$$u(r) = 1 - \left(\frac{r}{R}\right)^2.$$

The answer may depend on the parameters μ and R.

35. A supersonic jet that is traveling at a speed of Mach 2 (i.e., twice the speed of sound) is led by an oblique shock wave with a wave angle of $\beta = \pi/4$ rads (45°), as shown in Figure 2.29. (The wave angle is the angle between the shock wave and the centerline through the aircraft.) A shock wave is a plane that represents a sudden discontinuity in atmospheric pressure, density, and temperature. The *density ratio*, which is equal to the density of the air in region 2 divided by the density of the air in region 1, is given by

$$f(M) = \frac{(\gamma+1)M^2\sin^2\beta}{(\gamma-1)M^2\sin^2\beta+2},$$

where γ is the *adiabatic index* of the fluid ($\gamma = 1.4$ for air) and M is the Mach number.

(a) Determine the limit

$$\lim_{M\to\infty} f(M).$$

Explain the physical significance of your answer.

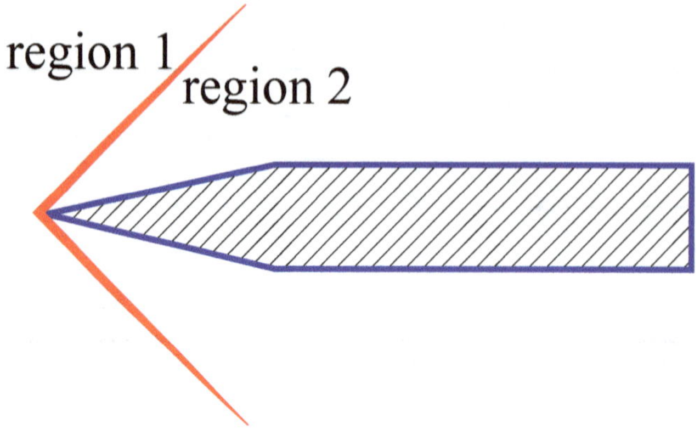

Figure 2.29: Shock wave on a supersonic aircraft; Problem 35.

(b) Suppose that the wave angle β is actually a function of the Mach number and that

$$\lim_{M\to\infty} \beta(M) = \frac{\pi}{12}.$$

(*Note*: $\pi/12$ rads = 15° = angle of the jet's leading edge.) Does this affect your answer for part (a)? Explain why or why not.

(c) The *pressure ratio*, which is equal to the air pressure in region 2 divided by the air pressure in region 1, is given by

$$g(M) = 1 + \frac{2\gamma}{\gamma+1}\left(M^2 \sin^2\beta - 1\right).$$

Use the limit definition of the derivative to determine $g'(2)$ (assume $\beta = \pi/4$).

(d) Calculate $g(2)$ and $g(2.1)$. Now approximate $g(2.1)$ using the formula

$$g(2.1) \approx g(2) + g'(2)(2.1 - 1).$$

Explain why this approximation formula works.

36. At a fixed temperature T, the pressure of a fluid $p(\rho)$ is a function of the density ρ. It can be shown that the *speed of sound* c is given by

$$c = \sqrt{\gamma \frac{dp}{d\rho}},$$

where γ is a constant, known as the *adiabatic index*, that depends on the gas.

(a) Suppose that superheated R-134a (a common refrigerant), when at constant temperature 0°C, has a density of 2.28 kg/m^3 when its pressure is 50 kPa and a density of 4.62 kg/m^3 when its pressure is 100 kPa.

Approximate the speed of sound in R-134a at 0°C. The adiabatic index of superheated R-134a is $\gamma = 1.106$.

(b) For an ideal gas, with specific real gas constant R, at fixed, absolute temperature T, the pressure is related to the density by the *ideal gas law*:

$$p(\rho) = \rho RT.$$

Use the limit definition of the derivative to show that $p'(\rho) = RT$. Conclude that the speed of sound in an ideal gas with adiabatic index γ, specific gas constant R, and temperature T is $c(T) = \sqrt{\gamma RT}$.

(c) One day, the temperature at a ski resort (elevation 2365m) in Snowbird, Utah was 283 Kelvin (K), and the temperature at the top of *Hidden Peak* (elevation 3353m) was 276 K. Assuming that air is an ideal gas with real gas constant $R = 287$ J/(kg K) and adiabatic index $\gamma = 1.4$, estimate the rate of change of the speed of sound, *with respect to elevation*, at the Snowbird ski resort.

37. A spacecraft is directly above the pole of the axis of rotation of a planet with rings. The gravitational acceleration on the spacecraft when it is at a distance z from the center of a planet is

$$a(z) = \frac{Gmz}{(R^2 + z^2)^{3/2}} + \frac{GM}{z^2},$$

where R is the radius of the planet's rings, m is the total mass of the rings, M is the mass of the planet, and G is the gravitational constant. For simplicity, let's set $G = 1$, $m = 1$, $M = 1$, and $R = 1$.

(a) Write down the limit definition for $a'(1)$.

(b) Approximate the value of this limit using the graphical and experimental approach. Explain the meaning of the sign.

38. The gravitational force at a distance r from the center of the Earth is given by

$$F(r) = \begin{cases} \dfrac{GMr}{R^3} & \text{for } 0 \le r \le R \\ \dfrac{GM}{r^2} & \text{for } R < r \end{cases},$$

where G is the gravitational constant, M is the mass of the Earth, and R is the Earth's radius. Is $F(r)$ continuous at $r = R$? Is $F(r)$ differentiable at $r = R$?

39. The strength of a magnetic field at a distance r from the center of a wire of radius R is given by

$$B(r) = \begin{cases} \dfrac{rB_0}{R} & \text{for } 0 \le r \le R \\ \dfrac{RB_0}{r} & \text{for } R < r \end{cases}.$$

What is the physical significance of the constant B_0? Is $B(r)$ continuous at $r = R$? Is $B(r)$ differentiable at $r = R$?

2.5 Real-World Interpretation of the Derivative

It is important to be able to describe mathematical equations "in practical terms." In this section, we take a closer look at derivatives, their meaning, their dimensions, and related constructs involving derivatives and inverse functions. We will also introduce what is meant by the *local linearization* of a function near a point.

2.5.1 Derivatives in Practical Terms

We begin with a basic review of the general elements of the derivative. The limit definition of the derivative, as given by (2.10), may be alternatively expressed in the form

$$\frac{dy}{dx} = \lim_{\Delta x \to 0} \frac{\Delta y}{\Delta x}. \tag{2.14}$$

Here, $\Delta x = h$, and $\Delta y = f(x+h) - f(x)$. Thus, the limit definition of the derivative captures something that we first examined in Section 2.1: the instantaneous rate of change of a function, as captured by the derivative, is a limit of the average rate of change, as the interval shrinks to one of zero width. Thus, the derivative tells us how much the y-variable should change given a unit change in the x-variable. The *units* of the derivative are therefore equal to the units of y divided by the units of x, as we stated in Proposition 2.2.1.

Consider now the derivative $f'(a)$ as evaluated at a point $x = a$. This gives us an approximation for the quantity $\Delta y / \Delta x$, at the point $(a, f(a))$. Hence,

$$\Delta y \approx f'(a)\Delta x, \tag{2.15}$$

if Δx is sufficiently small. This is no more than the point-slope equation for a line, and it describes the tangent line to the function. Thus, we may describe the derivative $f'(a)$ in words by following a form similar to the following:

> The value $f'(a)$ approximates how much the y variable will increase (or decrease, if $f'(a)$ is negative) if the variable x is increased by 1.

A few things to note when writing out such a thing in words:

1. It is vital to state that $f'(a)$ is an *approximation*; $f'(a)$ describes the *instantaneous* rate of change of y with respect to x, and is therefore only accurate *for an instant*. For any finite Δx, no matter how small, the equation (2.15) is only an approximation; this approximation becomes more accurate the smaller the value of Δx.
2. If the function $y = f(x)$ describes a physical situation, we would want to replace "the y variable" and "the x variable" with an appropriate real-world description of what they describe.
3. The units of the dependent and independent variables should always be clearly stated when describing the meaning of the derivative *in practical terms*.

■ **Example 2.13** Let $f(x)$ represent the loudness (in decibels) of a rocket launch at a distance of x miles. Describe the meaning of the quantity $f'(5)$ and of the equation $f'(5) = -1.74$ in practical terms.

The quantity $f'(5)$ represents the approximate amount by which the loudness (dB) will increase, when an observer is 5 miles from the launch site, if she were to instead stand one additional mile farther away.

The equation $f'(5) = -1.74$ means that when an observer is standing 5 miles from the launch site, the loudness will decrease by *approximately* 1.74 dB if she were to instead stand one mile farther away.

One can also get creative and come up with alternative ways of describing derivatives; for instance, consider the following alternative: the equation $f'(5) = -1.74$ means that the sound will be approximately 1.74 dB less loud at a distance of six miles than at five miles. ■

2.5.2 Derivatives and Inverse Functions

Next, we examine some more challenging constructs, obtained when one composes the operations of differentiation and function inversion.

In the language of mathematics, quantities such as $(f^{-1})'(5)$ are *nouns*, whereas equations such as $(f')^{-1}(3) = -12$ are complete *sentences*. When interpreting the physical meaning of *quantities*, we therefore should use them as part of a larger sentence; i.e., "The quantity $(f^{-1})'(5)$ represents..."

The key to understanding quantities and equations like the ones mentioned above is to understand what each quantity corresponds to symbolically. For example, if you are asked to interpret the mathematical statement $(f')^{-1}(3) = -12$, it is of monumental importance to be able to recognize that (if we call the dependent variable y and the independent variable x, so that $y = f(x)$) the quantity 3 corresponds to dy/dx and the quantity -12 corresponds to x. Not only is the knowledge of this correspondence the key to interpreting the statement $(f')^{-1}(3) = -12$, it is also the key to getting the units of each quantity correct; e.g., the units of 3 should equal the units of y divided by the units of x, and the units of -12 should equal the units of x.

Proposition 2.5.1 — Strategies for Interpretation. When interpreting mathematical quantities consisting of a function's derivative or inverse function or a particular composition of the two, implement the following approach:

1. **Assign Variables.** If variable names are not already assigned in the problem, assign a variable name to both the independent and dependent variables. Choose variable names that will remind you of the meaning of the variable, e.g., D for the number of daylight hours, L for distance, etc.
2. **Write the D&I Map.** The derivative-and-inverse (D&I) map is defined below. Once variables have been assigned, write out the corresponding D&I map for the given problem, following the outline below.
3. **Identify.** For each mathematical statement or quantity, identify which

numbers correspond to which symbols in the D&I map.

4. **Determine Units.** Using this correspondence, determine the units of each number following the standard rules.
5. **Form a Picture.** Form a picture in your mind of what is happen. Try to unravel the information to deduce the meaning of the symbols.
6. **Interpret.** Interpret the quantity or equation in practical terms, using guidance from the symbols and their units.

R In your final interpretation:

1. be sure to use the word *approximately* when dealing with derivatives;
2. be sure to include appropriate units wherever possible;
3. avoid using the phrase "the rate of change," as this can imply a rate of change with respect to *time* to the uninitiated. Instead, describe a derivative in terms of the approximate increase in y for a unit increase in x.

Definition 2.5.1 Given a function $y = f(x)$, its corresponding *derivatives-and-inverses map* (D&I map) is the following:

$$\begin{array}{ccc} y = f(x) & \Rightarrow & x = f^{-1}(y) \\ \Downarrow & & \Downarrow \\ \dfrac{dy}{dx} = f'(x) & & \dfrac{dx}{dy} = \left(f^{-1}\right)'(y) \\ \Downarrow & & \Downarrow \\ x = \left(f'\right)^{-1}\left(\dfrac{dy}{dx}\right) & & y = \left(\left(f^{-1}\right)'\right)^{-1}\left(\dfrac{dx}{dy}\right) \end{array} \tag{2.16}$$

The arrows give the order in which each equation is derived, starting with the root $y = f(x)$.

It is important to recognize that f, f', $(f')^{-1}$ are all functions. The function $(f')^{-1}$ represents the inverse function of the derivative. Following the map, one will never be lost with units. Similarly, f^{-1} and $(f^{-1})'$ are also functions. The function $(f^{-1})'$ is the derivative of the inverse function of f. From the D&I map, one can easily see that the functions $(f^{-1})'$ and $(f')^{-1}$ are *different*! The meaning of their inputs and outputs are not the same. Finally, the function $((f^{-1})')^{-1}$ is just the inverse function of $(f^{-1})'$, so it is obtained by swapping inputs for outputs (this is the strategy whenever computing the inverse function). We shall omit the case $((f')^{-1})'$ in order to eschew unnecessary complexity, but the prodigious student may take up this construction as a challenge.

■ **Example 2.14** Let $f(x)$ represent the *loudness* (measured in deciBels (dB)) of a jet engine at a distance x meters from the engine. Interpret, in practical terms, the meaning of each of the following equations.

(a) $f(50) = 140$.
(b) $f^{-1}(80) = 50,000$.
(c) $f'(50) = -0.1737$.
(d) $(f')^{-1}(-0.1086) = 80$.
(e) $(f^{-1})'(140) = -5.7565$.

Each of the preceding equations can be understood by writing down the D&I map and identifying each number to its corresponding variable expression. To begin, since the dependent variable has not been assigned in the problem, let us use the variable L; i.e., we can assign $L = f(x)$. Using these variables, we can write down our D&I table, following the scheme given in equation (2.16).

$$\begin{array}{ccc} L = f(x) & \Rightarrow & x = f^{-1}(L) \\ \Downarrow & & \Downarrow \\ \dfrac{dL}{dx} = f'(x) & & \dfrac{dx}{dL} = \left(f^{-1}\right)'(L) \\ \Downarrow & & \Downarrow \\ x = \left(f'\right)^{-1}\left(\dfrac{dL}{dx}\right) & & L = \left(\left(f^{-1}\right)'\right)^{-1}\left(\dfrac{dx}{dL}\right) \end{array}$$

Everything can be understood by studying these expressions and relating them back to the expressions we are trying to interpret. Let us examine carefully how this works, beginning with a reiteration of how the D&I maps are derived. We begin with the function f:

$$L = f(x). \tag{2.17}$$

Differentiating with respect to x, we obtain

$$\frac{dL}{dx} = f'(x). \tag{2.18}$$

Upon inverting this expression, we see that

$$x = (f')^{-1}\left(\frac{dL}{dx}\right). \tag{2.19}$$

Alternatively, if we start by inverting equation (2.17), we obtain

$$x = f^{-1}(L). \tag{2.20}$$

Differentiating this function with respect to L, we have

$$\frac{dx}{dL} = (f^{-1})'(L). \tag{2.21}$$

The five equations (2.17)–(2.21) constitute the D&I maps for the problem. To begin, let us compare expression (a) with equation (2.17); we immediately identify $x = 50$ and $L = 140$. Using the appropriate units, we have $x = 50$ m and $L = 140$ dB. The appropriate interpretation should therefore be something along the lines of the following: "*At a distance of 50 meters from the jet engine, the loudness is 140 deciBels.*" Notice that we *start* by describing what is *inside* the parentheses. The equation tells us that $f(50)$ is equal to 140; we therefore start by stating that the distance is 50 m, and then explaining the consequence of this fact.

Expression (b) is of the form of equation (2.20); by comparing, we can identify $L = 80$ dB and $x = 50{,}000$ m. Just like before, we will start on the inside of whatever function is present, i.e., on the inside of the parenthesis. The physical interpretation of expression (b) would be the statement, "*When the loudness of the jet engine is 80 deciBels, you are standing 50,000 meters away from the engine.*"

Expression (c) is of the form of equation (2.18); by comparing we can therefore identify $x = 50$ and $\frac{dL}{dx} = -0.1737$. The units of x are of course meters. The units of dL/dx are the units of L divided by the units of x; i.e.,

$$\text{untis}\left(\frac{dL}{dx}\right) = \frac{\text{units}(L)}{\text{units}(x)} = \frac{\text{dB}}{\text{m}};$$

Thus, expression (c) tells us that $x = 50$ m and $dL/dx = -0.1737$ dB/m. In practical terms, we therefore interpret expression (c) as follows: "*When standing at a distance of 50 meters from the jet engine, the loudness will decrease by approximately 0.1737 decibels if an observer were to instead stand one meter farther from the jet engine.*" Notice that we used the word *approximately*. The derivative is an instantaneous rate of change, so we can only use it to approximate the change of the dependent variable when we increment the independent variable by one unit. Notice also that we say that the loudness *decreases* by approximately 0.1737 dB, instead of saying that it *increases* by −0.1737 dB. The point is to put the equation into *practical* terms that anyone could understand. And once we have the idea of what is going on, we may massage our wording into an even more practical sounding sentence. For instance, we could alternatively state something along the lines of, "*When you are standing 51 meters from a jet engine, the loudness of the engine is approximately 0.1737 deci-Bels less than when you are standing 50 meters from the engine.*" If you made that statement on a cable talk show, anybody would be able to understand it. That is the point of giving a *practical* interpretation to *mathematical* sentences like $f'(50) = -0.1737$. Also notice we avoided using the phrase *rate of change*, which would often connote a rate of change with respect to *time*, in the minds of many, instead of a rate of change with respect to *distance*.

Expression (d) involves *f–prime–inverse*. (Not *f*–inverse–prime!) We therefore summon the powers of equation (2.19). Identifying the quantity −0.1086 with dL/dx and the quantity 80 with x, we can immediately interpret the statement as such: "*If the loudness of the jet engine were to decrease by approximately 0.1086 deciBels if you were to stand one meter farther away from the engine than you are, then you must be standing 80 meters from the jet engine.*" Notice that in this interpretation, we again start our sentence at the inside of the parenthesis.

Finally, for expression (e), we will use equation (2.21). We can therefore identify L with 140 dB and dx/dL with -5.7565. The units of dx/dL are meters per deciBel, i.e., m/dB. The practical statement goes something like this: "*At the spot where the loudness of the jet engine is 140 deciBels, the loudness is approximately 1 deciBel less at a spot that is 5.7565 meters farther away from the engine.*" Of course, we could alternatively end this sentence by stating, "...the loudness is approximately 1 deciBel *greater* at a spot that is 5.7565 meters *closer* to the engine."

In summary, the five mathematical statements and their interpretations are tabulated below.

(a) $f(50) = 140$.
- At a distance of 50 meters from the jet engine, the loudness is 140 deciBels.

(b) $f^{-1}(80) = 50,000$.
- When the loudness of the jet engine is 80 deciBels, you are standing 50 kilometers away from the engine.

(c) $f'(50) = -0.1737$.
- When standing at a distance of 50 meters from the jet engine, the loudness will decrease by approximately 0.1737 decibels if an observer were to instead stand one meter farther from the jet engine.

(d) $(f')^{-1}(-0.1086) = 80$.
- If the loudness of the jet engine is approximately 0.1086 deciBels at a distance of one meter farther away, then you must be standing 80 meters from the jet engine.

(e) $(f^{-1})'(140) = -5.7565$.
- At the spot where the loudness of the jet engine is 140 deciBels, the loudness is approximately 1 deciBel less at a spot that is 5.7565 meters farther away from the engine.

It is important to rearticulate that expressions like $(f^{-1})'(140) = -5.7565$ represent entire mathematical sentences, whereas expressions like $(f^{-1})'(140)$ are just mathematical *nouns*. You could be asked to interpret the meaning of $(f^{-1})'(140)$. After using (2.21) to identify that 140 is the loudness and $(f^{-1})'$ is dx/dL, the correct response should be something like:

- The quantity $(f^{-1})'(140)$ represents the distance, in meters, you must walk away from the engine in order for the loudness to increase by one deciBel, if you are standing at the spot where the loudness is 140 deciBels.

If the quantity $(f^{-1})'(140)$ is negative (which it happens to be), the above statement is still correct; one must simply walk a *negative* distance *away* from the engine in order to increase the loudness by one deciBel, which equivalent to walking a *positive* distance *towards* the engine. ∎

2.5.3 Local Linearization and Differentials

In this paragraph, we discuss the *local linearization* of a function near a point, which is also known as the *tangent line approximation*, since it essentially uses the tangent line in order to approximate value of the given function nearby the given point.

Local Linearization

We begin by recalling equation (2.14), which can be equivalently expressed for the derivative *at a point* $x = a$ as

$$f'(a) = \lim_{\Delta x \to 0} \frac{\Delta y}{\Delta x},$$

where $\Delta y = f(a+\Delta x) - f(a)$, and where Δx is measured from $x = a$. Since the ratio $\Delta y/\Delta x$ must approach the derivative $f'(a)$ as $\Delta x \to 0$, it follows that the ratio $\Delta y/\Delta x$ must approximate the derivative $f'(a)$ if Δx is small enough. If the value of the derivative is known, this observation leads us to the concept of local linearization.

Definition 2.5.2 Given a function f that is differentiable at $x = a$, with derivative $f'(a)$, the *local linearization* of the function f near the point $x = a$ is given by

$$\Delta y \approx f'(a)\Delta x, \tag{2.22}$$

where $\Delta y = f(x) - f(a)$ and $\Delta x = x - a$. This is valid if x is sufficiently close to a, i.e., if Δx is sufficiently small.

The local linearization of a function near a point is intimately related to the tangent line, given by equation (2.12). We may rearrange the tangent-line equation to form the relation

$$T(x) - f(a) = f'(a)(x-a).$$

Thus, the quantity $f'(a)\Delta x$ represents the change in the function's tangent line. The local linearization therefore uses the tangent line to approximate the actual value of the function f at a point x near the point a. This is illustrated in Figure 2.30.

Notice that Δy represents the *actual* change in the function f, whereas the quantity $f'(a)\Delta x$ represents the change in the function's tangent line.

■ **Example 2.15** Use local linearization to approximate the value of $\sqrt{9.1}$, *without* using the square-root function on a calculator.

To begin, let us consider the function f defined by $f(x) = \sqrt{x}$, and consider the local linearization about the point $a = 9$. To continue, we will kindly "borrow" a fact from Chapter 3, that

$$\frac{d}{dx}\left[\sqrt{x}\right] = \frac{1}{2\sqrt{x}}.$$

(If you are curious as to why this is true, read on! In particular, feel free to take an early look at the first few sections of Chapter 3.)

The local linearization using $a = 9$ therefore tells us that

$$\Delta y \approx f'(9)\Delta x.$$

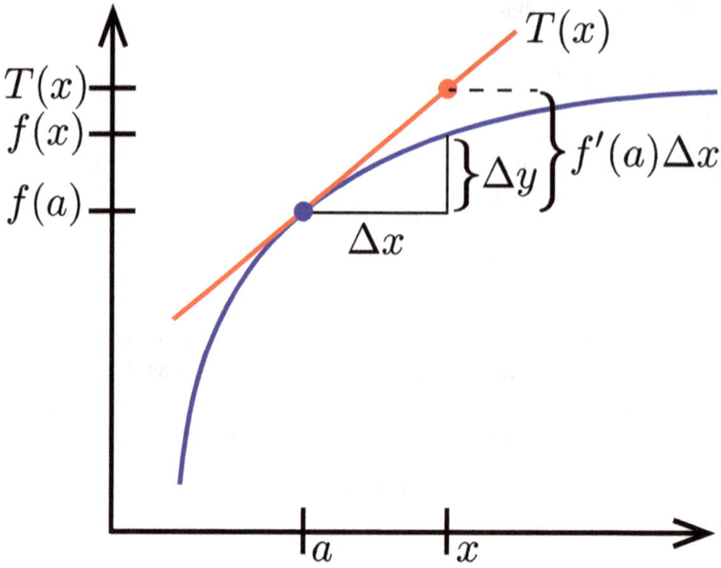

Figure 2.30: Illustration of the local linearization.

The Δx here is going to be $\Delta x = 0.1$, since we are trying to approximate the value of $f(9.1) = \sqrt{9.1}$. Using our borrowed fact from the next chapter, we also know that

$$f'(9) = \frac{1}{2\sqrt{9}} = \frac{1}{6}.$$

Hence,

$$\Delta y \approx \frac{0.1}{6}.$$

Since $\Delta y = f(9.1) - f(9) = \sqrt{9.1} - \sqrt{9} = \sqrt{9.1} - 3$, we can rearrange to find that

$$\sqrt{9.1} \approx 3 + \frac{1}{60} = \frac{181}{60} = 3.01\overline{66}.$$

The actual value, to nine decimal places, is $\sqrt{9.1} = 3.016620625\cdots$. Using the local linearization, aka the tangent line approximation, we therefore approximated $\sqrt{9.1}$ accurate to five significant figures. ■

The local linearization is therefore nothing more than using the tangent line—which can be computed when you know the value of the function's derivative—to approximate the actual function. This is valid as long as you are "close enough" to the point at which you are basing your approximation on.

Differentials

Next, we will take a look at a delicate yet useful notion—that of *differentials*. Essentially, a differential is an infinitesimally small change in a variable. The differential of the dependent variable, however, depends on the differential of the independent variable, as stated in the following definition.

Definition 2.5.3 Let $y = f(x)$, where the function f is differentiable. The *differentials* dx and dy are fictitious quantities that are infinitesimally small, such that their ratio yields the value of the derivative, i.e., such that they are in the proportion

$$dy = f'(x)dx. \tag{2.23}$$

Recall the local linearization of f, as given by equation (2.22). As Δx becomes smaller and smaller, the local linearization becomes a better and better approximation. In the limit as $\Delta x \to 0$, this equations defines the derivative, and therefore becomes exact. Thus, we may view equation (2.23) as what becomes of the local linearization in the limit as $\Delta x \to 0$. The differentials dx and dy are fictitious, because it's not *really* possible to have something that is infinitely small, except in our imagination. They are, however, useful constructs in a variety of situations, as we shall see throughout our journey. (They are particularly useful when we discuss the theory of integration.)

Taking the differential of an equation means that it should be differentiated in the form of equation (2.23). In a way, taking the differential is a way to differentiate an equation, but not with respect to any variable in particular. For example, the differential of the equation $y = x^2$ is

$$dy = 2x\, dx.$$

(The derivative of $f(x) = x^2$ is $f'(x) = 2x$, as we will see shortly.) Similarly, the differential of the equation for the circle $x^2 + y^2 = r^2$, where r is a constant, is

$$2x\, dx + 2y\, dy = 0.$$

If this is confusing, do not fret. Our aim at present is only to introduce the idea of differentials. This notation should be avoided for the time being, however, until we summon them again.

Exercises

For Exercises 1–10, use the given function and the value of its derivative at a point to approximate the indicated quantity using local linearization.

1. $f(x) = x^3$, $f'(2) = 12$; approximate $(2.1)^3$.
2. $f(x) = \sqrt[3]{x}$, $f'(27) = \frac{1}{27}$; approximate $\sqrt[3]{27.5}$.
3. $f(x) = e^x$, $f'(1) = e$; approximate $e^{1.2}$.
4. $f(x) = \sqrt{x}$, $f'(100) = \frac{1}{20}$; approximate $\sqrt{101}$

5. $f(x) = \frac{1}{\sqrt{x}}$, $f'(4) = \frac{-1}{16}$; approximate $\frac{1}{\sqrt{4.05}}$.

6. $f(x) = \sin(x)$, $f'(0) = 1$; approximate $\sin(0.0527)$.

7. $f(x) = \sin(x)$, $f'(\pi/3) = 0.5$; approximate $\sin(5\pi/12)$.

8. $f(x) = \tan(x)$, $f'(\pi/4) = 2$; approximate $\tan(3\pi/10)$.

9. $f(x) = \tan(x)$, $f'(\pi/3) = 4$; approximate $\tan(34\pi/99)$.

10. $f(x) = \log(x)$, $f'(100) = \frac{1}{100\ln(10)}$; approximate $\log(102)$.

Problems

11. Let $E = f(t)$ be the number of eggs in a certain octopus nest on the t-th day of the octopus mating season. Translate each of the following mathematical statements into a single english statement. Be sure to use units.
    (a) $f(0) = 0$
    (b) $f'(6) = 12$
    (c) $(f^{-1})'(112) = 3$
    (d) $(f')^{-1}(14) = 8$

12. Let $f(x)$ be the sound intensity, measured in Watts/meter^2(W/m^2), of a jet engine at a distance x meters from the engine. Give a physical interpretation to each of the following mathematical statements.
    (a) $f(50) = 100$.
    (b) $f^{-1}(25) = 100$.
    (c) $f'(50) = -4$.
    (d) $(f')^{-1}(-0.5) = 100$.
    (e) $(f^{-1})'(100) = -0.25$.

13. Let $f(v)$ be the *relativistic mass*, measured in kg, of a particle traveling at a speed v m/s. Let c denote the speed of light. Give a physical interpretation to each of the following mathematical statements.
    (a) $f\left(\frac{c}{3}\right) = \sqrt{\frac{9}{8}}$.
    (b) $f'\left(\frac{c}{3}\right) = \frac{0.3977}{c}$.
    (c) $f^{-1}(3) = \sqrt{\frac{8}{9}}c$.
    (d) $(f^{-1})'(3) = 0.0393c$.

14. Let $D = f(L)$ be the number of daylight hours during the summer solstice on the prime meridian at a latitude of L degrees. Give a physical interpretation to each of the following mathematical statements.
    (a) $f(37) = 14.7$.
    (b) $f'(37) = 0.1$.
    (c) $f^{-1}(20) = 62.47$.
    (d) $(f^{-1})'(20) = 2$.

(e) $(f')^{-1}(1.4) = 65$.

15. Suppose Sally takes out a \$400,000 loan from a bank in order to buy a home. If she wants to pay back the loan in 30 years, her monthly payment rate A, measured in dollars (\$), will be a function $A = f(R)$ of the annual interest rate R, measured in per cents (%). Interpret each of the following mathematical statements using words. Be sure to include units for all quantities.
    (a) $f(4) = 1893.21$.
    (b) $f'(4) = 221.47$.
    (c) $f^{-1}(2000) = 4.4762$.
    (d) $(f^{-1})'(2000) = 0.004409$.

16. Let $V = f(t)$ be the volume, measured in cm^3, of air contained in a certain balloon, t seconds after air began being pumped into the balloon. Write a mathematical statement that is equivalent to each of the following reports.
    (a) Five seconds after air began being pumped into the balloon, the volume of the balloon was 12 cm^3.
    (b) Five seconds after air began being pumped into the balloon, the volume of the balloon is expected to increase by 1.4 cm^3 over the next second.
    (c) The moment when the volume of the balloon is being increased by approximately 2 cm^3 over one second occurs 3 seconds after air began being pumped into the balloon.
    (d) The volume of the balloon is 20cm^3 exactly 10 seconds after air began being pumped into the balloon.
    (e) The volume of the balloon is 20cm^3 exactly when it requires approximately 1.4 seconds for the volume of the balloon to increase by an additional cubic centimeter.

17. In physics, a *blackbody* is an object that reflects none of the radiation that strikes it, so that it appears completely black. Blackbodies nonetheless have been shown to radiate thermal energy at various wavelengths. Consider a blackbody whose temperature is 300 degrees Kelvin. Let $I = f(\lambda)$ be the *total* thermal radiation intensity, measured in W/m^2, due to all wavelengths at our below λ μm (micrometers). Various values of the function appear in Table 2.16 below.

λ (micrometers)	5	5.01	9.8173609	9.8176805
I (W/m^2 per micrometer)	5.91908	6.00143	120	120.01

Table 2.16: Cumulative Intensity Function of a blackbody at temperature 300K

For parts (a)–(d), estimate or determine each of the following quantities given below from the table. Then explain intuitively what the given quantity represents.
    (a) $f(5)$.
    (b) $f'(5)$.

(c) $f^{-1}(120)$.

(d) $(f^{-1})'(120)$.

(e) It is determined that

$$\lim_{\lambda\to\infty} f(\lambda) = 459.782.$$

Explain what this statement means intuitively. Include units.

(f) The *radiancy*, $R(\lambda)$ is defined by the relation

$$R(\lambda) = \frac{dI}{d\lambda}.$$

What are the units of radiancy? If the total intensity of the all the radiation whose wavelengths are between 14 μm and 14.01 μm is 0.2341 W/m^2, estimate the radiancy of the blackbody at the wavelength 14 micrometers; i.e., what is $R(14)$?

(g) Given the limit in part (e), what must be true of the radiancy as $\lambda \to \infty$, i.e., what is the value of

$$\lim_{\lambda\to\infty} R(\lambda)?$$

Explain your reasoning.

18. Suppose a bacterial culture in a petri dish is growing at a rate proportional to the number of bacteria; i.e., if $N(t)$ is the number of bacteria after t minutes, then

$$N'(t) = kN(t),$$

for some positive constant $k > 0$. If the culture is growing at a rate of 3,000 bacteria per minute at the time when there are 12,000 bacteria in the dish, how many bacteria will be in the dish at the time when the culture is growing at a rate of 10,000 bacteria per minute?

19. Newton's law of heating and cooling governs how the temperature u of an object is changing if the ambient temperature R is constant. For example, $u(t)$ might be the temperature of a hot cup of coffee, t minutes after it is fresh. *Newton's law of heating and cooling* states that that the rate of change of an object's temperature is proportional to the temperature difference between the object and its surroundings. Assuming the constant of proportionality k is positive, i.e. $k > 0$, write a formula for $u'(t)$ as a function of u. (The expression should also include the parameters k and R.)

20. Let $L = f(x)$ be the loudness of a jet engine, measured in decibels, at a distance of x meters from the engine.

(a) What are the units of $\left((f')^{-1}\right)'(-0.2)$?

(b) What are the units of $\left((f^{-1})'\right)^{-1}(-6)$?

(c) What are the units of $\left(\left((f^{-1})'\right)^{-1}\right)'(-6)$?

Hint: Expand the D&I map given in the example.

21. Cosmologists measure a quantity $T(t)$ that represents the temperature of the universe (in degrees Kelvin, K) as a function in the number of gigayears (Gyrs) after the big bang. Note that 1 Gyr equals one billion years and that the Kelvin scale is an absolute temperature scale, meaning that zero degrees Kelvin *is* the absolute zero temperature; no lower temperature is physically possible.
Suppose that currently $t = 13.6$, $T(13.6) = 2.4$, and $T'(13.6) = -12$.
   (a) For each of the following statements, state whether or not it is reasonable to agree or disagree with the conclusion and justify your reasoning.
      i. In the next one billion years, the temperature of the universe will drop by approximately 12 degrees Kelvin.
      ii. In the next year, the temperature of the universe will drop by approximately $\frac{12}{1{,}000{,}000{,}000}$ degrees Kelvin.
   (b) Assuming that $T(t)$ is decreasing and has no inflection points on the interval $[13.6, \infty)$, do you expect $T(t)$ to be concave up or concave down on this interval? Justify your answer using physical reasoning.

Chapter 3

Techniques of Differentiation

3.1 The Five Basic Derivatives

In this section, we examine several differentiation formulas that will aid us in efficiently computing derivatives of complex functions. These formulas are the lifeblood of calculus, and they will permeate the remainder of our adventures; to succeed, their implementation must become second nature.

3.1.1 The Five Basic Derivatives

We begin with our five basic derivative rules.

Theorem 3.1.1 — The Five Basic Derivatives. The five basic derivatives are

$$\frac{d}{dx}\left[x^n\right] = nx^{n-1}, \tag{3.1}$$

$$\frac{d}{dx}[\sin x] = \cos x, \tag{3.2}$$

$$\frac{d}{dx}[\cos x] = -\sin x, \tag{3.3}$$

$$\frac{d}{dx}\left[e^x\right] = e^x, \tag{3.4}$$

$$\frac{d}{dx}[\ln|x|] = \frac{1}{x}. \tag{3.5}$$

These rules may be naturally extending using the linearity property of differentiation, i.e., using Theorem 2.4.2.

■ **Example 3.1** Consider, again, the function $f(x) = 3x^2 + 2x + 7$. We saw, in Example 2.11, that the derivative is given by $f'(x) = 6x + 2$. This result can immediately be obtained using the derivative rule (3.1), coupled with the linearity property captured by equation (2.13). To wit,

$$\frac{d}{dx}\left[3x^2+2x+7\right]=3\frac{d}{dx}\left[x^2\right]+2\frac{d}{dx}[x]+\frac{d}{dx}[7]=3(2x)+2=6x+2.$$

This confirms the result. ∎

∎ **Example 3.2** Compute the second derivative of the function $f(x)=\ln|x|+\sin(x)$. To begin, the first derivative is obtained using linearity coupled with the differentiation rules (3.2) and (3.5), from whence we find

$$\frac{d}{dx}[\ln|x|+\sin(x)]=\frac{1}{x}+\cos(x).$$

The second derivative is then computed using this result and the fact that $\frac{1}{x}=x^{-1}$. We find,

$$\frac{d}{dx}\left[\frac{1}{x}+\cos(x)\right]=\frac{-1}{x^2}-\sin(x).$$

Hence, the second derivative of $f(x)$ is given by

$$\frac{d^2}{dx^2}[\ln|x|+\sin(x)]=\frac{-1}{x^2}-\sin(x).$$

∎

∎ **Example 3.3** Consider the function $f(x)=\sqrt{x}+\sin(x)$. To compute the derivative, we recognize $\sqrt{x}=x^{1/2}$. Using equation (3.1), we therefore obtain

$$\frac{d}{dx}\left[\sqrt{x}\right]=\frac{d}{dx}\left[x^{1/2}\right]=\frac{1}{2}x^{-1/2}=\frac{1}{2\sqrt{x}}.$$

This final equality follows since $x^{-1/2}=1/x^{1/2}=1/\sqrt{x}$. Using linearity, we then conclude that

$$\frac{d}{dx}\left[\sqrt{x}+\sin(x)\right]=\frac{1}{2\sqrt{x}}+\cos(x).$$

∎

3.1.2 The Five Basic Derivatives *Plus*

The five basic derivatives, though fundamental, are noticeably limited. How would one find the derivative of functions like

$$\sin(\sin(x)),\qquad e^{\cos(x)},\qquad \text{or}\qquad \sin(12\pi x+17)?$$

Notice that each of these is a composite function. The answer to this question involves a powerful rule, which is used for constructing derivatives, that known as the chain rule. We have incorporated the chain rule automatically into the five basic derivatives, forming the following, so-called *five basic derivatives plus*. For each of these rules, we start with a *known* function $u(x)$, which itself has a *known* derivative $u'(x)$.

Theorem 3.1.2 — The Five Basic Derivatives *Plus.* Suppose the function u has a known derivative u'. Then

$$\frac{d}{dx}\left[(u(x))^n\right] = n(u(x))^{n-1}u'(x), \tag{3.6}$$

$$\frac{d}{dx}[\sin(u(x))] = \cos(u(x))u'(x), \tag{3.7}$$

$$\frac{d}{dx}[\cos(u(x))] = -\sin(u(x))u'(x), \tag{3.8}$$

$$\frac{d}{dx}\left[e^{u(x)}\right] = e^{u(x)}u'(x), \tag{3.9}$$

$$\frac{d}{dx}[\ln|u(x)|] = \frac{u'(x)}{u(x)}. \tag{3.10}$$

In the preceding theorem, the function u is sometimes referred to as the *inside* or *inner* function, as it represents some simpler function contained *inside* a basic, *outer* function. (The five outer functions of equations (3.6)–(3.10) are simply the five functions of equations (3.1)–(3.5), respectively.)

■ **Example 3.4** Consider the inner function $u(x) = 5x^3 + \sin(x)$. Its derivative, as obtained using the basic rule (3.1)–(3.5), is given by $u'(x) = 15x^2 + \cos(x)$. We may therefore use this function, which has a known derivative, in any of the variety of contexts enumerated in equations (3.6)–(3.10), thereby obtaining the five following examples:

$$\begin{aligned}
\frac{d}{dx}\left[\left(5x^3+\sin(x)\right)^{10}\right] &= 10\left(5x^3+\sin(x)\right)^9\left(15x^2+\cos(x)\right), \\
\frac{d}{dx}\left[\sin\left(5x^3+\sin(x)\right)\right] &= \cos\left(5x^3+\sin(x)\right)\left(15x^2+\cos(x)\right), \\
\frac{d}{dx}\left[\cos\left(5x^3+\sin(x)\right)\right] &= -\sin\left(5x^3+\sin(x)\right)\left(15x^2+\cos(x)\right), \\
\frac{d}{dx}\left[e^{5x^3+\sin(x)}\right] &= e^{5x^3+\sin(x)}\left(15x^2+\cos(x)\right), \\
\frac{d}{dx}\left[\ln\left|5x^3+\sin(x)\right|\right] &= \frac{15x^2+\cos(x)}{5x^3+\sin(x)}.
\end{aligned}$$

Notice that, for each of the preceding five examples, we write the derivative of the outer function (i.e., the derivative of u^{10}, $\sin(u)$, $\cos(u)$, e^u, and $\ln|u|$) as normal, but then we multiply this result by the derivative of the function u.

Let us consider an additional example, using the same inner function u, for good measure:

$$\frac{d}{dx}\left[\sqrt{5x^3+\sin(x)}\right] = \frac{15x^2+\cos(x)}{2\sqrt{5x^3+\sin(x)}}.$$

This result follows from equation (3.6), using $n = 1/2$. (Recall, again, that $u^{-1/2} = 1/\sqrt{u}$.) ■

■ **Example 3.5** Consider next the random assortment of examples utilizing a potpourri of inner functions.

$$\begin{aligned}\frac{d}{dx}\left[\sin(e^x)\right] &= \cos(e^x)e^x,\\ \frac{d}{dx}[\sin(\sin(x))] &= \cos(\sin(x))\cos(x),\\ \frac{d}{dx}\left[e^{\cos(x)}\right] &= -e^{\cos(x)}\sin(x),\\ \frac{d}{dx}[\ln|\sin(x)|] &= \frac{\cos(x)}{\sin(x)},\\ \frac{d}{dx}\left[\frac{1}{\sin(x)}\right] &= \frac{-\cos(x)}{\sin^2(x)}.\end{aligned}$$

The final example uses equation (3.6) with $n=-1$. ■

Exercises

For Exercises 1–35, compute the derivative of the indicated function.

1. $f(x) = 8x^2 + 9x + 10$.
2. $f(x) = 12x^{10} + 9x^9 + 6x^5$.
3. $f(x) = 10x^3 + 3x + 7$.
4. $f(x) = \sin(x) + \cos(x)$.
5. $f(x) = e^x + \ln|x| + \pi$.
6. $f(x) = \sqrt{x}$.
7. $f(x) = \frac{1}{x}$.
8. $f(x) = \frac{1}{\sqrt{x}}$.
9. $f(x) = e^{3x^2+7}$.
10. $f(x) = e^{\sin(x)+e^x}$.
11. $f(x) = e^{\ln|x|}$.
12. $f(x) = \ln|e^x|$.
13. $f(x) = \frac{1}{\sqrt{x^3}}$.
14. $f(x) = \pi^3$.
15. $f(x) = \ln|8x^4 + 13x^3 + 19|$.
16. $f(x) = \ln|\cos(x)|$.
17. $f(x) = \ln|\sec(x)|$.
18. $f(x) = \cos(\cos(x))$.
19. $f(x) = \sin(2\pi x + 7)$.
20. $f(x) = e^{\sin(x)+x^2}$.
21. $f(x) = (e^x + \ln|x|)^3$.
22. $f(x) = \left(e^{3x} + \sin(2x)\right)^{10}$.
23. $f(x) = e^{5\sin(7x+2)}$.
24. $f(x) = e^{e^x}$.
25. $f(x) = e^{\sqrt{x}}$.
26. $f(x) = \sin^3(x)$.
27. $g(t) = e^{a^2+t^2}$.
28. $h(t) = \sin(3\pi\sqrt{t} + 4)$.
29. $g(w) = \frac{1}{\sin(w)+\ln(w)}$.
30. $h(t) = \frac{1}{\sqrt{\sin(x)+\cos(x)}}$.
31. $f(\theta) = \sin^2(\theta) + \cos^2(\theta)$.
32. $g(z) = \sin(e^z + \cos(z))$.
33. $v(u) = \ln|\sin(3u + \pi) + e^u|$.
34. $f(x) = e^{\sin(e^x)}$.
35. $g(t) = \ln\sqrt{t}$.

For Exercises 36–40, determine the values of the parameters a and b that will guarantee that the given piece-wise defined function is differentiable at every point.

36. $f(x) = \begin{cases} \sin(x) & \text{for } x < 0, \\ ax+b & \text{for } x \geq 0 \end{cases}.$

37. $f(x) = \begin{cases} e^{2x} & \text{for } x < 0, \\ ax+b & \text{for } x \geq 0 \end{cases}.$

38. $f(x) = \begin{cases} ax+b & \text{for } x < 1, \\ \ln(x) & \text{for } x \geq 1 \end{cases}.$

39. $f(x) = \begin{cases} ax+b & \text{for } x < 0, \\ \sqrt{x+3} & \text{for } x \geq 0 \end{cases}.$

40. $f(x) = \begin{cases} ax+b & \text{for } x < 1, \\ 1/x & \text{for } x \geq 1 \end{cases}.$

For Exercises 41–45, determine a point on the given interval that satisfies the mean-value theorem, i.e., for the given function f and interval $[a,b]$, find a point $c \in [a,b]$ such that $f'(c) = [f(b) - f(a)]/(b-a)$.

41. $f(x) = e^x$, $[0,1]$.

42. $f(x) = \sin x$, $\left[0, \frac{\pi}{2}\right]$.

43. $f(x) = x^2$, $[0,1]$.

44. $f(x) = x^3$, $[0,1]$.

45. $f(x) = \ln|x|$, $[1,e]$.

Problems

46. Let $h(t)$ represent the height of an object of mass m that is acted on by a constant, downward gravitational acceleration g. If the object initially has a vertical position of h_0 and an initial vertical velocity v_0, its height as a function of time is given by the relation

$$h(t) = h_0 + v_0 t - \frac{1}{2} g t^2. \tag{3.11}$$

(a) Compute the derivative of the particle's height in order to obtain a formula for the particle's vertical velocity $v(t) = h'(t)$.
(b) Compute the derivative of the particle's vertical velocity in order to obtain a formula for its vertical acceleration $a(t) = h''(t)$.
(c) The Willis tower (formerly the Sears tower) in Chicago has a height of 442 m. Assuming a downward gravitational acceleration of $g = 9.8\ \text{m/s}^2$, determine the amount of time it will take a penny released from the top of the Willis tower to hit the ground. What will its velocity be when the penny reaches the ground? (Ignore air resistance.)

47. The gravitational force F between two bodies with masses M and m as a function of the distance r between their centers of mass is given by *Newton's law of gravity* by the formula

$$F(r) = \frac{GMm}{r^2}, \tag{3.12}$$

where G is the *gravitational constant*. ($G = 6.67259 \times 10^{-11}\ \text{m}^3\text{kg}^{-1}\text{s}^{-2}$.)
(a) Determine the rate of change of the gravitational force with respect to distance. Your answer may depend on the parameters G, M, and m.

(b) The mass and radius of the Earth are $M = 5.97 \times 10^{24}$ kg and $R =$ 6,371 km, respectively. Determine the rate of change of the gravitational force with respect to distance for a 1 kg mass located on the surface of the Earth. Use this value to approximate the weight of the object at an altitude of 100 km, expressed as a fraction of the object's weight at sea level.

48. The period T of a pendulum with length l is given by the formula

$$T(l) = 2\pi\sqrt{\frac{l}{g}}, \tag{3.13}$$

where g represents the acceleration due to gravity.

(a) Determine a formula for the derivative $T'(l)$.

(b) In particular terms, what does the sign of $T'(l)$ indicate?

(c) At sea level, acceleration due to gravity is approximately $g = 9.8$ m/s^2. What length l^* must a pendulum in a grandfather clock have in order to have a period of 2 seconds?

(d) If the length of the pendulum in part (b) were to increase by 1 mm, by approximately how much time would the grandfather clock be off by each hour? Use the derivative $T'(l^*)$ to approximate your answer.

3.2 Power Functions and Polynomials

Our next task is to take a closer look at the basic derivative rules and examine *why* they are true. For now, we will focus on the basic derivative rules and only keep in practice with *using* the advanced ones, which will be justified in a single swoop in Section 3.5. Before beginning our dissection of the derivatives of power functions and polynomials, we first introduce an important tool of advanced mathematics known as mathematical induction.

3.2.1 An Aside on Mathematical Induction

Mathematical induction is a technique that is commonly employed whenever one has an *infinite* string of statements to prove, each of which is either true or false. Let us begin by naming these statements P_1, P_2, P_3, and so on, so that P_1 represents statement number one, which is either true or false; P_2 represents statement number two; P_3 represents statement number three, and on and on. (These "statements," for example, might be individual equations.) In general, we refer to P_n as the nth statement. These statements will, invariably, depend on n in some way. Mathematical induction gives us a tool with which we can prove that they are *all* true.

Proposition 3.2.1 — Principle of Mathematical Induction. Consider an infinite set of ordered statements $\{P_1, P_2, P_3, \ldots\}$, such that each statement is either true or false. If both of the following conditions are satisfied:

1. the first statement is true (i.e., P_1 is true), *and*
2. the truth of the kth statement implies the truth of the $(k+1)$st statement (i.e., $P_k \Longrightarrow P_{k+1}$),

then *all* of the statements $P_1, P_2, \ldots$ are true.

Thus, in order to *prove* that a string of statements is true, we must (1) show that the first statement is true, and then (2) show that *if* the kth statement is true, *then* the $(k+1)$st statement must also be true. The first step is called the *basis step*; when it is complete, one says that there is *a basis for mathematical induction*. The second step is called the *induction step*.

A cautionary word for the wary: when implementing the induction step, we do not know, nor are we saying, that an arbitrary statement P_k is true. (Here, k just represents any natural number.) We must *assume* that the kth statement is true and, based on that assumption, try to prove that the $(k+1)$st statement is also true. Since we only assumed that the kth statement is true, the effect of this maneuver is to show that *if* the kth statement is true, *then* the $(k+1)$st statement must also be true. That is, the induction step could potentially work without any of the statements actually being true! With the induction step, we are only showing that there is a logical implication: *if* any of the statements is true, *then* the following statement must (automatically) also be true.

The effect of the basis step and induction step taken together is like knocking over a long chain of dominos: once the first is tipped over, the next one falls automatically, which causes the third domino to fall, which causes the forth, and so on. The basis step tells us that P_1 is true. The induction step tells us that if *any* statement is true, the immediate following statement must also be true. Thus, statement P_1 tips over statement P_2; statement P_2 tips over statement P_3; P_3 tips P_4; and so on, as it continues indefinitely.

■ **Example 3.6** Use mathematical induction to show that

$$1+2+\cdots+n=\frac{n(n+1)}{2}. \tag{3.14}$$

In this example, the nth statement is simply the equation (3.14), which, *a priori*, may either be true or false. It is our task to prove that these statements are true for every natural number n, i.e., for every $n \in \mathbb{N}$.

The basis step. For $n = 1$, equation (3.14) is simply:

$$1=\frac{1(1+1)}{2}.$$

Clearly, this is true. Thus, we have a basis for induction.

The induction step. Next, let us *assume* that the kth statement is true, i.e., let us assume that

$$1+2+\cdots+k=\frac{k(k+1)}{2}. \tag{3.15}$$

We will use this assumption to show that the equation (3.14) must *also* be true for $n = k+1$. Since the $(k+1)$st statement has one extra term on the left-hand side, let us add $(k+1)$ to both sides of equation (3.15), which we are assuming to be true; we obtain

$$1+2+\cdots+k+(k+1)=\frac{k(k+1)}{2}+(k+1).$$

This is true, assuming that equation (3.15) is true. The advantage of adding $(k+1)$ to each side, is that the left-hand side of our new equation now looks like the left hand side of statement P_{k+1}, i.e., the left hand side of equation (3.14) for $n = k+1$. Next, we only need to find a common denominator on the right hand side and simplify, to obtain

$$1+2+\cdots+(k+1)=\frac{k(k+1)}{2}+\frac{2k+2}{2}=\frac{(k+1)(k+2)}{2}.$$

But this is statement P_{k+1}! We have therefore proved the implication: if equation (3.14) is true for $n = k$, then it must also be true for $n = k+1$.

Conclusion. We conclude by the principle of mathematical induction that the formula (3.14) is true for every natural number n. ■

3.2.2 Derivatives of Polynomials

Due to the general linearity property of derivatives, as given by Theorem 2.4.2, we must only learn how to differentiate monomials; the general case of a polynomial with n terms will follow. We therefore begin with the following.

Lemma 3.2.2 The derivative of the monomial $f(x) = x^n$, where $n \in \mathbb{N}$, is given by

$$\frac{d}{dx}\left[x^n\right] = nx^{n-1}. \tag{3.16}$$

Proof. We proceed by mathematical induction. Here, P_n simply represents equation (3.16). For example,

P_1	represents the equation	$\frac{d}{dx}\left[x^1\right] = 1,$
P_2	represents the equation	$\frac{d}{dx}\left[x^2\right] = 2x,$
P_3	represents the equation	$\frac{d}{dx}\left[x^3\right] = 3x^2,$

and so forth.

In order to prove the basis step, i.e., that statement P_1 is true, we apply the definition of the derivative, as given in equation (2.11). We obtain

$$\frac{d}{dx}[x] = \lim_{h\to 0}\frac{(x+h)-x}{h} = \lim_{h\to 0} 1 = 1.$$

This agrees with equation (3.16) for $n = 1$; we have thus established a basis for induction.

Next, we proceed to the induction step. We begin by assuming that equation (3.16) is true for an arbitrary natural number k. Using the definition of the derivative, our assumption is equivalent to the following

$$\lim_{h\to 0}\frac{(x+h)^k - x^k}{h} = kx^{k-1}. \tag{3.17}$$

Remember, this is our *assumption*. Using this assumption, we wish to show that equation (3.16) would also be true for $n = k+1$. To proceed, let us write the definition of the derivative out for the function $f(x) = x^{k+1}$:

$$\frac{d}{dx}\left[x^{k+1}\right] = \lim_{h\to 0}\frac{(x+h)^{k+1} - x^{k+1}}{h}.$$

Our aim is to simplify this in a way such that we can use the assumption that equation (3.17) is true. Simplifying, we find

$$\begin{aligned}
\frac{d}{dx}\left[x^{k+1}\right] &= \lim_{h\to 0}\frac{(x+h)^{k+1}-x^{k+1}}{h} \\
&= \lim_{h\to 0}\frac{(x+h)(x+h)^k-xx^k}{h} \\
&= \lim_{h\to 0}\frac{x(x+h)^k+h(x+h)^k-xx^k}{h} \\
&= \lim_{h\to 0}\left[x\left(\frac{(x+h)^k-x^k}{h}\right)+\frac{h(x+h)^k}{h}\right] \\
&= x\left(\lim_{h\to 0}\frac{(x+h)^k-x^k}{h}\right)+\lim_{h\to 0}(x+h)^k.
\end{aligned}$$

This is a true equation, as we have yet to rely on any assumptions in stating it. Now, *assuming* that equation (3.17) is true, the preceding equation simplifies as

$$\frac{d}{dx}\left[x^{k+1}\right] = x\left(kx^{k-1}\right)+x^k = kx^k+x^k = (k+1)x^k.$$

Therefore, if equation (3.16) is true for $n = k$, then it must also be true for $n = k+1$.

By the principle of mathematical induction, we conclude that equation (3.16) is true for every natural number n. ■

Theorem 3.2.3 Consider the nth degree polynomial

$$P(x) = a_nx^n + a_{n-1}x^{n-1} + \cdots + a_2x^2 + a_1x + a_0,$$

for some constants $a_0, \ldots, a_n$, with $a_n \neq 0$. Its derivative is

$$P'(x) = na_nx^{n-1} + (n-1)a_{n-1}x^{n-2} + \cdots + 2a_2x + a_1.$$

Proof. The result follows by combining Lemma 3.2.2 with the linearity property of derivatives (Theorem 2.4.2). ■

▪ **Example 3.7** Compute the derivative of $f(x) = 6x^6 + 3x^3 + 27x^2 + 9x + 11$.

To accomplish this, we differentiate term-by-term. Since the coefficients 6, 3, 27, 9, and 11 are constant, they are simply carried along for the ride. For example, the derivative of x^6 is $6x^5$; therefore the derivative of $6x^6$ is $36x^5$, and so on. Proceeding in a similar fashion with each term, we obtain

$$f'(x) = 36x^5 + 9x^2 + 54x + 9.$$

▪

▪ **Example 3.8** Use the theory of differentiation to determine the intervals on which the function

$$f(x) = 8x^3 + 23x^2 - 18x + 15$$

is increasing and decreasing. Then determine the intervals on which the function is concave up and concave down.

We begin by computing the derivative

$$f'(x) = 24x^2 + 46x - 18.$$

Since the sign of the derivative controls whether the original function is increasing or decreasing, let us look for zeros (i.e., roots) of the derivative. Setting the derivative equal to zero and applying the quadratic formula, we obtain

$$x = \frac{-46 \pm \sqrt{46^2 + 4(24)(18)}}{48} = \frac{-23 \pm 31}{24} = \left\{-2.25, \frac{1}{3}\right\}.$$

Since the derivative is a quadratic function with a positive leading coefficient, we know that it is concave up. Therefore, $f'(x)$ will be negative for $-2.25 < x < 1/3$, and positive for $x < -2.25$ and $x > 1/3$. We conclude that the original function f is decreasing on the interval $(-2.25, 1/3)$, and increasing otherwise.

Next, we compute the second derivative:

$$f''(x) = 48x + 46.$$

This is a linear function with a positive slope, meaning it will be positive for x-values that are greater than the x-intercept. To compute the x-intercept, we set this equal to zero and solve for x, obtaining $x = -23/24$. We conclude that the original function f is concave down for $x < -23/24$ and concave up for $x > -23/24$. ■

3.2.3 Power Functions

The rule for taking the derivative of general power functions is identical to the rule for monomials.

Theorem 3.2.4 The derivative of the power function $f(x) = x^k$, for $k \in \mathbb{R}$, is given by

$$\frac{d}{dx}\left[x^k\right] = kx^{k-1}. \tag{3.18}$$

We will omit the proof of this theorem, noting only that it also derives from the limit-definition of the derivative.

■ **Example 3.9** Compute the derivative of $f(x) = \sqrt{x}$.

Here, we recognize the power as $1/2$, as $\sqrt{x} = x^{1/2}$. Therefore

$$f'(x) = \frac{1}{2}x^{-1/2} = \frac{1}{2\sqrt{x}}.$$

■

■ **Example 3.10** Compute the derivative of

$$f(x) = \frac{1}{\sqrt{x}}.$$

Here, the power is $-1/2$. Applying equation (3.18), with $k = -1/2$, we obtain

$$f'(x) = \frac{-1}{2}x^{-3/2} = \frac{-1}{2\sqrt{x^3}}.$$

■

3.2.4 The Power-Down Rule

Our next theorem combines the result from Theorem 3.2.4 with the power of the chain rule. Its proof follows immediately by applying the chain rule to the derivative rule for power functions; we will therefore save the proof until our full discussion of the chain rule in Section 3.5.

Theorem 3.2.5 — Power-Down Rule. Let $u(x)$ be a function with a known derivative $u'(x)$, and consider the function $f(x) = [u(x)]^k$, for some number k. Then

$$\frac{d}{dx}\left[[u(x)]^k\right] = k\,[u(x)]^{k-1}\,u'(x). \tag{3.19}$$

The function $u(x)$ is often referred to as the *inside function*, whereas the function u^k is referred to as the *outside function*.

■ **Example 3.11** Compute the derivative of

$$f(x) = \sqrt{4x^5 + 5x^4 + 6x^3 + 7x^2 + 8x + 9}.$$

In this example, our known function, i.e., the inside function, is $u(x) = 4x^5 + 5x^4 + 6x^3 + 7x^2 + 8x + 9$, which has its own derivative of

$$u'(x) = 20x^4 + 20x^3 + 18x^2 + 14x + 8.$$

The outer function is $\sqrt{u} = u^{1/2}$, so that, upon differentiating, we obtain

$$f'(x) = \frac{u'(x)}{2\sqrt{u(x)}} = \frac{10x^4 + 10x^3 + 9x^2 + 7x + 4}{\sqrt{4x^5 + 5x^4 + 6x^3 + 7x^2 + 8x + 9}}.$$

(For the derivative of $\sqrt{u}$, recall Example 3.9.) ■

■ **Example 3.12** Compute the derivative of

$$f(x) = \frac{1}{x^3 + 7}$$

. This is equivalent to $f(x) = (x^3 + 7)^{-1}$, so that we identify the inner function as $u(x) = x^3 + 7$ and the outer function as u^{-1}. The power-down rule therefore tells us that

$$f'(x) = -(x^3 + 7)^{-2}(3x^2) = \frac{-3x^2}{(x^3 + 7)^2}.$$

■

Exercises

For Exercises 1–10, determine the derivative of the indicated function.

1. $f(x) = x^{88} + 88x^8 + 8.$
2. $f(x) = 9x^{100} + 100x^9.$
3. $f(x) = 4x^4 + 3x^3 + 2x^2 + x + \pi.$
4. $f(x) = \dfrac{1}{\sqrt[3]{x^5}}.$
5. $f(x) = \sqrt{x^3 + 3x + 7}.$
6. $f(x) = \dfrac{1}{8x^9 + 9x^8}.$
7. $f(x) = \dfrac{1}{\sqrt{8x^9 + 9x^8}}.$
8. $f(x) = \dfrac{1}{\left(x^7 + 7x^6 + 6x^5 + 5x^4\right)^{\pi}}.$
9. $f(x) = \dfrac{9}{\sqrt[3]{(x^4 + 9x^3 + 7)^4}}.$
10. $f(x) = \dfrac{1}{\sqrt[13]{(x^3 + \pi)^{99}}}.$

For Exercises 11–15, determine the derivative of the indicated function using the limit definition of the derivative.

11. $f(x) = \dfrac{1}{x}.$
12. $f(x) = \dfrac{1}{x^2}.$
13. $f(x) = \dfrac{1}{x^3}.$
14. $f(x) = \sqrt{x}$. Use the *binomial series*, which says that
$$(1+z)^p \approx 1 + pz + \frac{p(p-1)}{2} z^2,$$
for small z. *Hint*: set $z = h/x$.
15. $f(x) = \dfrac{1}{\sqrt{x}}$. *Hint*: Use the binomial series again.

For Exercises 16–17, use mathematical induction to prove the indicated statement for all natural numbers n.

16. $1^2 + 2^2 + \cdots + n^2 = \frac{n(n+1)(2n+1)}{6}.$
17. $1 + r + r^2 + \cdots + r^{n-1} = \dfrac{r^n - 1}{r - 1}.$ This is known as a *finite geometric series*.

Problems

18. Compute each of the following derivatives. How do (b) and (c) compare? Show that their derivatives are, in fact, the same. Explain why this is true.
    (a) $y = (x+1)^{100}.$
    (b) $y = (x^2 + 2x + 1)^{100}.$
    (c) $y = (x+1)^{200}.$

19. Let $f(x)$ represent the number of minutes required to run a mile, when running at a pace of x mph.
    (a) Determine an expression for $f(x)$.
    (b) Compute the derivative $f'(x)$.
    (c) Evaluate $f(6)$ and $f'(6)$, and explain what these quantities represent in practical terms.
    (d) Use the results from part (c) and local linearization to approximate the value of $f(6.5)$. How does your approximation compare with the actual value?

20. The speed of sound in dry air, measured in m/s, is proportional to the square root of the absolute temperature of the air, measured in degrees Kelvin. The speed of sound is 343m/s when the temperature is 293K (20° C).
    (a) Write an expression for $f(T)$, the speed of sound as a function of the temperature. Find the value of the constant of proportionality k. What does the graph of $f(T)$ look like?
    (b) Evaluate $f'(T)$ at $T = 293K$. Use this to approximate the speed of sound at $298K$.
    (c) Based on the concavity of the graph of $f(T)$, would you expect the estimate of $f(298)$ to be an overestimate or an underestimate? Explain your reasoning.
    (d) Evaluate $f(T)$ at $T = 298K$ and compare with the value obtained from part (b). Is the approximation from part (b) an underestimate or an overestimate? (An *underestimate* is an estimate that is less than the actual value; an *overestimate* is the opposite.)
21. The sound intensity $I = f(x)$, measured in Watts per square meter (W/m^2), at a distance x m from a jet engine is given by the formula

$$f(x) = \frac{250,000}{x^2}.$$

    (a) Determine a formula for the derivative $f'(x)$.
    (b) Determine a formula for the inverse of the derivative, $x = (f')^{-1}(dI/dx)$.
    (c) Determine a formula for the inverse function $f^{-1}(I)$.
    (d) Determine a formula for the derivative of the inverse $dx/dI = (f^{-1})'(I)$.
    (e) Confirm that the stated equations in problem Problem 12 from Section 2.5 are correct, i.e., confirm that
        i. $f(50) = 100$.
        ii. $f^{-1}(25) = 100$.
        iii. $f'(50) = -4$.
        iv. $(f')^{-1}(-0.5) = 100$.
        v. $(f^{-1})'(100) = -0.25$.
    (f) Interpret each of the above statements using words.
22. The *ideal gas law* relates the pressure P, specific volume v, and absolute temperature T of an ideal gas with gas constant R, via the formula

$$Pv = RT. \tag{3.20}$$

    The *specific volume* is the volume per unit mass, i.e., the reciprocal of the density. The parameter R is a constant that depends on the given gas.
    (a) Assuming the temperature is a constant, compute the derivative $\frac{dP}{dv}$.
    (b) Assuming the pressure is a constant, compute the derivative $\frac{dv}{dT}$.
    (c) Assuming the specific volume is a constant, compute the derivative $\frac{dT}{dP}$.

(d) Combining the results from parts (a), (b), and (c), show that

$$\frac{dP}{dv}\frac{dv}{dT}\frac{dT}{dP} = -1.$$

23. A balloon is being inflated so that its radius r, measured in centimeters, can be modeled as a function of time t, measured in seconds, by the equation

$$r(t) = 13(1 - e^{-0.2t}).$$

(a) What is the balloon's maximum radius?
(b) At what time will the balloon's radius equal half of its maximum?
(c) At what rate is the radius changing at the moment when its radius is half its maximum?
(d) Determine the balloon's volume as a function of time.
(e) At what rate is the balloon's volume changing at the moment when its radius is half its maximum?

3.3 Exponential and Logarithmic Functions

In this section, we examine the rules for differentiation of exponential functions.

3.3.1 Derivatives of Exponential Functions

The basic exponential function $f(x) = e^x$ with base e is sometimes called *the indestructible function*, as it is its own derivative. Unlike a polynomial, e^x can therefore be differentiated many times without ever changing or eventually becoming the zero function.

Theorem 3.3.1 The derivative of the function $f(x) = e^x$ is itself, i.e.,

$$\frac{d}{dx}\left[e^x\right] = e^x. \tag{3.21}$$

Proof. We again call on the limit-definition of the derivative, which tells us that the derivative of e^x must be

$$\begin{aligned} f'(x) &= \lim_{h\to 0} \frac{e^{x+h} - e^x}{h} \\ &= \lim_{h\to 0} \frac{e^x e^h - e^x}{h} \\ &= \lim_{h\to 0} \frac{e^x(e^h - 1)}{h} \\ &= e^x \lim_{h\to 0} \frac{e^h - 1}{h}. \end{aligned}$$

We were allowed to pull a factor of e^x outside of the limit, since the limit is with respect to the variable h. The limit therefore regards x as though it were a constant. For the purpose of calculating the limit, x *is* a constant; it is only after the limit has taken its effect do we allow x to vary and consider the result as a function of x.

In order to complete the proof, we must study the behavior of the function

$$g(h) = \frac{e^h - 1}{h}$$

as $h \to 0$. The function $g(h)$ is plotted in Figure 3.1.

Though the function g is undefined at $h = 0$, it is clear from the graph that

$$\lim_{h\to 0} \frac{e^h - 1}{h} = 1.$$

This can be confirmed by evaluating a sequence of points that are close to $h = 0$. Since this limit equals 1, we may conclude that $f'(x) = e^x$, which completes the proof. ∎

Next, we state the extended version of Theorem 3.3.1 that automatically incorporates the chain rule.

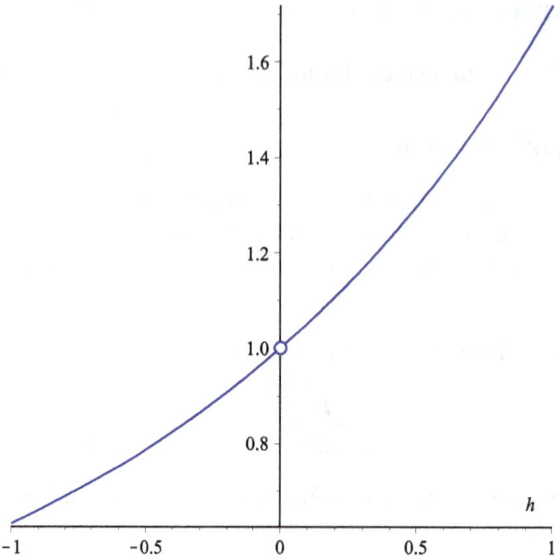

Figure 3.1: Graph of the function $g(h) = (e^h - 1)/h$ on the interval $[-1, 1]$.

Theorem 3.3.2 Let $u(x)$ be a function with a known derivative $u'(x)$. The derivative of the function $e^{u(x)}$ is given by

$$\frac{d}{dx}\left[e^{u(x)}\right] = e^{u(x)}u'(x). \tag{3.22}$$

The function $u(x)$ is known as the *inside function.*

When an exponential function, with base e, has an exponential other than x by itself, we therefore rewrite it and then multiply by the derivative of the exponent. As stated previously, this theorem follows directly from the chain rule, which we will discuss in Section 3.5, so we will omit its proof for the present.

■ **Example 3.13** Compute the derivative of $f(x) = e^{3x}$.

The insider function for this problem is $u(x) = 3x$, with derivative $u'(x) = 3$. The derivative of the function f is therefore given by

$$f'(x) = 3e^{3x}.$$

Note that we promoted the constant 3 to the front of the equation as a matter of style: coefficients precede functions. ■

■ **Example 3.14** Compute the derivative of $f(x) = e^{4x^5+9x^2+\pi}$.

The inside function in this case is $u(x) = 4x^5 + 9x^2 + \pi$, with derivative $u'(x) =$

$20x^4 + 18x$. The derivative of the function f is therefore

$$f'(x) = e^{4x^5+9x^2+\pi}\left(20x^4 + 18x\right).$$

■

We next proceed to the case of exponential functions with an arbitrary base.

Theorem 3.3.3 Given an exponential function $f(x) = a^x$, with base $a > 0$, its derivative is given by

$$\frac{d}{dx}\left[a^x\right] = a^x \ln(a). \tag{3.23}$$

Proof. It turns out that the exponential function $f(x) = a^x$ can be recast in the form of equation (3.22) with a simple algebraic twist:

$$a^x = e^{\ln(a^x)} = e^{x\ln(a)}.$$

We may now differentiate $f(x) = e^{x\ln(a)}$ following the rule given in equation (3.22):

$$f'(x) = e^{x\ln(a)}\ln(a) = a^x \ln(a).$$

This completes our proof. ■

Finally, we can spice up equation (3.23) using the chain rule (which, as usual, will follow from our results of Section 3.5, so that we can here omit the proof).

Theorem 3.3.4 Given a function $u(x)$ and its known derivative $u'(x)$, and a positive base $a > 0$, the derivative of the function $f(x) = a^{u(x)}$ is given by

$$\frac{d}{dx}\left[a^{u(x)}\right] = a^{u(x)}\ln(a)u'(x). \tag{3.24}$$

As usual, the function u is referred to as the *inside function*.

■ **Example 3.15** Compute the derivative of $f(x) = 2^{5x}$.

Here, the inside function is $u(x) = 5x$, so the derivative is simply

$$f'(x) = 5(2)^{5x}\ln(2).$$

■

■ **Example 3.16** Compute the derivative of $f(x) = \pi^{2x^7}$.

We obtain

$$f'(x) = 14x^6\pi^{2x^7}\ln(\pi),$$

which follows directly from equation (3.24). ■

Example 3.17 Compute the derivative of

$$f(x) = \pi^{(8x^2+3x+7)^{100}}.$$

Here, the inside function is $u(x) = (8x^2+3x+7)^{100}$; the derivative can be obtained by using the power-down rule (Theorem 3.2.5), which results in

$$u'(x) = 100(8x^2+3x+7)^{99}(16x+3).$$

Thus,

$$f'(x) = \underbrace{\pi^{(8x^2+3x+7)^{100}}}_{\text{original: } \pi^{u(x)}} \cdot \underbrace{100(8x^2+3x+7)^{99}(16x+3)}_{\text{derivative of exponent: } u'(x)} \cdot \underbrace{\ln(\pi)}_{\ln \text{ of base}}.$$

■

Notice that Theorems 3.3.1 and 3.3.2 *are* consistent with Theorems 3.3.3 and 3.3.4; in fact, the first two may be regarded as special cases of the latter two. If the base of the exponential function is the number e, then multiplying by the natural logarithm of the base results in multiplication by $\ln(e) = 1$.

These rules also reveal why Euler's number e is such a special number: of all exponential functions of the form a^x, base e is the *only* one that results in the derivative being *equal* to the original function, whereas for other choices of the base, the derivative is directly *proportional to* the original function.

3.3.2 Derivatives of Logarithmic Functions

Next, we turn our focus on the inverse of the exponential function: the logarithmic function.

Theorem 3.3.5 Consider the function $f(x) = \ln|x|$. Its derivative is

$$\frac{d}{dx}[\ln|x|] = \frac{1}{x}. \tag{3.25}$$

Proof. Using the limit-definition of the derivative, we write

$$f'(x) = \lim_{h\to 0} \frac{\ln|x+h| - \ln|x|}{h} = \lim_{h\to 0}\left(\frac{1}{h}\ln\left|\frac{x+h}{x}\right|\right) = \lim_{h\to 0}\left(\frac{1}{h}\ln\left|1+\frac{h}{x}\right|\right).$$

It is important to recall that, for the purpose of calculating the limit, the variable x is treated as a constant. This is because the limit asks, for a fixed value of x, what happens in the limit as h tends to zero. Only after we take the limit do we then allow x to vary, resulting in the derivative as a function of x.

We next introduce a new variable w, defined as $w = h/x$. For a fixed value of x, the variable w will go to zero proportionately as h goes to zero. Changing the variable h into the variable w, we therefore find an equivalent expression:

$$f'(x) = \lim_{w\to 0}\left(\frac{1}{wx}\ln|1+w|\right) = \frac{1}{x}\lim_{w\to 0}\ln|(1+w)^{1/w}|.$$

(Recall that x is treated as a constant for the purpose of computing the limit, which is why we may extract it from the limit.)

Recall from equation (1.12) that

$$\lim_{w\to 0}(1+w)^{1/w} = e;$$

therefore,

$$\lim_{w\to 0}\ln\left|(1+w)^{1/2}\right| = \ln(e) = 1.$$

We conclude that

$$f'(x) = \frac{1}{x},$$

which completes the proof. ■

Theorem 3.3.6 Let $u(x)$ be a function with a known derivative $u'(x)$. Then

$$\frac{d}{dx}[\ln|u(x)|] = \frac{u'(x)}{u(x)}. \tag{3.26}$$

The function $u(x)$ is referred to as the *inside function.*

This theorem follows by applying the chain rule to Theorem 3.3.5. (So once we learn the chain rule, the result will again be automatic.)

■ **Example 3.18** Compute the derivative of

$$f(x) = \ln(x^4 + 12x^2 + 20).$$

We first identify the inside function as $u(x) = x^4 + 12x^2 + 20$, which has a derivative of $u'(x) = 4x^3 + 24x$. The derivative of the function f is therefore

$$f'(x) = \frac{4x^3 + 24x}{x^4 + 12x^2 + 20}.$$

■

■ **Example 3.19** Compute the derivative of $f(x) = \ln(e^x)$.

The inside function is $u(x) = e^x$, which has the derivative $u'(x) = e^x$. Therefore

$$f'(x) = \frac{u'(x)}{u(x)} = \frac{e^x}{e^x} = 1.$$

This makes sense, as $f(x) = \ln(e^x) = x\ln(e) = x$, and so its derivative *should* equal 1. ■

■ **Example 3.20** Compute the derivative of $f(x) = \ln(\sqrt{x})$.

Here, the inside function is $u(x) = \sqrt{x}$, with derivative

$$u'(x) = \frac{1}{2\sqrt{x}}.$$

Therefore, we obtain

$$f'(x) = \frac{u'(x)}{u(x)} = \frac{1}{\sqrt{x}} \cdot \frac{1}{2\sqrt{x}} = \frac{1}{2x}.$$

Observe that $f(x) = \ln x^{0.5} = 0.5 \ln |x|$, and so this result is to be expected. ▪

3.3.3 Hyperbolic Trigonometric Functions

Finally, we introduce several important functions that are combinations of exponential functions.

Definition 3.3.1 The *hyperbolic sine*, *hyperbolic cosine*, and *hyperbolic tangent* functions are defined by the relations

$$\sinh(x) = \frac{e^x - e^{-x}}{2}, \quad (3.27)$$

$$\cosh(x) = \frac{e^x + e^{-x}}{2}, \quad (3.28)$$

$$\tanh(x) = \frac{e^x - e^{-x}}{e^x + e^{-x}}, \quad (3.29)$$

respectively.

These definitions of hyperbolic trig functions *are* actually related to the ordinary sine, cosine, and tangent functions; these relations require some background in complex variables to understand, so we will postpone such an analysis until we have developed the necessary tools. For now, however, we may simply regard the functions sinh, cosh, and tanh as specially defined functions as prescribed by equations (3.27)–(3.29).

Our first observation is the following.

Proposition 3.3.7 The following equality holds:

$$\tanh(x) = \frac{\sinh(x)}{\cosh(x)}. \quad (3.30)$$

Proof. Dividing $\sinh(x)$ by $\cosh(x)$, using equations (3.27) and (3.28), we obtain

$$\frac{\sinh(x)}{\cosh(x)} = \frac{\frac{e^x - e^{-x}}{2}}{\frac{e^x + e^{-x}}{2}} = \left(\frac{e^x - e^{-x}}{2}\right)\left(\frac{2}{e^x + e^{-x}}\right) = \frac{e^x - e^{-x}}{e^x + e^{-x}}.$$

This last expression we immediately identify as $\tanh(x)$, and the proof is complete. ■

Recall that the same property holds for regular trig functions: sine over cosine is equal to tangent.

Next, we may examine the graphs of the functions sinh and cosh, as shown in Figure 3.2. As to be expected, these two graphs grow close to each other as $x \to \infty$, as they only differ in the sign of their e^{-x} term. (And this term decays as $x \to \infty$.)

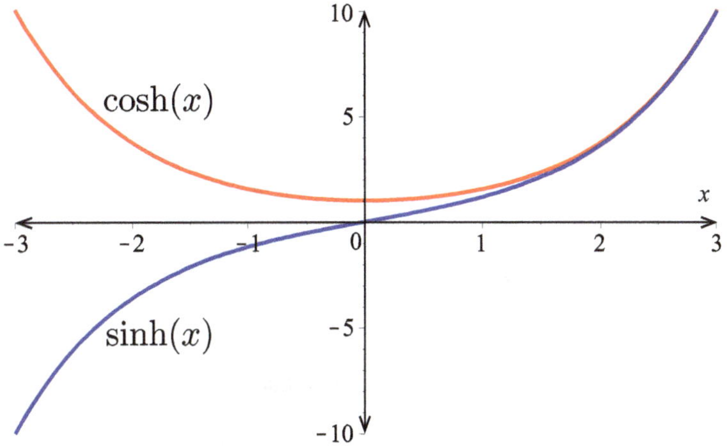

Figure 3.2: Graph of sinh(x) (blue) and cosh(x) (red).

Similarly, as $x \to -\infty$, the functions sinh and cosh approach a mirror image of each other as a reflection about the x-axis. This is because the e^x term decays as $x \to -\infty$, and sinh(x) therefore approaches $-\cosh(x)$ for large negative x values.

Since the hyperbolic tangent function can be described as the ratio of the hyperbolic sine to the hyperbolic cosine, we might predict the following behaviors: since the functions sinh and cosh approach each other for large x values, the values of the tanh function should approach 1 as $x \to \infty$. Similarly, since sinh and cosh approach each other's negative for large, negative x, we might similarly expect that

$$\lim_{x \to -\infty} \tanh(x) = -1.$$

These hypotheses are born out in reality, as is confirmed by the graph of tanh(x), shown in Figure 3.3. The hyperbolic tangent function therefore has the behavior of a smooth "switch," transitioning from one state (i.e., $y = -1$) to an alternate state (i.e., $y = +1$) in a smooth fashion (i.e., differentiable).

(We hope to instill in the reader a sense of first seeking to *think* about what a function might look like before turning to the crutches of technology—being able to intuit a function's behavior is an important skill, as it builds a *picture* and offers deeper *understanding*.)

Hyperbolic trigonometric functions share other similarities with the ordinary trig functions. For instance, recall that $\sin^2(x) + \cos^2(x) = 1$. The corresponding identity for hyperbolic trigonometric functions is given by the following.

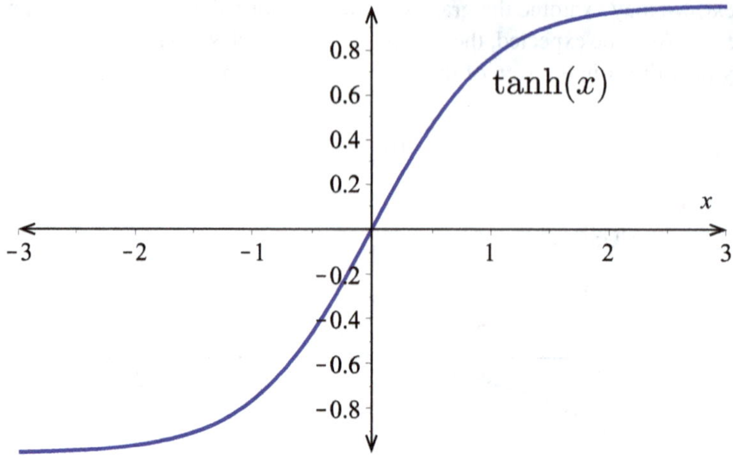

Figure 3.3: Graph of tanh(x).

Proposition 3.3.8 The hyperbolic sine and cosine functions satisfy the identity

$$\cosh^2(x) - \sinh^2(x) = 1. \tag{3.31}$$

We leave the proof of this identity as an exercise.

R Consider the *parametric* equations

$$x(\theta) = \cos(\theta) \qquad \text{and} \qquad y(\theta) = \sin(\theta).$$

Recall from Definition 1.6.1 on page 63 that these equations reproduce the unit circle as one varies θ from zero to 2π. Moreover, we may rewrite the identity (1.39) as

$$x^2 + y^2 = 1,$$

which is precisely the equation of the unit circle.

Alternatively, let us consider a similar construct using the hyperbolic trigonometric functions. If we write

$$x(\theta) = \cosh(\theta) \qquad \text{and} \qquad y(\theta) = \sinh(\theta),$$

we may rewrite the identity (3.31) as

$$x^2 - y^2 = 1.$$

This is the equation of a *hyperbola*.

Finally, let us examine the derivative rules for these special functions.

Theorem 3.3.9 The derivatives of the hyperbolic trigonometric functions are as follows:

$$\frac{d}{dx}[\sinh(x)] = \cosh(x), \tag{3.32}$$

$$\frac{d}{dx}[\cosh(x)] = \sinh(x), \tag{3.33}$$

$$\frac{d}{dx}[\tanh(x)] = \text{sech}^2(x), \tag{3.34}$$

where the hyperbolic secant function is defined as one would expect:

$$\text{sech}(x) = \frac{1}{\cosh(x)}.$$

The proof is reserved as an exercise; the proof of equation (3.34) will be postponed until Section 3.6.

Similar to the sine and cosine functions, the hyperbolic sine and hyperbolic cosine functions alternate back and forth as one takes successive derivatives: sinh—cosh—sinh—cosh—etc. The difference, however, is that the ordinary trig functions also alternate sign every two derivatives: sine—cosine—*negative* sine—*negative* cosine—sine—cosine—etc. Whereas the derivative of cosine is *negative* sine, the derivative of hyperbolic cosine is (positive) hyperbolic sine.

There are additional similarities between hyperbolic and regular trig functions other than these; in this short introduction, we hope to have a flavor for some of the ways in which the two classes of function are alike and some of the ways in which they are different.

Exercises

For Exercises 1–10, compute the derivative of the indicated function.

1. $f(x) = e^{8\pi x+7}$.
2. $f(x) = e^{x^e}$.
3. $f(x) = e^{e^x}$.
4. $f(x) = 12^{3x^7+7x^3}$.
5. $f(x) = 8^{e^x+\pi}$.
6. $f(x) = \ln|3x^3 + 9x^2 + 17|$.
7. $f(x) = \ln|5^x + x^5|$.
8. $f(x) = \ln|\ln|x||$.
9. $f(x) = \ln\left|e^{3x^2} + \ln|5x|\right|$.
10. $f(x) = \ln\left|x^3 + \sqrt{x} + \frac{1}{x}\right|$.

Problems

11. Proceed in a similar fashion as in the proof of Theorem 3.3.3 in order to determine the derivative of the functions:
    (a) $f(x) = x^x$; and
    (b) $f(x) = x^{x^x}$.
    For part (b), you may use the result from part (a).

12. Recalling the fact that

$$\log(x) = \frac{\ln(x)}{\ln(10)},$$

determine the derivative of the function $f(x) = \log(x)$.

13. Prove the identity (3.31).
14. Prove the derivative rules (3.32) and (3.33).
15. Let $P(h)$ represent the atmospheric pressure, measured in kiloPascals (kPa), at an altitude of h meters. The atmospheric pressure may be modeled by the formula

$$P(h) = 101.3e^{-h/7000}.$$

(a) According to this model, what is the atmospheric pressure at sea level?
(b) Compute $P'(0)$ and interpret the result in practical terms.
(c) Use the results from parts (a) and (b) and local linearization to approximate the atmospheric pressure at an altitude of 500 m. Compare this approximation with the exact value for $P(500)$.

16. Use the derivative rule (3.22) and the definitions (3.27) and (3.28) to generalize the derivative rules (3.32) and (3.33), i.e., show that

$$\frac{d}{dx}[\sinh(u(x))] = \cosh(u(x))u'(x), \tag{3.35}$$

$$\frac{d}{dx}[\cosh(u(x))] = \sinh(u(x))u'(x). \tag{3.36}$$

17. A *catenary* is the shape that a hanging cable or wire makes when it is only supported at its two endpoints. The equation of a catenary is given by the hyperbolic cosine function. Suppose that the height (above sea level) of the cables of the Golden Gate Bridge are modeled by the transformed catenary equation

$$h(x) = 75\cosh\left(\frac{x}{361}\right),$$

for $x \in [-640, 640]$, as shown in Figure 3.4.

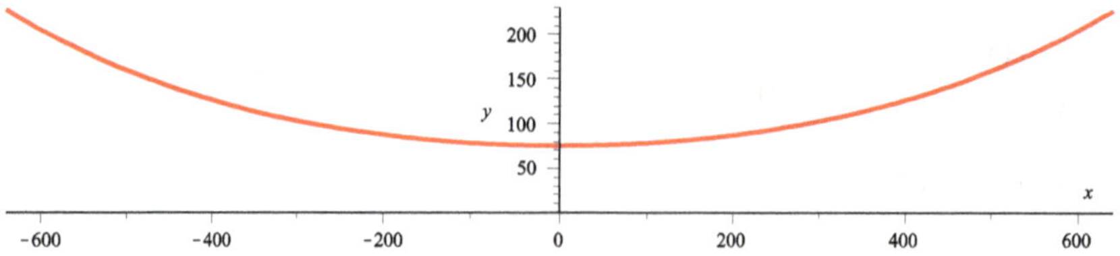

Figure 3.4: The height of the cable on the Golden Gate Bridge.

(a) Determine the derivative $h'(x)$. (You may use equation (3.36).)
(b) Evaluate $h(400)$ and $h'(400)$.

(c) Use local linearization and the results from part (b) to approximate the height of the cable at $x = 410$. How does this compare with the exact answer?

18. A frozen turkey is placed into an oven with a temperature of 180° C, so that its temperature u after t hours may be described by the function

$$u(t) = 180\left(1 - e^{-kt}\right),$$

where k is a positive parameter.

(a) If the turkey's temperature is 60° C after one hour, determine the exact value of the parameter k.

(b) Compute $u'(t)$ and evaluate $u'(0)$ and $u'(1)$. Interpret each value in practical terms.

(c) Use local linearization and the results from parts (a) and (b) to approximate the turkey's temperature after 10 minutes and after 70 minutes. Use the formula for $u(t)$ to determine the actual values of these temperatures.

ambient temperature

19. *Newton's law of heating and cooling* governs the rate at which the temperature u of an object is changing if it is placed in an environment with constant temperature R. The parameter R is called the *ambient temperature*; it is simply the temperature of the surrounding environment, i.e., the temperature of the ambience. Newton's law states that the rate of change of the object's temperature is proportional to the temperature difference between the object and its surroundings. This law is described by the *differential equation*

$$\frac{du}{dt} = -k(u - R), \tag{3.37}$$

where $k > 0$ is a positive constant of proportionality. (A differential equation is simply an equation that relates one or more of a function's derivatives to the function value.)

(a) Verify that the sign of equation (3.37) is correct. (I.e., if the object's temperature is hotter than its environment, is the object cooling? What if the object is cooler?)

(b) What are the units of the parameter k in terms of the units of temperature and the units of time?

(c) The general solution to the differential equation (3.37) is given by

$$u(t) = R - (R - u_0)e^{-kt}, \tag{3.38}$$

where u_0 is the initial temperature of the object at time $t = 0$. Verify that a function of this form indeed satisfies the differential equation (3.37); also verify that it satisfies $u(0) = u_0$.

(d) Determine the limit

$$\lim_{t\to\infty} u(t),$$

and explain why the result makes sense in practical terms.

(e) As the object heats or cools (this is driven by the temperature difference between the object and its surroundings), it exchanges thermal energy with its environment, and yet, the parameter R is treated as a constant. What assumptions must be made on the environment in order for this model to be valid?

3.4 Sinusoidal Functions

In this section, we will extend our repertoire of derivative rules to encompass sinusoidal functions: the sine and cosine functions, in particular. We will save the discussion of derivatives of inverse trigonometric functions for Section 3.8.

3.4.1 The Sine Function

Let us begin by recalling the shape of the graph of the sine and cosine functions, as shown in Figure 1.25. These graphs are reproduced in Figure 3.5, annotated with corresponding derivative properties (sign, trend, concavity). Notice that the cosine

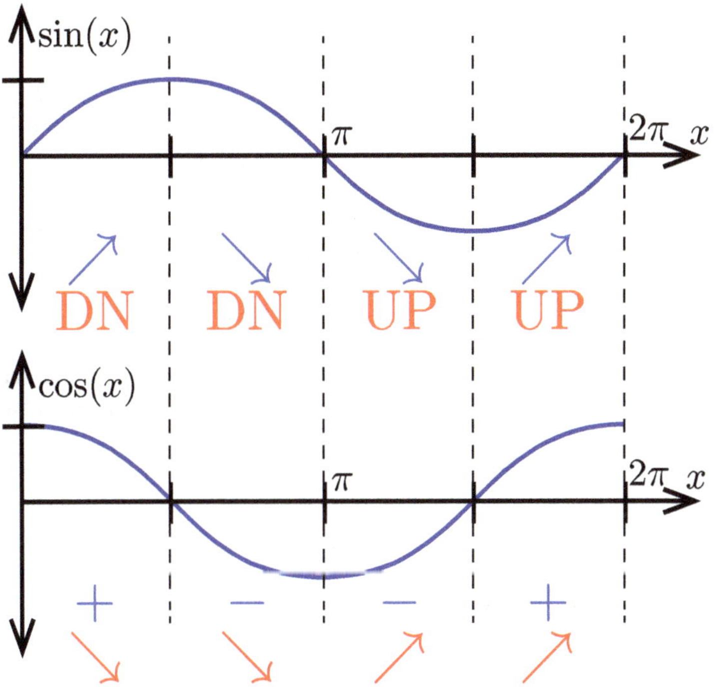

Figure 3.5: The graph of the sine and cosine functions, vertically aligned for comparison.

function is a perfect candidate to be the derivative of the sine function! The sine function is increasing on the intervals $\left[0, \frac{\pi}{2}\right]$ and $\left[\frac{3\pi}{2}, 2\pi\right]$; these are the identical intervals on which the cosine function is positive. Similarly, the sine function is decreasing on the interval $\left[\frac{\pi}{2}, \frac{3\pi}{2}\right]$, which is the same interval on which the cosine function is negative. Thus, the cosine function passes our first test to be viable as a derivative for the sine function: the sine function is increasing exactly when the cosine function is positive; the sine function is decreasing exactly when the cosine function is negative. Next, we must see if the concavity properties also match correctly.

The sine function is concave down on the interval $[0,\pi]$, which coincides exactly with the interval on which the cosine function is decreasing. Similarly, the sine function is concave up on the interval $[\pi,2\pi]$; again, this exactly matches the interval on which the cosine function is increasing. Therefore, the concavity of the sine function matches the trend of the cosine function in the correct ways if cosine is to be a viable candidate for the derivative of the sine function.

Verifying these properties does not *prove* that the derivative of sine is cosine: it is but a good indication. Our suspicions, however, turn out to be accurate, as is revealed by our next theorem. First, we have a lemma.

Lemma 3.4.1 The following equations hold:

$$\lim_{h\to 0}\frac{\cos(h)-1}{h} = 0$$

$$\lim_{h\to 0}\frac{\sin(h)}{h} = 1.$$

Proof. These limits can easily be verified using technology. The graph of the function

$$f(h)=\frac{\cos(h)-1}{h}$$

is shown in Figure 3.6. Though undefined at the origin, it is clear that the graph "passes through" the origin, so that the limit as $h\to 0$ of $f(h)$ should equal 0, as stated in the lemma.

Similarly, the graph of the function

$$g(h)=\frac{\sin(h)}{h}$$

is shown in Figure 3.7. Again, $g(0)$ is undefined, yet the limit Again, $g(0)$ is undefined, yet the limit of $g(h)$ as h approaches zero is clearly equal to 1, as we are trying to show. ■

We note that our proof of Lemma 3.4.1 is not a rigorous one, as it does not utilize the ε–δ definition of the limit; it does, however, suffice for our present purpose, as we will use this lemma in proving our next two theorems.

Theorem 3.4.2 The derivative of the sine function is

$$\frac{d}{dx}[\sin(x)]=\cos(x). \tag{3.39}$$

R **Warning!** Theorem 3.4.2 is *only valid* if the sine and cosine functions are interpreted in *radian mode*! (In fact, this is the *only* way to think about sine

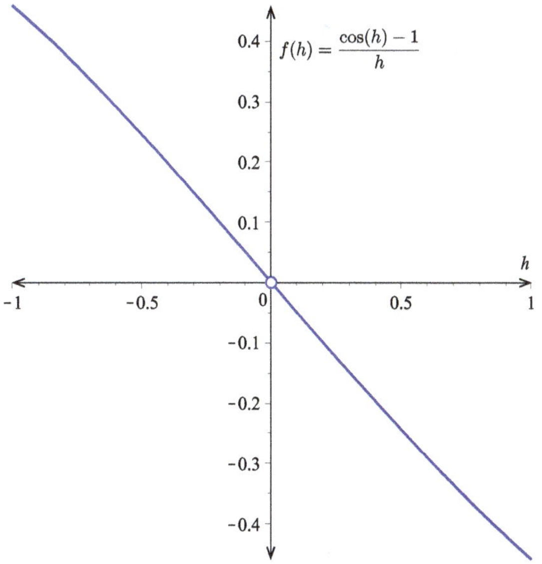

Figure 3.6: Graph of the function $f(h) = (\cos(h) - 1)/h$.

and cosine—exclusively in radians. Radians are actually dimensionless—they are not units like meters and seconds and kilowatts. This way, the sine and cosine of so many radians similarly returns a dimensionless result.)

Proof. To prove this derivative rule, we must again turn to the limit definition of the derivative. Setting $f(x) = \sin(x)$, we may write

$$\begin{aligned} f'(x) &= \lim_{h\to 0} \frac{\sin(x+h) - \sin(x)}{h} \\ &= \lim_{h\to 0} \frac{\sin(x)\cos(h) + \cos(x)\sin(h) - \sin(x)}{h} \\ &= \sin(x) \lim_{h\to 0} \frac{\cos(h) - 1}{h} + \cos(x) \lim_{h\to 0} \frac{\sin(h)}{h} \\ &= \sin(x). \end{aligned}$$

The second equality follows from the trigonometric identity

$$\sin(a \pm b) = \sin(a)\cos(b) \pm \cos(a)\sin(b).$$

The final equality in our equation chain follows from Lemma 3.4.1. ■

We have thus proved that the derivative of sine is cosine. We next turn to the derivative of the cosine function, but first, a brief example.

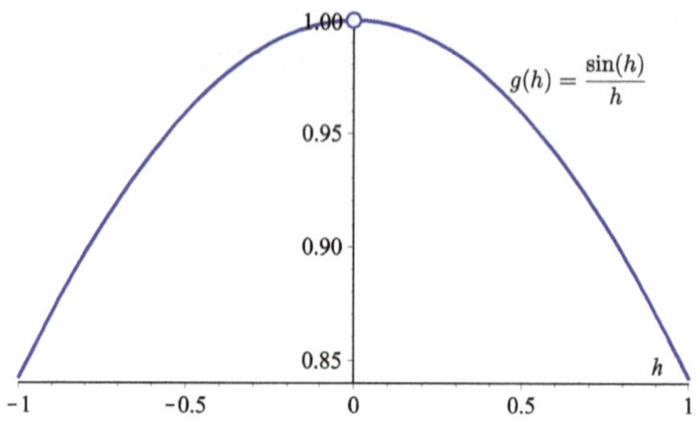

Figure 3.7: Graph of the function $g(h) = \sin(h)/h$.

■ **Example 3.21** Determine the derivative of the function

$$f(x) = e^{\sin(x)}.$$

For this function, we use the rule given by equation (3.22). The inside function is $u(x) = \sin(x)$, which has a known derivative of $u'(x) = \cos(x)$. Therefore

$$f'(x) = e^{\sin(x)} \cos(x).$$

■

3.4.2 The Cosine Function

If we draw the graph of the cosine function, we will see that the sine function is not a possible candidate for its derivative, as everything is off. (For example, when cosine is increasing, sine is negative, and so forth. Not a good candidate for the derivative.) By comparing the graphs of cosine and *negative* sine (which is just the sine function flipped up-side-down), one again sees the appropriate correlations between sign, trend, and concavity that one would expect between a function and its derivative. This observation is validated in our following theorem.

Theorem 3.4.3 The derivative of the cosine function is

$$\frac{d}{dx}[\cos(x)] = -\sin(x). \tag{3.40}$$

(R) **Warning!** Theorem 3.4.3 is *only valid* if the sine and cosine functions are interpreted in *radian mode*!

Proof. To prove this derivative rule, we again invoke the limit definition of the derivative. Setting $f(x) = \cos(x)$, we write

$$\begin{aligned} f'(x) &= \lim_{h\to 0} \frac{\cos(x+h) - \cos(x)}{h} \\ &= \lim_{h\to 0} \frac{\cos(x)\cos(h) - \sin(x)\sin(h) - \cos(x)}{h} \\ &= \cos(x) \lim_{h\to 0} \frac{\cos(h) - 1}{h} - \sin(x) \lim_{h\to 0} \frac{\sin(h)}{h} \\ &= -\sin(x). \end{aligned}$$

The second equality follows from the trigonometric identity

$$\cos(a \pm b) = \cos(a)\cos(b) \mp \sin(a)\sin(b);$$

the final equality follows again from Lemma 3.4.1. ■

■ **Example 3.22** We may differentiate the function $f(x) = \cos^2(x)$ using equation (3.19) with $k = 2$ and $u(x) = \cos(x)$. We obtain

$$f'(x) = -2\cos(x)\sin(x).$$

We promoted the minus sign to the front of the expression simply as a matter of style. ■

3.4.3 Generalized Forms

We next incorporate the chain rule into the rules for differentiating sine and cosine, to obtain the following result.

Theorem 3.4.4 Let $u(x)$ be a function with a known derivative $u'(x)$. Then

$$\frac{d}{dx}[\sin(u(x))] = \cos(u(x))u'(x), \tag{3.41}$$

$$\frac{d}{dx}[\cos(u(x))] = -\sin(u(x))u'(x). \tag{3.42}$$

The function u is known as the *inside function.*

This theorem follows from the chain rule, which we will discuss in Section 3.5, so we do not need to prove it here.

■ **Example 3.23** Compute the derivative of the function

$$f(x) = \sin\left(4x^5 + 9x^3 + 17x + \pi\right).$$

The inside function, in this example, is $u(x) = 4x^5 + 9x^3 + 17x + \pi$, with known derivative $u'(x) = 20x^4 + 27x^2 + 17$. Using equation (3.41), we therefore obtain

$$f'(x) = \cos\left(4x^5 + 9x^3 + 17x + \pi\right)\left(20x^4 + 27x^2 + 17\right).$$

Hence, we simply replaced sine with cosine and multiplied by the derivative of the inside function. ■

Example 3.24 Compute the derivative of the function

$$f(x) = \cos(\ln|x|).$$

We first identify the inside function as $u(x) = \ln|x|$, which has a known derivative $u'(x) = 1/x$. The derivative of the function f is therefore given by

$$f'(x) = \frac{-\sin(\ln|x|)}{x}.$$

Hence, we replaced cosine with negative sine, and then multiplied by the derivative of the inside. ■

3.4.4 Derivatives of Reciprocal Sinusoidal Functions

As an immediate corollary to the rules established so far, we can easily derive the derivative rules for the secant and cosecant functions, which, as you will recall, are defined by the relations

$$\sec(x) = \frac{1}{\cos(x)} \qquad \text{and} \qquad \csc(x) = \frac{1}{\sin(x)},$$

respectively.

(R) Recall that the notation $\cos^{-1}$ and $\sin^{-1}$ is reserved exclusively for the *inverse trigonometric functions* (from Definition 1.6.4 on page 75). This is why the reciprocals of sine and cosine are given special names: cosecant and secant.

Corollary 3.4.5 The derivatives of the secant and cosecant functions are given by the formulas

$$\frac{d}{dx}[\sec(x)] = \sec(x)\tan(x), \tag{3.43}$$

$$\frac{d}{dx}[\csc(x)] = -\csc(x)\cot(x), \tag{3.44}$$

respectively.

Proof. We will prove equation (3.43) and leave the proof of equation (3.44) as an exercise. The secant function may be equivalently expressed as

$$f(x) = \sec(x) = \frac{1}{\cos(x)} = [\cos(x)]^{-1}.$$

This falls under the guise of rule (3.19), with inside function $u(x) = \cos(x)$ and its derivative $u'(x) = -\sin(x)$. Therefore, following equation (3.19), we obtain

$$f'(x) = -1\,[\cos(x)]^{-2}(-\sin(x)) = \frac{\sin(x)}{\cos^2(x)} = \frac{1}{\cos(x)}\frac{\sin(x)}{\cos(x)}.$$

This is equivalent to the stated result; our proof is therefore complete. ■

The derivative rules (3.43) and (3.44) can also be generalized using the power of the chain rule; the result is stated in the following.

Theorem 3.4.6 Let $u(x)$ be a function with a known derivative $u'(x)$. Then

$$\frac{d}{dx}[\sec(u(x))] = \sec(u(x))\tan(u(x))u'(x), \tag{3.45}$$

$$\frac{d}{dx}[\csc(u(x))] = -\csc(u(x))\cot(u(x))u'(x). \tag{3.46}$$

The function u is known as the *inside function.*

The preceding theorem is yet another application of the chain rule; we will therefore not consider its proof, aside from our discussion of the chain rule in our next section.

■ **Example 3.25** Compute the derivative of $f(x) = \sec(3x^2+7)$.

Here, the inside function is $u(x) = 3x^2+7$, with derivative $u'(x) = 6x$. The derivative of the function f is therefore given by

$$f'(x) = 6x\sec(3x^2+7)\tan(3x^2+7).$$

■

■ **Example 3.26** Compute the derivative of $f(x) = \csc(e^{7x})$.

In this example, the inside function is $u(x) = e^{7x}$, with known derivative $u'(x) = 7e^{7x}$ (following rule (3.22)). The derivative of the function f is therefore

$$f'(x) = -7e^{7x}\csc(e^{7x})\cot(e^{7x}).$$

■

■ **Example 3.27** Compute the derivative of the function $f(x) = \sec^2(x)$.

In this example, it is $u(x) = \sec(x)$ that is the inside function. (The outside function is u^2.) This situation calls for rule (3.19), using $k = 2$. Since the derivative of secant is $u'(x) = \sec(x)\tan(x)$, we obtain

$$f'(x) = 2\underbrace{\sec(x)}_{u(x)^1}\underbrace{\sec(x)\tan(x)}_{u'(x)} = 2\sec^2(x)\tan(x).$$

■

Exercises

For Exercises 1–25, compute the indicated derivative, using the rules from Sections 3.2–3.4.

1. $f(x) = \sin(x^{100})$.
2. $f(x) = \cos(e^x)$.
3. $f(x) = \sin(\sin(x))$.
4. $f(x) = \cos(\sqrt{x})$.
5. $f(x) = \cos\left(\frac{1}{\sqrt{x}}\right)$.

6. $f(x) = \sin(x^2 + 1)$.
7. $f(x) = 80 + 20\cos\left(6\pi x + \frac{\pi}{4}\right)$.
8. $x(t) = M + A\cos(\omega t + \varphi_0)$.
9. $y(t) = 17\sin(16\pi t)$.
10. $h(w) = \sin(w^2 + k^2)$.
11. $f(x) = e^{\sin(x)+\cos(x)}$.
12. $f(x) = \sin^2(x) + \cos^2(x)$.
13. $f(x) = \sin^3(x) + \cos^3(x)$.
14. $f(x) = \sqrt{\sin(x)}$.
15. $f(x) = \ln|\cos(x)|$.
16. $f(x) = \pi^{\cos(x)}$.
17. $f(x) = \frac{3}{\sqrt{\sin(x)}}$.
18. $f(x) = \sqrt{\sin(8x)}$.
19. $f(x) = e^{\sin(3x+7)}$.
20. $f(x) = \sin^2(5x+8)$.
21. $f(x) = \sec(8\pi x + 9)$.
22. $f(x) = \sec^2(x) + \csc^2(x)$.
23. $f(x) = e^{\csc(x)}$.
24. $f(x) = \ln|\sec(x)|$.
25. $f(x) = \sec(\cos(x))$.

Problems

26. Prove equation (3.44).
27. Show that the functions $y_1(x) = \sin(\omega x)$ and $y_2(x) = \cos(\omega x)$ each satisfy the differential equation

$$y'' + \omega^2 y = 0.$$

(I.e., substitute y_1 and y_2 into this equation, by first computing the second derivative of each, and show that the equation is satisfied.)

28. Show that the functions $y_1(x) = \sinh(\omega x)$ and $y_2(x) = \cosh(\omega x)$ each satisfy the differential equation

$$y'' - \omega^2 y = 0.$$

(I.e., substitute y_1 and y_2 into this equation, by first computing the second derivative of each, and show that the equation is satisfied.)

29. On the summer and winter solstices, Fairbanks, Alaska will have 21 hours and 49 minutes of daylight (June 21, 2014) and 3 hours and 41 minutes of daylight (December 21, 2014), respectively. Let $S(d)$ represent the number of hours of daylight in Fairbanks, Alaska on the dth day of the year in 2014. The summer solstice occurred on day $d = 172$, and the winter solstice occurred on day $d = 355$.
    (a) Assuming the number of daylights to be a sinusoidal function, determine a formula for $S(d)$. Express this function in the standard form given by equation (1.46) on page 69.
    (b) Compute a formula for the derivative $S'(d)$.
    (c) Determine the day d_{eq} of the autumnal equinox according to this model, assuming that it is the day on which there are exactly 12 hours and the number of daylight hours is decreasing.
    (d) Compute $S'(d_{eq})$. Use this to approximate how many fewer hours, minutes, and seconds of daylight there will be one day later.

30. The voltage V, measured in volts, supplied by an electrical outlet is given as a function of time t, measured in seconds, by the relation

$$V(t) = V_0 \sin(\omega t),$$

where the parameters V_0 and ω are constants.
   (a) Determine the period T of the oscillations?
   (b) At what point on the interval $[0, T]$ is the voltage the greatest?
   (c) Determine a formula for $V'(t)$.
   (d) at what point on the interval $[0, T]$ is the voltage increasing at the greatest rate?

3.5 The Chain Rule

In this section, we finally write down and formalize the chain rule, and gain an understanding of why it is true. We first review each of the applications of the chain rule that we have already discussed, in order to identify the pattern that each of these rules has in common with the others.

3.5.1 Patterns

We have already seen the chain rule in action, as, with each new derivative rule, we introduced the corresponding chain-rule version of the rule, which had the chain rule already built in. Thus, we already know *how* to use the chain rule, at least for our current repertoire of functions, even though we haven't explored *what* the chain rule is. To begin, examine each of the basic rules along side their chain-rule generalizations, as summarized in Table 3.1.

(eqn)	basic rule	(eqn)	advanced rule
(3.18)	$\frac{d}{dx}\left[x^k\right] = kx^{k-1}$	(3.19)	$\frac{d}{dx}\left[u(x)^k\right] = k(u(x))^{k-1}u'(x)$
(3.21)	$\frac{d}{dx}\left[e^x\right] = e^x$	(3.22)	$\frac{d}{dx}\left[e^{u(x)}\right] = e^{u(x)}u'(x)$
(3.23)	$\frac{d}{dx}\left[a^x\right] = a^x \ln(a)$	(3.24)	$\frac{d}{dx}\left[a^{u(x)}\right] = a^{u(x)}\ln(a)u'(x)$
(3.25)	$\frac{d}{dx}[\ln\lvert x\rvert] = \frac{1}{x}$	(3.26)	$\frac{d}{dx}[\ln\lvert u(x)\rvert] = \frac{1}{u(x)}u'(x)$
(3.39)	$\frac{d}{dx}[\sin(x)] = \cos(x)$	(3.41)	$\frac{d}{dx}[\sin(u(x))] = \cos(u(x))u'(x)$
(3.40)	$\frac{d}{dx}[\cos(x)] = -\sin(x)$	(3.42)	$\frac{d}{dx}[\cos(u(x))] = -\sin(u(x))u'(x)$

Table 3.1: Recap of basic derivatives and their advanced versions.

For each of these rules, the advanced version operates the same way as the basic version, except the independent variable x in the basic version is replaced by an *inside function* $u(x)$ in the advanced version, and then, once the basic derivative rule has been executed—with respect to u—the result is multiplied by a single factor of $u'(x)$ in order to correct.

3.5.2 The Chain Rule

Each of the advanced derivative rules, given by equations (3.19), (3.22), (3.24), (3.26), (3.41), and (3.42) follow a pattern that can be formalized according to the following theorem.

Theorem 3.5.1 — The Chain Rule. Suppose that the function f is a composite of two differentiable functions v and u, so that $f = v \circ u$, i.e., $f(x) = v(u(x))$. Then

$$f'(x) = v'(u(x))u'(x). \tag{3.47}$$

Alternatively,

$$\frac{d}{dx}[v(u(x))] = v'(u(x))u'(x), \tag{3.48}$$

or, using Leibniz notation,

$$\frac{dv}{dx} = \frac{dv}{du}\frac{du}{dx}. \tag{3.49}$$

The logic behind the chain rule can be easily appreciated by examining equation (3.49): one can heuristically think of the "du" in the denominator of the derivative dv/du as canceling with the "du" in the numerator of the derivative du/dx. Such a cancellation, however, should only be regarded as a metaphorical one, because the individual terms in a derivative, be it dv/du or du/dx, are not independent algebraic agents; it is the whole expression dv/du or du/dx that must be regarded as a single entity. Nevertheless, this metaphorical cancellation indeed turns out to be the case, so one may use it to enliven the imagination.

(R) Suppose we have a double composite, i.e., suppose that $f = w \circ v \circ u$, so that $f(x) = w(v(u(x)))$. Then

$$f'(x) = w'(v(u(x)))v'(u(x))u'(x). \tag{3.50}$$

This result follows by double application of the chain rule. This process may be continued ad infinitum.

Proof. To prove Theorem 3.5.1, we rely on the limit-definition of the derivative, i.e., Definition 2.4.2. Given a function $f(x) = v(u(x))$, its derivative is given by equation (2.11), yielding

$$f'(x) = \lim_{h \to 0} \frac{f(x+h) - f(x)}{h} = \lim_{h \to 0} \frac{v(u(x+h)) - v(u(x))}{h}.$$

The trick, here, is to massage this expression into a familiar form, in particular, something involving the derivative of v with respect to u, and something involving the derivative of u with respect to x. To achieve this, we define the parameter k by the relation

$$k = u(x+h) - u(x).$$

Due to the continuity of the function u, we know that $k \to 0$ as $h \to 0$. (Observe that, for fixed x, k is a function of h.) The advantage here, however, is that we may express $v(u(x+h))$ as $v(u(x)+k)$. Therefore,

$$f'(x) = \lim_{h\to 0}\left(\frac{v(u(x)+k)-v(u(x))}{k}\cdot\frac{u(x+h)-u(x)}{h}\right).$$

(Note that

$$\frac{u(x+h)-u(x)}{k} = 1,$$

due to the definition of k.) As long as each separate limit exists, we may therefore use equation (1.60) to obtain

$$f'(x) = \left(\lim_{k\to 0}\frac{v(u(x)+k)-v(u(x))}{k}\right)\left(\lim_{h\to 0}\frac{u(x+h)-u(x)}{h}\right) = v'(u(x))u'(x).$$

This completes the proof. ■

We have already seen the chain rule in action, as the chain rule is precisely the mathematical device that makes the five basic derivatives plus, as recited in equations (3.6)–(3.10), work. Understanding the theoretical language of the chain rule, however, is by no means a redundancy, as the five basic derivatives plus are mere illustrations of the rule, and are by no means an exclusive listing of its applications. The chain rule allows one to apply its efficacy to new situations (suppose, for instance, you learn the derivative of $\tan(x)$, and wish to compute the derivative of $\tan(3\pi x+12)$?) and, in particular, situations in which there might not be a simple expression for the object you are studying (e.g., suppose you are analyzing complex stock data or rocket telemetry data that no simple expression describes).

■ **Example 3.28** Compute the derivative of $f(x) = \sin(3x^2+7x+\pi)$.

First, we identify the outer and inner functions. The outer function is given by $v(u) = \sin(u)$, and the inner function is given by $u(x) = 3x^2+7x+\pi$. With annotations, the function f is

$$f(x) = \sin(\underbrace{3x^2+7x+\pi}_{\text{inner, } u(x)}).$$

The chain rule tells us that

$$f'(x) = v'(u(x))u'(x).$$

Thus, we need to compute the derivative of the outer function v and the inner function u. These derivatives are

$$v'(u) = \cos(u) \qquad \text{and} \qquad u'(x) = 6x+7.$$

Next, we evaluate v' at $u(x)$, thereby obtaining

$$f'(x) = \underbrace{\underbrace{\cos}_{v'}\underbrace{(3x^2+7x+\pi)}_{\text{inner, } u(x)}}_{v'(u(x))} \cdot \underbrace{(6x+7)}_{u'(x)}.$$

And so it goes. ■

■ **Example 3.29** Use the chain rule to compute the derivative of the function $f(x) = \sqrt{\sin(x)+x}$.

Here, the inner function is $u(x) = \sin(x)+x$, and the outer function is $v(u) = \sqrt{u}$. The derivatives of the inner and outer function are

$$u'(x) = \cos(x)+1 \qquad \text{and} \qquad v'(u) = \frac{1}{2\sqrt{u}},$$

respectively. (Recall that $\sqrt{u} = u^{1/2}$. Therefore its derivative is $(1/2)u^{-1/2}$.) Thus, the derivative of the composite function f is given by

$$f'(x) = \underbrace{\frac{1}{2\sqrt{\underbrace{\sin(x)+x}_{\text{inner, } u(x)}}}}_{v'(u(x))} \cdot \underbrace{(\cos(x)+1)}_{u'(x)},$$

as obtained by applying the chain rule. ■

■ **Example 3.30** A *reduced-gravity aircraft* is an astronaut-training aircraft that flies at a 45° upward angle at nearly the speed of sound, and then cuts its engine and free falls along a parabolic trajectory for approximately 25 seconds, simulating a zero-gravity environment, as shown in Figure 3.8. For one such flight, the aircraft cuts its engines as it passes through an altitude of 9500 m, with an ascent rate of 122.5 m/s. Its altitude x, as measured in meters, is given as a function of time t, as measured in seconds, by the relation

$$x(t) = 9500 + 122.5t - 4.9t^2,$$

for $t \in [0, 25]$. Meanwhile, atmospheric pressure at various altitudes is given in Table 3.2. A gauge in the aircraft shows the ambient atmospheric pressure.

altitude (m)	9500	9600	9700	9800	9900
pressure (kPa)	28.523	28.095	27.673	27.255	26.843
altitude (m)	10000	10100	10200	10300	10400
pressure (kPa)	26.436	26.034	25.636	25.244	24.857

Table 3.2: Atmospheric pressure (kPa) at various altitudes (m).

How quickly was this gauge changing when the aircraft was 10 seconds into its zero-gravity maneuver?

Figure 3.8: Mercury astronauts in a C-131 aircraft in a simulated zero-gravity environment in 1959. Photograph courtesy NASA.

To solve this, we first recognize that we have two functions: the plane's altitude x is given as a function of time t, and the pressure p is given as a function of altitude x. Let f represent the ambient atmospheric pressure of the aircraft as a function of the number of seconds t into the zero-gravity maneuver, so that $f = p \circ x$, i.e., $f(t) = p(x(t))$.

Next, we use the chain rule to write

$$f'(t) = p'(x(t))x'(t).$$

Ten seconds into the flight, the altitude of the aircraft is given by $x(10) = 10,235$ m. The vertical velocity of the aircraft is given by

$$x'(t) = 122.5 - 9.8t;$$

the aircraft's ascent rate after ten seconds is therefore given by $x'(10) = 24.5$ m/s. Since we only have atmospheric data for the air pressure, we must approximate the derivative p' at an altitude of 10,235 m. The best we can do is

$$p'(10235) \approx \frac{p(10300) - p(10200)}{100} = \frac{25.244 - 25.636}{100} = -0.00392 \text{ kPa/m}.$$

Therefore, the pressure gauge in the cockpit should be changing at a rate of

$$f'(10) = p'(10235)x'(10) \approx (-0.00392 \text{ kPa/m})(24.5 \text{ m/s}) = -0.09604.$$

Thus, the pressure is decreasing at a rate of approximately 96.04 Pascals per second, when the aircraft is ten seconds into its zero-velocity maneuver. ∎

The preceding example illustrates how a composite function may arise in application, in which each constituent of the composite is given by a different form (e.g., equation and data table). It also demonstrates how one must integrate these various pieces of information into a coherent ensemble, to which the chain rule may be applied.

3.5.3 Quadratic Approximations

In Section 2.4, we first discussed the tangent line to the graph of a function f at a point $x = a$, which is given by equation (2.12). This equation is related to the point–slope formula for a line, as is revealed in the following annotated version of the tangent-line equation:

$$\underbrace{T(x)}_{y} = \underbrace{f(a)}_{y_0} + \underbrace{f'(a)}_{m}\underbrace{(x-a)}_{\Delta x}.$$

As we explored in Section 2.5, the tangent line provides a linear approximation to the function f for points near $x = a$. In this paragraph, we will discuss the best-fit *quadratic* function that approximates the function f near the point $x = a$. But first, we formalize two important properties of the tangent line.

Proposition 3.5.2 Consider a differentiable function f and the tangent-line function T to the graph at the point $x = a$, as given by equation (2.12). The tangent-line function T is the unique linear function that satisfies the conditions

$$T(a) = f(a), \quad (3.51)$$
$$T'(a) = f'(a). \quad (3.52)$$

(R) Proposition 3.5.2 tells us that the tangent line is the only linear function that intersects the graph of the function f at the point $x = a$ *and* has a slope equal to the derivative of f at the point a.

The preceding proposition follows is easily proven. Using equation (2.12), we observe that

$$T(a) = f(a) + f'(a)(a-a) = f(a).$$

Moreover,

$$T'(x) = f'(a);$$

in particular, $T'(a) = f'(a)$.

When the tangent line is viewed in terms of the conditions (3.51) and (3.52), a natural question arises: why stop at just a linear approximation to the graph? Indeed, we can generalize these ideas to determine a *best-fit quadratic* function to the graph.

Definition 3.5.1 Given a function f that is twice differentiable at a point $x = a$, then the function T_2 defined by

$$T_2(x) = f(a) + f'(a)(x-a) + \frac{f''(a)}{2}(x-a)^2 \tag{3.53}$$

is called the *second-order Taylor polynomial* for $f(x)$ *centered at* $x = a$.

(The tangent-line function may similarly be called the *first-order Taylor polynomial* for $f(x)$ centered at $x = a$. We may also add a subscript 1, writing T_1 in place of T, for clarity.)

(R) Notice that the functions f, f', and f'' are *evaluated* at the point $x = a$ in order to generate the coefficients of the constant, linear $(x-a)$, and quadratic $(x-a)^2$ terms of the polynomials. The function (3.53) truly is a quadratic function in the variable x, regardless of how exotic and untamable the function f.

Proposition 3.5.3 Given a function f that is twice differentiable at a point $x = a$, then its second-order Taylor polynomial T_2 is the unique quadratic polynomial that satisfies the conditions

$$\begin{aligned} T_2(a) &= f(a), & (3.54)\\ T_2'(a) &= f'(a), & (3.55)\\ T_2''(a) &= f''(a), & (3.56) \end{aligned}$$

i.e., the function T_2 is the *best-fit* quadratic approximation to f near the point $x = a$.

Proof. To prove the proposition, we must simply evaluate the function (3.53) and its first two derivatives at the point $x = a$ to show that these values agree with equations (3.54)–(3.56). First, $T_2(a) = f(a)$, as the second and third terms vanish when evaluated at $x = a$. Next, consider the derivative

$$T_2'(x) = f'(a) + f''(a)(x-a).$$

Notice that this derivative is obtained using the chain rule, with $u(x) = (x-a)$. We have $u'(x) = 1$, so that

$$\frac{d}{dx}\left[(x-a)^2\right] = 2(x-a).$$

Evaluating the derivative at $x = a$, we immediately confirm that $T_2'(a) = f'(a)$. Taking a second derivative, we obtain

$$T_2''(x) = f''(a).$$

Since the second derivative is a constant function, its value at $x = a$ clearly matches the requirement (3.56). Since a general quadratic equation has at most three independent parameters (the coefficients), and since we have placed three conditions on such a quadratic, our given quadratic function T_2 must be unique. The proof is therefore complete. ■

(R) Notice that the first two terms of the quadratic approximation T_2 are precisely the linear approximation given by the tangent-line function T_1. Thus, the quadratic approximation T_2 *is* the linear approximation *plus* a quadratic term:

$$T_2(x) = \underbrace{f(a) + f'(a)(x-a)}_{T_1(x)} + \frac{f''(a)}{2}(x-a)^2.$$

■ **Example 3.31** Determine the tangent-line and quadratic approximations to the graph of the function $f(x) = e^{1.2x}$ at the point $x = 0$. Use use to approximate the value of $e^{0.12}$. Compare the approximations with the (approximate) "exact" value as obtained from a calculator (i.e., compare with the first 10 digits of the exact value).

To utilize equations (2.12) and (3.53), we require the following ingredients:

$$f(0), \qquad f'(0), \qquad \text{and} \qquad f''(0).$$

Since

$$f(x) = e^{1.2x}, \qquad f'(x) = 1.2e^{1.2x}, \qquad \text{and} \qquad f''(x) = 1.44e^{1.2x},$$

we obtain

$$f(0) = 1, \qquad f'(0) = 1.2, \qquad \text{and} \qquad f''(0) = 1.44.$$

Using these numbers (along with $a = 0$) in the expressions (2.12) and (3.53), we obtain the results

$$\begin{aligned} T_1(x) &= 1 + 1.2x, \\ T_2(x) &= 1 + 1.2x + 0.72x^2. \end{aligned}$$

The function f is shown along with its first- and second-order Taylor polynomials in Figure 3.9. Notice that the function T_2 is a better approximation to $f(x)$ near the point $x = 0$.

Next, we approximate the value of $e^{0.12}$ by evaluating our linear and quadratic approximations at $x = 0.1$; we obtain

$$T_1(0.1) = 1.12 \qquad \text{and} \qquad T_2(0.1) = 1.1272,$$

respectively. The actual value of $e^{0.12}$ (to 16 significant figures) is

$$e^{0.12} \approx 1.127496851579376.$$

Clearly, the quadratic approximation is closer to the exact value than the linear approximation is. ■

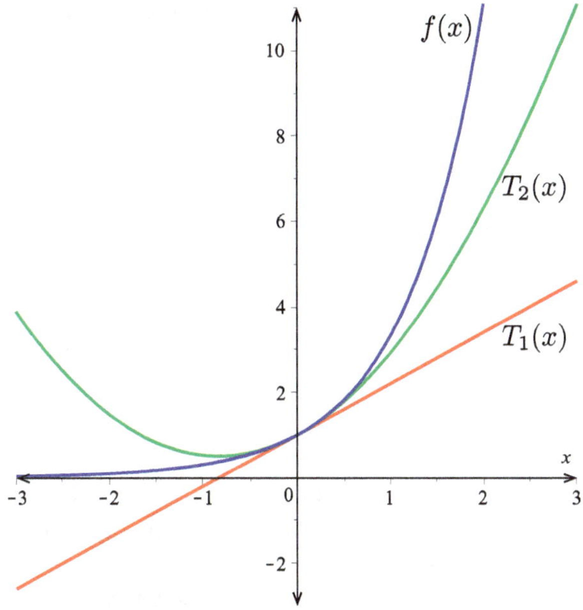

Figure 3.9: The function $f(x) = e^{1.2x}$ and its linear T_1 and quadratic T_2 approximations.

Exercises

For Exercises 1–10, use the chain rule to compute derivative of the given function. Identify the inner function u and the outer function v for each.

1. $f(x) = \sin(\sqrt{x})$.
2. $f(x) = \left(\sin(x) + \frac{1}{x}\right)^{80}$.
3. $f(x) = \ln|e^x|$.
4. $f(x) = \sec(x)$.
5. $f(x) = \sin^2(x)$.
6. $f(x) = \sin(e^x)$.
7. $f(x) = \sin(\sin(x))$.
8. $f(x) = e^{\cos(x)}$.
9. $f(x) = 8\sin(3\pi x + 2)$.
10. $f(x) = -\ln|\cos(x)|$.

For Exercises 11–20, compute the derivative of the given function.

11. $f(x) = \left(\sin(3x) + x^3\right)^{100}$.
12. $f(x) = e^{8\sin(8x)+2}$.
13. $f(x) = \sqrt{\sin(6x+7)}$.
14. $f(x) = \dfrac{1}{(\sin(x)+\cos(x))^2}$.
15. $f(x) = \dfrac{1}{\sin(8x+9)}$.
16. $f(x) = \sin^3\left(\sqrt{3x^2+2x+6}\right)$.
17. $f(x) = \ln\left|\dfrac{1}{\sqrt{x}}\right|$.
18. $f(x) = \sin(\sin(\sin(x)))$.
19. $f(x) = e^{e^x}$.
20. $f(x) = e^{\pi+1}$.

For Exercises (21)–(30), determine the quadratic approximation T_2 for the given function about the point $x = a$. Graph the function concurrently with its linear and quadratic approximations.

21. $f(x) = \sin(x)$; $a = 0$.
22. $f(x) = \cos(x)$; $a = 0$.
23. $f(x) = x^4 + 5x^2 + 6x + 7$; $a = 0$.
24. $f(x) = \sqrt{x}$; $a = 9$.
25. $f(x) = \sqrt[3]{x}$; $a = 8$.
26. $f(x) = \frac{1}{x}$; $a = 2$.
27. $f(x) = \cosh(3x)$; $a = 0$.
28. $f(x) = \tan(x)$; $a = \frac{\pi}{4}$.
29. $f(x) = \ln(x)$; $a = 1$.
30. $f(x) = e^{1.2x}$; $a = 1$.

Problems

31. Consider the function $f(x) = x^4$.
    (a) Calculate $f'(x)$ the normal way.
    (b) Rewrite $f(x) = (x^2)^2$ and identify the inner function as $u(x) = x^2$. Compute the derivative using the chain rule and show that you recover the same result as in part (a).
32. Consider the functions u and v, as plotted concurrently in Figure 3.10. Define the functions

$$f(x) = u(v(x)), \qquad g(x) = v(u(x)), \qquad \text{and} \qquad h(x) = u(x) + v(x).$$

    Estimate the value of each of the following derivatives.
    (a) $f'(-2)$.
    (b) $f'(2)$.
    (c) $g'(1)$.
    (d) $g'(4)$.
    (e) $h'(-1)$.
33. Determine an equation for the tangent line to the function

$$f(x) = \sin\left(\frac{\pi e^{0.5x}}{3}\right)$$

    at the point $x = 0$. Where does this tangent line intersect the x-axis?
34. In this problem, we will compare the quadratic approximations for the cosine and hyperbolic cosine functions.
    (a) Determine the quadratic approximations for the functions $\cos(x)$ and $\cosh(x)$ about $x = 0$. How are they alike? How are they different?
    (b) Approximate $\cos(0.1)$ and $\cosh(0.1)$. Determine the error for each. (The error is the absolute value of the difference between the true value and the approximation; e.g., $|\cos(0.1) - T_2(0.1)|$.)
    (c) Approximate $\cos(0.01)$ and $\cosh(0.01)$. Determine the error for each. How does the error difference than the error from part (b)?
35. The hydrostatic pressure P (kPa) at a depth of x m below the surface of the ocean is described by the function

$$P(x) = P_0 + \rho g x,$$

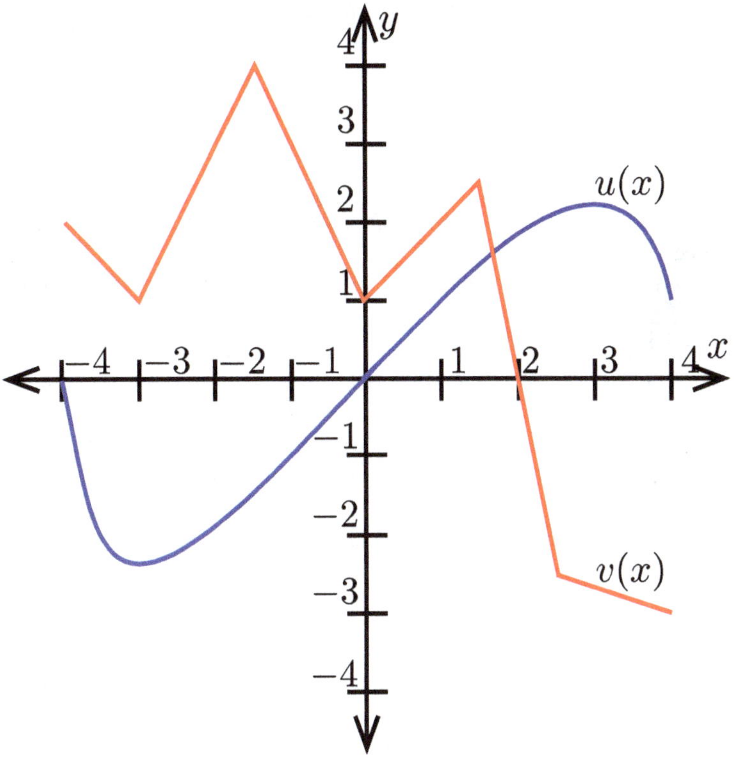

Figure 3.10: Graphs of the functions u and v; Problem 32.

where ρ is the density of water ($\rho = 1000$ kg/m^3), g is acceleration due to gravity ($g = 9.8$ m/s^2), and P_0 is the atmospheric pressure at sea level ($P_0 = 101$ kPa). A deep-sea diver is located at depth

$$x(t) = 3500(1 - e^{-0.0028t}) \text{ m},$$

at t s after the start of her dive.

(a) What is the diver's depth after five minutes?
(b) What is the hydrostatic pressure at that depth?
(c) At what rate is the diver's depth increasing five minutes into the dive?
(d) At what rate is the hydrostatic pressure, as experienced by the diver, changing at that moment in time?

36. The voltage drop E (volts) across a resistor with resistance R (Ohms) is given as a function of the current I (amps) by *Ohm's law*:

$$E = IR.$$

Suppose that the current across a 5200 Ohm resistor is given by $I(t) =$

$23(1 - e^{-0.2t})$ amps. Determine the rate at which the voltage drop is initially increasing.

37. In the special theory of relativity, the *relativistic energy* E of a particle with mass m is given by

$$E(\beta) = \frac{mc^2}{\sqrt{1-\beta^2}}, \qquad (3.57)$$

where C is the speed of light and β is the particle's speed, as a fraction of the speed of light (i.e., $\beta = v/c$).

   (a) Determine a formula for $E'(\beta)$. What does this represent in practical terms?
   (b) Suppose that a particle with mass m has a velocity of $v = c/2$ and is accelerating at a rate of $0.01c$ per second. Determine the rate in which the particle's relativistic energy is increasing at that moment in time.
   (c) Determine a formula for $E''(\beta)$.
   (d) Determine the quadratic approximation for the function $E(\beta)$ about $\beta = 0$.

38. Determine the values of the constants c_0, c_1, c_2, and c_3 so that the cubic equation

$$T_3(x) = c_0 + c_1(x-a) + c_2(x-a)^2 + c_3(x-a)^3$$

satisfies equations (3.54)–(3.56) as well as the condition $f'''(a) = T_3'''(a)$, for a given thrice-differentiable function f and a point $x = a$. The function T_3 is called the *third-order Taylor polynomial* for f about $x = a$.

39. Use the quadratic approximation about $x = 0$ to derive the first few terms of the *binomial series expansion*

$$(1+x)^p = 1 + px + \frac{p(p-1)}{2}x^2 + \cdots.$$

3.6 The Product and Quotient Rules

With the aid of a few simple techniques, we have learned how to compute the derivative of a vast array of seemingly complex functions. Remaining elusive to our powers, however, are such simple functions as

$$xe^x, \qquad e^x \sin(x), \qquad \text{and} \qquad \frac{\sin(x)}{\cos(x)}.$$

None of our known derivative rules are amenable here. In these several examples, we have functions that are constructed by taking the product or quotient of simpler functions, which we already know the derivative of. A natural question arises: can we, in a simple way, use these simpler functions and their known derivatives to construct the derivatives of these products and quotients? Such is the topic of the current paragraph. In this section, we learn two new, albeit closely related, rules for differentiating products and quotients of functions.

3.6.1 The Product Rule

Our first rule teaches us how to differentiate products of functions.

Theorem 3.6.1 — The Product Rule. Suppose the functions u and v are differentiable, and suppose the function f is given by the product $f(x) = u(x)v(x)$. Then the derivative of the product is given by

$$f'(x) = u(x)v'(x) + u'(x)v(x). \tag{3.58}$$

Alternatively, in Leibniz notation,

$$\frac{d}{dx}[u(x)v(x)] = u(x)v'(x) + u'(x)v(x).$$

R Theorem 3.6.1 tells us that the derivative of the product of two functions—the *first* and the *second* function—is given by, "the first times the derivative of the second, plus the second times the derivative of the first." This useful mnemonic can aid in remembering the product rule.

Proof. We present a novel formal proof of the product rule. Suppose we are given a function $f(x) = u(x)v(x)$. Taking the logarithm of both sides, we obtain

$$\ln|f(x)| = \ln|u(x)v(x)| = \ln|u(x)| + \ln|v(x)|.$$

(Notice we used the logarithmic property, given by equation (1.19), that allows us to split this product into a sum of logarithms.) Next, we may compute the derivative of each of these individual terms using equation (3.10) (which itself is the derivative rule for $\ln(x)$ coupled with the chain rule), obtaining

$$\frac{f'(x)}{f(x)} = \frac{u'(x)}{u(x)} + \frac{v'(x)}{v(x)}.$$

Recognizing that $f(x) = u(x)v(x)$, and multiplying both sides by this quantity, we recover our result. ■

■ **Example 3.32** Compute the derivative of the function $f(x) = xe^x$. Identifying the first and second function (which is which is rather irrelevant), we may annotate the function as

$$f(x) = \underbrace{x}_{u(x)}\underbrace{e^x}_{v(x)}.$$

Following the product rule, the derivative is given by the first times the derivative of the second, plus the second times the derivative of the first, i.e., by

$$f'(x) = \underbrace{x}_{u(x)}\underbrace{e^x}_{v'(x)} + \underbrace{1}_{u'(x)}\underbrace{e^x}_{v(x)}.$$

Sans annotation, the derivative is simply $f'(x) = xe^x + e^x$, or, equivalently, $f'(x) = e^x(x+1)$. ■

■ **Example 3.33** Next, let us consider the function $f(x) = e^{3x}\sin(7x)$. We may identify $u(x) = e^{3x}$ and $v(x) = \sin(7x)$. Thus, the derivative is given by

$$f'(x) = \underbrace{e^{3x}}_{u(x)}\underbrace{7\cos(7x)}_{v'(x)} + \underbrace{3e^{3x}}_{u'(x)}\underbrace{\sin(7x)}_{v(x)},$$

i.e., by $f'(x) = 7e^{3x}\cos(7x) + 3e^{3x}\sin(7x)$. ■

3.6.2 The Quotient Rule

A closely related rule, the quotient rule teaches us how to find the derivative of the quotient of two functions.

Theorem 3.6.2 — The Quotient Rule. Suppose the functions u and v are differentiable and that the function f is given as the quotient $f(x) = \frac{u(x)}{v(x)}$. Then the derivative of the quotient function f is given by

$$f'(x) = \frac{u'(x)v(x) - u(x)v'(x)}{v(x)^2}, \tag{3.59}$$

whenever $v(x) \neq 0$. In Leibniz notation,

$$\frac{d}{dx}\left[\frac{u(x)}{v(x)}\right] = \frac{u'(x)v(x) - u(x)v'(x)}{v(x)^2}.$$

(R) Theorem 3.6.2 tells us that the derivative of the quotient of two functions—the *top* and the *bottom* function—is given by, "the bottom times the derivative of the top, minus the top times the derivative of the bottom, all over the bottom squared." This useful mnemonic can aid in remembering the quotient rule.

Proof. The quotient rule is not really a new rule: it can be obtained by using the produce rule coupled with the chain rule. We will proceed in this fashion in our proof. Let us arrange the function f as a product, in the fashion of

$$f(x) = \frac{u(x)}{v(x)} = u(x) \cdot \frac{1}{v(x)}.$$

Since we are already using the functions u and v, let us use the names u and w to represent the first and second factors in this product, i.e.,

$$f(x) = \underbrace{u(x)}_{u(x)} \underbrace{\frac{1}{v(x)}}_{w(x)}.$$

(Conveniently, the numerator u of the quotient u/v is also the first function of the product uw.)

Now, let us observe, a la the chain rule, that the derivative of $w(x) = v(x)^{-1}$ is given by

$$w'(x) = -v(x)^{-2} v'(x) = \frac{-v'(x)}{v(x)^2}.$$

Therefore, the derivative of the quotient $f = u/v$ (which we are treating as a product $f = uw$), is given by the product rule ($\frac{d}{dx}[uw] = uw' + u'w$) as

$$f'(x) = \underbrace{u(x)}_{u(x)} \underbrace{\left(\frac{-v'(x)}{v(x)^2} \right)}_{w'(x)} + \underbrace{u'(x)}_{u'(x)} \underbrace{\frac{1}{v(x)}}_{w(x)}.$$

Alternating the order of these two terms and simplifying, we see that

$$f'(x) = \frac{u'(x)}{v(x)} - \frac{u(x)v'(x)}{v(x)^2} = \frac{u'(x)v(x)}{v(x)^2} - \frac{u(x)v'(x)}{v(x)^2}.$$

Combining both terms, which now have a common denominator, as a single fraction, we recover our result. ■

Our first example of the quotient rule constitutes an important derivative rule in its own right, and is therefore given by the following theorem.

Theorem 3.6.3 The derivative of the tangent function is given by

$$\frac{d}{dx}[\tan(x)] = \sec^2(x), \tag{3.60}$$

for all points on its domain.

Proof. To differentiate the function $f(x) = \tan(x)$, we first write it as a quotient

$$f(x) = \frac{\sin(x)}{\cos(x)}.$$

We identify the top and bottom functions of the quotient as $u(x) = \sin(x)$ and $v(x) = \cos(x)$. Next, applying the quotient rule, we obtain

$$f'(x) = \frac{\underbrace{\cos(x)}_{u'(x)}\underbrace{\cos(x)}_{v(x)} - \underbrace{\sin(x)}_{u(x)}\underbrace{(-\sin(x))}_{v'(x)}}{\underbrace{\cos^2(x)}_{v(x)^2}} = \frac{\cos^2 x + \sin^2 x}{\cos^2 x} = \sec^2 x.$$

The last equality follows due to the identity $\sin^2 x + \cos^2 x = 1$ and due to the definition $\sec(x) = \frac{1}{\cos(x)}$. ■

Corollary 3.6.4 The derivative of the function $f(x) = \tan(u(x))$ is given by

$$\frac{d}{dx}[\tan(u(x))] = \sec^2(u(x))u'(x). \tag{3.61}$$

Proof. This corollary follows immediately by applying the chain rule (as given in Theorem 3.5.1) to Theorem 3.6.3. ■

Exercises

For Exercises 1–20, compute the derivative of the given function.

1. $f(x) = \sin(x)\cos(x)$.
2. $f(x) = x\sin(x)$.
3. $f(x) = x\ln|x|$.
4. $f(x) = \dfrac{x}{\sin(x)}$.
5. $f(x) = \dfrac{\sin(x)}{x}$.
6. $f(x) = \dfrac{\ln|x|}{x}$.
7. $f(x) = x^8\sin\left(\sqrt{x}\right)$.
8. $f(x) = e^{3x+7}\sin(2\pi x)$.
9. $f(x) = x^6\sin(x)$.
10. $f(x) = x\ln|x|$.
11. $f(x) = \sin^4(x)\cos^8(x)$.
12. $f(x) = \sin^4(3x+7)\cos^8(9x+2)$.
13. $f(x) = xe^{e^x}$.
14. $f(x) = \dfrac{\sqrt{8x+3}}{\sin(x)}$.
15. $f(x) = \dfrac{\sin(4\pi x+12)}{\ln|x^8|}$.
16. $f(x) = \dfrac{e^{\sin(9x)}}{x^3}$.
17. $f(x) = \sin(\sin(x))\sin(x)$.
18. $f(x) = \dfrac{\tan(x)}{\sin(x)}$.
19. $f(x) = \sin^2(x) + \cos^2(x)$.
20. $f(x) = \ln\left|x^6\sin^2(3x+7)\right|$.

Problems

21. Prove the derivative rule (3.34) on page 205.
22. Determine the derivative of $f(x) = \cot(x)$. Then use the chain rule to derive a formula for the derivative of the function $f(x) = \cot(u(x))$.

23. A power surge can be modeled by the function

$$P(t) = ate^{-bt}.$$

(a) Draw a sketch of the function P for $t \geq 0$. How do the parameters a and b affect the shape of the graph?
(b) Compute $P'(t)$.
(c) Set $P'(t) = 0$ and solve for t. What point does this correspond to on the graph? What is the value of the function P at this point? What does that value correspond to in practical terms?

24. The position of a certain damped oscillator[1] (e.g., a block–spring system with damping or an automobile with shock absorbers) is given by

$$x(t) = e^{-t} \sin\left[\left(2+\sqrt{3}\right)t\right].$$

(Consider using an abbreviation $x(t) = e^{-t}\sin(\omega t)$, where $\omega = 2+\sqrt{3}$, if desired.)
(a) Determine a formula for the derivative $x'(t)$.
(b) Set the derivative x' equal to zero and solve for t. Determine all roots of the equation $x'(t) = 0$ on the interval $t \in [0,5]$. *Hint*: you may find the fact

$$\tan\left(\frac{5\pi}{12}\right) = 2+\sqrt{3}$$

useful in this endeavor. There are six roots on the interval $[0,5]$.
(c) What type of tangent line does the graph of x have at the t values corresponding to the roots of x' found in part (b)?
(d) Evaluate the function x at the six t-values found in part (b). Use these values to sketch a graph of the function x on the interval $[0,5]$.
(e) *Optional*: Use technology to plot the function $x(t)$ on the interval $[0,5]$ and compare with your result from part (d).

25. Suppose that the number of people N who buy a ticket to a certain concert is a function of the ticket price x.
(a) Explain why the revenue R from ticket sales can be modeled by the function $R(x) = xN(x)$.
(b) Show that $R'(x) = 0$ if and only if the condition

$$N'(x) = \frac{-N(x)}{x}$$

is satisfied.
(c) Suppose that $N(x) = 400 - 0.04x^2$. Determine the domain of the function N. What value of the ticket price x will result in the condition from part (b) being satisfied?

[1]Solutions of this form arise from the differential equation $mx'' + cx' + kx = 0$, where m is the mass of the block, c is the damping coefficient, and k is the spring constant. This differential equation may be derived from Newton's second law of motion.

26. Suppose that the fuel economy η (mpg) and the rate of fuel consumption R (gph) of a certain automobile is given as a function of its speed v (mph).
    (a) Explain why the relation

$$\eta(v) = \frac{v}{R(v)}$$

    holds. Complete a dimensional analysis to show that the units work out correctly.
    (b) Compute a formula for the derivative $\eta'(v)$. (This formula may involve the variable v and the functions R and R'.)
    (c) Show that $\eta'(v) = 0$ if and only if

$$R'(v) = \frac{R(v)}{v}.$$

    (d) Assume that the fuel consumption rate increases linearly with speed, so that $R(v) = av + b$.
27. Derive a formula for the *triple product rule*, i.e., suppose that $f(x) = u(x)v(x)w(x)$ and show that

$$f'(x) = uvw' + uv'w + u'vw. \qquad (3.62)$$

    Hint: First use the regular product rule to compute the derivative of the function $g(x) = v(x)w(x)$. Then apply the product rule again to the function $f(x) = u(x)g(x)$, using the derivative of the function g.
28. Use equation (3.62) to determine the derivative of the function

$$A(t) = t^3 e^{-0.5t} \sin(\pi t).$$

29. Consider a function in the form of

$$f(x) = u(x)^{v(x)}.$$

    (a) We will proceed in the same fashion as in the proof of Theorem 3.3.3. Show that the function f is equivalent to

$$f(x) = e^{v(x)\ln(u(x))}.$$

    (b) Use equation (3.22) to differentiate. Simplify your answer to show that

$$\frac{d}{dx}\left[u^v\right] = vu^{v-1}u' + u^v \ln|u| v'. \qquad (3.63)$$

    (c) Write out equation (3.63) for the special cases (i) $u(x)$ is a constant, and (ii) $v(x)$ is a constant. Which derivative rules did you uncover as a special case?

3.7 Implicit Differentiation

Our next topic for exploration introduces a new way of examining certain types of curves—curves, in particular, for which neither variable can be written as a function of the other variable. (At least, there is no *global* representation of these curves as the graph of a *function*.) They are, however, described by certain equations, and upon those equations we can perform calculus. Far from being merely an esoteric exercise for curve-loving erudite academics, *implicit differentiation* is a practical technique of calculus that loans itself to a variety of applications; in particular, it sets the stage for the algebraic ways in which we will handle related-rates problems in Section 4.3.

3.7.1 Explicit and Implicit Equations

We begin with some nomenclature.

> **Definition 3.7.1** An *explicit equation* is an equation in which one variable can be solved (without ambiguity) as an *explicit function* of the other variable.
> An *implicit equation* is an equation relating two (or more) variables, such that neither variable can be solved uniquely as a function of the other variable.

(R) The English-language definitions of the words explicit and implicit offer additional clarity:

1. *Explicit*—fully and clearly expressed; unequivocal.
2. *Implicit*—implied, rather than expressly stated.

Thus, an explicit equation is one for which the dependent variable may be fully and unequivocally expressed as a direct (explicit) function of the independent variable; e.g., in the relation

$$y = f(x),$$

the variable y is an *explicit function* of the variable x. There is no ambiguity: given x, we may uniquely calculate y.

In contrast, an implicit equation is a relation between the variables in which neither variable can be expressed explicitly as a function of the other; e.g., the equation

$$R(x,y) = 0$$

is an implicit equation. Here, R is a *function of two variables*, so that it represents some combination of the variables x and y.

> **Definition 3.7.2** The *graph* of an equation is the set of points that satisfy the given equation.

For example, given an explicit function $y = f(x)$, its graph is the set of coordinate pairs (x,y) such that $y = f(x)$; i.e., it is the set of points $(x, f(x))$ that is obtained as one varies x across the domain of the function f.

On the other hand, the graph of an implicit function $R(x,y) = 0$ is simply the set of coordinate pairs (x,y) such that each given pair satisfies the underlying equation.

R When considering implicit equations, it is common for mathematicians to say that y is an "implicit function" of x, or vice versa. They do not mean that y is *globally* a function of x, but rather that it may be described *locally* as a function of x. That is, different pieces of the curve might look like the graph of a function, even though the entire curve itself fails the vertical and horizontal line tests. For the avoidance of confusion, we shall eschew such vocabulary here and throughout, referring only to implicit *equations* and graphs of implicit equations, but never to "implicit functions," which aren't *really* functions at all.

As we have previously studied an arsenal of explicit functions, let us next consider graphs of a few example implicit equations.

■ **Example 3.34** The unit circle is described by the implicit equation

$$x^2 + y^2 = 1. \tag{3.64}$$

(But can't we solve this for y as a function of x, obtaining $y = \sqrt{1-x^2}$? No. This represents only *half* of the circle; the other half is described by the equation $y = -\sqrt{1-x^2}$.) The graph of equation (3.64) is shown in Figure 3.11. Notice that the

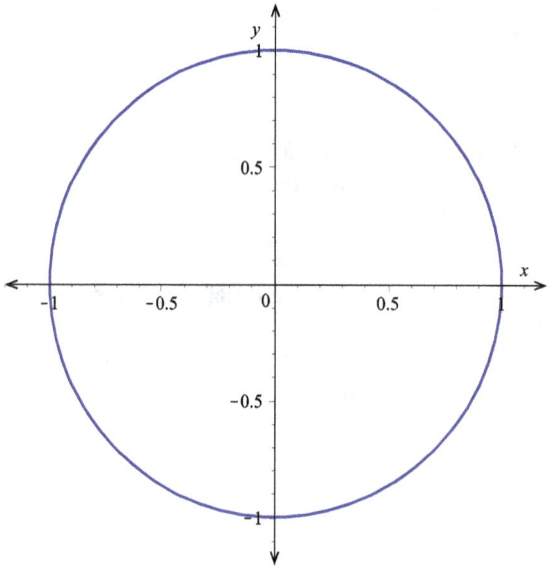

Figure 3.11: Graph of the implicit equation $x^2 + y^2 = 1$.

graph fails both the horizontal and vertical line tests. For this function, in particular, each x-value in the interval $(-1,1)$ corresponds to *two* y-values. Therefore—and as is the case for general implicit functions—an x-value alone cannot specify a unique point on the curve; points on the curve can only be specified, uniquely, by an (x,y) pair.

If one were to replace equation (3.64) with the equation

$$x^2 + y^2 = r^2,$$

for $r > 0$, one would instead obtain a circle of radius r. Otherwise, the picture is exactly the same. ∎

∎ Example 3.35 Next, we consider the general equation for an ellipse

$$\frac{x^2}{a^2} + \frac{y^2}{b^2} = 1, \tag{3.65}$$

where a and b are positive parameters. The x-intercepts occur at $x = \pm a$, whereas the y-intercepts occur at $x = \pm b$. The particular ellipse

$$\frac{x^2}{9} + \frac{y^2}{4} = 1 \tag{3.66}$$

is graphed in Figure 3.12.

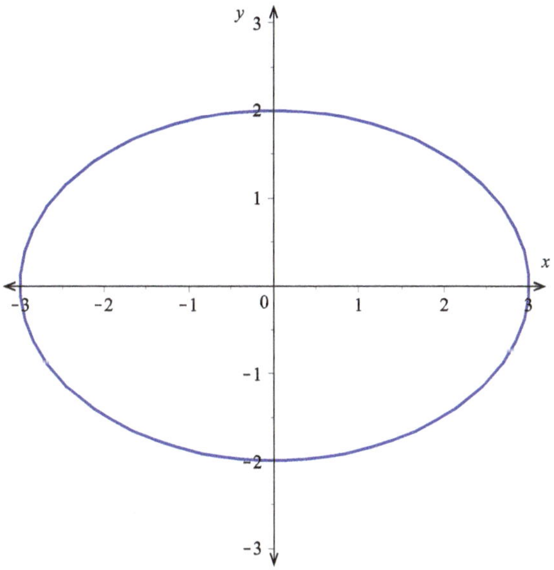

Figure 3.12: Graph of the implicit equation $x^2/9 + y^2/4 = 1$.

For this example ellipse, the *major axis* corresponds to the x-axis and the *minor axis* corresponds to the y-axis. (The major and minor axes of an ellipse are the long and short axes, respectively.) In general, the x-axis corresponds to the major axis whenever $a > b$, and it corresponds to the minor axis whenever $b < a$. ∎

■ **Example 3.36** Aside from circles and ellipses, there are many exotic curves that have been studied by mathematicians. One such example is known as a *cardioid*. A cardioid is any curve that is described by the implicit equation

$$\left(x^2+y^2-2ax\right)^2=4a^2(x^2+y^2), \tag{3.67}$$

for a given parameter a. (The word *cardioid* derives from the Greek word *καρδια* for "heart.") An example cardioid, for $a = 1$, is graphed in Figure 3.13. Notice that the graph of this cardioid exhibits a *cusp* at the origin. ■

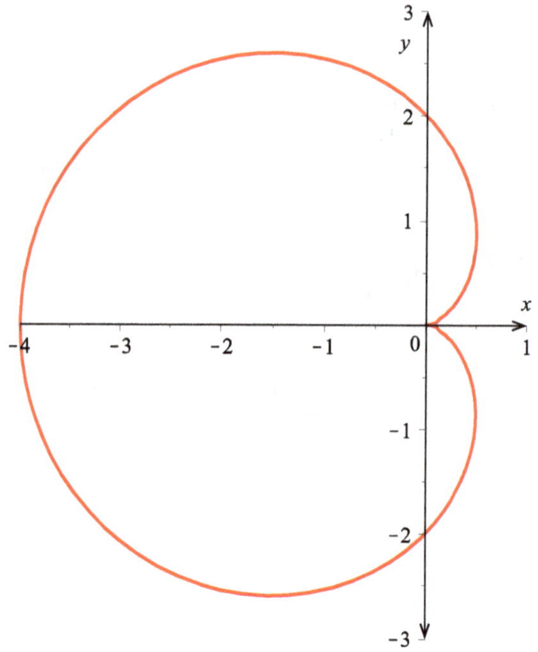

Figure 3.13: Graph of a cardioid; $a = 1$.

These few introductory examples should give the reader a feel for *what* an implicit equation is—it is an equation that involves both x and y, where neither variable can be solved explicitly in terms of the other. It requires little imagination to fathom the use of such creatures: why should, in real life, every relation between two variables be given to us where one variable is simply a straight-up function of the other? In a variety of applications, there is only some complicated relationship between two variables that must be studied. Even when one variable in an equation *can* be solved as an explicit function of the other, it is not always easiest to proceed in such a fashion.

3.7.2 Implicit Differentiation

In the study of differential calculus, we are interested in derivatives; such is the case here. But how can we go about computing the derivative of an implicit equation? We next take up the task of answering this question.

Given an implicit equation and its corresponding graph, the concept of a derivative is still applicable. The derivative at a point on the graph corresponds to a small change in the y-variable divided by a small change in the x-variable, in the limit as the change in x goes to zero. Thus, the ratio of infinitesimal quantities dy/dx may still be defined for points on the graph (except, of course, for points with a cusp or a vertical tangent line). Since the underlying equation is implicit, there will be, in general, multiple y-values associated to every x-value. The derivative therefore cannot depend on the value of x alone, as the value of x alone fails to specify a unique point on the curve. However we end up obtaining the derivative, it should depend on *both x and y*.

To proceed, we require the following useful observation: while y is not a function of x *globally*, when we consider a restricted *arc* of the curve, y can *locally* be thought of as a function of x. In order to differentiate an implicit equation, we may therefore employ the useful fiction that $y = y(x)$, and then differentiate as normal.

Definition 3.7.3 — Implicit Differentiation. Let $R(x,y) = 0$ be an implicit equation. To *differentiate* this equation *implicitly* with respect to the variable x, proceed as follows:

1. Think of the variable y as a secret function of x; i.e., in your mind, think of $y = y(x)$. Even though y is not really a function of x, this is a useful fiction.
2. Differentiate the equation using the ordinary derivative rules. The variable x, whenever encountered, is differentiated as normal. The variable $y = y(x)$ is differentiated a la the chain rule: differentiate the variable y as normal, and then multiply by a factor of y', which corresponds to the "derivative of the inside function."
3. The result is a linear algebraic equation for y'. Solve for y' in terms of the variables x and y.

In order to *differentiate* the equation $R(x,y) = 0$ *implicitly* with respect to the variable y, instead view x as a secret function of y, so that $x = x(y)$, and then differentiate as normal. When the variable y is encountered, differentiate the expression containing y as normal. Whenever the variable x is encountered, differentiate as normal and then multiply by the derivative of the inside, i.e., multiply by a factor of x', where x' represents dx/dy. Then solve for x' as a function of x and y.

These rules are best understood through means of example.

Example 3.37 Let us consider again the equation for the circle from Example 3.34. First, let's compute a formula for the derivative $y' = y'(x,y)$. Second, let us derive an equation for the tangent line to the circle at the point $(\sqrt{3}/2, 1/2)$.

We begin by differentiating equation (3.64) implicitly with respect to the variable x. The first term on the left, x^2, is simply a function of x; we therefore only need to take its derivative. The second term on the left-hand side, y^2 is a function of y; we must therefore regard $y = y(x)$ and compute the derivative using the chain rule. We obtain

$$2x + 2yy' = 0.$$

(Compare this with equation (3.19) on page 192: replace u with y and k with 2 to obtain the derivative of y^2.) Solving for the derivative $y' = dy/dx$, we obtain

$$y'(x,y) = \frac{-x}{y}.$$

The derivative for a point (x,y) on the unit circle can therefore be computed using this formula. In particular, consider the point $(\sqrt{3}/2, 1/2)$. (The reader will check that this point satisfies the equation of the circle $x^2 + y^2 = 1$.) The derivative at this point is equal to

$$y'\left(\frac{\sqrt{3}}{2}, \frac{1}{2}\right) = \frac{-\sqrt{3}/2}{1/2} = -\sqrt{3}.$$

The point–slope equation for a line may now be employed in order to obtain the equation for the tangent line at the point $(\sqrt{3}/2, 1/2)$, which is given by

$$T(x) = 2 - \sqrt{3}x.$$

Equation (3.64) is plotted along with this tangent line in Figure 3.14. ∎

∎ **Example 3.38** Consider the ellipse from equation (3.66). Differentiating implicitly with respect to x, we obtain

$$\frac{2x}{9} + \frac{y}{2}y' = 0.$$

Solving for the derivative, we have

$$y' = \frac{-4x}{9y}.$$

Next, let us look at the equation for the tangent line at the point $(1.5, \sqrt{3})$. To compute this, we first need to evaluate the derivative at this point:

$$y'\left(\frac{3}{2}, \sqrt{3}\right) = \frac{-12/2}{9\sqrt{3}} = \frac{-2}{3\sqrt{3}}.$$

Using the point–slope formula for a line, we obtain the equation for our tangent line:

$$T(x) = \frac{4}{\sqrt{3}} - \frac{2}{3\sqrt{3}}x.$$

The ellipse is plotted concurrently with this tangent line in Figure 3.15. ∎

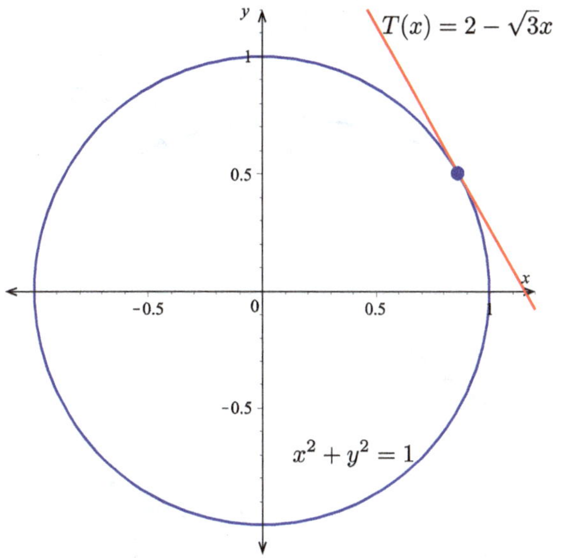

Figure 3.14: Graph of the circle $x^2 + y^2 = 1$ and its tangent line at the point $(\sqrt{3}/2, 1/2)$.

Next, let us consider some examples that focus solely on the computational aspects of implicit differentiation.

■ **Example 3.39** Differentiate the function

$$x^4 + x^3 y^5 + y^7 = 42$$

implicitly with respect to the variable x. (For purposes of satiating curiosity, the graph of this implicit equation is shown in Figure 3.16.)

Let us proceed term-by-term. Differentiating the first term with respect to x, we obtain simply

$$\frac{d}{dx}\left[x^4\right] = 4x^3.$$

Momentarily skipping ahead to the third term, we find

$$\frac{d}{dx}\left[y(x)^7\right] = 7y(x)^6 y'(x).$$

Normally, writing $y(x)$ instead of y takes place mentally; on paper we simply write

$$\frac{d}{dx}\left[y^7\right] = 7y^6 y'.$$

The second term is a product of x^3 and y^5, so we must use the product rule (along with the chain rule on anything that contains y):

$$\frac{d}{dx}\left[x^3 y^5\right] = 3x^2 y^5 + 5x^3 y^4 y'.$$

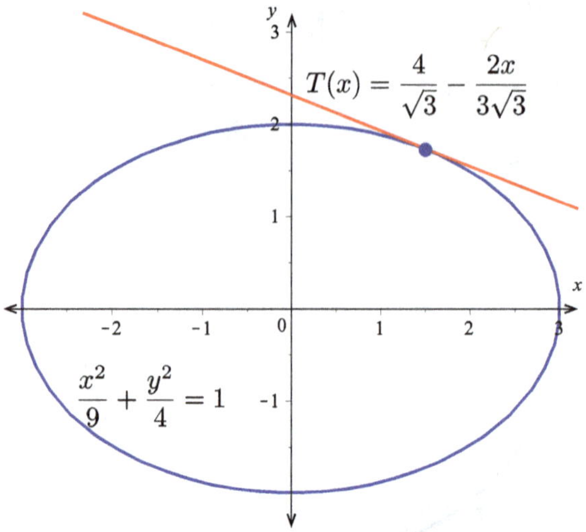

Figure 3.15: Graph of an ellipse and its tangent line; Example 3.38.

Of course, we are only writing out each term one at a time for learning purposes, so the reader might closely examine what is happening one derivative at a time. In practice, we just start with the equation $x^4 + x^3y^5 + y^7 = 42$, and then write that its implicit derivative with respect to x is given by

$$4x^3 + 3x^2y^5 + 5x^3y^4y' + 7y^6y' = 0.$$

As is always the case, the result is linear in the derivative y'; it can therefore be solved for y' using ordinary algebra:

$$y' = \frac{-(4x^3 + 3x^2y^5)}{5x^3y^4 + 7y^6}.$$

The result is the derivative $y' = dy/dx$, which depends on both variables x and y. ■

■ **Example 3.40** Consider the implicit equation

$$\sin(xy) = \frac{1}{2},$$

the graph of which is plotted in Figure 3.17. (Why does the graph look like this?)

Differentiating implicitly with respect to the variable x, we obtain

$$\cos(xy)\left(xy' + y\right) = 0.$$

This follows using the chain rule: the derivative of sine is cosine, but then we must multiply by the derivative of the inside. The inside function here is the product xy.

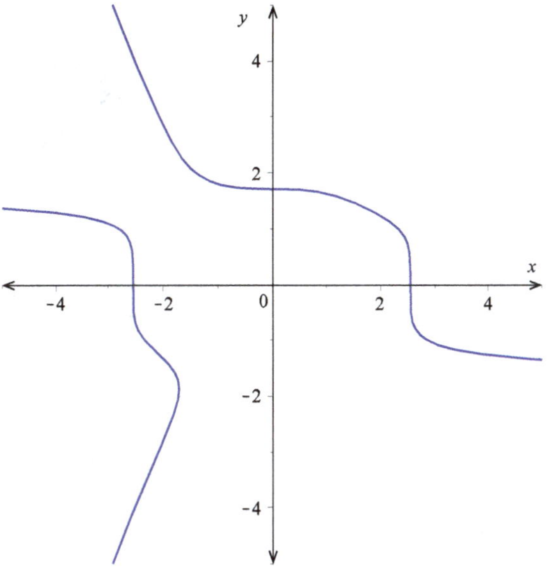

Figure 3.16: Graph of the function from Example 3.39.

Differentiating this product implicitly with respect to x, using the product rule, we obtain $xy' + y$.

Solving this previous equation for y', we obtain

$$y' = \frac{-x}{y}$$

(as long as $\cos(xy) \neq 0$, for which points the derivative is not defined). ■

3.7.3 Differentials

In this paragraph, we continue our discussion of differentials, which were introduced in Definition 2.5.3 on page 175. Given an implicit equation, one may *compute its differential*. In a way, this procedure is tantamount to taking its derivative, but without specifying which variable with respect to which we are differentiating. Not only do we need to use the chain rule when differentiating a term involving the variable y, but we must also use the chain rule when differentiating terms involving the variable x. With the following caveat: instead of multiplying by the derivative of the inside function, we must multiply by the differential of the inside function.

Definition 3.7.4 Given an equation between two or more variables, one may *take the differential* by doing the following:

1. regard each variable as an inside function (i.e., for the chain rule);

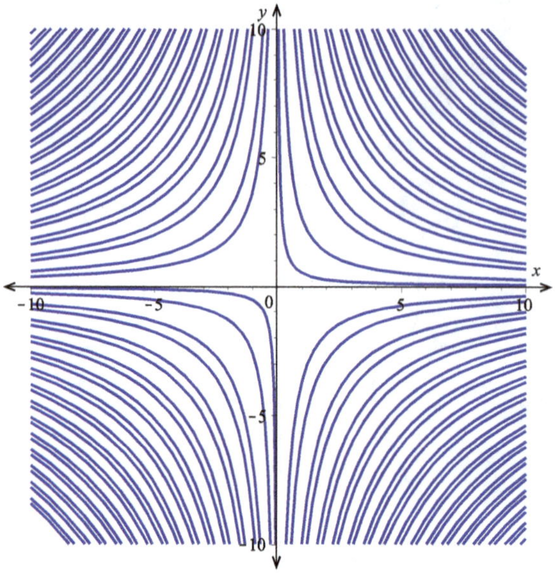

Figure 3.17: Graph of the implicit function $\sin(xy) = 0.5$, from Example 3.40.

2. differentiate the expression, one variable at a time, using the chain rule for each derivative;
3. except: instead of multiplying by the derivative of a given variable, multiply by its differential.

Recall that, given variables x and y, their differentials are the (fictitious) infinitesimally small quantities dx and dy, such that their ratios are correctly maintained in the sense of calculus.

■ **Example 3.41** Compute the differential of the equation for the circle

$$x^2 + y^2 = 1.$$

To proceed, we differentiate as normal, using the chain rule on *both* variables:

$$2x\,dx + 2y\,dy = 0.$$

This equation is the differential.

The differential of an equation has the following advantage: we may use the differential to directly obtain any of the implicit derivatives. For example, to obtain the implicit derivative of the equation of the circle with respect to x, we only need to divide its differential by the differential of x:

$$2x\frac{dx}{dx} + 2y\frac{dy}{dx} = 0, \qquad \text{or} \qquad 2x + 2yy' = 0,$$

as we previously obtained. Alternatively, we could divide the differential by dy, to instead obtain the implicit derivative with respect to the variable y:

$$2x\frac{dx}{dy} + 2y\frac{dy}{dy} = 0, \qquad \text{or} \qquad 2xx' + 2y = 0.$$

Here, x' is understood to represent $x' = dx/dy$. This is why the differential may be thought of as a non-committed implicit derivative. ■

■ **Example 3.42** Compute the differential of the implicit equation

$$\sin(xy) = 1.$$

Again, we proceed using our modified differential form of the chain rule. Taking the differential, we obtain

$$\cos(xy)(x\,dy + y\,dx) = 0.$$

If we divide this result by the differential dx, we again obtain the implicit derivative with respect to x. ■

Proposition 3.7.1 Consider an equation relating a set of variables and their differentials. If the various differentials are replaced with small finite differences, then this equation becomes an approximation. This approximation is called the *local linearization.*

■ **Example 3.43** Consider a can, which has the shape of a right-circular cylinder. Its volume is given by the formula

$$V = \frac{1}{3}\pi r^2 h.$$

The differential of this equation is given by

$$dV = \frac{\pi r^2}{3}\,dh + \frac{2\pi rh}{3}\,dr.$$

The corresponding local linearization is therefore given by

$$\Delta V \approx \frac{\pi r^2}{3}\Delta h + \frac{2\pi rh}{3}\Delta r.$$

Given a variance in the radius Δr and a variance in the height Δh, one may use this equation to compute the corresponding resulting variance in the can's volume ΔV. For example, if the material used to construct the can has a certain thickness, resulting in a certain value for Δr and Δh, one can use this equation to compute the approximate volume required to construct such a can.

In this example, since the volume depends on both r and h, which may vary independently of one another, the finite differences Δr and Δh may also be set independently of one another. Once Δr and Δh are specified, however, the finite difference ΔV must follow the prescribed formula. ■

■ **Example 3.44** Consider the equation for the area of a circle

$$A = \pi r^2.$$

Its differential is given by

$$dA = 2\pi r\, dr.$$

Converting the differentials dr and dA into their corresponding finite differences Δr and ΔA, we obtain the associated local linearization, give by the approximation

$$\Delta A \approx 2\pi r\, \Delta r.$$

Notice that the change of volume is approximately equal to the circumference times the change in radius. The reason why this is true can be understood by examining Figure 3.18. By incrementing the radius by a small amount Δr, the

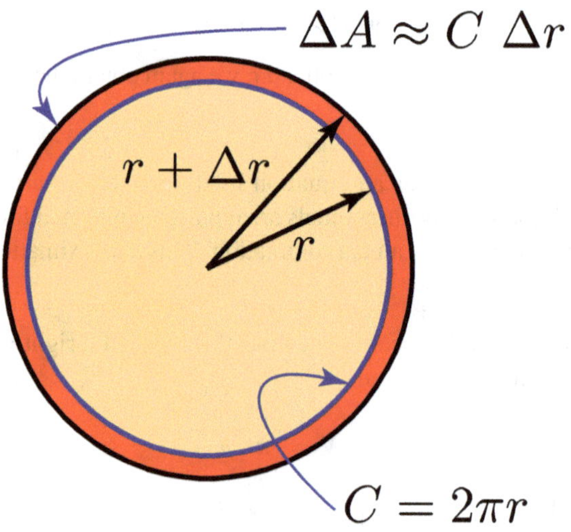

Figure 3.18: Illustration of the linearization $\Delta A \approx 2\pi r \Delta r$.

additional area is essentially a thin, circular ring with width Δr and length equal to the circumference of the circle. (Of course, the inner and outer circumferences are slightly different; there will be a correction that is quadratic in Δr, which is an order of magnitude smaller than the *linear* approximation, so that we may neglect it.)

This example also explains why

$$\frac{dA}{dr} = 2\pi r,$$

i.e., why the derivative of the area of a circle is equal to its circumference: given an infinitesimally small increase in the radius dr, the area will increase in an amount $dA = 2\pi r\, dr$ equal to the circumference times the infinitesimal change in radius. ■

Exercises

For Exercises 1–10, differentiate the given equation implicitly with respect to x. Solve for dy/dx. Determine the equation for the tangent line to the graph at the indicated point.

1. $x^3+y^3=9$; at $(1,2)$.
2. $x\sin(y)=2$; at $\left(4,\frac{\pi}{6}\right)$.
3. $4\cos(x)\sin(y)=1$; at $\left(\frac{\pi}{3},\frac{\pi}{6}\right)$.
4. $x^4+y^4=1$; at $\left(\sqrt[4]{0.5},\sqrt[4]{0.5}\right)$.
5. $y^x=4$; at $(2,2)$.
6. x^2-xy+y^2; at $(1,1)$.
7. $x^3y^2=32$; at $(2,2)$.
8. $xe^{xy}=3$; at $(3,0)$.
9. $\sqrt{x}+\sqrt{y}=4$; at $(2,2)$.
10. $x^2-y^2=8$; at $(3,1)$.

For Exercises 11–12, compute the differential of the given equation.

11. $z^3=x^3+y^3$.
12. $w=x^5y^7$.
13. $x^3y^5+x^5y^3=42$.
14. $\tan(x+y)=1$.
15. $x^2+8xy+y^2=9$.
16. $\sec(xy)=2$.
17. $x^3y^5=7$.
18. $\sin(xy)=\cos(x^3+y)$.
19. $Pv=RT$, where R is a constant.
20. $K=\dfrac{mv^2}{2}$, where m is a constant.

Problems

21. Consider the function
$$f(x)=\sqrt{1-x^2}.$$
   (a) How does this compare with the implicit equation (3.64)? (Their graphs *are* different. How?)
   (b) Show that $f(\sqrt{3}/2)=1/2$.
   (c) Compute the derivative $f'(x)$.
   (d) Evaluate $f'(\sqrt{3}/2)$ and determine an equation for the tangent line to the graph at the point $(\sqrt{3}/2,1/2)$.
   (e) Compare the result with Example 3.37.
22. Explain why the graph in Figure 3.17 makes sense, given the implicit equation $\sin(xy)=0.5$ from Example 3.40.
23. Consider the implicit equation, which represents a circle (not centered at the origin):
$$x^2+y^2-5x+y=2.5.$$
   (a) Differentiate implicitly with respect to the variable x. Compute a formula for dy/dx.
   (b) Set $dy/dx=0$ and solve for x. This is the point at which the tangent line is horizontal.
   (c) Differentiate implicitly with respect to the variable y. Compute a formula for dx/dy.

(d) How are the formulas for dy/dx and dx/dy similar? How are they different?
(e) Set $dx/dy = 0$ and solve for y. This is the point at which the tangent line is vertical.
(f) At what point is this circle centered? What is its radius?
(g) Complete the square to rewrite the original equation in the form

$$(x-x_0)^2+(y-y_0)^2=r^2.$$

Use this to confirm the answer to part (f).

24. Consider the implicit equation

$$ax^2+bxy+cy^2=1.$$

(a) Differentiate implicitly and solve for dy/dx.
(b) Determine the x values at which $y'(x) = 0$.

25. A *lemniscate* is the curve defined by the implicit equation

$$3(x^2+y^2)^2=25(x^2-y^2).$$

The graph of this equation is shown in Figure 3.19. Determine the equation for the tangent line at the point $(1,1)$.

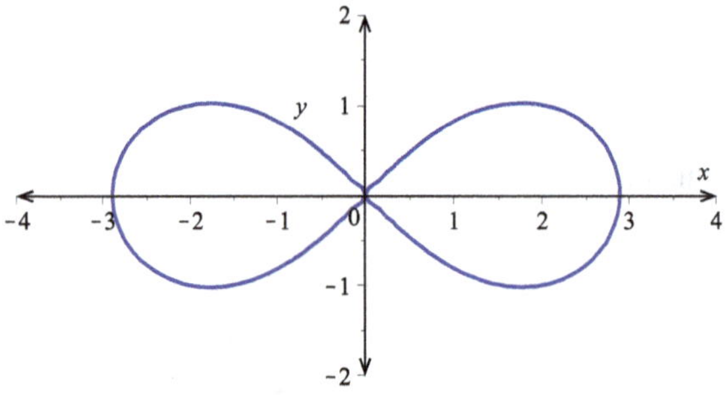

Figure 3.19: Graph of a lemniscate.

26. Consider the volume equation for a sphere, as given by the familiar formula

$$V=\frac{4}{3}\pi r^3.$$

(a) Compute the differential of the volume formula.
(b) Show that $dV = S\ dr$, where S represents the surface area of a sphere.
(c) Explain why the local linearization $\Delta V \approx 4\pi r^2\ \Delta r$ makes practical sense. Draw a picture of this local linearization (analogous to Figure 3.18).

27. A company wishes to manufacture a can that has the shape of a right-circular cylinder, with volume

$$V=\frac{1}{3}\pi r^2 h,$$

where r is the radius of the cross-sectional circles and h is the height.

(a) Compute the differential of the volume equation.
(b) Write down the associated local linearization.
(c) Suppose that the radius of the can is 5 cm and the height is 8 cm. If the tin on the sides must be 0.03 cm thick and the tin on the top and bottom must be 0.04 cm thick, use the local linearization to approximate the total volume of tin required to produce the can.

28. A *heat engine* is a mechanical system that converts thermal energy (i.e., heat) into work (e.g., work done on a turbine to produce electricity). The *thermodynamic efficiency* η of a heat engine is the fraction of the input heat energy that is actually converted into work. The second law of thermodynamics prevents a heat engine from being able to convert all of the incoming heat energy into work; the maximum (theoretical) thermodynamic efficiency is produced by a *Carnot cycle*, and is given by

$$\eta = 1 - \frac{L}{H},$$

where H is the high temperature at which heat enters the system and L is the low temperature at which excess heat energy (that was unable to be converted to work) is lost to the environment.
(a) Compute the differential of the equation for the maximum thermodynamic efficiency of a Carnot cycle.
(b) Write down the associated local linearization.
(c) Suppose that a heat engine is operating between the temperatures 240 K (output) and 350 K (input). Use local linearization to approximate the increase in the maximum thermodynamic efficiency if the high temperature increases by 3 K and the low temperature increases by 2 K.
(d) (Part (c), continued.) If the high temperature increases by 3 K, by how much must the low temperature increase to result in no net change in the maximum thermodynamic efficiency, according to the local linearization?

29. The ideal gas law states that, for an ideal gas,

$$Pv = RT,$$

where P is the pressure of the gas, v is its specific volume (volume per unit mass), and T is its absolute temperature. The constant R is the specific gas constant, which depends on the particular gas. For air, $R = 287\ \text{J}\cdot\text{kg}^{-1}\cdot\text{K}^{-1}$.
(a) Compute the differential of the ideal gas law.
(b) Write down the associated local linearization.
(c) Suppose now that 0.5 kg of air is trapped in a piston and cylinder, and that the piston/cylinder has a cross-sectional area of 0.21 m^2. The air is pressurized to 500,000 Pa and has an absolute temperature of 300 K.
  i. Determine the specific volume of the gas.
  ii. If the piston moves into the cylinder by 0.8 cm, thereby reducing the total volume in the chamber and resulting in a 0.5 K increase in temperature, approximate the resulting pressure.

3.8 Derivatives of Inverse Functions

In this section, we explore derivatives of inverse functions. Given a function $y = f(x)$, its inverse function may be described as $x = f^{-1}(y)$. The derivative of the inverse function is therefore given by

$$\frac{dx}{dy} = \left(f^{-1}\right)'(y).$$

Essentially, the result we will obtain states that

$$\frac{dx}{dy} = \frac{1}{dy/dx},$$

where dy/dx is the derivative of the original function, i.e., $dy/dx = f'(x)$. In other words, the derivative of the inverse function is the inverse of the derivative. Our task in the following pages will be to formalize this concept and to apply the result to the study of the main inverse trigonometric functions.

3.8.1 General Inverse Functions

Before proceeding to the general formula, let us examine several examples.

▪ **Example 3.45** Compute the inverse function of the function $f(x) = 3x + 8$. Then compare the derivatives of the two functions.

To obtain the inverse function, we first write the general relations

$$y = f(x) \qquad \text{and} \qquad x = f^{-1}(y).$$

Next, we write

$$y = 3x + 8$$

and solve for x, obtaining

$$x = \frac{y-8}{3}.$$

We conclude that

$$f^{-1}(y) = \frac{y-8}{3}.$$

Alternatively, we may replace y with x and write

$$f^{-1}(x) = \frac{x-8}{3}.$$

(The objects f and f^{-1} are each *functions*, i.e., they do not care what we call their independent variable.)

Computing the derivatives of the functions f and f^{-1}, we obtain

$$f'(x) = 3 \qquad \text{and} \qquad \left(f^{-1}\right)'(x) = \frac{1}{3},$$

respectively. We observe that the derivatives of the (linear) functions f and f^{-1} are reciprocals of each other. This makes sense, as the derivatives f' and $(f^{-1})'$ represent

$$\frac{dy}{dx} \quad \text{and} \quad \frac{dx}{dy},$$

respectively. ■

■ **Example 3.46** Consider the invertible function $f(x) = x^3 + 8$. Compute the inverse function $f^{-1}(x)$, and then compute and compare the derivatives of the functions f and f^{-1}.

We start by writing

$$y = x^3 + 8.$$

Solving for x, we obtain

$$x = (y-8)^{1/3} = \sqrt[3]{y-8}.$$

(The expression on the right-hand side must equal $f^{-1}(y)$.) Replace y with x to obtain

$$f^{-1}(x) = (x-8)^{1/3}.$$

(As an exercise, verify that $f^{-1} \circ f(x) = x$, i.e., $f^{-1}(f(x)) = x$.)

Next, we compute the derivatives as normal, obtaining

$$f'(x) = 3x^2 \quad \text{and} \quad \left(f^{-1}\right)'(x) = \frac{1}{3(x-8)^{2/3}}.$$

The reason that the derivatives f' and $(f^{-1})'$ are not straight-up reciprocals of each other, is that they are only reciprocals for a given *point on the graph*. Recall that the input variable of the function f' is the independent variable x, whereas the input variable of the function $(f^{-1})'$ is the (original) dependent variable y. For a given x value, the corresponding y value is $y = f(x)$. When we evaluate the function $(f^{-1})'$ at the point $f(x)$, we obtain

$$\left(f^{-1}\right)'(f(x)) = \frac{1}{3\left[\left(x^3+8\right)-8\right]^{2/3}} = \frac{1}{3x^2} = \frac{1}{f'(x)}.$$

This explains the relation between the two derivatives. ■

Given a function f, we have observed that the derivatives f' and $(f^{-1})'$ are not simply the straight-up reciprocals of one another. Rather, the following can be said: given a point (x,y) on the graph of the function f, we have

$$\left(f^{-1}\right)'(y) = \frac{1}{f'(x)}.$$

This concept is formalized (using a slightly different form) in the following theorem.

Theorem 3.8.1 Given an invertible function f, the derivative of its inverse is given by

$$\left(f^{-1}\right)'(x) = \frac{1}{f'(f^{-1}(x))}. \tag{3.68}$$

Proof. First, let us remark that the variable x in equation (3.68) corresponds to the *output* variable of the original function f. Starting with $y = f(x)$, we can write

$$\frac{dy}{dx} = f'(x).$$

The inverse function tells us how to compute $x = f^{-1}(y)$. Therefore

$$(f^{-1})'(y) = \frac{dx}{dy} = \frac{1}{f'(x)} = \frac{1}{f'(f^{-1}(y))}.$$

Switching the variable y with the variable x—so that the input of the inverse function f' is now the variable x—we obtain our result. ■

Corollary 3.8.2 Let the function f be invertible and differentiable. Given a point (x, y) on the graph of the function f, the relation

$$\left(f^{-1}\right)'(y) = \frac{1}{f'(x)} \tag{3.69}$$

holds.

Proof. Rewrite equation (3.68) with x replaced with y. Recognizing the relation $x = f^{-1}(y)$ completes the proof. ■

▪ **Example 3.47** Consider the function $f(x) = e^x$, with inverse function $f^{-1}(x) = \ln(x)$, which is defined on the domain $x \in (0, \infty)$. (This domain coincides with the *range* of the exponential function e^x.)

We know that $f'(x) = e^x$. The denominator on the right-hand side of equation (3.68) therefore simplifies as

$$f'(f^{-1}(x)) = f'(\ln(x)) = e^{\ln(x)} = x.$$

It follows from equation (3.68), that the derivative of the logarithmic function $f^{-1}(x) = \ln(x)$ is given by

$$\left(f^{-1}\right)'(x) = \frac{1}{x},$$

which agrees with our experience. ▪

3.8.2 Inverse Trigonometric Functions

Next, we apply the results from Theorem 3.8.1 to derive the formulas for the derivatives of the inverse trigonometric functions.

Theorem 3.8.3 The following rules give the derivatives of the inverse trigonometric functions

$$\frac{d}{dx}\left[\sin^{-1}(x)\right] = \frac{1}{\sqrt{1-x^2}}, \tag{3.70}$$

$$\frac{d}{dx}\left[\cos^{-1}(x)\right] = \frac{-1}{\sqrt{1-x^2}}, \tag{3.71}$$

$$\frac{d}{dx}\left[\tan^{-1}(x)\right] = \frac{1}{1+x^2}. \tag{3.72}$$

Proof. First, let us consider the function $f(x) = \sin(x)$, whose inverse is given by the arcsin function $f^{-1}(x) = \sin^{-1}(x)$. Equation (3.71) tells us that

$$\frac{d}{dx}\left[\sin^{-1}(x)\right] = \frac{1}{\cos(\sin^{-1}(x))}.$$

We recall, however, that $\cos(y) = \sqrt{1-\sin^2(y)}$, since $\sin^2(y)+\cos^2(y) = 1$. It follows that $\cos(\sin^{-1}(x)) = \sqrt{1-x^2}$, which verifies identity (3.70). A more detailed approach, that doesn't rely on a direct application of equation (3.71), would be to start by writing out

$$y = \sin^{-1}(x).$$

Therefore, $x = \sin(y)$, and

$$\frac{dx}{dy} = \cos(y) = \sqrt{1-\sin^2(y)}.$$

Substituting back in $y = \sin^{-1}(x)$, we again obtain our result.

A similar technique can be used to derive equations (3.71) and (3.72) using the formula (3.68) and trigonometric identities. ∎

Exercises

For Exercises 1–10, use the chain rule and known derivatives in order to compute the derivative of the given function.

1. $f(x) = \tan(e^x)$.
2. $f(x) = \tan(3\pi x+7)$.
3. $f(x) = \sin^{-1}(3x)$.
4. $f(x) = \cos^{-1}(\sqrt{x})$.
5. $f(x) = \tan^{-1}\left(8x^{3/2}+9\right)$.
6. $f(x) = \tan^{-1}(\sin(x))$.
7. $f(x) = \sin^{-1}(\cos(x))$.
8. $f(x) = \sin^{-1}(x^2)$.
9. $f(x) = \tan^{-1}(\ln|x|)$.
10. $f(x) = \sin^{-1}(\sin(x))$.

Problems

11. Prove the derivative rules (3.71) and (3.72) using Theorem 3.8.1.
12. A ten foot tall statute casts a shadow that is x feet long. The ray of light traveling from the top of the statue to the ground, landing at the shadow's edge, makes an angle θ with the ground.
    (a) Show that the angle θ can be expressed as a function of the length of the shadow:
    $$\theta(x) = \tan^{-1}\left(\frac{10}{x}\right).$$
    (b) Determine an expression for $\theta'(x)$.
    (c) If the shadow is 12 feet long, use linearization to approximate by how much the angle will change if the shadow were to increase by an additional 0.5 feet.

Chapter 4

Applications of Differentiation

4.1 Maxima and Minima

Optimization is one of the cornerstone applications of calculus and has left its mark on a diverse panorama of disciplines. Before delving into the modeling aspects of optimization, as we soon shall do, we first introduce some language and basic theory that will assist us on our way.

4.1.1 Local Extrema

We discussed the relationship between whether a function is increasing or decreasing and whether its derivative is positive or negative in Proposition 2.3.2. Today, we will examine the peaks and valleys of differentiable functions: when does a function obtain its maximum and minimum output values?

Definition 4.1.1 — Local Extrema. A function f has a *local minimum* at the point $x = a$ if $f(a) \leq f(x)$, for all points in the domain of f that are sufficiently close to the point $x = a$.

A function f has a *local maximum* at the point $x = a$ if $f(a) \geq f(x)$, for all points in the domain of f that are sufficiently close to the point $x = a$.

(R) The phrase "sufficiently close" has a precise mathematical meaning: there must be some open interval containing the point $x = a$, so that the inequality condition of Definition 4.1.1 must hold for all points that are in both the open interval and the domain of the function f.

Example 4.1 Consider the function f that is graphed in Figure 4.1. Let us assume that the function f is only defined on the domain $[a,d]$, as shown. The point $x = b$ is clearly a local maximum, as the output $f(b)$ is greater than all of the nearby outputs. Similarly, the point $x = c$ is clearly a local minimum, as the output $f(c)$ is less than all of the nearby outputs.

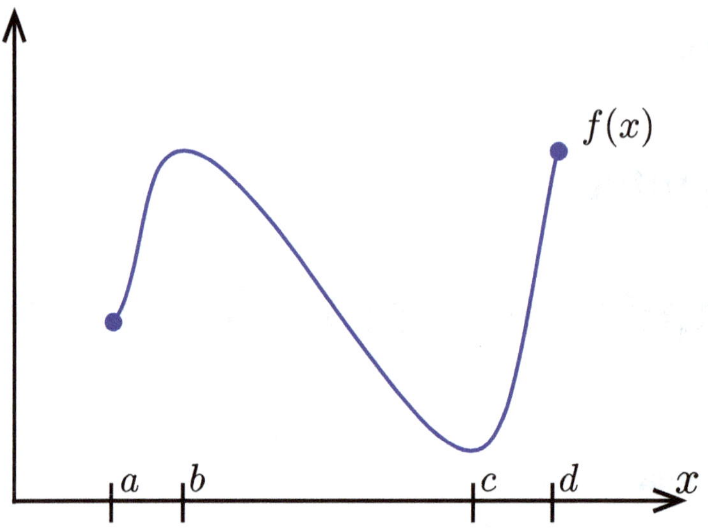

Figure 4.1: A function with several local extrema.

Perhaps less obvious, the *endpoints* of the domain also constitute local extrema: the point $x = a$ is a local minimum and the point $x = d$ is a local maximum. This follows since the outputs $f(a)$ and $f(d)$ are less than and greater than, respectively, the nearby points *on the domain*. Thus, the fact that the graph is not "level"—i.e., the fact that the graph does not have a horizontal tangent line—does not prevent the point from being a local extremum. ∎

∎ **Example 4.2** Consider the function f that is shown in Figure 4.2, defined on the domain $[a, c]$. This function has local maxima at the two endpoints $x = a$ and $x = c$ and a local minimum at the point $x = b$. Notice that the function fails to be differentiable at the point $x = b$, even though this point is a local minimum. ∎

Critical Points and Interior Points

So far, we have seen examples in which local extrema occur at points with a horizontal tangent line, endpoints of the domain, and at points in which the function fails to be differentiable. This motivates the following definition.

Definition 4.1.2 A point $x = a$ on the domain of a function f is said to be a *critical point* of the function f if either of the following two conditions holds:

1. the function f fails to be differentiable at the point $x = a$;
2. the function f is differentiable at the point $x = a$ and its derivative vanishes, i.e., $f'(a) = 0$.

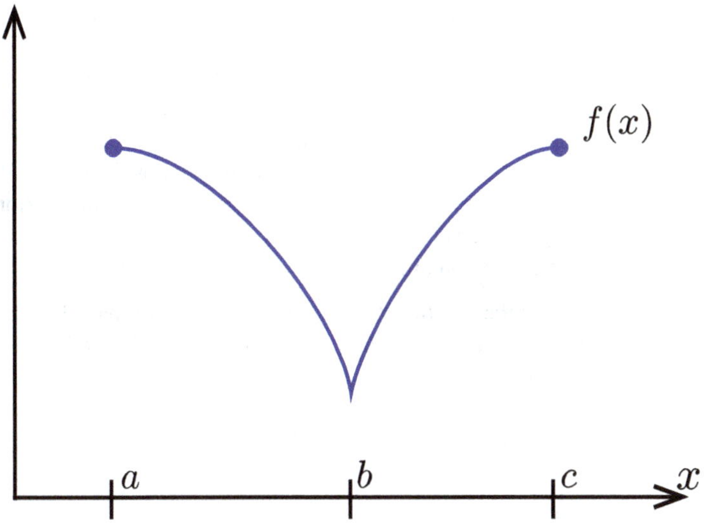

Figure 4.2: A function with several local extrema.

(R) Thus, the critical points of a function are the points at which the function has a discontinuity (and are therefore also nondifferentiable), the points at which the graph of the function has a sharp corner or edge (and are therefore nondifferentiable), and the points at which the function has a horizontal tangent line (and therefore have a vanishing derivative).

Example 4.3 In Figure 4.1, the points $x = b$ and $x = c$ are critical points, due to the fact that the function f has a horizontal tangent line at these points. The points $x = a$ and $x = d$, however, are not: the function f neither fails to be differentiable at these points nor does it have a horizontal tangent line.

In Figure 4.2, the point $x = b$ is a critical point, as the function f fails to be differentiable at this point. The two endpoints $x = a$ and $x = c$ are *likely* critical points as well, as it appears as though the function meets its ends with a horizontal tangent. Assuming that the graph indeed levels off to a horizontal tangent at its endpoints, we may conclude that the three points $x = a$, $x = b$, and $x = c$ each constitute a critical point. ∎

We will make use of one additional definition.

Definition 4.1.3 A point $x = a$ is an *interior point* of the function f's domain if there exists an open interval that contains the point $x = a$ and that is itself contained entirely within the domain of f.

If a point on a function's domain is not an interior point, it is called a *boundary point*.

(R) Intuitively, interior points are simply points that are *inside* the function's domain, i.e., neither endpoints nor isolated points of the domain.

■ **Example 4.4** Consider the function $f(x) = \frac{1}{x}$ defined on the domain $(0,1]$, i.e., defined for the x-values $0 < x \leq 1$. The interior points are precisely the points contained in the open interval $(0,1)$. For example, the point $x = 0.001$ is an interior point, because the open interval $(0.0001, 0.0011)$ both contains the point 0.001 and is itself contained in the domain $(0,1]$. The point $x = 1$ is *not* an interior point. This is because *any* open interval that contains the point $x = 1$ will also have some points that lie outside of the domain (since the domain ends sharply at $x = 1$). ■

■ **Example 4.5** Consider the function $f(x) = \frac{1}{x}$ defined on the domain $\mathbb{N} = \{1,2,3,4,\ldots\}$, i.e., defined on the set of natural numbers. Then every point in the function's domain is a boundary point, and none of the points is an interior point.

Consequently, each point in this function's domain is both a local minimum and a local maximum, as each point satisfies these definitions. ■

The Local-Extremum Theorem

We have seen examples in which an interior point (not an endpoint or isolated point) of a function coincidentally constitutes both a local extremum and a critical point of that function. One may wonder: is it possible for an interior point to be a local extremum, but not a critical point? Is it possible for an interior point to be a critical point, but not a local extremum? As it turns out, the answer to the first question is no—as we shall recite in our next theorem—and the answer to the second question is yes.

Theorem 4.1.1 — The Local-Extremum Theorem. If an interior point $x = a$ is a local extremum of a function f, then the point $x = a$ must also be a critical point of f.

(R) The converse of Theorem 4.1.1 (i.e., the statement that if a point is a critical point, then it must be a local extremum) is not necessarily true. In this theorem, the logical implication only acts one way:

$$\text{local extremum (and interior point)} \Longrightarrow \text{critical point},$$

but

$$\text{critical point} \not\Longrightarrow \text{local extremum}.$$

Thus, if we only know that a point is a critical point, it may or may not be a local extremum of the function f.

We begin with a counterexample of the converse, i.e., we begin with an example of a function that possesses a critical point that is not a local extremum.

■ **Example 4.6** Consider the function $f(x) = x^3$. The point $x = 0$ is a critical point of f, since

$$f'(x) = 3x^2,$$

and, therefore, $f'(0) = 0$. (Thus, the graph of f has a horizontal tangent at the point $x = 0$.) The graph of the function f is shown in Figure 4.3, plotted concurrently with the tangent line T to the graph at the point $(0,0)$.

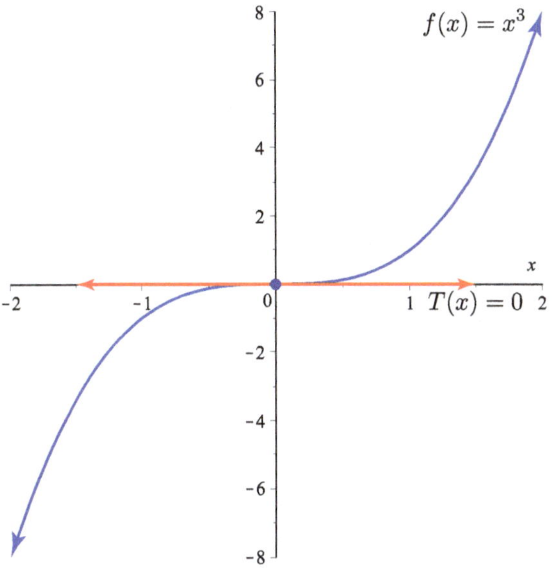

Figure 4.3: Graph of the function $f(x) = x^3$ with the tangent line T to the point $(0,0)$.

Since the line T passes through the point $(0,0)$ with a slope of $f'(0) = 0$, its equation is therefore given by $T(x) = 0$. One can readily see, by examining Figure 4.3, that this tangent line is horizontal. The function f, however, does *not* possess a local extremum at the point $x = 0$. We can understand this as follows: the function is increasing for the negative x-values, its rate of change "levels off" to zero slope, just for an instant, as it passes through the origin, and then it returns its ascent, increasing again for positive x-values. The point $x = 0$ is a critical point, yet there is not a local minimum or maximum at this point. ■

Though the critical-point condition does not guarantee that a given point is a local extremum, it is a useful indicator nevertheless that it might be. Since all (interior-point) local extrema are also critical points, one needs only search through the critical points (and the end points) on the search for local extrema. Our next example demonstrates the usefulness of this theorem.

■ **Example 4.7** Consider the function

$$f(x) = 16x^3 + 63x^2 - 232.5x - 200,$$

as shown in Figure 4.4. Determine the local extrema.

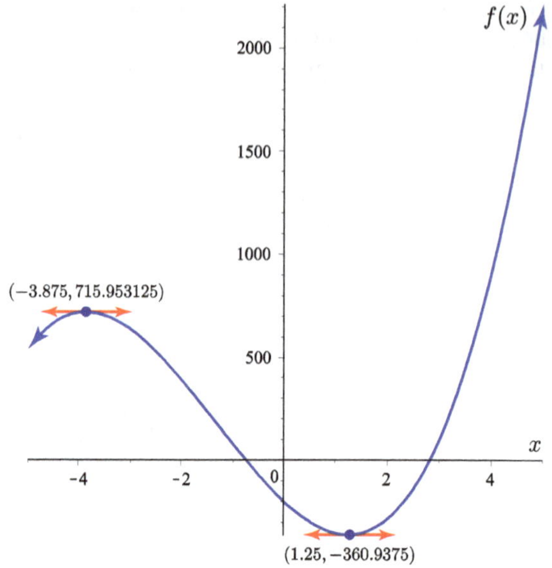

Figure 4.4: Graph of the function $f(x) = 16x^3 + 63x^2 - 232.5x - 200$.

Since a local extremum must be a critical point (or a boundary point), we begin by identifying the critical points. The derivative is computed as

$$f'(x) = 48x^2 + 126x - 232.5.$$

To find the critical points, we set $f'(x) = 0$ and solve for x. Using the quadratic formula, the two critical points of this function are

$$x = \frac{-126 \pm \sqrt{126^2 + 4(48)(232.5)}}{96} = \frac{-21 \pm 41}{16} = \{-3.875, 1.25\}.$$

Therefore, the two critical points of the function f are the points $x = -3.875$ and $x = 1.25$. The graph of the function f is shown in Figure 4.4. Indeed, the two points $x = -3.875$ and $x = 1.25$ are critical points, as evidenced by their horizontal tangent lines. Furthermore, they each constitute a local extremum: the point $x = -3.875$ is a local maximum and the point $x = 1.25$ is a local minimum. This means that *locally* they constitute the function's highest and lowest point within the neighboring vicinity. ∎

4.1.2 Global Extrema

As local extrema constitute the maximum or minimum values in their local neighborhood (i.e., in an open interval containing the point), *global* extrema constitute the maximum or minimum values *globally*, i.e., as compared with each point on the domain.

Definition 4.1.4 A function f has a *global minimum* at the point $x = a$ if $f(a) \leq f(x)$, for all x in the domain of f.

A function f has a *global maximum* at the point $x = a$ if $f(a) \geq f(x)$, for all x in the domain of f.

Global extrema are sometimes referred to as *absolute* extrema; we will use the former nomenclature throughout this text, however.

(R) It follows from the definitions that if a point is a global maximum (or global minimum), then it must also be a local maximum (or local minimum) as well. This is true because the inequality condition only needs to be valid *locally* for a point to be a local extremum, whereas it must be valid *globally*, i.e., for the entire domain, in order for the point to be a global extremum. Thus, Theorem 4.1.1 applies to global extrema as well as local extrema: if an interior point is a global extremum, then it must be a critical point.

■ **Example 4.8** Consider the function f that is shown in Figure 4.5. We first observe that there are discontinuities at the points $x = 5$, $x = 11$, and $x = 18$; these points are therefore critical points of the function f. (Recall: differentiability implies continuity; therefore being continuous is a necessary condition of differentiability, though not sufficient.) The points $x = 3$, $x = 6$, $x = 9$, $x = 12$, and $x = 16$ are also critical points, as the function f has a horizontal tangent line at these points. Finally, the point $x = 14$ is a critical point, since the graph of the function has a sharp corner at this point.

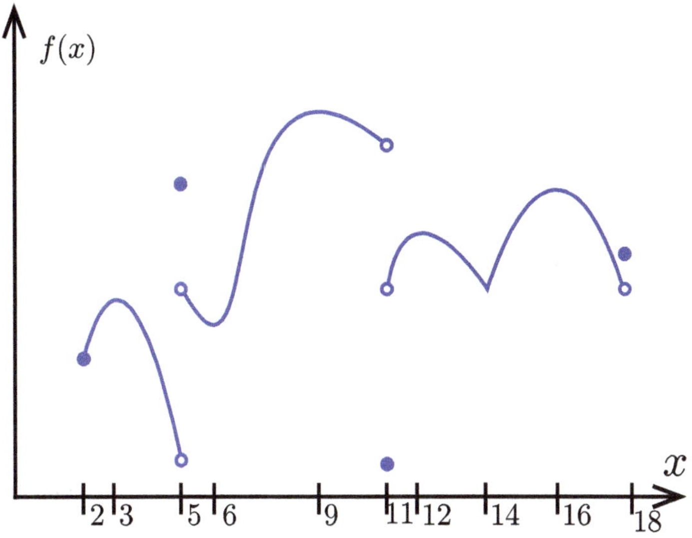

Figure 4.5: Function from Example 4.8.

The local minima occur at the points $x = 2$, $x = 6$, $x = 11$, and $x = 14$. The local maxima occur at the points $x = 3$, $x = 5$, $x = 9$, $x = 12$, $x = 16$, and $x = 18$.

The global minimum occurs at the point $x = 11$, and the global maximum occurs at the point $x = 9$. ∎

4.1.3 The Extreme-Value Theorem

It can be the case that a function does not have a global maximum or a global minimum. Our next theorem gives us a condition that guarantees the existence of the global extrema.

Theorem 4.1.2 — The Extreme-Value Theorem. If a continuous function f is defined on a closed interval I of finite length, then it actually achieves its global extrema for some points on I.

∎ **Example 4.9** Consider the function $f(x) = \frac{1}{x}$, defined on the domain $x \in (0, 1)$, as shown in Figure 4.6.

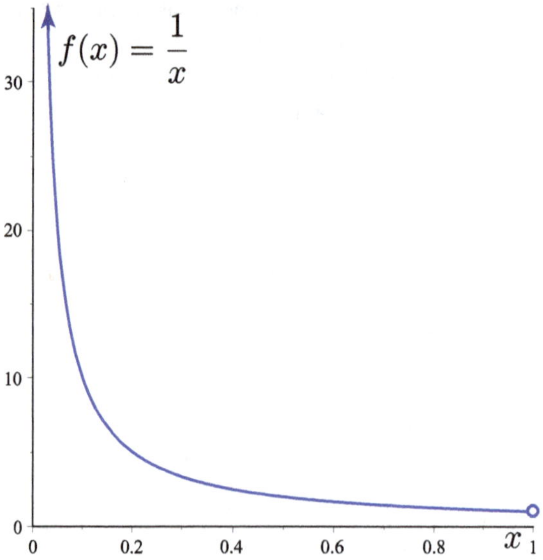

Figure 4.6: Function from Example 4.9.

This function does not have a global maximum, as this function achieves output values that are arbitrarily high. For example, suppose you were to say that $x = 0.1$ was the global maximum, with $f(0.1) = 10$. This cannot be true, since $f(0.01) = 100$. So next suppose that $x = 0.01$ is the global maximum. Again, this conclusion fails because $f(0.001) = 1000$. And so on.

Similarly, the function f does not have a global minimum, since the endpoint $x = 1$ is not included in the domain. The outputs become arbitrarily close to the output value of 1, but never reach it. Therefore, no single point acts as the global minimum.

These conclusions do not contradict Theorem 4.1.2, since our function f is not defined on a *closed* interval, but, rather, it is defined on the *open* interval $(0,1)$. ■

■ **Example 4.10** Consider the function

$$f(x) = \sin(16x^2)\cos(8x), \tag{4.1}$$

defined on the domain $[0,1]$. Since this function is continuous and is defined on a closed interval, the maximum-value theorem guarantees that it actually achieves its global minimum and global maximum at some point on the interval $[0,1]$. This function is plotted in Figure 4.7. Its local and global extrema are labeled. The

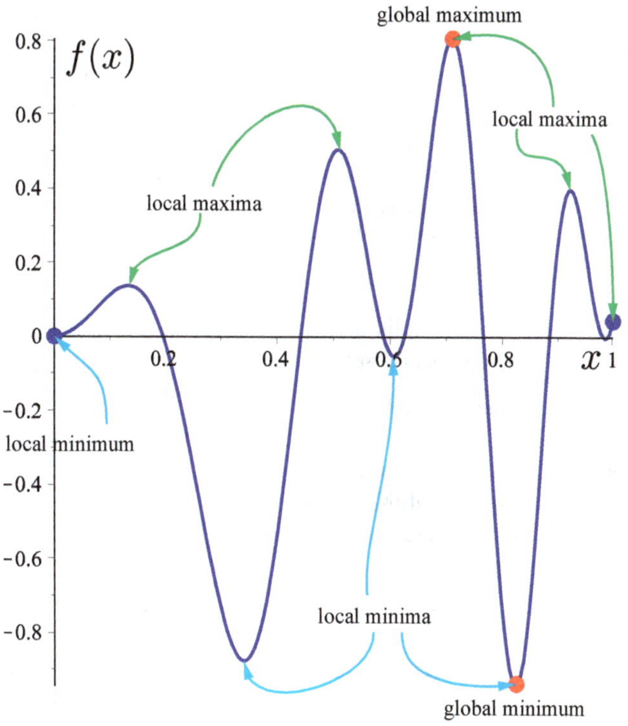

Figure 4.7: The function $f(x) = \sin(16x^2)\cos(8x)$, as given in Example 4.10. The local and global extrema are labeled.

graph confirms the result of Theorem 4.1.2: this function indeed achieves its global maximum and global minimum on the interval $[0,1]$. ■

(R) If the function defined in Example 4.10 were instead defined on the open interval $(0,1)$, then it would still possess the same global maximum and minimum points. Thus, Theorem 4.1.2 only tells us that a continuous function defined on a closed interval of finite length is a sufficient condition to conclude that it achieves its global extrema. Some functions, however, will still achieve their global extrema, even if they are defined on open intervals, or if they fail to be continuous.

4.1.4 Locating Global Extrema

Finally, we present a theorem that shows us how we can go about finding the global extrema of a function on a given domain.

Theorem 4.1.3 In order to locate the global extrema of a continuous function f on a given domain $\mathcal{D}$, we proceed as follows:

1. Locate all the critical points of the function f in the domain $\mathcal{D}$; these points are comprised of the points where $f'(x) = 0$ and where the derivative f' is not defined.
2. Locate the boundary points of the domain $\mathcal{D}$; these points consist of the endpoints of any contiguous interval as well as isolated points (i.e., points not connected to any interval) in the domain.
3. Evaluate the function f at each critical point and boundary point.
4. The point(s) that returns the greatest output value is (are) the global maximum. The point(s) that returns the least output value is (are) the global minimum.

In short, Theorem 4.1.3 tells us that we only need to check the value of a function at the critical points and at the boundary of the domain.

▪ **Example 4.11** Consider the function $f(x) = xe^{-0.5x}$, which is defined on the domain $\mathcal{D} = [1,5]$, as shown in Figure 4.8. Theorem 4.1.3 tells us that we need to check the endpoints of the domain, i.e., the points $x = 1$ and $x = 5$, as well as any critical points in the interior of the domain.

To locate the critical points, we compute the derivative of f using the product rule, thereby obtaining

$$f'(x) = e^{-0.5x} - 0.5xe^{-0.5x} = (1 - 0.5x)e^{-0.5x}.$$

Since the factor $e^{-0.5x}$ is never zero, the only critical point occurs when the factor $(1 - 0.5x) = 0$, i.e., when $x = 2$. Checking the output values of this critical point and the two endpoints, we find that

$$\begin{aligned} f(1) &= e^{-0.5} \approx 0.60653, \\ f(2) &= 2e^{-1} \approx 0.73576, \text{ and} \\ f(5) &= 5e^{-2.5} \approx 0.41042. \end{aligned}$$

Therefore, the global maximum occurs at the point $x = 2$ and the global minimum occurs at the point $x = 5$. ▪

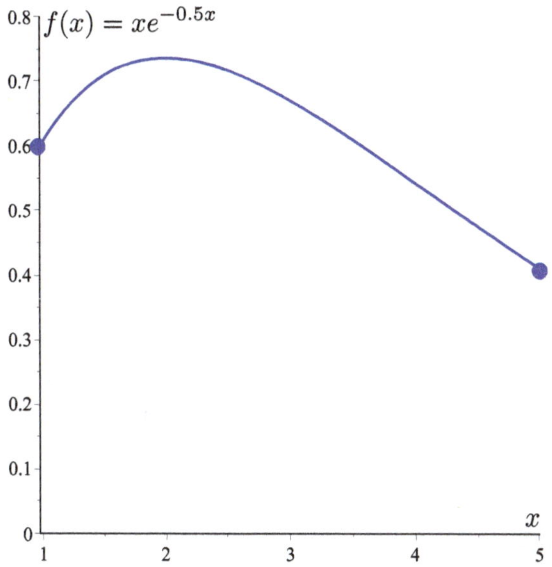

Figure 4.8: The function f, as given in Example 4.11.

4.1.5 The Second-Derivative Test

We have previously seen that the critical points of a function are candidates for the function's local minima and maxima. Using the second derivative, we next present a test that, in some cases, lets us know for sure.

Theorem 4.1.4 — The Second-Derivative Test. Let $x = a$ be a critical point on the domain of a function f at which the function is differentiable, i.e., $f'(a) = 0$.

1. If $f'(a) = 0$ and $f''(a) < 0$, then the point a is a local maximum.
2. If $f'(a) = 0$ and $f''(a) > 0$, then the point a is a local minimum.

The preceding theorem makes a good deal of sense: if the derivative at a point is equal to zero and the function is concave down at that point (i.e., $f''(a) < 0$), then the point must be a local maximum. Similarly, if the derivative is zero and the function is concave up (i.e., $f''(a) > 0$), then the point must be a local minimum.

Exercises

For Exercises 1–10, determine the critical points, the local minima and maxima, and the global minima and maxima on the indicated interval.

1. $f(x) = 2x^3 - 9x^2 - 4x + 5$; $x \in [0, 5]$.

2. $f(x) = 6x^3 - 30x^2 + 24x$; $x \in [0, 5]$.

3. $f(x) = 6x^3 - 12x^2 - 30x + 36;\ x \in [0,3]$.
4. $f(x) = -x^2 + x + 2;\ x \in [0,2]$.
5. $f(x) = 8 + 3\sin(4x+2);\ x \in [0,1]$.
6. $f(x) = \ln(3 + \sin(x));\ x \in [0,4]$.
7. $f(x) = \sin(3x - x^2);\ x \in [0,3]$.
8. $f(x) = 2x^5 - 25x^4 + 110x^3 - 200x^2 + 160x;\ x \in [0,5]$.
9. $f(x) = 3x^4 - 28x^3 + 96x^2 - 144x;\ x \in [1,4]$.
10. $f(x) = \dfrac{\ln(x)}{x};\ x \in [1,3]$.
11. Show that a point is a critical point of the function given by equation (4.1) only if that point satisfies the condition $4x = \tan(16x^2)\tan(8x)$.

Problems

12. A certain luxury car sells new for $36,000. In this problem, we will determine the optimal length of time to own the vehicle in order to minimize the average cost.
    (a) Suppose the vehicle depreciates 15% per year. Write a formula for the value V of the car after x years.
    (b) Suppose that the cumulative cost M of maintaining and repairing the vehicle during the first x years of the vehicle's life is given by the formula
    $$M(x) = 1200\left(1.35^x - 1\right).$$
    Use this to write a formula for the total cost C of owning the vehicle for x years. (*Hint*: C should include (i) the price of the vehicle less its resale value after x years and (ii) the total cost to maintain the repair the vehicle over the first x years of its life.)
    (c) Show that the average *cost per year* A of owning the vehicle for x years is given by
    $$A(x) = \frac{34800 - 36000(0.85)^x + 1200(1.35)^x}{x}.$$
    (d) Explain why one would want to minimize the average cost per year when deciding how long to own a vehicle.
    (e) Compute the derivative A'.
    (f) Show that the function A has a critical point when
    $$34800 = 1200(1.35)^x(x\ln(1.35) - 1) - 36000(0.85)^x(x\ln(0.85) - 1).$$
    Verify that this condition is satisfied when $x \approx 6.8257$. (This equation cannot be solved for x directly by hand. Technology is required.)
    (g) Conclude that, under these assumptions, it is optimal to purchase a new vehicle approximately every six years 10 months. If this strategy is implemented, what is the average annual cost of owning this vehicle?

4.2 Optimization and Modeling

In this section, we seek to apply the tools that we have learned to modeling problems. Our current focus is therefore on understanding how we can take a real-world problem and model it as a mathematical question that we can solve with the techniques of calculus.

Definition 4.2.1 An *optimization problem* occurs when one wishes to ascertain the maximum or minimum values of a certain quantity Q that depends on one or more physical variables.

A *constraint* is a relationship that must hold between two or more of the physical input variables on which the function Q depends.

Typically, whenever an optimization problem has more than one input variable, there will be corresponding constraints that will allow us to reduce Q to a function of a single variable. We will now recite some advice on approaching optimization problems that require mathematical modeling.

Proposition 4.2.1 — Strategies for Solving Optimization Problems. The following steps are commonly employed to solve an optimization problem:

1. **Read the problem.** The first step is to carefully read and understand the problem statement. It is also helpful to write down all of the information from the problem statement.
2. **Draw a picture.** If the problem setup can be depicted visually, draw a sketch of the situation. This is especially useful if the problem involves geometry.
3. **Assign variables.** Identify the quantity Q that you are trying to optimize and the quantity (or quantities) that are input variables. It is good practice to assign as few input variables as possible. For our purposes, we will never require more than two.
4. **Identify constraints.** If there are two (or more) input variables, identify any relationships that must hold between those variables. These are the constraint equations. Typically, you will have one fewer constraint equation than you have input variables.
5. **Reduce the variables.** Use the constraint equations to eliminate all but one input variable. Then express the quantity Q as a function of that single input variable.
6. **Identify the domain.** There may be physical restrictions on the input variables. Identify the domain on which the remaining input variable is valid.
7. **Determine the global extrema.** Use the techniques of Section 4.1 to find the global extrema on the domain in question.

■ **Example 4.12** Determine the maximum volume of a box that can be constructed out of a sheet of cardboard with dimensions 12" × 14", as shown in Figure 4.9.

The idea is that we can cut equal-sized squares out of each corner and then

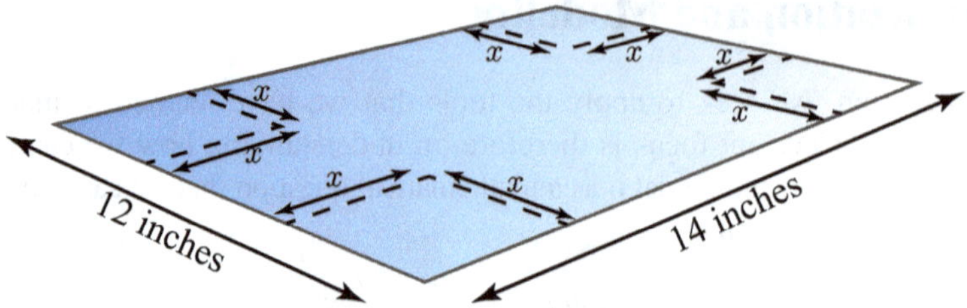

Figure 4.9: The piece of cardboard from Example 4.12.

create a box by folding the remaining flaps upward. A natural variable, therefore, is the side length of the corner squares, which we shall call x. The input variable x therefore represents the height of the resulting box, once the flaps are folded to construct a box. The physical dimension x is not arbitrary, but it must vary between 0" and 6". (At both of these endpoints, the box will have zero volume.)

Since a total length of $2x$ is removed from each side of the box, the box's final volume will be

$$V(x) = x(12-2x)(14-2x) = 4x^3 - 52x^2 + 168x.$$

The derivative of this function is given by

$$V'(x) = 12x^2 - 104x + 168.$$

Setting the derivative equal to zero, we obtain the following critical points for the function V:

$$x = \frac{104 \pm \sqrt{(104)^2 - 4(12)(168)}}{24} = \frac{13 \pm \sqrt{43}}{3} \approx \{2.147520,\ 6.519146\}.$$

The only critical point on the domain $[0,6]$ is therefore located at $x = \frac{13-\sqrt{43}}{3} \approx 2.147520$. Since $V(0) = V(6) = 0$, the maximum possible volume of the box is approximately $V(2.147520) = 160.5836612$ cubic inches. The function $V(x)$ is shown in Figure 4.10, which validates our conclusion. ■

■ **Example 4.13** Determine the dimensions of the right-circular cylinder that can be constructed using a fixed amount of surface area A and that has maximum volume.

First, let us draw a picture of the cylinder and assign variables. The cylinder is shown in Figure 4.11, in which we have defined the cylinder's height h and the cylinder's base radius r.

We note that there is a fundamental difference between the parameter A and the input variables r and h. The parameter A is to be regarded as a *fixed* constant, i.e., we will pretend like it is a fixed, known quantity. On the other hand, the variables r

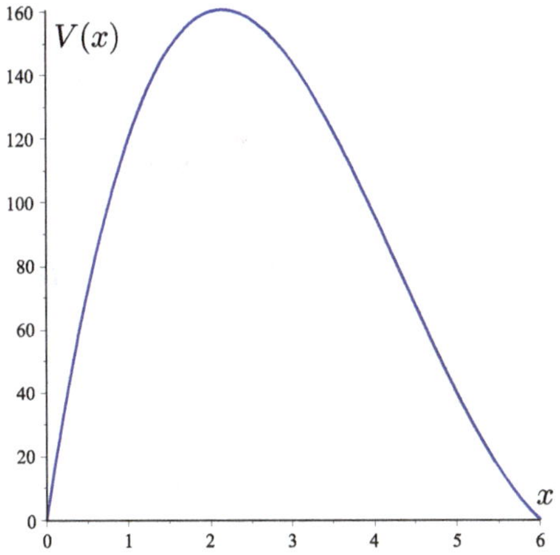

Figure 4.10: The volume V of the box as a function of the side length x of the cut-out square.

and h are design variables that are free to vary, subject to our constraint that the total surface area is the fixed amount A.

Next, we define the variable V to represent the volume of the cylinder, which is what we wish to optimize. The volume V is given in terms of the input variables by the relation

$$V = \pi r^2 h. \tag{4.2}$$

To write our constraint equation, we recognize that the area of the base and of the top of the cylinder are both equal to πr^2, and the area of the lateral side of the cylinder is given by the circumference of the base times the height, i.e., is given by $2\pi rh$. Thus, constraining the total surface area to equal the constant A yields the constraint equation

$$2\pi r^2 + 2\pi rh = A.$$

It is easier to solve for h in terms of r, rather than the other way around. Doing this, we obtain

$$h = \frac{A}{2\pi r} - r. \tag{4.3}$$

Clearly, the variable r must be positive. It can become arbitrarily close to zero at the cost of the cylinder's height h becoming arbitrarily large. On the other hand, there is also a maximum value for the variable r, which represents the cylinder's height approaching zero. If we set $h = 0$, we obtain

$$r = \sqrt{\frac{A}{2\pi}}.$$

We conclude that the domain for the variable r is given by $r \in \left(0, \sqrt{\frac{A}{2\pi}}\right)$.

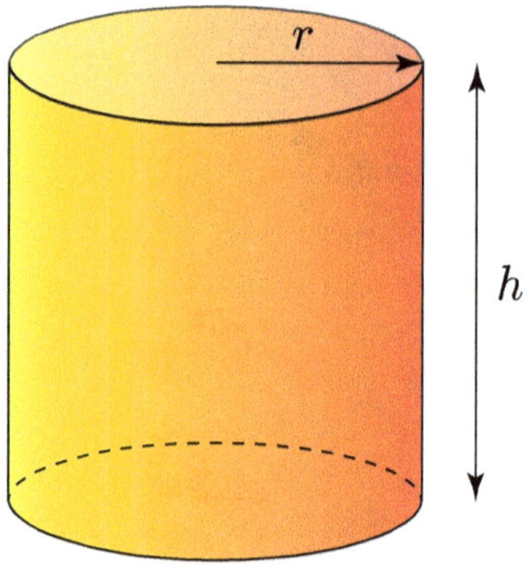

Figure 4.11: The cylinder from Example 4.13.

We may now use our constraint equation, in the form of equation (4.3), to eliminate the variable h in equation (4.2), thereby obtaining

$$V = \pi r^2 \left(\frac{A}{2\pi r} - r \right) = \frac{A}{2} r - \pi r^3.$$

Taking the derivative, we obtain

$$\frac{dV}{dr} = \frac{A}{2} - 3\pi r^2.$$

Setting this derivative equal to zero, we find that the only two critical points of the function V occur at the points

$$r = \pm\sqrt{\frac{A}{6\pi}}.$$

The only critical point in our domain is the positive root. Since the endpoints of the domain, $r = 0$ and $r = \sqrt{A/2\pi}$, correspond to zero volume, we conclude that the volume is maximized when the radius is

$$r = \sqrt{A/6\pi}. \tag{4.4}$$

(Also, note that $V'' = -6\pi r$. Therefore, $V'' < 0$ at this critical point and it is a local maximum according to the second-derivative test.) Substituting this back into the constraint equation (4.3), we find that the optimum height of the cylinder is

$$h = \left(\sqrt{3} - \frac{1}{\sqrt{3}} \right) \sqrt{\frac{A}{2\pi}}. \tag{4.5}$$

The maximum obtainable volume is therefore

$$V_{\max} = \frac{A}{6}\left(\sqrt{3} - \frac{1}{\sqrt{3}}\right)\sqrt{\frac{A}{2\pi}}.$$

To confirm this result, we can graph the function $V(r)$, with the parameter A set to the value of $A = 1$ (arbitrarily, just so we can use actual numbers). This function is shown in Figure 4.12, plotted on the domain $(0, 1/\sqrt{2\pi})$. The maximum volume

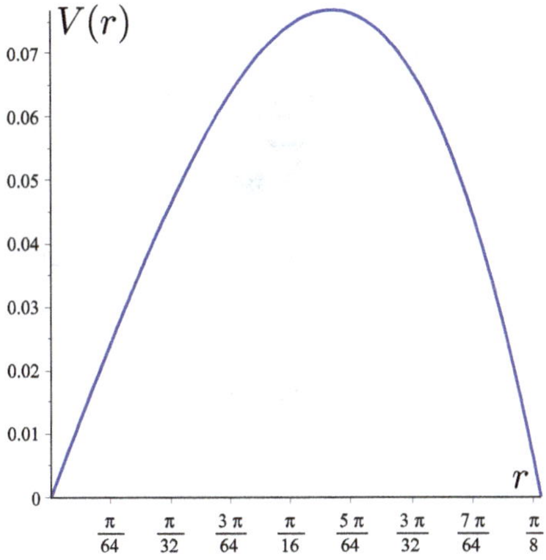

Figure 4.12: The function $V(r)$ in Example 4.13.

indeed occurs at the point $r = 1/\sqrt{6\pi} \approx 0.230329$, and the maximum volume is indeed $V = \left(\sqrt{3} - \frac{1}{\sqrt{3}}\right)\frac{1}{6\sqrt{2\pi}} \approx 0.076776$.

We see that, with calculus, we solved an entire class of problems just as easily as we may have solved a single problem with a given value of the parameter A. This is the usefulness of parameters: one does not have to settle for one particular value of the parameter. Perhaps you are tasked with optimizing 100 different cylinders, each with a different amount of total surface area A. We solved all 100 problems simultaneously. The optimal radius and height will be given by the formulas (4.4) and (4.5), regardless the value of the parameter A. ■

■ **Example 4.14** Two street lamps in a city park are arranged as shown in Figure 4.13. Each lamp is 1 unit tall and the distance between the two lamps is 2 units. Assuming that the intensity of the light at a point is inversely proportional to the distance between the point and the lamp squared, determine the points on the ground, between the two lamps, that are the brightest and the dimmest.

We begin by defining the x-axis as the horizontal axis along which the two street lamps lie. Let the point $x = 0$ represent the midpoint between the two street lamps, so that the street lamps are located at $x = -1$ and $x = 1$, as shown in Figure 4.14.

Figure 4.13: Two street lamps, as given in Example 4.14.

Furthermore, let r_1 and r_2 represent the distance between a point on the x-axis and

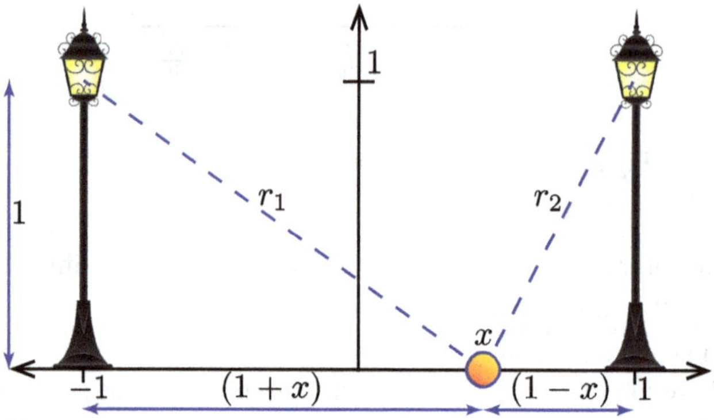

Figure 4.14: Schematic of the two street lamps.

the left and right street lamps, respectively, also shown in the Figure. Since the horizontal distance between a point on the x-axis and the two endpoints $x = -1$ and $x = 1$ is given by the formulas $(1+x)$ and $(1-x)$, respectively, the Pythagorean theorem tells us that r_1 and r_2 are equal to

$$\begin{aligned} r_1 &= \sqrt{(1+x)^2+1} = \sqrt{x^2+2x+2}, \\ r_2 &= \sqrt{(1-x)^2+1} = \sqrt{x^2-2x+2}. \end{aligned}$$

We therefore conclude that the total light intensity I at the point x is given by the relation

$$I(x) = \frac{a}{x^2+2x+2} + \frac{a}{x^2-2x+2},$$

where $a > 0$ represents the light intensity on the ground directly below either lamp. Computing the derivative using the chain rule, we find

$$\frac{dI}{dx} = \frac{-a(2x+2)}{(x^2+2x+2)^2} + \frac{-a(2x-2)}{(x^2-2x+2)^2}.$$

Completing the square, this, in turn, is equivalent to

$$\begin{aligned} \frac{dI}{dx} &= -a\frac{(2x+2)(x^2-2x+2)^2+(2x-2)(x^2+2x+2)^2}{(x^2+2x+2)^2(x^2-2x+2)^2} \\ &= -4a\frac{x^5+4x^3-4x}{(x^2+2x+2)^2(x^2-2x+2)^2}. \end{aligned}$$

Thus, a point is a critical point only if $I'(x) = 0$, which can only occur when the numerator $(x^5+4x^3-4x) = 0$. Thus, there are critical points when $x = 0$ and when $(x^4+4x^2-4) = 0$. This latter relation can be solved for x^2 using the quadratic formula:

$$x^2 = \frac{-4 \pm \sqrt{16+16}}{2} = -2 \pm 2\sqrt{2}.$$

Since the negative root results in a negative number for x^2, and hence an imaginary number for x, we can omit this. The two remaining real critical points are therefore

$$x = \pm\sqrt{2\left(\sqrt{2}-1\right)} \approx \pm 0.9101797.$$

Evaluating the intensity function at the three critical points and at the endpoints $x = -1$ and $x = 1$, we find

$$\begin{aligned} I(\pm 1) &= 1.2a, \\ I\left(\pm\sqrt{2\left(\sqrt{2}-1\right)}\right) &= 1.20710678a, \\ I(0) &= a. \end{aligned}$$

We conclude that the dimmest point is the point on the ground exactly between the two lamps, and the two brightest points are the points $x = \pm\sqrt{2\left(\sqrt{2}-1\right)} \approx \pm 0.9101797$. This is verified in the graph of the function $I(x)$, as shown in Figure 4.15. ∎

∎ **Example 4.15** In our final example, we seek to determine how far away from the statue of liberty a boat should be positioned in order to optimize its viewing angle of the monument. Let us assume that the dock of the boat is on level with the base of the pedestal, which is 20 m in height, and that the statue is 45 m in height, as shown in Figure 4.16.

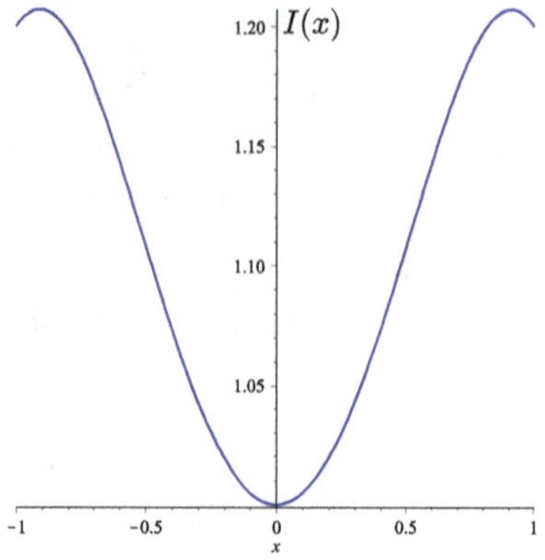

Figure 4.15: Light intensity between the two street lamps as a function of x.

We begin by defining our variables: let x represent the distance (in meters) between the boat and the statue; let θ represent the viewing angle of the statue, as shown in the figure; and let φ represent the angle between the lower line of sight and the horizon. Finally, let us define ψ as the total angle between the horizontal and the upper line of sight, so that $\psi = \varphi + \theta$.

To proceed, we use trigonometry to express the angles φ and ψ in terms of the input variable x, obtaining the relations

$$\tan(\varphi) = \frac{20}{x} \qquad \text{and} \qquad \tan(\psi) = \frac{65}{x}.$$

It follows that

$$\varphi = \tan^{-1}\left(\frac{20}{x}\right) \qquad \text{and} \qquad \psi = \tan^{-1}\left(\frac{65}{x}\right).$$

Therefore, since $\theta = \psi - \varphi$, we may express the viewing angle θ as a function of the boat's distance x using the relation

$$\theta(x) = \tan^{-1}\left(\frac{65}{x}\right) - \tan^{-1}\left(\frac{20}{x}\right).$$

Using equation (3.72) coupled with the chain rule, the derivative of the viewing angle θ with respect to the distance x is given by

$$\begin{aligned}
\theta'(x) &= \frac{-\frac{65}{x^2}}{1+\left(\frac{65}{x}\right)^2} - \frac{-\frac{20}{x^2}}{1+\left(\frac{20}{x}\right)^2} \\
&= \frac{20}{x^2+400} - \frac{65}{x^2+4225} \\
&= \frac{58500-45x^2}{(x^2+400)(x^2+4225)}.
\end{aligned}$$

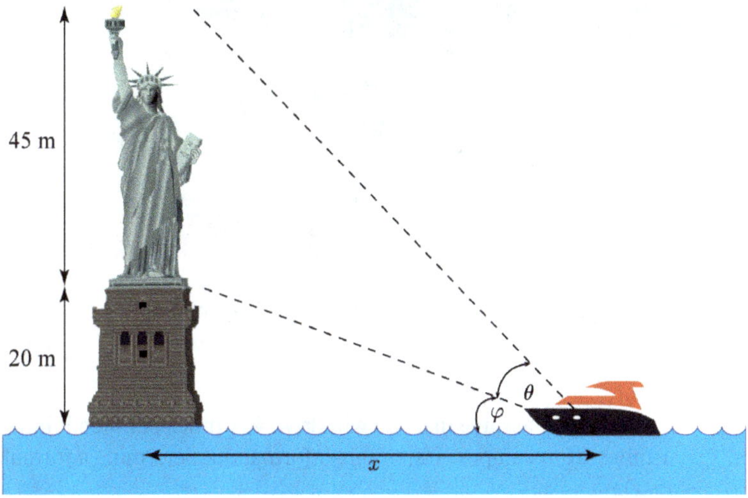

Figure 4.16: Viewing angle for the Statue of Liberty.

Therefore, the only two critical points of the function $\theta(x)$ are at $x = \pm\sqrt{1300}$. We conclude that the boat should be approximately 36 m away from the statue in order to achieve the maximum viewing angle. ■

Problems

1. Determine the dimensions of the rectangle with largest area that can be inscribed in a unit circle.
2. A right-circular cylinder is to be constructed to have a fixed volume V. Determine the optimal dimensions (in terms of the parameter V) that minimize the total surface area.
3. An attic wall is the shape of a right triangle with base measurement 8 feet and height 12 feet. This shape is equivalent to the area under the graph of the function $y = 12 - 1.5x$ in the first quadrant. Determine the measurements of the largest rectangular painting that can be placed against this wall.
4. A farmer wants to create an area that is enclosed by a fence and a brick wall, as shown in Figure 4.17. If the farmer has 100 feet of fencing, what dimensions should the farmer use in order to maximize the enclosed area?
5. Let $R(x)$ be the typical rent of a one-bedroom apartment located x miles away from the central campus of the University of Michigan, and let $G(x)$ represent the total monthly cost of gasoline one incurs by commuting if one is to live x miles away from campus. Let $C = R + G$ represent the total monthly cost associated with living x miles from campus.
   (a) Assume that Jon's car has an average gas mileage of 20 mpg, and that gasoline can be purchased at the average rate of \$2.50 per gallon.

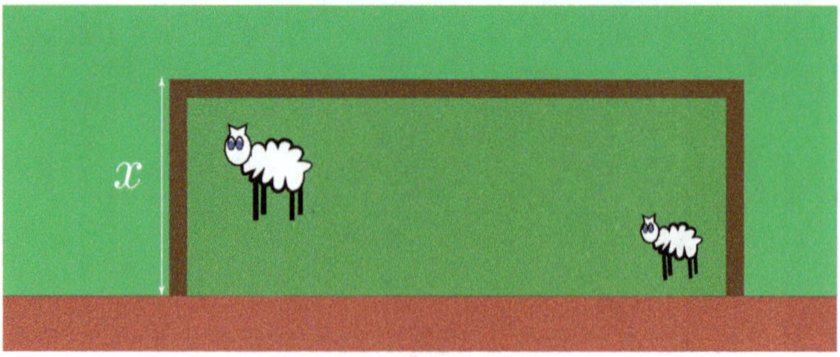

Figure 4.17: Fence and a brick wall; Problem 4.

Furthermore, assume that Jon will drive $40x$ miles per month if he lives x miles from campus. Using this information, determine a formula for $G(x)$.

(b) Let d represent the optimal distance from campus that one should live in order to minimize the total cost. Using the previous information, determine the value of $R'(d)$.

(c) Suppose that the average rent price can be modeled by the formula

$$R(x) = \frac{x^3}{16} - \frac{15x^2}{8} + 750.$$

Assuming that Jon wishes to live no farther than 20 miles from campus, determine how far away he should live in order to minimize his total cost.

6. Consider a water tank with height H. If one were to make a small hole in the side of the tank at a height of x, the distance D away from the tank that the water would land is given by the formula

$$D = \sqrt{4x(H-x)}.$$

(This equation can be derived from Bernouilli's principle from fluid mechanics.) The situation is illustrated in Figure 4.18. Determine the height x that the hole should be placed in order to maximize the initial distance D that the water will squirt.

7. A box with a square bottom and no top is to be constructed such that its total volume is 25 ft^3. Determine the dimensions of such a box that minimize the surface area used in its construction.
8. Repeat Problem 7 if the top is to be included in the box.
9. A right-circular cone is to be constructed from a circular sheet of paper of radius R by removing a wedge and gluing the two cut-out edges together. Determine the maximum volume of such a cone. Your answer may depend on the parameter R.

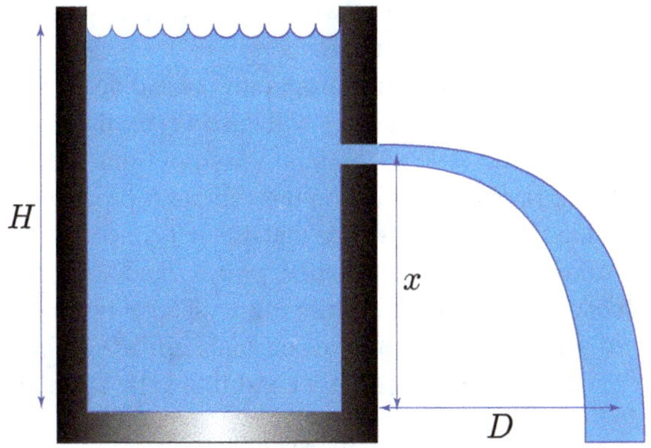

Figure 4.18: A leaking water tank; Problem 6.

10. A projectile that is launched at an angle of θ will land a distance of

$$D = \frac{2v_0^2 \sin\theta \cos\theta}{g}, \tag{4.6}$$

where v_0 is the speed of the projectile at launch and g is the acceleration due to gravity. The situation is shown in Figure 4.19. Determine the angle θ that will maximize the distance D of the projectile. What is the optimal distance D. Your answer may be in terms of the parameters g and v_0. (*Physics bonus*: derive equation (4.6).)

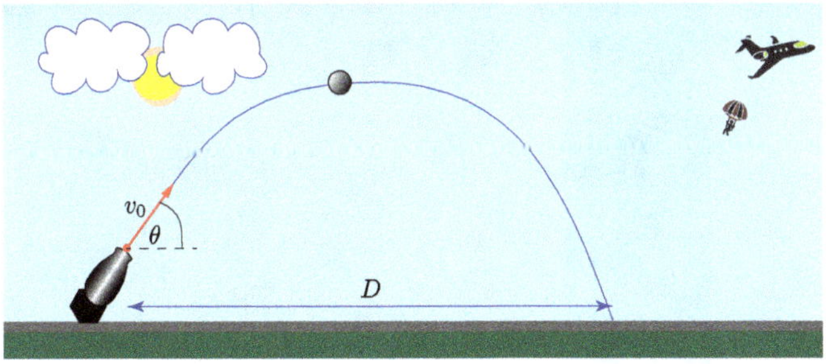

Figure 4.19: Launch of a projectile; Problem 10.

11. In this problem, we will study the optimal cruising speed that maximizes the fuel economy of an automobile. Let v represent the (steady) cruising speed of the vehicle, measured in mph; let $R = R(v)$ represent the rate of fuel consumption at speed v, measured in gph (gallons per hour); and let $\eta = \eta(v)$ represent the fuel efficiency obtained when cruising at a speed v, measured in mpg (miles per gallon).

(a) Determine a relationship between v, R, and η. In particular, express the function η in terms of v and R.

(b) Differentiate the expression from part (a) and determine the condition under which the fuel efficiency will have a critical point. $\left(Ans.\ \frac{dR}{dv} = \frac{R}{v}\right)$.

(c) Next, let us add several modeling assumptions: first, let us assume that the air resistance acting on the vehicle is proportional to the speed squared. Second, let us assume that the fuel consumption rate is directly proportional to the air resistance, *plus* a fixed amount that represents the fuel consumption rate while idling. These assumptions imply that $R = av^2 + b$, for certain positive constants $a, b > 0$. Using the result from part (b), show that, under these conditions, the optimal fuel efficiency is achieved when cruising at a speed $v^* = \sqrt{\frac{b}{a}}$.

(d) Draw a sketch of the graph of $R(v)$. Study the meaning of the optimality condition from part (b) graphically. (*Hint*: what is the slope of the tangent line at the optimal cruising speed? What is the slope of the line that connects the origin to a point on the graph (v, R)?) Explain why this may be so.

12. A company wishes to ship goods between two cities: A and B. City A is a port city, located on the coast, and City B is located 100 miles north of City A and 50 miles inland, as shown in Figure 4.20. The company will ship

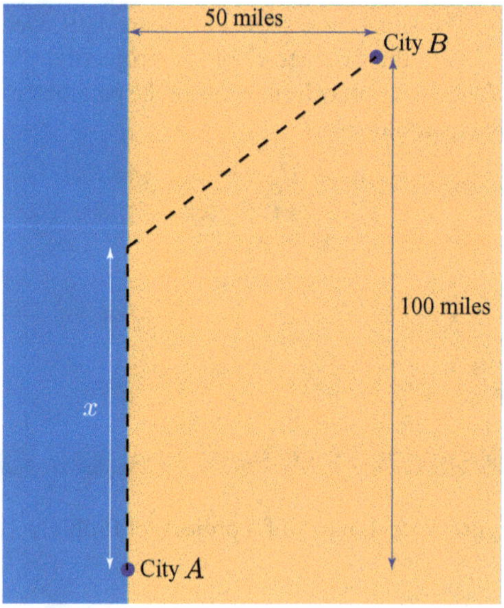

Figure 4.20: Shipping by land and by sea; Problem 12.

the goods by sea a distance of x miles up the coast, where the goods will be transferred to a truck and shipped the remaining distance by land. If a shipment costs \$0.70 per mile to ship by sea and \$1.20 per mile to ship by land, determine the location x where the goods should be transferred in order to minimize the shipping costs.

13. Assuming the setup from Problem 12, if shipping by land costs a factor of α more than shipping by sea, determine the optimal location x where the goods should be transferred, i.e., determine x as a function of α. (For Problem 12, $\alpha = 12/7$.)
14. A crate with mass m is being pulled across a level surface by an attached rope that makes an angle θ with the horizontal, as shown in Figure 4.21. If the crate is acted upon by a constant downward gravitational acceleration g and has a coefficient of friction μ, then the force required to pull the crate at a constant speed is equal to

$$F = \frac{\mu m g}{\cos(\theta) + \mu \sin(\theta)}.$$

Determine the angle θ that minimizes the force required to pull the crate.

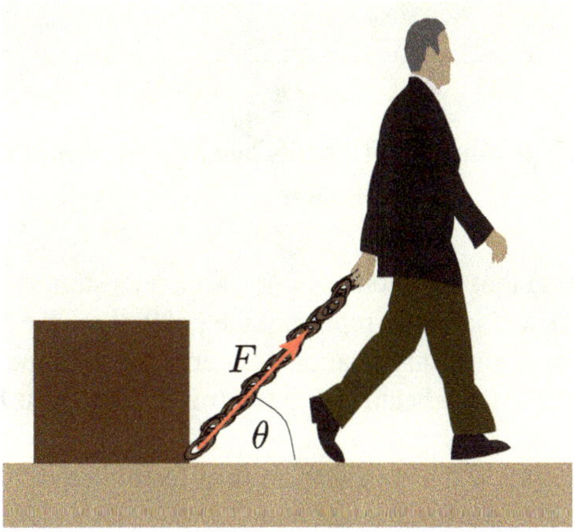

Figure 4.21: A man pulling a crate; Problem 14.

15. Photons of light travel at different speeds through different mediums (e.g., through vacuum, air, water, etc.). *Fermat's principle* states that a photon of light that travels between two points P_1 and P_2 will take the path that minimizes its time. In this problem, we will study the path of a light beam through two different mediums, as shown in Figure 4.22. We will normalize distance so that the horizontal distance between P_1 and P_2 is one unit. Suppose that the medium interface is halfway between the vertical separation of the two points, and that the vertical distance from either point to the medium is a units. The angles θ_1 and θ_2 are as shown. Finally, suppose that a photon can travel with speed v_1 through medium 1 and with speed v_2 through medium 2.

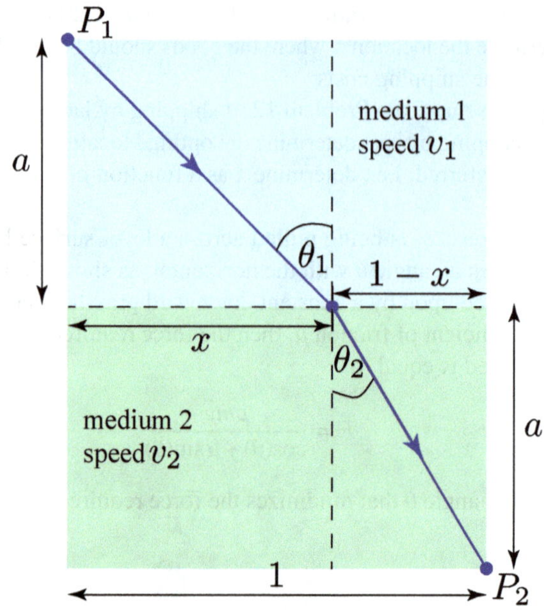

Figure 4.22: Illustration of Snell's law of refraction; Problem 15.

(a) Assuming that a photon of light travels in a straight line through either medium, and letting x represent the point along the interface where the light beam suddenly changes direction, determine a formula for the total time T for a beam of light to travel from point P_1 to point P_2 as a function of x.

(b) Differentiate $T(x)$, and determine a formula that the variable x must satisfy in order for T to have a critical point at that value of x.

(c) Use the result from part (b) to show that Fermat's principle implies the condition

$$\frac{v_1}{v_2} = \frac{\sin\theta_1}{\sin\theta_2}.$$

This relation is known as *Snell's law.*

4.3 Related Rates

In our next application, we apply the tools of calculus and further our modeling skills to the analysis of situations in which multiple variables are each changing in time in a coordinated fashion.

Definition 4.3.1 A *related-rates problem* is one in which multiple variables are changing in time, and where those variables and their rates of changes are connected by a relation known as the *related-rates equation*.

As was the case with optimization problems, there are typically one fewer related-rates equations than variables. We now offer some advice on how to set up, model, and solve related-rates problems.

Proposition 4.3.1 — Strategies for Solving Related-Rates Problems. The following steps are commonly employed when solving related-rates problems:

1. **Read the problem.** The first step is to read and understand the problem statement. It is also helpful to write down all of the information from the problem statement.
2. **Draw a picture.** If the problem setup can be depicted visually, draw a sketch of the situation. This is especially useful if the problem involves geometry.
3. **Identify the quantities that are changing in time.** In a related-rates problem, certain quantities will be changing in time and others will be fixed constants. Identify the variables that are changing in time. In particular, be sure to distinguish between numbers that are given that represent constants and numbers that are given that only represent the instantaneous value of a time-changing quantity at some moment.
4. **Assign variables.** Assign a variable name to each quantity that is changing in time. It is always best to achieve this with as few variables as possible.
5. **Identify constraints.** Determine any relationship that must hold between the dynamic quantities (i.e., between the time-dependent variables).
6. **Derivate the related-rates equation.** Differentiate any constraint equation to obtain the related-rates equation. Solve for the unknown.

■ **Example 4.16** Consider a ladder with length L being pulled away from a wall at a constant rate of 2 m/s, as shown in Figure 4.23. Suppose that the ladder is 5 m long. At the moment in time when the base of the ladder is exactly 3 m from the wall, at what rate is the height of the top of the ladder above the ground changing?

Since a picture has already been given, let us begin by introducing variables. The distance between the bottom of the water and the wall and the distance between the top of the ladder and the floor are both changing in time; let us denote these two quantities as x and y, respectively. Since the length of the ladder is given to us as $L = 5$—a quantity that is constant in time—the dynamic variables are related via the Pythagorean theorem, which gives us the constraint equation

$$x^2 + y^2 = 25. \tag{4.7}$$

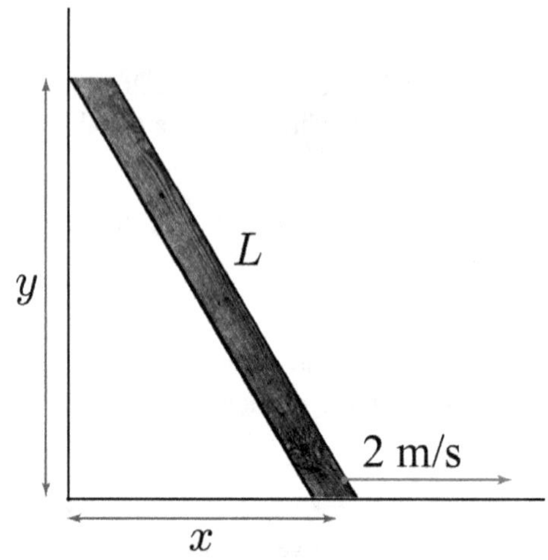

Figure 4.23: A ladder being pulled away from a wall.

Differentiating our constraint equation with respect to time, we obtain

$$2x\frac{dx}{dt} + 2y\frac{dy}{dt} = 0. \tag{4.8}$$

Notice that we used the chain rule. This can be done since both variables x and y depend on time, i.e., $x = x(t)$ and $y = y(t)$. At the moment in question, $x = 3$. Therefore, from the constraint equation (4.7), we know that $y = 4$ at that same moment. Moreover, we are told that $\frac{dx}{dt} = 2$. Therefore, the related-rates equation (4.8) tells us that

$$\frac{dy}{dt} = -\frac{x}{y}\frac{dx}{dt} = -\frac{3}{2}.$$

Thus, the height of the ladder is decreasing at a rate of 1.5 m/s at the moment when the foot of the ladder is 3 m from the wall. ■

■ **Example 4.17** Consider two aircrafts, as shown in Figure 4.24. The first is located 60 miles east of San Francisco Airport (SFO), traveling due east with a groundspeed of 380 mph. The second is located 50 miles north of Los Angeles Airport (LAX), traveling due south with a groundspeed of 300 mph. SFO is located 195 miles west and 230 miles north of LAX. Determine the rate at which the distance between these two aircrafts is changing at this moment in time.

To begin, we assign variables as follows: let x be the distance (in miles) between the first plane and SFO and let y be the distance (in miles) between the second plane and LAX, as shown in Figure 4.25. Let D represent the distance (in miles) between the two aircraft. The Pythagorean theorem tells us that the variable D is related to the variables x and y by the equation

$$D^2 = (195 - x)^2 + (230 - y)^2. \tag{4.9}$$

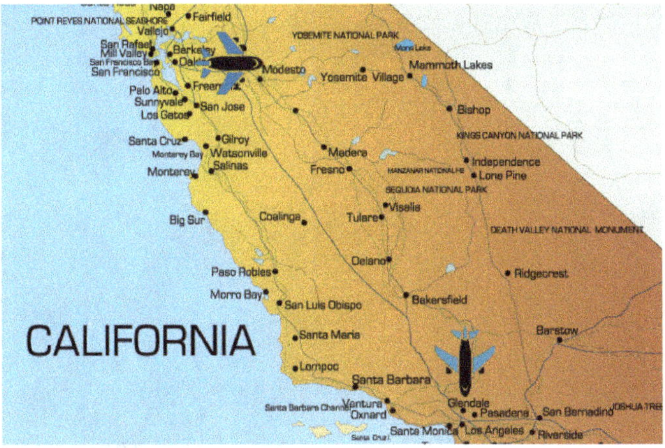

Figure 4.24: A map of California showing the two planes from Example 4.17.

At the moment in question, we are given that

$$x = 60, \qquad y = 50, \qquad \frac{dx}{dt} = 380, \qquad \text{and} \qquad \frac{dy}{dt} = -300.$$

One can use the constraint equation (4.9) to determine that the total distance between the two planes at that instant works out (luckily enough) to be exactly 225 miles, so that $D = 225$ (at that single moment).

Next, differentiating the constraint equation (4.9) with respect to time gives us the related rates equation

$$2D\frac{dD}{dt} = -2(195 - x)\frac{dx}{dt} - 2(230 - y)\frac{dy}{dt}.$$

Substituting in the instantaneous values of the variables at that moment, we obtain

$$\frac{dD}{dt} = \frac{-(135)(380) + (180)(300)}{225} = 12.$$

Therefore, the distance between the two planes is increasing at a rate of 12 mph at that moment in time. ■

■ **Example 4.18** The hour hand and minute hand of Big Ben have a length of 2.7 m and 4.2 m, respectively. The clock face is shown in Figure 4.26. At what rate is the distance between the tips of the two clock hands changing when the time is 4:00?

First, we assign variables: let D and θ represent the distance (meters) and angle measurement (radians) between the tips of the two clock hands, respectively. These two quantities are related due to the law of cosines, which tells us

$$D^2 = 2.7^2 + 4.2^2 - 2(2.7)(4.2)\cos(\theta). \tag{4.10}$$

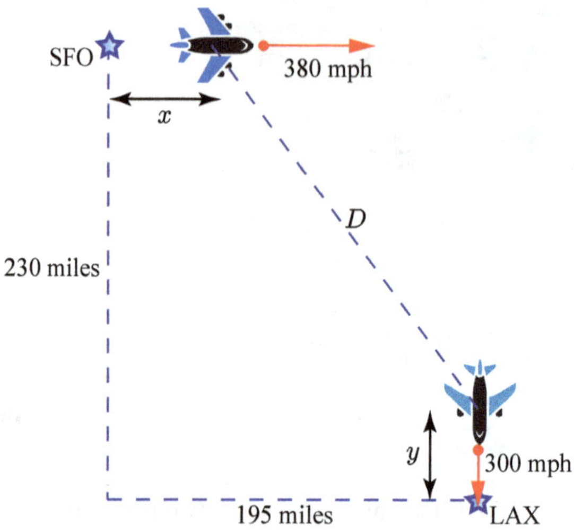

Figure 4.25: A schematic of the aircraft in Example 4.17.

This is our constraint equation. At four o'clock, the angle θ is equal to $\theta = 2\pi/3$. Thus, the distance between the two clock hands at four o'clock can be determined using the constraint equation (4.10) as $D = \sqrt{36.27}$ m.

Next, we differentiate the constraint equation with respect to time in order to produce the related-rates equation; we obtain

$$2D\frac{dD}{dt} = 22.68\sin(\theta)\frac{d\theta}{dt}.$$

therefore, at four o'clock, the distance between the tips of the two clock hands is changing at a rate of

$$\frac{dD}{dt} = \frac{11.34}{\sqrt{36.27}}\sin\left(\frac{2\pi}{3}\right)\frac{d\theta}{dt} = \frac{5.67}{\sqrt{12.09}}\frac{d\theta}{dt}.$$

All that remains is to ascertain the rate of change of the angle θ at four o'clock. To do this, we note that the minute hand has an angular rate of 2π radians per hour, and the hour hand has an angular rate of $\pi/6$ radians per hour. The rate of change of the angle between the two hands is therefore $\frac{d\theta}{dt} = \frac{11\pi}{6}$ radians per hour. We conclude that the rate of change of the distance between the tips of the two hands is given by

$$\frac{dD}{dt} = \frac{5.67}{\sqrt{12.09}}\frac{11\pi}{6} = \frac{10.395\pi}{\sqrt{12.09}},$$

or approximately 9.392068 m/hr, or about 2.6 mm/s. ∎

Figure 4.26: The clock hands of Big Ben at 4:00.

■ **Example 4.19** In this example, we will analyze the workings of an automobile engine, which approximates a certain thermodynamic process known as the *Otto cycle*. During the expansion stroke, the compressed gas trapped in the piston-cylinder assembly expands, pushing the piston outward. This phase of the engine cycle is modeled as an *isentropic process*, which means the total *entropy* in the engine is conserved during expansion. For an isentropic process, the pressure P and volume V of the expanding gas are constrained by the relation $PV^{\gamma} = c$, where c is a constant and where γ is the heat capacity ratio (the parameter γ has a value of $\gamma = 1.4$ for air). The end of the piston is attached to a rod that is connected to the crankshaft, as shown in Figure 4.27. Thus, the expansion and compression of the gas in the piston-cylinder assembly give rise to a torque that is applied to the crankshaft that generates rotation.

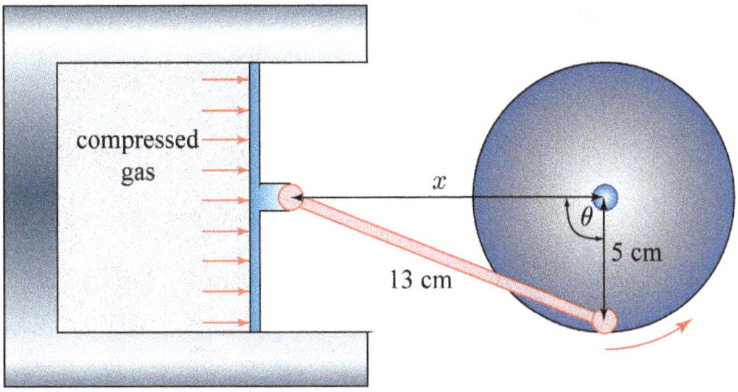

Figure 4.27: Schematic of a piston-cylinder and camshaft.

In a certain engine, the cylinder has a bore radius of 5 cm, the rod has a length of 13 cm, and the crankshaft has a radius of 5 cm. At one particular moment during the expansion stroke, the pressure in the piston-cylinder assembly has a value of $P = 0.876$ MPa (megaPascals) and a volume of $V = 580\ \text{cm}^3$, and the pressure is decreasing at a rate of 182 MPa/s. At this moment, the crank angle θ, as shown in the figure, is 90°. Determine the angular rate at which the crankshaft is rotating at this moment.

This problem requires a three-part analysis.

Part I: Isentropic Expansion. We are told that the expansion stroke is an isentropic process and therefore follows the thermodynamic relation

$$PV^{\gamma} = c.$$

Differentiating this relation with respect to time, we obtain our first related-rates equation:

$$\gamma PV^{\gamma-1}\frac{dV}{dt} + V^{\gamma}\frac{dP}{dt} = 0.$$

Solving for the volume rate of change, we obtain

$$\frac{dV}{dt} = -\frac{V}{\gamma P}\frac{dP}{dt} \approx 86{,}073\ \text{cm}^3/\text{s}.$$

Part II: Volume and Position. Let h represent the length of the open space in the piston-cylinder assembly. Since the radius of the cylinder is 5 cm, the volume of the chamber is related to its length by the relation

$$V = 25\pi h.$$

Differentiating this with respect to time, we obtain

$$\frac{dV}{dt} = 25\pi\frac{dh}{dt}.$$

Next, if we let the variable x represent the distance between the center of the camshaft and the point where the rod is connected to the cylinder, we recognize that the quantity $h + x$ must be a constant. Therefore

$$\frac{dx}{dt} = -\frac{dh}{dt}.$$

Combining the previous two results, we find that

$$\frac{dx}{dt} = -\frac{1}{25\pi}\frac{dV}{dt} \approx -1096\ \text{cm/s}.$$

Part III. Camshaft rotation. Finally, to relate the motion of the piston to the rotation of the camshaft, we use the law of cosines, which tells us that $13^2 = 5^2 + x^2 - 10x\cos(\theta)$, or, more simply,

$$144 = x^2 - 10x\cos(\theta).$$

Differentiating this with respect to time, we obtain

$$0 = 2x\frac{dx}{dt} - 10\cos(\theta)\frac{dx}{dt} + 10x\sin(\theta)\frac{d\theta}{dt}.$$

Therefore:

$$\frac{dx}{dt} = \frac{-10x\sin(\theta)}{2x - 10\cos(\theta)}\frac{d\theta}{dt}.$$

Since $\theta = \pi/2$ and $x = 12$ at this moment, we have

$$\frac{d\theta}{dt} = -\frac{1}{5}\frac{dx}{dt} \approx 219 \text{ rad/s}.$$

Thus, the camshaft is rotating at approximately 2100 revolutions per minute (rpm) at this moment in time. ∎

Problems

1. The *kinetic energy* K of an object, measured in Joules (J), is given by the relation
$$K = \frac{1}{2}mv^2,$$
where m is the mass of the object (kg) and v is the velocity of the object (m/s). If a 3 kg block is accelerating downward at a rate of 9.8m/s^2, at what rate is its kinetic energy changing at the moment when it falling downward at a rate of 20 m/s?
2. The *total mechanical energy* E of an object in a uniform gravity field, with gravitational acceleration g, is given by
$$E = \frac{1}{2}mv^2 + mgh,$$
where m is the mass of the object, v is the object's velocity, and h is the object's height above a reference point (e.g., the ground). If the total mechanical energy of an object is constant, prove that its acceleration must also be a constant, and, moreover, it must be given by $\frac{dv}{dt} = -g$.
3. A spherical snowball melts at a rate proportional to its surface area. Show that its radius must be changing at a constant rate.
4. A sailboat is being pulled to a boat dock with the aid of a pulley, as shown in Figure 4.28. The top of the pulley is 5 ft above the water level. Determine the rate at which the rope is being pulled in at the moment when the boat is 12 ft from the dock and moving toward the dock at a rate of 0.5 ft/s.
5. A skydiver is falling at the rate of 16 mph as a car with a speed of 70 mph is driving toward the area just below, as shown in Figure 4.29. At what rate is the distance between the car and the skydiver changing at the moment when the skydiver is $1/3$ miles above the ground and the car is 2 miles away from the point on the road directly below the skydiver?

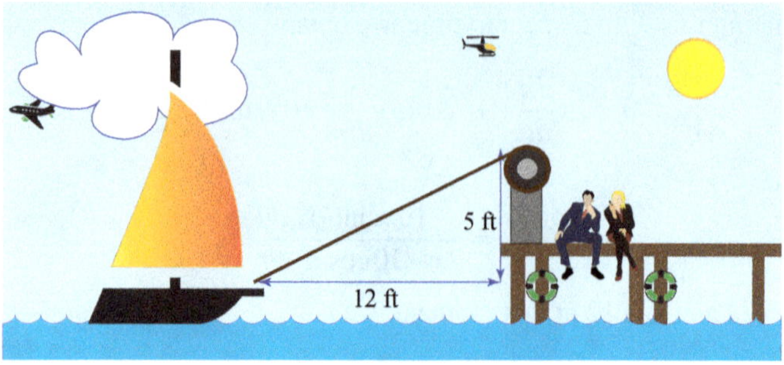

Figure 4.28: Sailboat for Problem 4.

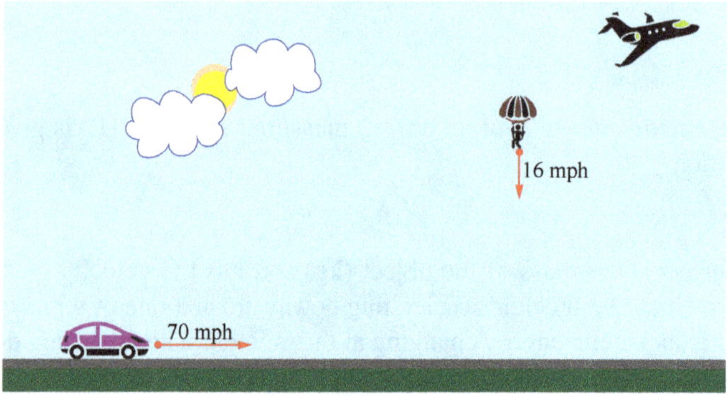

Figure 4.29: Skydiver and automobile for Problem 5.

6. A martini glass is in the shape of an inverted right-circular cone, with base radius 5 cm and height 5 cm. A cocktail is being poured into the glass at the rate of 25 cm^3/s. At what rate is the level of the liquid changing at the moment when the liquid level is 3 cm high.
7. A swimming pool is in the shape of a hemisphere with radius 5 m. The swimming pool is being filled at a rate of 4 m^3/min. At what rate is the water level changing at the moment when the water level is 3 m?
8. A spherical balloon is deflating at the rate of 5 cm^3/s. At what rate is its radius changing the moment when its radius is 12 cm?
9. Sand is falling onto a pile in the shape of a right-circular cone whose height is twice its base radius, accumulating at the rate of 18 cm^3/s. At what rate is the height of the pile increasing the moment the height is 10 cm?
10. A six-foot tall businessman is walking away from a ten-foot tall streetlamp, as shown in Figure 4.30. At the moment he is two feet away from the lamp, at what rate is the length of his shadow changing?

Figure 4.30: A businessman walking under a streetlamp for Problem 10.

11. The ideal gas law states that

$$Pv = RT,$$

where P represents the pressure of the gas, v represents the *specific volume* (i.e., the volume per unit of mass: literally the reciprocal of the density), R represents the gas constant of the particular gas (for air, $R = 287 \text{ J kg}^{-1} \text{ K}^{-1}$), and T represents the *absolute* temperature of the gas (i.e., the temperature as measured from absolute zero). Air with a temperature of 400 K and a pressure of 5 MPa is trapped in a box and is cooling at a rate of 5 K/s. At what rate is the pressure changing at this moment?

12. A certain snowman is comprised of three spherical snowballs, whose radii are in a 1:2:3 proportionality, which remains fixed as the snowman melts, as shown in Figure 4.31. Let r represent the radius of the top snowball (the snowman's head), so that the middle and bottom snowballs have radii $2r$ and $3r$, respectively. At noon on a certain day, $r = 6$ inches, and the snowman's height is decreasing at a rate of 1 inch/hr. Determine the rate at which the snowman's volume is changing.

13. A certain comet can be modeled as a spherical chunk of ice that is in orbit about the sun. The sun's intensity I at a distance of s from the center of the sun is inversely proportional to the distance squared, i.e.,

$$I(s) = \frac{k}{s^2},$$

for some constant $k > 0$. Suppose that the rate of change of the comet's

Figure 4.31: The snowman from Problem 12.

surface area A is proportional to the intensity of the sunlight, so that

$$\frac{dA}{dt} = cI(s),$$

for some constant $c > 0$. Determine a formula for the rate of change of the comet's volume in terms of the comet's radius r and its distance to the sun s and in terms of the parameters k and c.

14. A Boeing 747 aircraft is making its descent into Los Angeles International Airport, as shown in Figure 4.32. At one moment, its altitude h is 29,500 ft, its ground distance D to the airport is 253,440 ft, and its ground speed is 440 ft/s. At this moment, its guidance angle θ stops increasing and starts decreasing. Determine the airplane's rate of descent at that moment.
15. A spherical balloon is being inflated so that its radius r after t seconds is given by the relation

$$r(t) = R_{\max}(1 - e^{-\alpha t}),$$

where $\alpha > 0$ is a positive constant.
    (a) Compute the rate of change of the balloon's volume at time t. Your answer may depend on the parameters $R_{\max}$ and α.
    (b) Determine the time t_0 at which the volume is increasing at the greatest rate. What is the value of this rate (in terms of the parameters)?
    (c) Suppose that $R_{\max} = 8$ cm and $\alpha = \frac{1}{3}$. Determine t_0 and $\frac{dV}{dt}(t_0)$.
16. The *loudness* L of a sound (decibels; dB) with *sound intensity* I (kW/m^2) is given by the relation

$$L(I) = 10\log\left(\frac{I}{I_0}\right), \qquad (4.11)$$

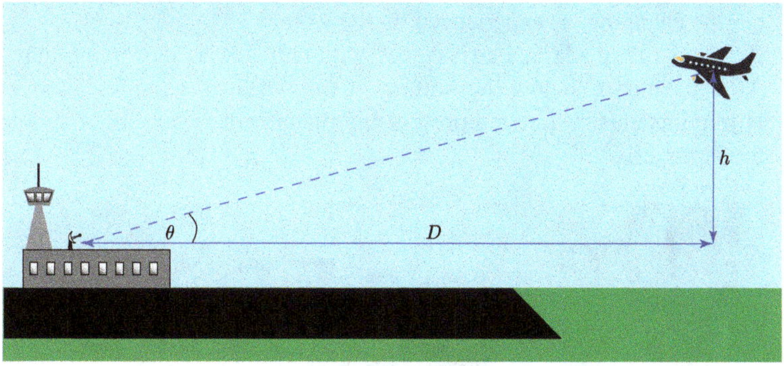

Figure 4.32: An aircraft on approach; Problem 14.

where I_0 is a benchmark sound intensity, which is set to $I_0 = 10^{-12}$ W/m^2 by convention (this approximates the threshold of human hearing). The sound intensity I is inversely proportional to the square of the distance r to the source of the sound, so that

$$I(r) = \frac{k}{r^2},$$

for some constant k.

A jet engine has a loudness of 140 dB at a distance of 30 m. A woman is walking away from the engine at a rate of 1.2 m/s, as shown in Figure 4.33. At what rate is the loudness changing at the moment she is 50 m from the engine?

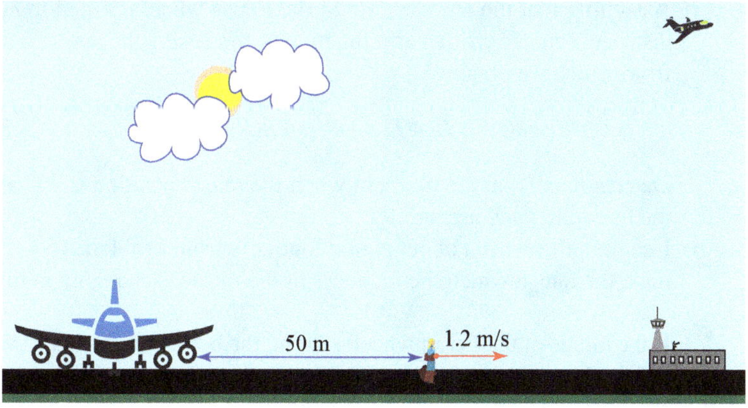

Figure 4.33: A woman walking away from an aircraft; Problem 16.

17. Consider the situation of Problem 16. Suppose now that the aircraft is also moving forward at a speed of 8 m/s. At what rate is the loudness of the jet

engine changing, as experienced by the woman?

18. A man is riding a large Ferris wheel, with radius R, as shown in Figure 4.34. At a distance D below the bottom of the wheel is a loudspeaker, which is playing music. Let x represent the distance between the man and the loudspeaker.

Figure 4.34: Ferris wheel for Problem 18.

(a) Assuming that the angular rate of the Ferris wheel is equal to ω, the distance x can be written as a function of time using the law of cosines, from which we obtain

$$x(t)^2 = (R+D)^2 + R^2 - 2R(R+D)\cos(\omega t).$$

Determine $x'(t)$ at the moment when the man's position is 45° above the horizontal and ascending.

(b) Use the information about sound loudness from Problem 16 to determine the rate at which the loudness of the music is changing in time at that moment.

(c) If the radius of the Ferris wheel is 20 m, the bottom of the Ferris wheel is 5 m above the loudspeaker, the Ferris when goes around once every sixty seconds, and if the sound from the speakers registers at 90 dB from 5 m away, determine the rate at which the volume of the music is changing, in dB/s, as perceived by the man.

19. A woman is walking her poodle using a retractable leash, as shown in Figure 4.35. The leash is attached to the poodle's collar at a height of 18 inches above the ground, and the woman is holding the leash at her waist-

level, which is 32 inches above the ground. The woman is walking at a rate of 4 ft/s when the dog starts running forward at a rate of 12 ft/s. How fast is the length of the leash changing at this instant if the dog is 4 ft in front of the woman?

Figure 4.35: A woman walking her poodle; Problem 19.

4.4 Parametric Equations

In this section, we examine a method that can be used for studying geometric curves or a particle's trajectory. Instead of representing such curves by an implicit equation (see Section 3.7), a method which is not always practical, here we will define a new, hidden variable that controls the location along the curve in question. This variable, which describes our location along a given curve, is called a parameter. Regrettably, this type of parameter has nothing to do with the type of parameters we encountered in modeling situations, which were constants of the problem. Despite this slight ambiguity in language, the context usually makes clear what type of parameter is being used—an intrinsic physical constant of the system or a secret variable that describes the position along a parametric curve.

4.4.1 Planar Curves

In this paragraph, we will be studying planar curves: a *planar curve* is a set of points on the Euclidean plane that can be traced without lifting the pencil from the page. Line segments, lines, circles, parabolas, ellipses, branches of hyperbolas, triangles, squares, pentagons, and graphs of continuous functions are all examples of planar curves. Often, we will name our curves, just as we name functions. The Greek letter γ (gamma) is a common choice that is often used for this purpose. To be precise, the curve's name refers to the *set* of points that comprise of the curve; one must exercise care not to use the curve's name in an equation unless it is used in the context of set notation.

Definition 4.4.1 A curve γ is *bounded* (i.e., it is a *bounded curve*) if there exists a positive number $R > 0$ such that

$$x^2 + y^2 < R^2,$$

for all $(x, y) \in \gamma$. A curve that is not bounded is said to be *unbounded*.

(R) Intuitively, bounded curves are curves that do not stretch to infinity. Recall that, for a given point $(x, y) \in \gamma$ on the curve, the quantity $\sqrt{x^2 + y^2}$ represents the *distance* between the given point (x, y) and the origin. Thus, Definition 4.4.1 states that there is some upper bound on the distance between the points on the curve and the origin.

An example of a bounded curve is shown in Figure 4.36. Every point on the curve falls within a distance R from the origin; therefore, no point on the curve lies outside of the circle, centered at the origin, with radius R. It does not matter what value R has, just so long as there is some positive number R that has this property.

The curve shown in Figure 4.36 is also an example of an *open curve*, i.e., it is a curve that has endpoints, which do not meet. We make the distinction between open and closed curves in our next definition.

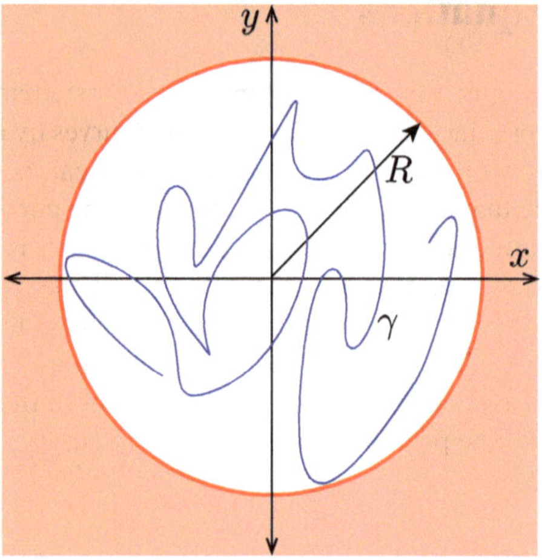

Figure 4.36: An example of a bounded curve γ; the points on the curve stay within a distance R from the origin.

Definition 4.4.2 A *closed curve* is a bounded curve with no endpoints, i.e., if one were to trace the curve, one would eventually arrive back at the starting point. An *open curve* is a curve that is not closed.

A closed curve is like a loop: circles, triangles, and ellipses are all examples of closed curves. Line segments, lines, parabolas, and hyperbolas are examples of open curves. An open curve is either unbounded, or bounded with two endpoints. In addition to the distinction of closed and open curves, some curves have a "twist," so that they cross themselves, whereas other curves don't. We formalize this notion in our next definition.

Definition 4.4.3 A *simple curve* is a curve that does not intersect itself.

An example of a simple closed curve and a non-simple closed curve is shown in Figure 4.37.

The following result is intuitive, though its proof lies within the scope of a field of mathematics known as *topology*, which we shall not delve into at this juncture.

Theorem 4.4.1 — Jordan Curve Theorem. A simple, closed curve γ divides the plane $\mathbb{R}^2$ into two separate regions: an *interior region* that consists of all the points bounded by the curve and an *exterior region* that consists of all points that lie outside of the curve. Any point $(x, y) \in \mathbb{R}^2 \setminus \gamma$, i.e., any point in the plane, but not on the curve γ, must belong to either the interior region or the exterior region.

The Jordan curve theorem is illustrated in Figure 4.38. Notice how the sim-

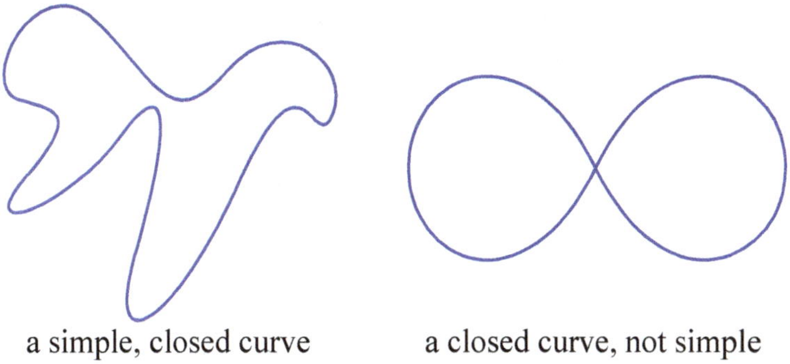

Figure 4.37: An example of a simple, closed curve (left) and a closed curve that is not simple (right).

ple closed curve γ divides the plane into a bounded region (the interior) and an unbounded region (the exterior).

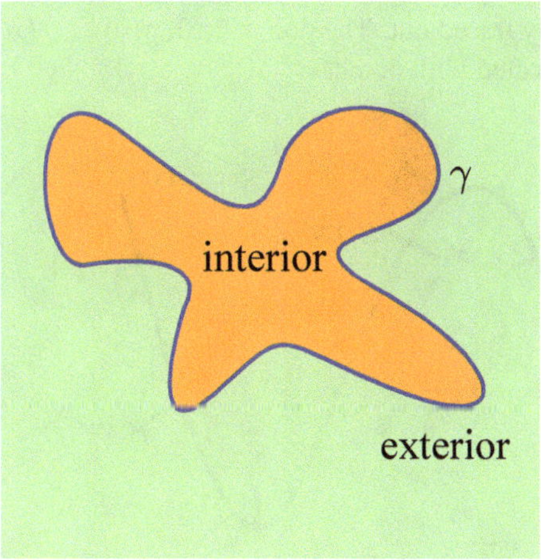

Figure 4.38: A simple closed curve γ and its interior (orange) and exterior (green) regions.

4.4.2 Parametric Equations

Now that we have established some basic terminology relating to planar curves, we next introduce the concept of a parameterization of a curve.

Definition 4.4.4 Given a planar curve γ, a *parameterization* is a pair of two strictly increasing functions

$$x = x(s) \qquad \text{and} \qquad y = y(s), \qquad s \in I,$$

known as *parametric equations*, where $I \subset \mathbb{R}$ is an interval, such that as the independent variable s varies along its domain I, the outputs $(x(s), y(s))$, taken as a coordinate pair, trace out the curve γ. The variable s is referred to as a *parameter* for the curve.

R Basically, a parameter (in this context) is a variable that is defined along a curve and that can be used to measure the position of a point along a curve.

R The requirement that the parametric equations $x(s)$ and $y(s)$ are given by strictly increasing functions, i.e., $x'(s) > 0$ and $y'(s) > 0$, is used to guarantee that the parameterization doesn't "stop" or retrace a portion of the curve.

Definition 4.4.4 is illustrated in Figure 4.39. Here, we find a curve parameterized on an interval $I = [0,7]$. As the parameter s varies from $s = 0$ to $s = 7$, the curve γ is completely traced out. The points $(x(0), y(0)), \ldots, (x(7), y(7))$, for integer values of s, are indicated with notches.

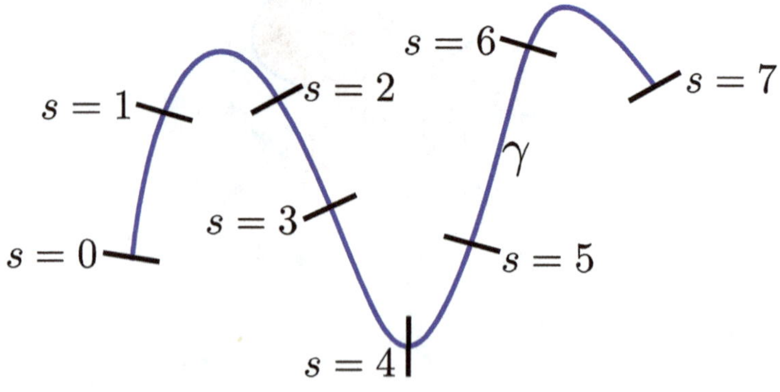

Figure 4.39: A parameterized curve γ, with parameter s.

Our first example should be quite familiar.

■ **Example 4.20** The unit circle $x^2 + y^2 = 1$ may be parameterized by the equations

$$x(\theta) = \cos(\theta), \tag{4.12}$$
$$y(\theta) = \sin(\theta), \tag{4.13}$$

for $\theta \in [0, 2\pi]$.

This result follows directly from the definition of the sine and cosine functions, as contained in Definition 1.6.1 on page 63. The parameter here is the angle θ. The unit circle is an example of a simple closed curve; the two endpoints are identical, as $x(0) = x(2\pi)$ and $y(0) = y(2\pi)$. ■

■ **Example 4.21** Determine a parameterization for the line ℓ that passes through the point (x_0, y_0) with slope m.

Since the result is a line, it is simplest to choose a parameterization that is a linear function of the parameter s. We may also set up the parameterization so that it passes through the given point (x_0, y_0) when $s = 0$, so that

$$x(0) = x_0 \qquad \text{and} \qquad y(0) = y_0.$$

If the parametric equations $x(s)$ and $y(s)$ are to be linear in s, we now only need to determine the coefficients of s that appear in these equations. We may use the slope toward these ends. The slope tells us that every time the x-variable is incremented by 1, the y-variable will be incremented by an amount equation to m. (Recall, $\Delta y/\Delta x = m$, so that $\Delta y = m\Delta x$.) Therefore, the correct parametric equations for this line are given by

$$x(s) = x_0 + s, \tag{4.14}$$

$$y(s) = x_0 + ms, \tag{4.15}$$

for $s \in \mathbb{R}$. (The parameter s varies across all real numbers, as the line ℓ extends to infinity in both directions.)

As an example of this, let us consider the line ℓ that passes through the point $(1, 1)$ with slope $m = 2$. The parametric equations are

$$x(s) = 1 + s \qquad \text{and} \qquad y(s) = 1 + 2s. \tag{4.16}$$

Various points on this line are plotted in Figure 4.40. One immediately confirms that these sample points indeed lie on the line ℓ. For instance, evaluating our parametric equations at $s = -2$, $s = -1$, $s = 0$, and $s = 1$, we obtain

$$(x(-2), y(-2)) = (-1, -3), \qquad (x(-1), y(-1)) = (0, -1),$$
$$(x(0), y(0)) = (1, 1), \quad \text{and} \quad (x(1), y(1)) = (2, 3).$$

By figuring these sample points, we gain an understanding of why the parametric equations (4.16) work. The initial position is the given point, so that $x(0) = 1$ and $y(0) = 1$. Further, as we increment the parameter s, the variables x and y are incremented in precisely a $1 : m$ ratio, maintaining a constant slope for the line. ■

(R) A parameterization of a curve γ is essentially a one-to-one mapping from an interval I (possibly an infinite or semi-infinite interval) onto the curve γ. A parameter for a curve can be thought of like a "highway marker" along the curve: each value of the parameter uniquely specifies a point on the curve.

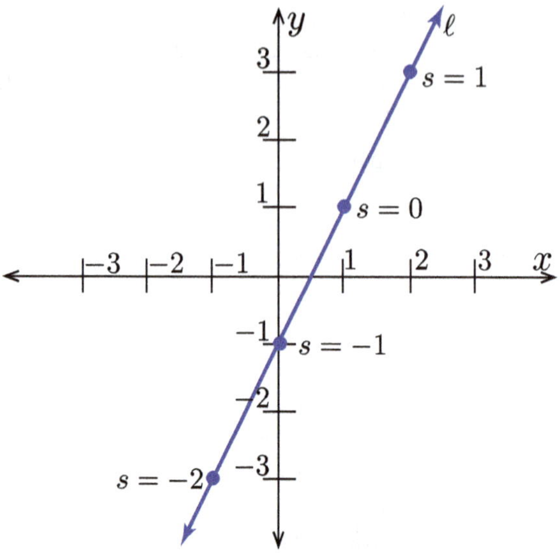

Figure 4.40: A parameterized line ℓ.

When we studied curves that were represented as the graph of an implicit equation, we devised a method for determining its derivative at a given point. We shall do the same here.

Theorem 4.4.2 Given a curve γ with a parameterization $x = x(s)$ and $y = y(s)$, for $s \in I$, the derivative of y with respect to x is given by

$$\frac{dy}{dx} = \frac{y'(s)}{x'(s)}, \tag{4.17}$$

whenever this quantity is defined.

We omit a technical proof; intuitively, however, the result is clear:

$$\frac{y'(s)}{x'(s)} = \frac{dy/ds}{dx/ds} = \frac{dy}{ds}\frac{ds}{dx} = \frac{dy}{dx}.$$

This equation is not fully rigorous, as dy/ds and dx/ds are themselves quantities properly defined by means of the limit definition of the derivative; one may not simply "cancel" the competing ds's. That said, the rigorous proof is just an elaborate version of the preceding line, as it avoids a direct cancellation of ds's by means of some algebraic gymnastics and the spice known to mathematicians as *trickery*.

■ **Example 4.22** Let us again consider the parametric equations (4.16) for the line segment ℓ given in Example 4.21. Substituting these equations into (4.17), we obtain

$$\frac{dy}{dx} = \frac{y'(s)}{x'(s)} = \frac{2}{1} = 2, \tag{4.18}$$

which agrees with the given slope. ■

■ **Example 4.23** Consider the parametric equations

$$x(s) = \sin(2\pi s) \qquad \text{and} \qquad y(s) = \cos(3\pi s),$$

for $t \in [0,2]$. The resulting parametric curve is plotted in Figure 4.41. Our task is to compute an equation for the tangent line at the point $(0.5, 1/\sqrt{2})$.

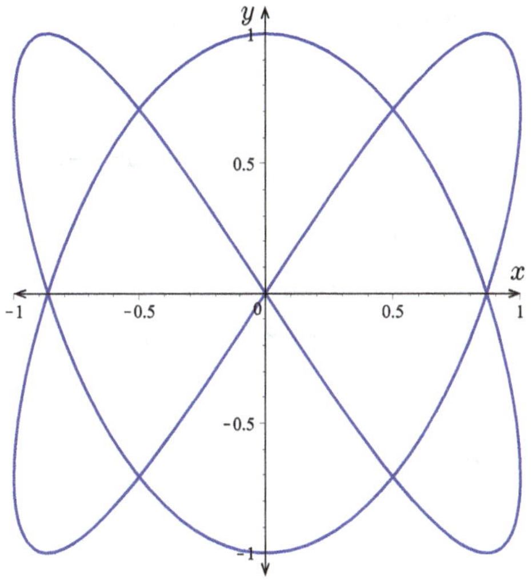

Figure 4.41: The parametric curve of Example 4.23.

We recognize that the point $(0.5, 1/\sqrt{2})$ corresponds to the parameter value $s = 1/12$, since

$$\sin\left(\frac{2\pi}{12}\right) = 0.5 \qquad \text{and} \qquad \cos\left(\frac{3\pi}{12}\right) = \frac{1}{\sqrt{2}}.$$

Applying equation (4.17), we obtain

$$\frac{dy}{dx} = \frac{y'(s)}{x'(s)} = \frac{-3\pi\sin(3\pi s)}{2\pi\cos(2\pi s)}.$$

Evaluating the derivative for $s = 1/12$, we uncover the slope of the tangent line, which is given by

$$\left.\frac{dy}{dx}\right|_{s=1/12} = \frac{-3\sin(\pi/4)}{2\cos(\pi/6)} = -\sqrt{1.5}.$$

Finally, the point-slope formula is applied to obtain

$$T(x) = \frac{1}{\sqrt{2}} + \sqrt{\frac{3}{2}}\,(x - 0.5).$$

This completes the example. ■

4.4.3 Dynamics: Speed and Velocity

In addition to describing static curves, parametric equations are also commonly employed to study the dynamics of a particle's trajectory. (A trajectory is simply the path followed by a moving particle.) In this context, the trajectory is ubiquitously parameterized by *time*. The parametric equations

$$x = x(t) \qquad \text{and} \qquad y = y(t),$$

for $t \in I$, therefore tell us the particle's position on the x–y plane as a function of time.

As was the case in one-dimensional particle motion, given a particle's trajectory, we are interested in determining its velocity.

Definition 4.4.5 Suppose a particle follows a trajectory γ, which is parameterized by time as

$$x = x(t) \qquad \text{and} \qquad y = y(t).$$

Then the particle's *x-velocity* v_x (or the *x-component of the velocity*) and *y-velocity* v_y (or the y-component of the velocity) are given by the derivatives

$$v_x(t) = x'(t), \tag{4.19}$$

$$v_y(t) = y'(t). \tag{4.20}$$

The *velocity* $\mathbf{v}$ of the particle is the ordered pair

$$\mathbf{v}(t) = \langle v_x(t), v_y(t) \rangle. \tag{4.21}$$

The *speed* v of the particle is defined by

$$v(t) = \sqrt{v_x(t)^2 + v_y(t)^2}. \tag{4.22}$$

(R) The velocity of a particle is a *vector*, that is, it captures information pertaining to both how fast the particle is traveling and in what direction. The velocity has two components: the x-component (defined by equation (4.19)) and the y-component (defined by equation (4.20)). Both of the velocity components are required to fully state its velocity. Given a particle's x-velocity and y-velocity, one can determine the direction the particle is traveling in by constructing a right triangle, constructed so that its legs are parallel to the coordinate axes with lengths v_x and v_y.

On the other hand, the speed of a particle is a *scalar*, that is, it is just a number. In fact, a particle's speed is always a positive number. Whereas the velocity captures the particle's direction, the speed only tells us how fast the particle is traveling.

To illustrate the preceding set of definitions, consider Figure 4.42, which shows a particle (represented by the blue dot), its velocity components (orange), and its velocity (red).

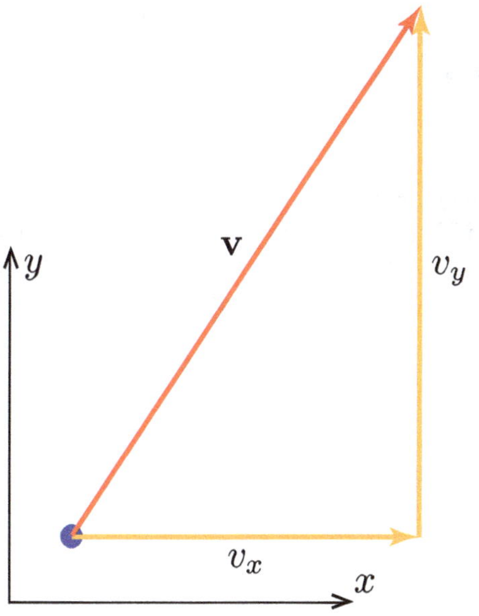

Figure 4.42: The velocity of a particle.

Basically, the x- and y-velocities tell us the instantaneous rate at which the particle is moving in the x- and y-directions, respectively. The direction of the particle's actual motion is represented by the red velocity vector $\mathbf{v}$, as shown in Figure 4.42. The speed of the particle is simply the *length* of its velocity vector $\mathbf{v}$. Since the x- and y-components of the particle's velocity form perpendicular legs to a right triangle, the speed is calculated using Pythagorean's theorem, resulting in equation (4.22). The particle's actual path might look something like the dashed grey line in Figure 4.42; in any case, the particle's velocity (the red arrow) will lie tangent to its path at all moments in time.

In addition to velocity, we are also interested in a particle's acceleration. (This is particularly useful, as the laws of Newtonian mechanics relate the forces acting on the particle to its acceleration.) Like the velocity, acceleration is also a vector quantity; we will therefore speak of both the x- and y-components of the acceleration.

Definition 4.4.6 Suppose a particle follows a trajectory γ, which is parameterized by time as

$$x = x(t) \qquad \text{and} \qquad y = y(t).$$

Then the particle's *x-acceleration* a_x (or the *x-component of the acceleration*) and *y-acceleration* a_y (or the y-component of the acceleration) are given by the derivatives

$$a_x(t) = x''(t), \tag{4.23}$$

$$a_y(t) = y''(t). \tag{4.24}$$

The *acceleration* **a** of the particle is the ordered pair

$$\mathbf{a}(t) = \langle a_x(t), a_y(t)\rangle. \tag{4.25}$$

■ **Example 4.24** Suppose that a projectile is launches from a cannon located at the origin, as shown in Figure 4.19 on page 281. The projectile has an initial speed v_0 and is launched at an angle of θ_0 above the horizontal. The projectile will follow a parabolic trajectory, and its trajectory is given as a function of time t by the parametric equations

$$x(t) = v_0\cos(\theta_0)t, \tag{4.26}$$

$$y(t) = v_0\sin(\theta_0)t - \frac{gt^2}{2}, \tag{4.27}$$

where g represents acceleration due to gravity.

Computing the derivative of each equation, we obtain the x- and y-velocities

$$\begin{aligned} v_x(t) &= v_0\cos(\theta_0), \\ v_y(t) &= v_0\sin(\theta_0) - gt. \end{aligned}$$

Notice that the initial velocity is given by

$$\big\langle v_x(0), v_y(0)\big\rangle = \langle v_0\cos(\theta_0), v_0\sin(\theta_0)\rangle,$$

and the corresponding initial speed

$$v(0) = \sqrt{v_0^2\cos^2(\theta_0) + v_0^2\sin^2(\theta_0)} = v_0.$$

(Notice that the initial x- and y- velocities form the legs of a right triangle with angle θ_0.

The x- and y-accelerations are given by

$$\begin{aligned} a_x(t) &= 0, \\ a_y(t) &= -g. \end{aligned}$$

This confirms that the projectile is acted by a constant downward gravitational acceleration. By examining the particle's velocity and acceleration, we can confirm that the trajectory equations (4.34) and (4.35) satisfy all the requirements of the stated system. ■

■ Example 4.25 A particle travels around a circle of radius R with angular speed ω, so that its position is given by

$$x(t) = R\cos(\omega t), \tag{4.28}$$
$$y(t) = R\sin(\omega t). \tag{4.29}$$

The particle's path is shown in Figure 4.43. The velocity components are obtained

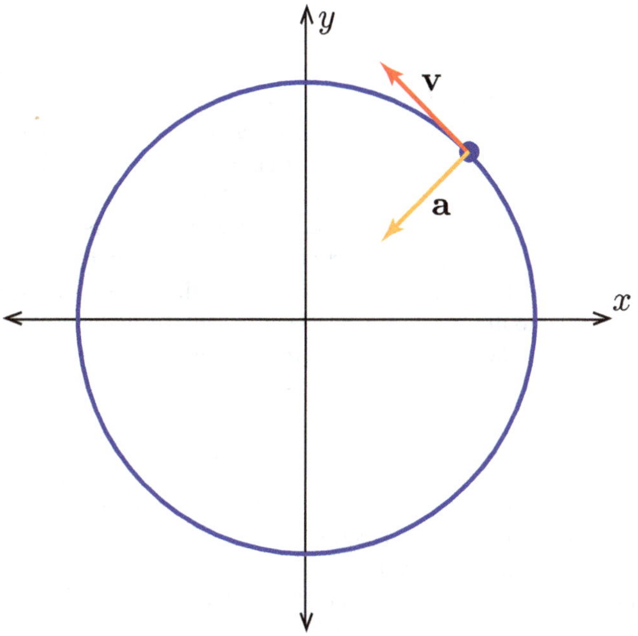

Figure 4.43: A particle traveling in a circle; shown with its velocity and acceleration.

by differentiating equations (4.36) and (4.37) with respect to time:

$$v_x(t) = -R\omega\sin(\omega t),$$
$$v_y(t) = R\omega\cos(\omega t).$$

The speed at which the particle traverses this circle is constant, for

$$v(t) = \sqrt{v_x^2 + v_y^2} = \sqrt{R^2\omega^2\sin^2(\omega t) + R^2\omega^2\cos^2(\omega t)} = R\omega. \tag{4.30}$$

Differentiating the velocity equations, we uncover the components of the particle's acceleration

$$a_x(t) = -R\omega^2\cos(\omega t),$$
$$a_y(t) = -R\omega^2\sin(\omega t).$$

Notice that the acceleration is equal to $-\omega^2$ times the particle's position. While the velocity of the particle is always tangent to the particle's circular path, the acceleration of the particle always lies in the opposite direction as the particle's position: the acceleration is always direct toward the center of the circle.

(R) Even though the speed of the particle is constant, the particle is nevertheless accelerating. This is because its velocity, which is a vector, is changing direction. The acceleration captures the rate at which the velocity vector is changing.

The acceleration in this problem, which is directed toward the center of the circle, is known as *centripetal acceleration*. (Centripetal means *center seeking*.) In order for a particle to remain on a circular path, an inward force must be applied so that the particle's motion accelerates toward the center of the circle. *Centrifugal force*, on the other hand, refers to the outward "force" the people can feel when traveling along a circular path. Centrifugal force, however, is not really a force at all: it is an effect of inertia. The particle has a tendency to remain in a straight line. As an automobile, for example, makes a sharp left turn, its passengers must also be accelerated to the left. Since their bodies wants to continue along a straight path, it will *feel* like they are being pushed to the right, in the reference frame of the vehicle. ■

Exercises

For Exercises 1–10, determine an equation for the tangent line to the given parametric curve at the indicated point.

1. $x(t) = \sin(2t)$, $y(t) = \cos(3t)$; $t = \frac{\pi}{6}$.
2. $x(t) = e^t$, $y(t) = e^{-t}$; $t = \ln(2)$.
3. $x(t) = t$, $y(t) = t^2$; $t = 2$.
4. $x(t) = \sin(t)$, $y(t) = \sin(2t)$; $t = \frac{\pi}{4}$.
5. $x(t) = e^{2t}$, $y(t) = 3 + 4e^{2t}$; $t = 5$.
6. $x(t) = 12\cos(t)$, $y(t) = 5\sin(t)$; $t = \frac{\pi}{4}$.
7. $x(t) = t\cos(t)$, $y(t) = t\sin(t)$; $t = \frac{9\pi}{4}$.
8. $x(t) = \cos(3t)$, $y(t) = 0.5t + \sin(4t)$; $t = \frac{\pi}{12}$.
9. $x(t) =$, $y(t) =$; $t =$.
10. $x(t) =$, $y(t) =$; $t =$.

Problems

11. In this problem, we consider a line segment ℓ that connects two points (x_0, y_0) and (x_f, y_f).
    (a) Determine a parameterization for the line segment ℓ. Be sure to state the domain of the parameter.
    (b) Show that one parameterization is given by

$$x(s) = x_0 + (x_f - x_0)s, \quad (4.31)$$
$$y(s) = y_0 + (y_f - y_0)s, \quad (4.32)$$

for $s \in [0,1]$. Explain why the interval $I = [0,1]$ is the correct domain for the parameter s.

(c) Use equations (4.31) and (4.32) to determine a parameterization for the line segment connecting the points $(-1,-3)$ and $(2,3)$. Evaluate $(x(s), y(s))$ for $s = 0$, $1/3$, $2/3$, and 1. Show that these points lie on the line segment.

(d) If the equations given in equation (4.16) are instead used to parameterize the line segment in part (c), determine the correct domain s for the parameterization. (This shows, by elementary example, that there is no unique way to parameterize a curve.)

12. Consider the parametric equations

$$x(s) = 1 + s^3 \qquad \text{and} \qquad 1 + 2s^3.$$

(a) Compare these equations with the parametric equations (4.16).

(b) Plot several points on the curve described by these parametric equations. What curve do they describe?

(c) Employ equation (4.17) to confirm your hypothesis.

13. We consider here the unit circle from Example 4.20.

(a) The unit circle can be parameterized by equations (4.12) and (4.13). Use this parameterization and equation (4.17) to compute the derivative dy/dx. Evaluate the derivative at the point $\theta = \pi/6$.

(b) The top half of the unit circle may also be described by the function $f(x) = \sqrt{1-x^2}$. Compute the derivative of this function and evaluate it at the point $x = 1/2$. How does this relate to the answer obtained from part (a)? Why?

14. Determine all values of the parameter s on the interval $[0,2]$ at which the parametric equations (4.23) result in a horizontal tangent line.

15. Explain why the graph of a continuous function $y = f(x)$ may be parameterized using the equations

$$x(s) = s \qquad \text{and} \qquad y(s) = f(s), \qquad s \in \text{domain}(f). \tag{4.33}$$

Construct an example of this, graph and confirm.

16. Consider again a particle traveling in a circular path, as discussed in Example 4.25.

(a) If the particle's angular speed is ω radians per second, determine the period T of the particle's motion, i.e., how long does it take for the particle to traverse the circle one complete time?

(b) Use the circumference of the circle and the equation $d = vT$ (distance equals rate times time) to determine the particle's constant velocity v. Confirm equation (4.30).

17. In this problem, we will design a *vertical loop* for a roller coaster. Consider the parametric equations

$$x(t) = at(t^2 - b^2), \tag{4.34}$$

$$y(t) = ce^{-dt^2}. \tag{4.35}$$

The quantities a, b, c, and d are constants, which may be chosen based on physical criteria. These equations trace the curve of a vertical loop, as shown in Figure 4.44.

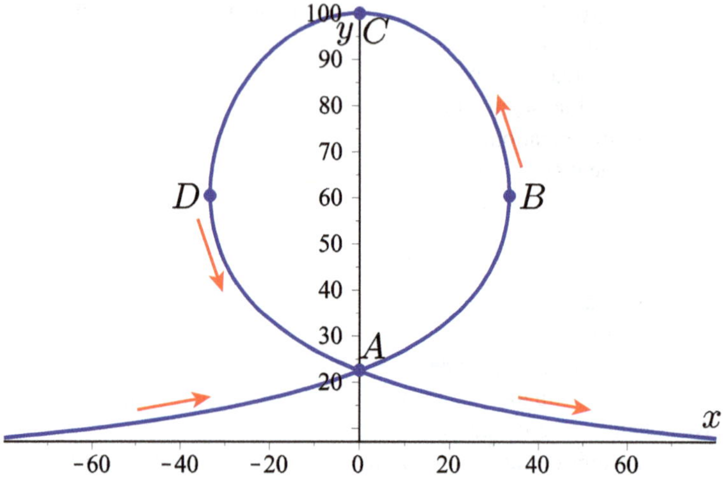

Figure 4.44: A vertical loop.

(a) Produce a generic graph of the functions $x(t)$ and $y(t)$. Label the intercepts where applicable.

(b) Show that the "loop time" T_ℓ, i.e., the time to complete the loop $ABCDA$, is given by $T_\ell = 2b$.

(c) Show that the maximum vertical height h that the roller coaster reaches is given by $h = c$.

(d) Derive a formula for $x'(t)$, $x''(t)$, $y'(t)$, and $y''(t)$.

(e) Show that the vertical acceleration when the particle passes through point C is given by $-2cd$.

(f) Show that the horizontal acceleration when the particle passes through point B is given by $\dfrac{-6ab}{\sqrt{3}}$.

(g) Show that the graph of $y(t)$ has an inflection point at the points $t = \dfrac{\pm 1}{\sqrt{2d}}$.

(h) Determine the values of the constants a, b, c, and d required to enforce the following physical criteria:

   i. the maximum vertical height is $h = 100$ ft;

   ii. the vertical acceleration at point C is equal to $-2g$, where the gravitational acceleration is $g = 32$ ft/s^2;

   iii. the graph of y has an inflection points at the points corresponding to points B and D in the particle's trajectory; and

   iv. the horizontal acceleration at point B is equal to $-g$.

18. In this problem, we will use a cubic polynomial to model the flight path of an airplane during its descent. Assuming the airplane maintains a constant ground speed during its descent, its flight path can be modeled using the following parametric equations

$$x(t) = v_0 t, \tag{4.36}$$

$$y(t) = \frac{at^3}{3} - \frac{aTt^2}{2} + b. \tag{4.37}$$

The variable x describes the distance along the ground traveled by the airplane, and the variable y describes the airplanes altitude as a function of time.

(a) Identify the groundspeed of the aircraft.
(b) Evaluate $y(0)$. What is the physical meaning of the constant b?
(c) Compute $y'(0)$, and show that

$$y'(0) = at(t-T).$$

At what two points in time does the airplane have zero vertical velocity? What two points in the airplane's flight path do these two points correspond to. (Instead of starting with an arbitrary cubic function of t, we selected the coefficients so that these conditions would be satisfied.)
(d) The time domain of the flight path is the interval $[0,T]$. To enforce the condition that $y(T) = 0$, show that T must depend on the constants a and B through the relationship

$$T = \sqrt[3]{\frac{6b}{a}}.$$

(e) Compute $y''(t)$ and evaluate $y''(0)$ and $y''(T)$. Suppose that the maximum vertical acceleration of the airplane is αg, where g represents acceleration due to gravity. Show that this condition may be enforced through the relation $aT = \alpha g$.
(f) Suppose that an airplane with initial cruising altitude h wishes to limit its vertical acceleration to a maximum value of αg. Use the results from parts (d) and (e) to show that the constants a and T can be solved in terms of the physical specifications given by the values of h and α, thereby obtaining the relations

$$T = \sqrt{\frac{6h}{\alpha g}} \quad \text{and} \quad a = \sqrt{\frac{\alpha^3 g^3}{6h}}.$$

(g) Suppose an aircraft has a cruising altitude of 35,000 ft, a constant ground speed of 420 mph (616 ft/s), and that $g = 32$ ft/s^2. If it wishes to limit its vertical acceleration during landing to $0.01g$, determine the flight path for its descent. How far away from the runway should the airplane begin its descent?

19. Starting with the general cubic polynomial

$$y(t) = c_3 t^3 + c_2 t^2 + c_1 t + c_0,$$

recover equation (4.37) by enforcing the conditions that
   (a) the vertical velocity y' equals zero at times $t = 0$ and $t = T$, for some constant T;
   (b) the vertical velocity is negative during the interval $(0, T)$.
20. In this problem, we will explore the concept of the *curvature* of a curve.
   (a) A planar curve with parametric equations $x(s)$ and $y(s)$ is said to be *parameterized with respect to arc length* if it is traced with unit speed, i.e., if

$$v(s) = \sqrt{x'(s)^2 + y'(s)^2} = 1.$$

   Explain why, in such a situation, the parameter s may be thought of as the "arc length" along the curve?
   (b) The *curvature* $\kappa(s)$ at a point s along a curve parameterized with respect to arc length is given by

$$\kappa(s) = \sqrt{x''(s)^2 + y''(s)^2}.$$

   Explain the following interpretation: the curvature of a planar curve is the magnitude of the particle's acceleration, if the particle traces the curve with a constant unit speed.
   (c) Determine the angular frequency ω so that the circle of radius R, given by the equations

$$x(s) = Rcos(\omega s) \qquad \text{and} \qquad y(s) = R\sin(\omega s),$$

   is parameterized with respect to arc length. Show that the curvature of this circle is $\kappa = 1/R$.

4.5 Polar Equations

In this section, we examine an alternate coordinate system to describe points on the Euclidean plane. Such coordinates can lead to simplifications and are particularly useful for curves that possess a certain symmetry.

4.5.1 Polar Coordinates

As an alternative to describing points on the Euclidean plane $\mathbb{R}^2$ using Cartesian coordinates (x, y), let us consider *polar coordinates.*

Definition 4.5.1 Any point P on the Euclidean plane may be described by an ordered pair (r, θ) known as *polar coordinates*, where $r \geq 0$. The coordinate r is called the *radial coordinate* or the *radius*; the coordinate θ is called the *angular coordinate*, the *polar angle*, or the *azimuth.*

The radial coordinate r measures the distance between the point P and the origin O. The angular coordinate θ is the angle measured (in radians) from the the positive x-axis, counterclockwise, to the ray $\overrightarrow{OP}$.

Polar coordinates are illustrated in Figure 4.43. The rays $\theta = 0$, $\pi/2$, π, $3\pi/2$,

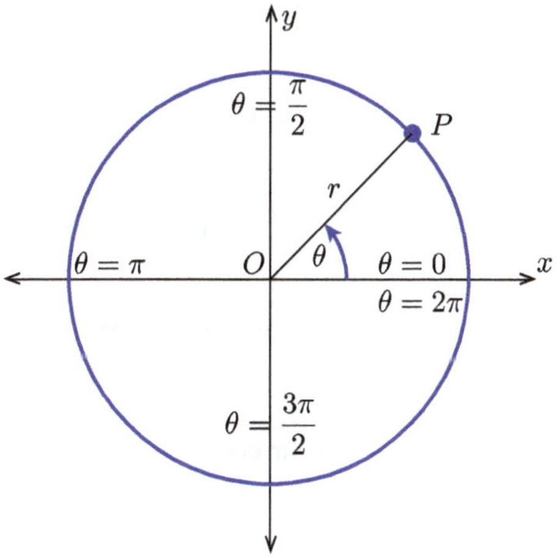

Figure 4.45: A point $P \in \mathbb{R}^2$ and its polar coordinates (r, θ).

and 2π, which represent the positive x-axis, the positive y-axis, the negative x-axis, and negative y-axis, and the positive x-axis (again), respectively, are also labeled in the figure.

(R) Polar coordinates are not unique, as incrementing the angle θ by an integer multiple of 2π results in the same point. To avoid ambiguity, a range of θ-values is often specified. Popular choices include $\theta \in [0, 2\pi)$ and $\theta \in (-\pi, \pi]$.

Since both Cartesian and polar coordinates describe points on the Euclidean plane, we should be able to convert from one coordinate system to the other. This is the result of our next proposition.

Proposition 4.5.1 Consider a point $P \in \mathbb{R}^2$ and its representation in Cartesian coordinates (x,y) and polar coordinates (r,θ). To convert from polar coordinates to Cartesian coordinates, the following equations may be used:

$$x = r\cos(\theta), \tag{4.38}$$

$$y = r\sin(\theta). \tag{4.39}$$

Similarly, to convert from Cartesian coordinates to polar coordinates, the following equations may be used:

$$r = \sqrt{x^2+y^2}, \tag{4.40}$$

$$\tan(\theta) = \frac{y}{x}. \tag{4.41}$$

These relations follow from simple trigonometry, as illustrated in Figure 4.46.

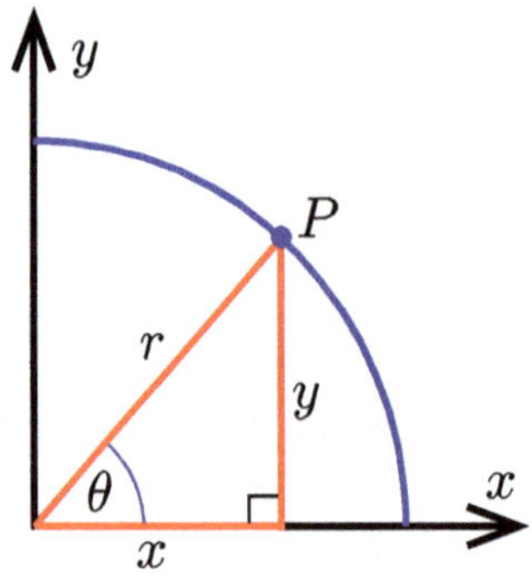

Figure 4.46: The relation between polar coordinates (r,θ) and Cartesian coordinates (x,y).

One must, of course, take care when executing equation (4.41), as the arctangent function only yields values on the interval $\left(\frac{-\pi}{2},\frac{\pi}{2}\right)$. To remedy this situation, let us determine the polar angle $\theta \in (-\pi,\pi]$ given a point P expressed in Cartesian coordinates (x,y). First, if the point P lies on one of the coordinate axes (i.e., if either x or y is equal to zero), assign the polar angle accordingly:

positive x-axis	$x > 0,\ y = 0$	$\theta = 0,$
positive y-axis	$x = 0,\ y > 0$	$\theta = \frac{\pi}{2},$
negative x-axis	$x < 0,\ y = 0$	$\theta = \pi,$
negative y-axis	$x = 0,\ y < 0$	$\theta = \frac{-\pi}{2}.$

(The polar angle is undefined at the origin.) Otherwise, identify which quadrant the point (x,y) is located in and compute the arctangent of the ratio y/x, shifting the end result if necessary:

$$\begin{array}{lll} \text{Quadrant I} & x>0,\ y>0 & \theta=\tan^{-1}\left(\frac{y}{x}\right), \\ \text{Quadrant II} & x<0,\ y>0 & \theta=\tan^{-1}\left(\frac{y}{x}\right)+\pi, \\ \text{Quadrant III} & x<0,\ y<0 & \theta=\tan^{-1}\left(\frac{y}{x}\right)-\pi, \\ \text{Quadrant IV} & x>0,\ y<0 & \theta=\tan^{-1}\left(\frac{y}{x}\right). \end{array}$$

To understand why these rules work, let us examine Figures 4.47 and 4.48. The first figure shows how the four quadrants partition the domain of the polar angle θ: quadrant I corresponds to $\theta \in (0,\pi/2)$; quadrant II corresponds to $\theta \in (\pi/2,\pi)$; quadrant III corresponds to $\theta \in (-\pi/2,-\pi)$; quadrant IV corresponds to $\theta \in (0,-\pi/2)$.

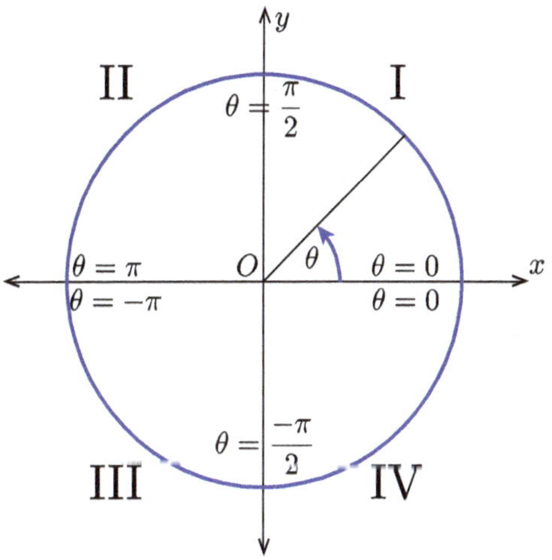

Figure 4.47: The four quadrants and the polar angle $\theta \in (-\pi,\pi]$.

Figure 4.48 is a graph of the tangent function on the domain $\theta \in (-\pi,\pi)$. Recall that the arctangent function maps each y-value to a θ value within the domain $\theta \in (-\pi/2,\pi/2)$. Therefore, the arctangent function returns the correct angle if the point P lies in quadrant I or IV, but the incorrect angle if the point P lies in quadrant II or III. If the point lies in quadrant II, the θ-value returned by the arctangent function must be incremented by $+\pi$, as is apparent in the figure. This will yield the correct θ-value in the domain $(\pi/2,\pi)$. Similarly, if the point lies in quadrant III, the θ-value must be incremented by $-\pi$, in order to obtain the correct θ-value in the domain $(-\pi,-\pi/2)$.

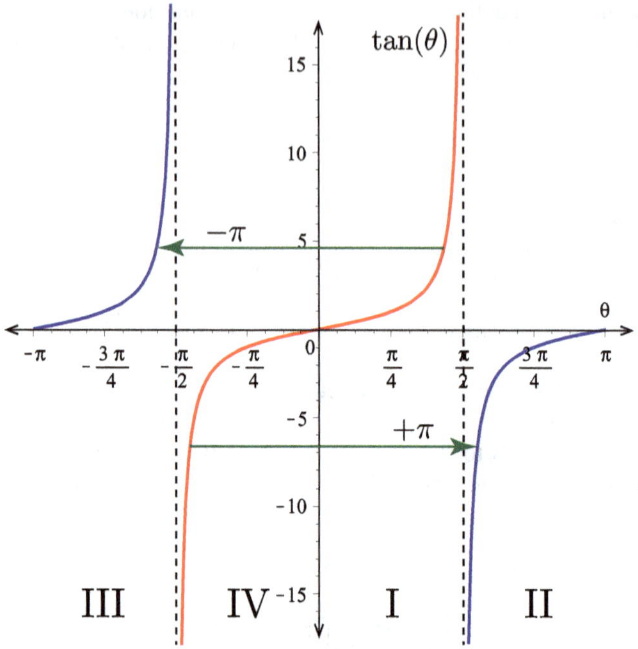

Figure 4.48: Graph of the function $\tan(\theta)$; each quadrant is labeled.

■ **Example 4.26** Compute the polar coordinates for the point P, given in Cartesian coordinates as $(-3,6)$.

The radius is obtained by equation (4.40):

$$r = \sqrt{(-3)^2 + 6^2} = 3\sqrt{5}.$$

To employ equation (4.41), we must first identify the quadrant. Since $x < 0$ and $y > 0$, the point P belongs to quadrant II. Therefore, we must increment the arctangent function by $+\pi$, thereby obtaining

$$\theta = \tan^{-1}\left(\frac{6}{-3}\right) + \pi \approx -1.107148718 + \pi \approx 2.034443936.$$

Observe that $\theta = 2.034443936$ indeed lies in the second quadrant, between the values of $\pi/2$ and π. (We may further confirm that the tangent function evaluated at 2.034443936 yields a value of -2, as desired.) ■

4.5.2 Polar Functions

Next, we examine how polar coordinates can be used in plotting certain functions.

Definition 4.5.2 A function f is a *polar function* if it determines the radius as a function of the polar angle, i.e., if

$$r = f(\theta),$$

for all $\theta \in \mathscr{D}$, where $\mathscr{D} \subset \mathbb{R}$ is the domain of f.

The graph of a polar function f is the set of all points in the plane corresponding to polar coordinates $(f(\theta), \theta)$, for all $\theta \in \mathscr{D}$.

(R) Each value of the polar angle θ determines a ray emanating from the origin. A polar function $r = f(\theta)$ tells us how far out along each ray to go to find the corresponding point on the graph.

Compare this with a Cartesian function. Each x-value corresponds to a vertical line. The Cartesian function $y = f(x)$ tells us where on each vertical line we find the corresponding point on its graph.

As we vary the polar angle θ, it is like a radar sweep: the distances r, as a function of the angle θ, are plotted as we sweep across the radar range (as expressed by the domain of the independent variable θ).

Example 4.27 Consider the polar function

$$f(\theta) = \theta,$$

defined on the domain $[0, 2\pi]$. The graph of this function is shown in Figure 4.49. In this plot, fixed values of θ are indicated by rays emanating from the origin. Similarly, fixed values of the radius are indicated by a sequence of concentric circles, each centered at the origin. Beginning with $\theta = 0$, we evaluate $f(0) = 0$. Therefore, along the ray $\theta = 0$ (which represents the positive x-axis), our graph intersects at $r = 0$. As we increase the polar angle θ, the radius increases linearly with θ. For example, when $\theta = \pi/4$, the radius is equal to $r = f(\pi/4) = \pi/4$. When $\theta = \pi/2$ (the positive y-axis), the radius is $r = f(\pi/2) = \pi/2$, and so forth. This results in the spiral curve shown in the figure. If we were to extend the domain, we would end up with additional loops around the circle. (For instance, if the domain were extended to $[0, 4\pi]$, we would obtain two complete loops around the circle.) ■

Example 4.28 Next, let us consider the polar function

$$f(\theta) = \cos(2\theta),$$

defined on the domain $[0, 2\pi]$. Before jumping directly into the polar plot of this function, let us first consider its Cartesian plot; the function $y = f(\theta) = \cos(2\theta)$ is shown in Figure 4.50. Moreover, each quadrant (as determined by the values of θ) is labeled. An odd thing, however, occurs with this function: for some values of the polar angle θ, the corresponding radius $r = f(\theta)$ is *negative*! To remedy this situation, we plot negative radius values along the antipodal polar-angle ray (i.e., the exact opposite polar-angle ray). For example, the negative r-values in the domain $\theta \in (\pi/4, \pi/2)$ end up being plotted (with a positive radius) in the sector $\theta \in (5\pi/4, 3\pi/2)$.

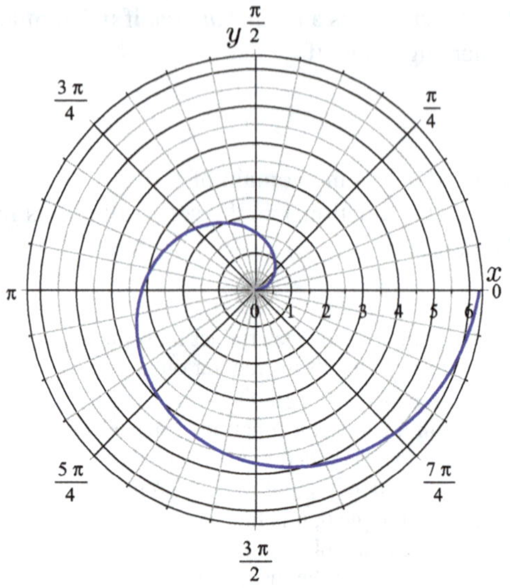

Figure 4.49: Graph of the polar function $f(\theta) = \theta$; Example 4.27.

The final polar plot is shown in Figure 4.51. Compare this plot with the one shown in Figure 4.50.

■

4.5.3 Derivatives of Polar Curves

Next, we examine the Cartesian derivative dy/dx of a polar function.

Theorem 4.5.2 Given a differentiable polar function $f(\theta)$, its Cartesian derivative is given, as a function of θ, by the relation

$$\frac{dy}{dx} = \frac{f'(\theta)\sin(\theta) + f(\theta)\cos(\theta)}{f'(\theta)\cos(\theta) - f(\theta)\sin(\theta)}. \tag{4.42}$$

Proof. Since $r = f(\theta)$, we may rewrite equations (4.38) and (4.39) in the form

$$\begin{aligned} x(\theta) &= f(\theta)\cos(\theta), \\ y(\theta) &= f(\theta)\sin(\theta). \end{aligned}$$

These equations, however, may be viewed as parametric equations in the parameter θ! The Cartesian derivative is therefore obtained by applying equation (4.17):

$$\frac{dy}{dx} = \frac{y'(\theta)}{x'(\theta)} = \frac{f'(\theta)\sin(\theta) + f(\theta)\cos(\theta)}{f'(\theta)\cos(\theta) - f(\theta)\sin(\theta)}.$$

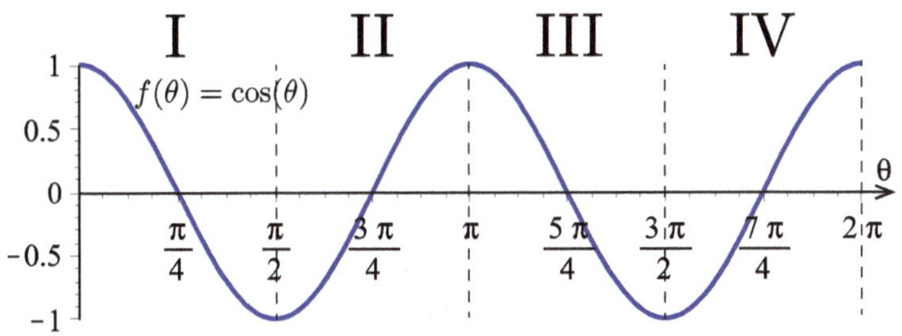

Figure 4.50: Graph of the function $f(\theta) = \cos(2\theta)$; Example 4.28.

(We used the product rule to differentiate the parametric functions $x(\theta)$ and $y(\theta)$.) But this recovers our result. ■

▪ **Example 4.29** Determine the equation for the tangent line to the polar function $f(\theta) = \theta$ at the point $\theta = \pi/2$. (Continuation of Example 4.27.)

Applying equation (4.42), we obtain

$$\frac{dy}{dx}(\theta) = \frac{\sin(\theta) + \theta\cos(\theta)}{\cos(\theta) - \theta\sin(\theta)}.$$

Evaluating at $\theta = \pi/2$, we obtain

$$\frac{dy}{dx}\left(\frac{\pi}{2}\right) = \frac{-2}{\pi}.$$

The point $\theta = \pi/2$, at which $r = f(\pi/2) = \pi/2$, is given in polar coordinates as $(\pi/2, \pi/2)$. In Cartesian coordinates, this is the point $(0, \pi/2)$. Using the point-slope formula, we obtain the equation for the tangent line as

$$T(x) = \frac{\pi}{2} - \frac{2}{\pi}x.$$

The tangent line is plotted concurrently with the polar plot of the function f in Figure 4.52. ▪

Exercises

For Exercises 1–5, convert the point in Cartesian coordinates into polar coordinates, using the domain $\theta \in [0, 2\pi)$. Express the polar angle to at least four significant figures.

1. $(3, 4)$.
2. $(-5, 12)$.
3. $(-3, -3)$.
4. $(2, -5)$.
5. $(0, -8)$.

For Exercises 6–10, convert the point in polar coordinates into Cartesian coordinates.

6. $\left(5, \frac{\pi}{4}\right)$.

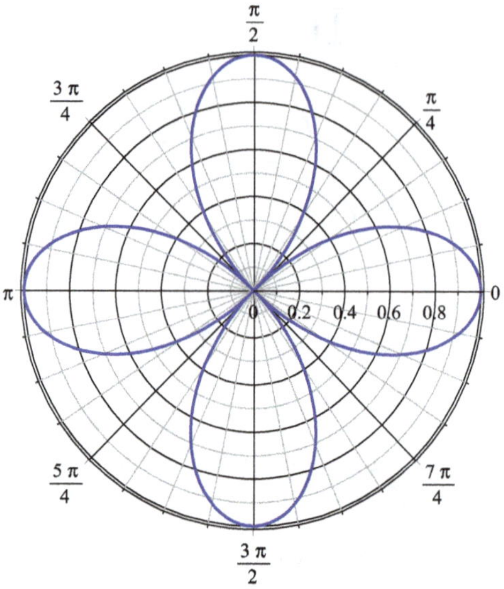

Figure 4.51: Graph of the polar function $r = \cos(2\theta)$; Example 4.28.

7. $\left(3, \frac{7\pi}{4}\right)$.
8. $\left(13, \frac{7\pi}{6}\right)$.
9. $\left(7, \frac{9\pi}{2}\right)$.
10. $\left(4, \frac{-3\pi}{4}\right)$.

Problems

11. Compare the graphs shown in Figures 4.50 and 4.51. Match the sections in both plots for each of the intervals: $[0, \pi/4]$, $[\pi/4, \pi/2]$, $[\pi/2, 3\pi/4]$, $[3\pi/4, \pi]$, $[\pi, 5\pi/4]$, $[5\pi/4, 3\pi/2]$, $[3\pi/2, 7\pi/4]$, $[7\pi/4, 2\pi]$.
12. In astrodynamics, an object's orbit about a primary object (e.g., a satellite's orbit about the Earth) can be described by the polar equation

$$r = \frac{p}{1 + e\cos(\theta)}.$$

For $0 < e < 1$, this represents an ellipse with a focus located at the origin. (The parameters p and e are referred to as the semi-latus rectum and the eccentricity of the orbit, respectively.) Determine an equation for the derivative dy/dx in terms of the constants p and e and the variable θ.
13. The *Fibonacci numbers* are the numbers formed via the recursion relation

$$a_{n+2} = a_{n+1} + a_n,$$

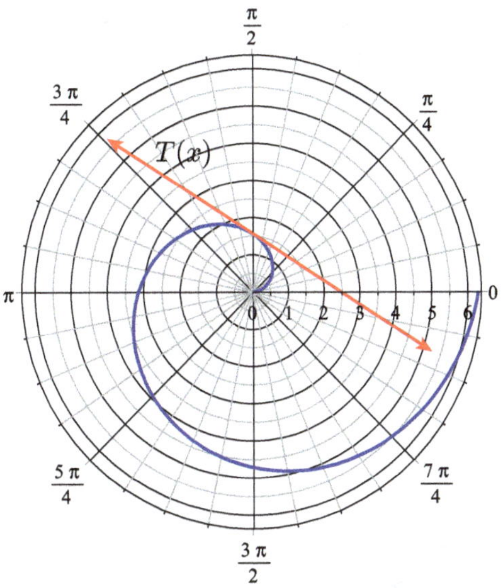

Figure 4.52: Graph of the polar function $r = \theta$ with its tangent line at $\theta = \pi/2$; Example 4.29.

starting with $a_1 = 1$ and $a_2 = 1$, i.e., they comprise the set

$$\mathcal{F} = \{1, 1, 2, 3, 5, 8, 13, 21, 34, 55, 89, 144, \ldots\}.$$

(Each number is formed by adding the previous two numbers.) As $n \to \infty$, the ratio a_{n+1}/a_n approaches the *golden ratio*

$$\Phi = \frac{1+\sqrt{5}}{2}.$$

A *golden spiral*, as shown in Figure 4.53, may be constructed as follows. Begin by drawing two unit squares, one on top of the other. Next to these two squares, draw a third square of side length 2. On top of the first three squares, draw a fourth square with side length 3, and so on, such that the nth square has a side length equal to the nth Fibonacci number. Produce a plot of this and reproduce the golden spiral shown in the Figure.

14. A *logarithmic spiral* is any curve resulting from a polar equation of the form

$$r = ae^{b\theta}.$$

Logarithmic spirals abound in nature: the path followed by a hawk as it approaches its prey, the arms of a spiral galaxy, the nerves of the cornea, nautilus shells, and the arms of a hurricane each constitutes a logarithmic spiral. An example of a logarithmic spiral is plotted in Figure 4.54.

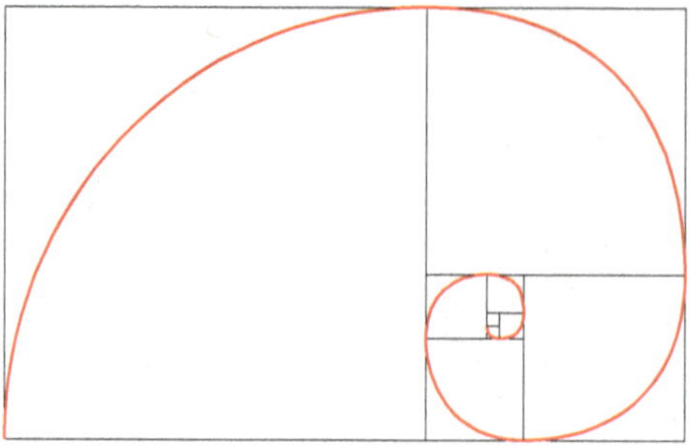

Figure 4.53: A golden spiral.

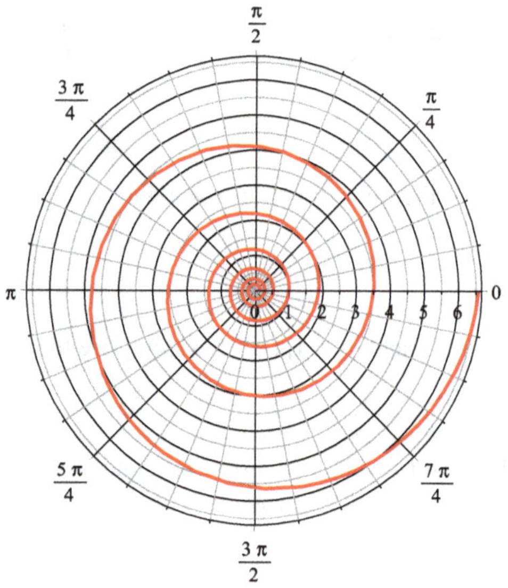

Figure 4.54: A logarithmic spiral with $a = 1, b = 0.1$, shown for $\theta \in [-6\pi, 6\pi]$.

(a) Show that the radius increases by a factor of $e^{2\pi b}$ with each revolutions.
(b) Solve for the parameter b if the radius' growth factor for each quarter revolution is the golden ratio Φ. This produces the *golden spiral*.
(c) Solve for the parameter b if the radius' increases by a factor equal to the golden ratio Φ for every half revolution. This approximates the curve of a *nautilus shell*.

4.6 Applications to Economics

Next, we discuss applications of calculus to business and economics. We have seen several such examples already, sprinkled throughout the text. We will make a more formal study, however, now. The main, powerful concepts that will permeate our discussion are revenue, cost, and profit.

4.6.1 Revenue, Cost, and Profit

We begin by introducing the revenue function.

> **Definition 4.6.1** The *revenue* $R(q)$ is a function that returns the total expected gross earnings if a quantity of q goods are sold.

■ **Example 4.30** Suppose that goods are sold at a constant price of p dollars per item. Then the revenue function will be

$$R(q) = pq,$$

i.e., the total revenue received from the sale of q goods, selling at p dollars per item, is the price times the number of items sold. ■

■ **Example 4.31** A tropical resort hotel is offering an online promotion in which a family can stay for five nights for the price \$500. If more than 20 families sign up for the deal, each will receive a 20% discount and will only be charged \$400.

Since the price per item depends on the number of families who sign up for the package, we will model the revenue as a piecewise-defined function. If 20 or fewer families sign up for the deal, the revenue will equal $500q$. However, if more than 20 families sign up, the revenue will equal $400q$. Thus, the revenue function may be represented as

$$R(q) = \begin{cases} 500q & \text{for } 0 \le q \le 20, \\ 400q & \text{for } 20 < q \end{cases}.$$

Notice that the revenue function is discontinuous at the point where the deal is actualized. ■

> **Definition 4.6.2** The *cost* $C(q)$ is a function that returns the total amount of capital invested in producing a quantity of q goods to be sold.
>
> The *fixed cost* is the vertical intercept $C(0)$. This represents the total startup capital required before the first good can be produced.

■ **Example 4.32** A small e-commerce company requires a startup capital of \$35,000 and pays \$18 for each item they sell. The cost function is given by

$$C(q) = 35000 + 18q,$$

where q represents the number of items sold. The fixed cost $C(0) = 35000$ is equal to the startup cost of the company, and the slope 18 represents the cost to acquire each good that is sold. ■

■ Example 4.33 A small e-commerce website sells goods by accepting credit-card payments over its site. The bank charges the company a monthly merchant account fee of $45. Each credit-card transaction is also charged an amount equal to 30 cents plus 2.9% of the transaction amount. Suppose that everything the company sells costs either $20 or $50. Each $20 transaction therefore incurs a fee of

$$0.30 + 0.029 \times 20 = 0.88.$$

Similarly, each $50 transaction incurs a fee of

$$0.30 + 0.029 \times 50 = 1.75.$$

If m $20 items are sold per month and n $50 items are sold per month, the total cost associated with being able to accept credit-card payments is therefore

$$C(m,n) = 45 + 0.88m + 1.75n$$

per month. Notice that the cost, in this example, depends on two variables. ■

Definition 4.6.3 The *profit* $\Pi(q)$ attained from the sale of q goods is equal to the total revenue minus the total cost, i.e.,

$$\Pi(q) = R(q) - C(q). \tag{4.43}$$

(R) Notice that we use the *capital* Greek letter pi (Π) to denote the profit function. In certain circles, however, the *lower-case* Greek letter π is used instead to represent the profit function. We shall eschew such heresy here and reserve the lower-case letter π to exclusively represent the ratio of a circle's area to the square of its radius.

■ Example 4.34 Suppose that the products from Example 4.32 sell for $39 each, so that the revenue function is $R(q) = 39q$. The profit function is therefore given by

$$\Pi(q) = 39q - (35000 + 18q) = 21q - 35000.$$

Note that 1,667 items must be sold in order for the company to break even. ■

4.6.2 Marginality

In business, it is typically the objective to maximize profit. In setting up the relations to achieve this, we will consider a few closely related concepts. For each, we present the formal definition juxtaposed with its intuitive counterpart.

Definition 4.6.4 The *marginal revenue* $MR(q)$, when q goods have already been sold, is defined as the derivative $MR(q) = R'(q)$.

(R) If q goods have already been sold, the *marginal revenue* is the amount of additional revenue that a company would bring in if one additional good were sold.

Definition 4.6.5 The *marginal cost* $MC(q)$, when q goods have already been produced, is defined as the derivative $MC(q) = C'(q)$.

R If q goods have already been produced, the *marginal cost* is the additional cost required to produce one additional good.

Definition 4.6.6 The *marginal profit* $M\Pi(q)$, when q goods have already been produced and sold, is defined as the derivative $M\Pi(q) = \Pi'(q)$.

R If q goods have already been produced and sold, the *marginal profit* is the additional profit that would be obtained if one additional good were produced and sold.

To maximize profit, one naturally considered the question: what if I were to produce and sell more goods? If the marginal profit is positive, it means that more profit can be earned by further scaling up the enterprise. If the marginal profit is negative, however, it means that producing and selling additional items actually costs more than can be gained, and that one should, perhaps, consider downsizing the operation.

Theorem 4.6.1 The maximum and minimum obtainable profits occur when the number of goods produced and sold is either

1. an endpoint of the domain;
2. a point where either the revenue or cost function is discontinuous; or
3. a point where the marginal revenue equals the marginal cost, i.e., $MR(q) = MC(q)$.

This theorem follows since the profit function is defined by the relation $\Pi(q) = R(q) - C(q)$. Therefore, its derivative is given by

$$\Pi'(q) = R'(q) - C'(q) = MR(q) - MC(q).$$

The derivative of the profit function is therefore equal to zero precisely when the marginal revenue is balanced with the marginal cost.

■ **Example 4.35** Suppose that the revenue and cost function for a certain company are as shown in Figure 4.55. First, we observe that if fewer than q_1 quantity of goods are produced and sold, the company will operate under a net loss, i.e., the revenue will be less than the cost. Thus, at least q_1 goods must be produced and sold in order for the company to be profitable. Second, observe that if the company produces and sells more than q_2 goods, the company again has a greater cost than revenue, and will operate under a loss. We conclude that the company is only profitable if it produces and sells between q_1 and q_2 quantity of goods.

In order to maximize profit, we look for the point at which the *difference* between the revenue and cost is as large as possible. This occurs when the tangent

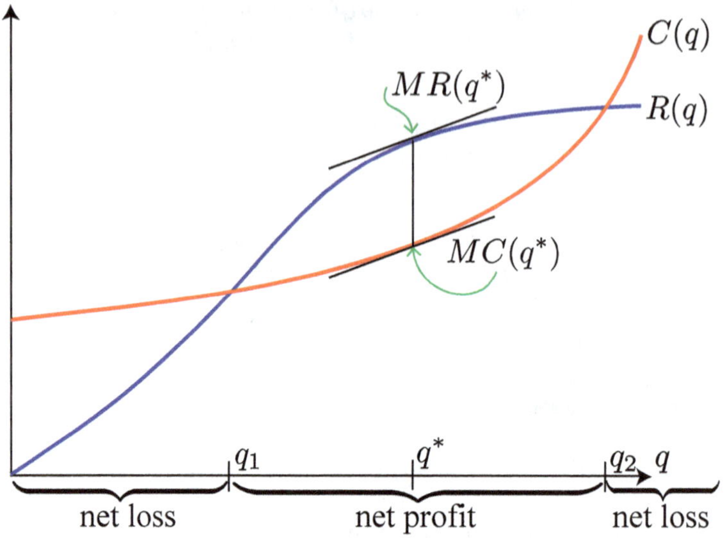

Figure 4.55: The revenue and cost functions for the company of Example 4.35.

line to the revenue function is parallel to the tangent line to the cost function. Recall that the slope of the tangent line to the revenue function represents the marginal revenue: this is how much additional revenue will be generated by producing one more item. Similarly, the slope of the tangent line to the cost function represents the marginal cost: this is how much it will cost to manufacture one more item. At the point q^*, we observe that $MR(q^*) = MC(q^*)$ and, thus, profit is maximized.

■

Problems

1. Paypal offers a discounted fee to high-volume merchants, depending on the total transaction amount of monthly sales. Paypal currently advertises the rates shown in Table 4.1 Suppose that an e-commerce website sells high-end

Monthly total sales	Transaction fee per sale
\$0 to \$3,000	\$0.30 plus 2.9%
\$3,000.01 to \$10,000	\$0.30 plus 2.5%
\$10,000.01 and up	\$0.30 plus 2.2%

Table 4.1: Paypal transaction fee based on monthly volume; Problem 1.

sweaters at a price of \$125 each. Write a piecewise-defined formula for the total paypal fee $P(q)$ if q sweaters are sold in a given month.

2. If $R'(800) = 90$ and $C'(800) = 75$, should the quantity of goods produced be increased or decreased from $q = 800$ in order to increase profit?
3. A certain laptop computer sells for \$1,295. Suppose that the cost of manufacturing q computers is $C(q) = 12000 + 3q^2$. Determine how many computers should be manufactured and sold in order to maximize profit.
4. A riverboat cruise travels a 30-mile section of river each night. Suppose that the company pays \$576 per hour for labor, and that fuel costs $(1/6)v^2$ dollars per mile, where v is the speed of the cruise ship, measured in mph. Determine the cruising speed v that will minimize the cost of each cruise.
5. Suppose that the quantity q of goods that are manufactured and sold depends on the price p per item at which the goods sell. Use the chain rule to show that
$$\frac{d\Pi}{dp} = \left(R'(q(p)) - C'(q(p))\right) q'(p).$$
6. Suppose that the *demand* $D(p)$ for a product at price p represents the number of people who would be willing to purchase the commodity if it is priced at \$p per item. Suppose that for a certain good, the demand function can be modeled by the equation
$$D(p) = q_0 e^{-\alpha p},$$
for some positive constant $\alpha > 0$.
   (a) Explain what the parameter q_0 represents in practical terms.
   (b) Explain why $R(p) = pD(p)$.
   (c) Explain why $C(p) = aD(p)$. Explain the meaning of the parameter a.
   (d) Determine the price p that will maximize the profit per item. The answer may depend on the parameters a and α.
7. Suppose that Geneuga's Ice Cream Cones wants to maximize their profits. The surface area S of their cone, i.e., the area of waffle needed to make one cone, is given by the formula
$$S = \pi r \sqrt{r^2 + h^2},$$
where r is the radius of the base of the cone and h is the cone's height. The volume V of ice cream required to fill a cone is given by the formula
$$V = \frac{1}{3}\pi r^2 h + \frac{2}{3}\pi r^3.$$
Suppose that the total cost to produce one ice cream cone is given by
$$C = 0.0029V + 0.0045S.$$
Also, suppose that the price p of one ice cream cone is proportional to its surface area, so that $P = 0.015S$. If Geneuga's only produces ice cream cones whose height is twice the radius, i.e., $h = 2r$, how much will an ice cream cone from Geneuga's cost?

8. A 300-seat theater will be featuring a new musical production. If the cost per ticket is \$80, the theater will be sold out. For every \$1 increase in ticket price, however, 2 fewer tickets will be sold. Determine what the ticket price should be in order to maximize the nightly revenue.
9. A certain 100-room hotel is offering an internet sale. The price of renting one room for five nights is normally \$1,000. If more than 20 rooms are rented, however, the hotel has agreed to subtract \$10 from everyone's bill for each room, in excess of 20 rooms, that are rented. For example, if 24 rooms are rented, each room will cost \$960 for five nights. Meanwhile, there is a fixed cost of \$25,000 for keeping the hotel running for these five days. In addition, there is a cost of \$200 for each room that is rented.
   (a) Write a formula for the cost $C(n)$ that would be incurred if n rooms are rented.
   (b) Write a formula for the revenue $R(n)$ that would be generated if n rooms are rented. The revenue $R(n)$ is, in this situation, a piecewise-defined function.
   (c) Use marginality to explain why the optimal profit will not be achieved if fewer than 20 rooms are rented.
   (d) How many rooms should be rented in order to maximize profit?
10. If the cost of producing q units of a certain commodity is given by the cost function $C(q)$, then the *average cost* per unit, if q units are manufactured, is given by the formula

$$a(q) = \frac{C(q)}{q}.$$

   (a) Show that when the average cost per item is minimized, that the average cost equals the marginal cost.
   (b) If it costs $C(q) = 12{,}000 + 8q + 0.02q^2$ to produce a total quantity q of a certain commodity, determine the quantity q^* of goods that should be produced in order to minimize the average cost per item.
11. The *Cobb–Douglas production function* can be used to model the total production level P (i.e., the total value of all goods produced in a year) as a function of the amount of labor L (the total number person-hours worked in a given year) and the amount of capital investment K (the total value invested in infrastructure, equipment, machines, etc.) by the relation

$$P(L,K) = cL^{\alpha}K^{1-\alpha},$$

where c and α are a positive parameters, with $0 < \alpha < 1$[1].
Suppose a small, Silicon Valley tech startup has enough startup capital so that infrastructure, etc., is not an issue, and so that we may treat the variable K as a constant. Further, suppose that the parameter α has a value of $\alpha = 0.25$. In this case, the Cobb–Douglas production formula will take the simplified

[1]Charles Cobb and Paul Douglas, the namesakes of this formula, published a study of the American economy from 1899–1922, in which they determined $c = 1.01$ and $\alpha = 0.75$. They normalized their data so that $P = L = K = 100$ in 1899, i.e., each quantity is reported as a percentage of their 1899 value.

form

$$P(L) = kL^{1/4} = k\sqrt[4]{L},$$

for a positive proportionality constant k.

(a) If each "good" produced (let us say that each item is a virtual product, such as a website membership) sells at price p, and each person-hour of labor costs the company h, show that the weekly profit of the company, if it employs a total labor force of L person-hours per week, is given by the formula

$$\Pi(L) = pkL^{0.25} - hL.$$

(b) Show that the maximum profit for this web startup is obtained if they employ a total labor force consisting of

$$L^* = \left(\frac{pk}{4k}\right)^{4/3} = \sqrt[3]{\frac{p^4k^4}{256h^4}}$$

person-hours per week.

12. Suppose that a small business sells a product, for which there is a demand of Q units per day. The business orders this product in batches, once every T days. The goal of this problem is to determine a formula for the length of time T that a business should wait in between orders so that they will minimize the combined shipping and storage cost of the product.

(a) Assume that the total shipping cost S of q items is given by the formula

$$S(q) = a + bq.$$

Explain the meaning of the parameters a and b in practical terms.

(b) Assume that the storage cost G incurred for renting enough storage space to store q items for a total of T days is given by

$$G(q,T) = cqT.$$

Explain the meaning of the parameter c in practical terms.

(c) If the company orders additional goods every T days, explain why $q = TQ$, where q is the quantity ordered with each shipment.

(d) Show that the average cost per day c for both shipping and storing the goods, if an order is placed every T days, is given by the formula

$$c(T) = \frac{a}{T} + bQ + cQT.$$

(e) Show that the average cost per day is minimized if an order is placed every

$$T^* = \sqrt{\frac{a}{cQ}}$$

days.

4.7 L'Hospital's Rule

In this section, we examine a method for evaluating certain kinds of limits.

4.7.1 Indeterminate Forms

Consider the limits

$$\lim_{x\to 0} \frac{\sin(x)}{x} \qquad \text{and} \qquad \lim_{x\to 0} \frac{\ln|x|}{\tan(x-\pi/2)}.$$

Neither function is defined for $x = 0$. Moreover, if we were to attempt to evaluate these functions at $x = 0$, we would obtain

$$\frac{0}{0} \qquad \text{and} \qquad \frac{\infty}{\infty},$$

respectively. (Of course, by these expressions, we merely mean that both the numerator and denominator is equal to either zero or infinity when $x = 0$.) Since the ratios $0/0$ and ∞/∞ are undefined, this result is *not* the value of the limit. These undefined ratios, however, tip us off that the limit is of indeterminate form, as we now define.

Definition 4.7.1 Suppose that either

$$\lim_{x\to a} f(x) = 0 \qquad \text{and} \qquad \lim_{x\to a} g(x) = 0, \tag{4.44}$$

or

$$\lim_{x\to a} f(x) = \pm\infty \qquad \text{and} \qquad \lim_{x\to a} g(x) = \pm\infty, \tag{4.45}$$

then we say that the limit of the ratio

$$\lim_{x\to a} \frac{f(x)}{g(x)}$$

is an *indeterminate form*. For the case of equation (4.44), we say that the indeterminate form is of *type* $0/0$. For the case of equation (4.45), we say that the indeterminate form is of *type* ∞/∞.

The same terminology applies if we replace each limit with either a left-hand limit, a right-hand limit, or a limit to $+\infty$ or $-\infty$.

R Notice that the limits

$$\lim_{x\to a} f(x) \qquad \text{and} \qquad \lim_{x\to a} g(x)$$

must *both* approach zero, or they must *both* approach $\pm\infty$.
In the event that one of these limits approaches zero and the other approaches infinity, and we are computing the limit of their product

$$\lim_{x\to a} f(x)g(x),$$

we may rewrite the limit as an indeterminate form by choosing either

$$\lim_{x\to a}\frac{f(x)}{1/g(x)} \qquad \text{or} \qquad \lim_{x\to a}\frac{g(x)}{1/f(x)}.$$

4.7.2 L'Hospital's Rule

There is a simple rule that can be used to evaluate indeterminate-form limits.

Theorem 4.7.1 — L'Hospital's Rule. Suppose that the limit as $x \to a$ of the ratio $f(x)/g(x)$ is an indeterminate form of type $0/0$ or type ∞/∞, then

$$\lim_{x\to a}\frac{f(x)}{g(x)} = \lim_{x\to a}\frac{f'(x)}{g'(x)}, \tag{4.46}$$

provided that the limit on the right-hand side exists. Here, $a \in \mathbb{R} \cup \{+\infty, -\infty\}$, i.e., a is either a real number or $\pm\infty$. The same result applies to right-hand and left-hand limits.

(R) If the limit on the right-hand side of equation (4.46) is also an indeterminate form of type $0/0$ or ∞/∞, then L'Hospital's rule may be applied again, until the value of the limit is recovered.

Proof. We will not prove the general case of L'Hospital's rule here. We will, however, provide a heuristic justification of why the result holds for the case of an indeterminate form of type $0/0$, in which the limiting point a is finite.

Without loss of generality, let us assume that $f(a) = g(a) = 0$. If this were not true, we could always define the functions

$$\begin{aligned} F(x) &= \begin{cases} f(x) & \text{for } x \neq a, \\ 0 & \text{for } x = a \end{cases}, \\ G(x) &= \begin{cases} g(x) & \text{for } x \neq a, \\ 0 & \text{for } x = a \end{cases}. \end{aligned}$$

Since

$$\lim_{x\to a} f(x) = 0 \qquad \text{and} \qquad \lim_{x\to a} g(x) = 0,$$

it follows that the functions F and G are continuous. (Recall Definition 1.7.3 on page 82.)

Next, let us assume that the first and second derivatives of the functions f and g at the point $x = a$ also exist, so that we may consider the quadratic approximations as given by equation (3.53). We may therefore represent the functions f and g to second order in $(x-a)$ using the approximations

$$\begin{aligned} T_2^f(x) &= f(a) + f'(a)(x-a) + \frac{f''(a)}{2}(x-a)^2, \\ T_2^g(x) &= g(a) + g'(a)(x-a) + \frac{g''(a)}{2}(x-a)^2, \end{aligned}$$

respectively. The values $f(a) = g(a) = 0$, so that

$$\lim_{x\to a} \frac{f(x)}{g(x)} = \lim_{x\to 0} \frac{f'(a)(x-a) + 0.5f''(a)(x-a)^2}{g'(a)(x-a) + 0.5g''(a)(x-a)^2}.$$

(The difference between f and T_2^f goes to zero even faster than the quadratic term $(x-a)^2$, so that these limits do actually equal each other.) Removing a factor of $(x-a)$ from the numerator and denominator, we obtain

$$\lim_{x\to a} \frac{f(x)}{g(x)} = \frac{f'(a) + 0.5f''(a)(x-a)}{g'(a) + 0.5f''(a)(x-a)} = \frac{f'(a)}{g'(a)}.$$

The $(x-a)$ terms, however, go to zero, as well as any of the higher order terms we left off in using the quadratic approximations for f and g. The limit therefore approaches the ratio of the derivatives. Now, if the functions f and g are undefined or discontinuous at $x = a$, their derivatives will likewise be undefined. As long as the limit of the ratio of these derivatives, i.e., the right-hand side of equation (4.46), exists, the result is still valid. ■

■ **Example 4.36** Determine the limit

$$\lim_{x\to 0} \frac{\sin(x)}{x}.$$

This limit is an indeterminate form of type $0/0$. We may therefore apply L'Hospital's rule, obtaining

$$\lim_{x\to 0} \frac{\sin(x)}{x} = \lim_{x\to 0} \frac{\cos(x)}{1} = 1.$$

■

■ **Example 4.37** Determine the limit

$$\lim_{x\to 0} \frac{\ln|x|}{\tan(x-\pi/2)}.$$

This limit is an indeterminate form of type ∞/∞. We may therefore apply L'Hospital's rule, obtaining

$$\lim_{x\to 0} \frac{\ln|x|}{\tan(x-\pi/2)} = \lim_{x\to 0} \frac{1/x}{\sec^2(x-\pi/2)} = \lim_{x\to 0} \frac{\cos^2(x-\pi/2)}{x} = \lim_{x\to 0} \frac{\sin^2(x)}{x}.$$

(The second equality follows solely from algebraic manipulation; the third equality from the trigonometric identity $\cos(x-\pi/2) = -\sin(x)$.) Our new expression is also an indeterminate form, this time of type $0/0$. We may therefore apply L'Hospital's rule a second time, obtaining

$$\lim_{x\to 0} \frac{\ln|x|}{\tan(x-\pi/2)} = \lim_{x\to 0} \frac{-\sin^2(x)}{x} = \lim_{x\to 0} \frac{-2\sin(x)\cos(x)}{1} = 0.$$

The final equality follows by evaluating $\sin(0)\cos(0) = 0$. Our result is confirmed by the graph of the function $f(x)/g(x)$, as shown in Figure 4.56. ■

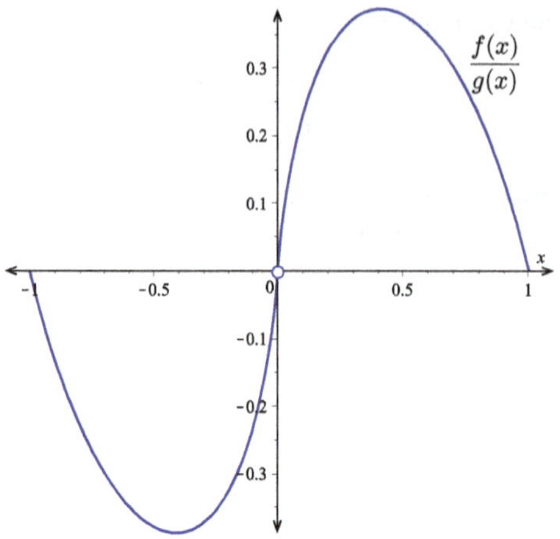

Figure 4.56: Graph of the function $f(x)/g(x)$, where $f(x) = \ln|x|$ and $g(x) = \tan(x - \pi/2)$.

■ **Example 4.38** Use L'Hospital's rule to evaluate the limit

$$\lim_{x\to\infty} \frac{5x^3 + 4x^2 + 8x + 15}{14x^3 + 16x^2 + 23x + 42}.$$

This limit is an indeterminate form of type ∞/∞. We may therefore apply L'Hospital's rule:

$$\begin{aligned}
\lim_{x\to\infty} \frac{5x^3 + 4x^2 + 8x + 15}{14x^3 + 16x^2 + 23x + 42} &= \lim_{x\to\infty} \frac{15x^2 + 8x + 8}{42x^2 + 32x + 23} \\
&= \lim_{x\to\infty} \frac{30x + 8}{84x + 32} \\
&= \lim_{x\to\infty} \frac{30}{84} = \frac{5}{14}.
\end{aligned}$$

Notice that each limit, up until the last, is an indeterminate form. In the preceding equation chain, we applied L'Hospital's rule three successive times before arriving at the answer. ■

■ **Example 4.39** Consider the limit

$$\lim_{x\to 1} \tan\left(\frac{\pi x}{2}\right) \ln(x)$$

Notice that

$$\lim_{x\to 1} \left|\tan\left(\frac{\pi x}{2}\right)\right| = \infty \qquad \text{and} \qquad \lim_{x\to 1} \ln(x) = 0.$$

In order to apply L'Hospital's rule, we must rewrite this limit so that it is an indeterminate form of type $0/0$ or ∞/∞. To do this, we write

$$\lim_{x\to 1}\tan\left(\frac{\pi x}{2}\right)\ln(x) = \lim_{x\to 1}\frac{\ln(x)}{\cot\left(\frac{\pi x}{2}\right)}.$$

Now our limit is an indeterminate form of type $0/0$. Applying L'Hospital's rule, we obtain

$$\lim_{x\to 1}\frac{\ln(x)}{\cot\left(\frac{\pi x}{2}\right)} = \lim_{x\to 1}\frac{1/x}{-\frac{\pi}{2}\csc^2\left(\frac{\pi x}{2}\right)} = \frac{-2}{\pi}.$$

The function $f(x) = \tan(\pi x/2)\ln(x)$ is plotted in Figure 4.57, which confirms our result. ∎

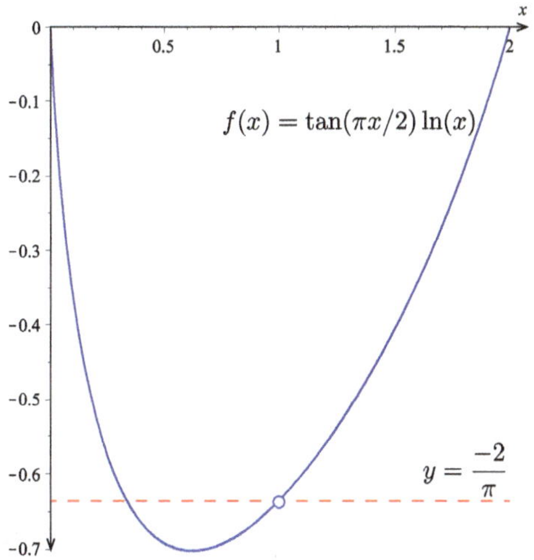

Figure 4.57: Graph of the function $f(x) = \tan(\pi x/2)\ln(x)$.

Exercises

For Exercises 1–10, use L'Hospital's rule to evaluate the given limit.

1. $\lim_{x\to\infty}\frac{x^2+e^{0.4x}}{5x^2+3e^{0.4x}}$.
2. $\lim_{x\to 1}\frac{\ln(x)}{(x-1)}$.
3. $\lim_{x\to 0}\frac{x^3+5x}{17x^4+9x}$.
4. $\lim_{x\to 0}\frac{2\cos(x)-2+x^2}{x^4}$.
5. $\lim_{x\to\pi}\frac{\sin^2(x)}{(x-\pi)^2}$.
6. $\lim_{x\to 0}\frac{x-\sin(x)}{x^3}$.
7. $\lim_{x\to 0}\frac{\sin(\alpha x)}{\sinh(\beta x)}$.
8. $\lim_{x\to 0}\frac{\arcsin(x)}{\sin(x)}$.
9. $\lim_{x\to 2}\frac{\sin(x^3-8)}{x-2}$.
10. $\lim_{x\to\infty}\frac{\ln(x)}{\sqrt{x}}$.

Chapter 5

Antiderivatives and Summation

5.1 Antiderivatives and Indefinite Integrals

In this section, we introduce the basic concept and notation of antiderivatives and indefinite integrals. These two concepts are closely related, as an antiderivative of a function is an instance of the indefinite integral, which represents the entire *family* of antiderivatives for the given function. These ideas will form the bedrock of our subsequent discussion on integration. Though they may at first appear abstract, or as a lofty academic exercise with no practical application, we will just as quickly ground these ideas in practicality when we use them to introduce integration in general—the definite integral, in particular.

5.1.1 Antiderivatives

An *antiderivative* of a function is just the opposite of the derivative. Computing antiderivatives is the first manifestation we shall encounter of the more general concept of *integration*, which we will discuss in depth over the course of the next several chapters. The act of differentiation diminishes where the act of integration amplifies. Without further ado, let us now state precisely what an antiderivative is.

Definition 5.1.1 An *antiderivative* F of a function f is a new function that has the property that f is the derivative of F; i.e., F is an antiderivative of f if $F' = f$.

(R) For the sake of visualization, I, personally, like to think of differentiation as *moving down* and integration as *moving up*. Think of a function and its derivatives occupying different floors of a building. Moving from one floor to the floor below is differentiation whereas moving from one floor to the floor above is integration. Thus, differentiation and integration are two opposing operators used to move up and down the rungs of a ladder, as depicted in Figure 5.1.

Just as subtraction undoes addition and division undoes multiplication, integration undoes differentiation.

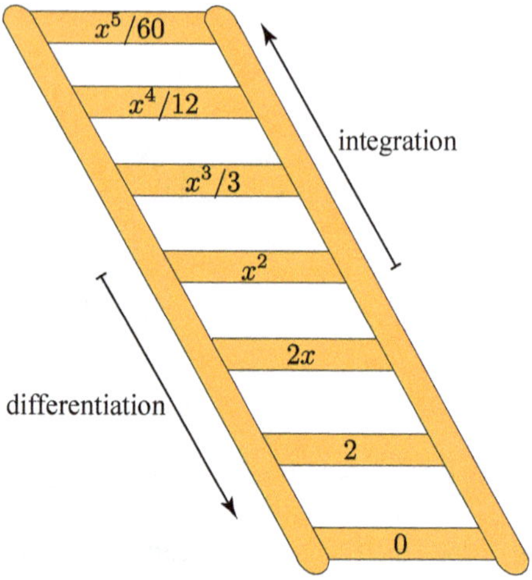

Figure 5.1: An illustration of differentiation and integration.

■ **Example 5.1** Let $f(x) = x^6$. Find an antiderivative for the function f.

To do this, we must think about the rules of differentiation, except *backwards*. That is to say, we must think of a function that, when differentiated, returns the given function $f(x) = x^6$.

When taking derivatives of a power function, such as this, we (1) bring the exponent down and place it out front, and (2) subtract one from the exponent. It makes sense, then, that to find an antiderivative, we might first *add one* to the exponent, and then, second, *divide* the result by the *new* exponent. Thus, we obtain

$$F(x) = \frac{x^7}{7}.$$

Finally, we always *check* our answer, by computing its derivative. If we now take the derivative of $F(x) = x^7/7$, we obtain $F'(x) = 7x^6/7 = x^6$. This confirms that $F'(x) = f(x)$, and, thus, the new function F is indeed an antiderivative for our starting function f. ■

■ **Example 5.2** Let $f(x) = \sin(3x)$. Find an antiderivative for the function f.

Again, we will examine the rules of differentiation backwards. When taking the derivative of sine or cosine, we apply the following rubric: the derivative of sine is cosine and the derivative of cosine is negative sine. Thus, what *should* the antiderivative of sine be? Well, the *derivative* of cosine is *negative* sine, so, going backwards, the antiderivative of sine should be negative cosine.

Finally, when differentiating sine or cosine of $3x$, we complete the process by multiplying by the derivative of the inside, i.e., by the number 3. For the purpose of computing an antiderivative, we therefore divide by the number 3. Altogether, this gives us the following antiderivative

$$F(x) = \frac{-\cos(3x)}{3}.$$

To check, we may compute the derivative of our answer:

$$F'(x) = --3\sin(3x)/3 = \sin(3x),$$

which indeed confirms that $F'(x) = f(x)$, as desired. ▪

The preceding examples illustrate a general rule: when faced with a function you wish to determine the antiderivative of, think about how the derivative rules work, and apply them, if you can, *backwards*. Also, always check your answer by computing its derivative and verifying that you recover the original function.

Next, please take notice of the language: we compute *an* antiderivative of a function or we compute *the* derivative of a function. This indicates that, whereas a function can have only one derivative, a function may have several antiderivatives. Indeed this is the case and, as it turns out, a function that has one antiderivative has also an infinite number of antiderivatives. Despite this apparently vastness, those antiderivatives are closely related, as is revealed by our next theorem.

Theorem 5.1.1 Suppose that the functions F and G are *both* antiderivatives of the function f. Then both antiderivatives F and G must be related to each other by an additive constant, i.e.,

$$F(x) = G(x) + C,$$

for some real number C.

Proof. According to the premise, we are to assume that both F and G are antiderivatives of the function f. By definition, this means that

$$F'(x) = f(x) \qquad \text{and} \qquad G'(x) = f(x).$$

Next, let us define a function H by the relation $H(x) = F(x) - G(x)$, i.e., H is the *difference* between these two antiderivatives (which, after all, we are trying to prove is constant). Computing the derivative of H, we obtain

$$H'(x) = F'(x) - G'(x) = f(x) - f(x) = 0.$$

Therefore, the function H has a derivative that vanishes everywhere. This is only possible if the function H is a constant function, i.e., if $H(x) = C$. The result follows. ■

■ **Example 5.3** Consider the function $f(x) = \sin(3x)$. Find several antiderivatives.

We have already computed one antiderivative, i.e., $F(x) = -\cos(3x)/3$. According to Theorem 5.1.1, the following functions are *also* antiderivatives of the function f:

$$G(x) = \frac{-\cos(3x)}{3} + 17, \quad H(x) = \frac{-\cos(3x)}{3} + \pi^8, \quad \text{and} \quad J(x) = \frac{-\cos(3x)}{3} + 9{,}999.$$

■

Theorem 5.1.1 has the following graphical consequence.

Corollary 5.1.2 Two functions F and G that are antiderivatives of a common function f must be vertical translations of one another.

In other words, the graph of two antiderivatives of a common function must simply be the same graph, except shifted up or down, as shown in Figure 5.2.

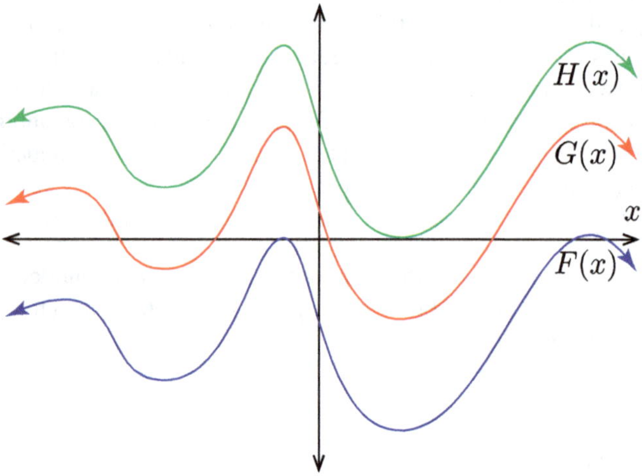

Figure 5.2: The graphs of three antiderivatives F, G, and H of a function f.

5.1.2 The Indefinite Integral

Next, we introduce the concept of the (indefinite) integral. We will study *definite* integrals in our next chapter; therefore, do not be alarmed that the modifier *indefinite* does not contrast with anything at the moment.

Definition 5.1.2 The *indefinite integral* of a function $f(x)$ *with respect to* x, which is denoted

$$\int f(x)\,dx,$$

represents the family of antiderivatives of the function f, when regarding x as the independent variable.

If $F(x)$ is a particular antiderivative for $f(x)$, one may write

$$\int f(x)\,dx = F(x) + C,$$

where C represents an arbitrary constant, as provided by Theorem 5.1.1, and which is known as the *constant of integration*.

(R) **[Anatomy of an Integral]** In the notation

$$\int f(x)\,dx,$$

the function $f(x)$ is called the *integrand*, i.e., the integrand is the function that is being integrated. The symbol dx is read, "with respect to x," and it indicates the variable with respect to which we are performing the integration. The d is always used, and the symbol for the independent variable always follows the d. One may also think of the dx as "closing" the integral, in the sense that if something appears *after* the dx, it should be regarded as being *outside* the integral. Though, style-wise, it is always preferred to put a function multiplying an integral *in front* of the integral, i.e., the notation

$$x\int x dx \quad \text{is preferred over} \quad \int x\,dx\,x,$$

as the latter may be confusing.

Example 5.4 Recalling our previous two examples, we may write

$$\begin{aligned} \int x^6\,dx &= \frac{x^7}{7} + C, \\ \int \sin(3x)\,dx &= \frac{-\cos(3x)}{3} + C. \end{aligned}$$

Notice that all three antiderivatives in Example 5.3 fall under the scheme of this second formula, only for different values of the constant C. ■

The linearity properties of the derivative operator imply the same linearity properties of the integral.

Theorem 5.1.3 Let f and g be functions that possess antiderivatives. Then

$$\int (af(x)+bg(x))\, dx = a\int f(x)\, dx + b\int g(x)\, dx, \qquad (5.1)$$

for any constants a and b.

■ **Example 5.5** One can easily show that the following two integral formulas hold:

$$\begin{aligned}\int x^3\, dx &= \frac{x^4}{4}+C,\\ \int x^5\, dx &= \frac{x^6}{6}+C.\end{aligned}$$

Therefore, from linearity, it follows that

$$\int \left(7x^3+11x^5\right)\, dx = \frac{7x^4}{4}+\frac{11x^6}{6}+C.$$

This may easily be verified by checking

$$\frac{d}{dx}\left[\frac{7x^4}{4}+\frac{11x^6}{6}\right] = 7x^3+11x^5,$$

which confirms the result. ■

Thus, as was the case during differentiation, during integration one may integrate a formula term-by-term. Constant multiples can be pulled out of integrals, as they simply get carried along for the ride.

Let us re-articulate the difference between antiderivatives and integrals. An antiderivative is a particular instance of the indefinite integral, i.e., it is a member of the family of functions that the indefinite integral represents. The (indefinite) integral, on the other hand, represents *all* antiderivatives of a particular function simultaneously, i.e., the integral is the *class* or *family* of antiderivatives.

5.1.3 Some Basic Integral Formulas

As we did with derivatives, we introduce a quick-and-easy reference enumerating a number of basic integrals.

Theorem 5.1.4 — The Five Basic Integrals. The five basic integrals are

$$\int x^n \, dx = \frac{x^{n+1}}{n+1} + C, \qquad \text{for } n \neq -1, \tag{5.2}$$

$$\int \sin(ax) \, dx = \frac{-\cos(ax)}{a} + C, \tag{5.3}$$

$$\int \cos(ax) \, dx = \frac{\sin(ax)}{a} + C, \tag{5.4}$$

$$\int e^{ax} \, dx = \frac{1}{a} e^{ax} + C, \tag{5.5}$$

$$\int \frac{1}{x} \, dx = \ln|x| + C. \tag{5.6}$$

▪ **Example 5.6** Compute the integral

$$\int \left(x^3 + 2x^2 + 3x + 4 + \frac{5}{x} \right) dx.$$

Using equation (5.2) and the linearity property of the integral, we conclude

$$\int \left(x^3 + 2x^2 + 3x + 4 + \frac{5}{x} \right) dx = \frac{x^4}{4} + \frac{2x^3}{3} + \frac{3x^2}{2} + 4x + 5\ln|x| + C.$$

Notice that an invisible "+0" in the integrand is always integrated to the "+C" in the output. ▪

▪ **Example 5.7** Using formulas (5.3)–(5.5), we may compute

$$\int \left(\sin(3x) + \cos(4x) + e^{8x} \right) dx = \frac{-\cos(3x)}{3} + \frac{\sin(4x)}{4} + \frac{e^{8x}}{8} + C.$$

▪

▪ **Example 5.8** One may use formula (5.2) even for non-integer values of n.

$$\int \left(\sqrt{x} + \frac{1}{x^2} \right) dx = \frac{2x^{3/2}}{3} - \frac{1}{x} + C.$$

Notice that the process is similar to what we have done in the past: we first recognize $\sqrt{x} = x^{1/2}$, then we *add* one to the exponent, obtaining $x^{3/2}$; finally, we divide by the new exponent, obtaining $2x^{3/2}/3$. Similarly, $1/x^2 = x^{-2}$; therefore, its integral is $-x^{-1} = -1/x$. ▪

One must be careful when the "inside function" is not a straight-out constant multiple of x. Our next example is of a function that does not have an antiderivative—at least, its antiderivative cannot be written in terms of elementary functions using a finite number of terms.

▪ **Example 5.9** Consider

$$\int e^{x^2} \, dx.$$

This is actually an important example in statistics and probability. Though we will reach a dead end here, we will return to this example later in the text, when we have tools to do more with it.

At first, one might suspect that one can compute the integral as follows:

$$\int e^{x^2}\,dx \stackrel{?}{=} \frac{e^{x^2}}{2x} + C.$$

Here, we simply did the opposite of how we would have computed the derivative: instead of rewriting the exponential function and then multiplying by the derivative of the exponent, we rewrote the exponential function and *divided* by the derivative of the exponent. It seems to make sense, but does this work? To check, we must take the derivative of our answer (we are, after all, pros at differentiation by now). To do so, we apply the quotient rule, thereby obtaining

$$\frac{d}{dx}\left[\frac{e^{x^2}}{2x}\right] = \frac{4x^2e^{x^2} - 2e^{x^2}}{4x^2} = e^{x^2} - \frac{e^{x^2}}{2x^2}.$$

This is clearly *not* equal to e^{x^2}, therefore our answer is incorrect:

$$\int e^{x^2}\,dx \neq \frac{e^{x^2}}{2x} + C.$$

As it turns out, this antiderivative, though it exists, cannot be written in terms of elementary functions (i.e., power, trig, exponential, logarithmic) in a finite number of terms. ■

The previous example shows where we must be careful in constructing antiderivatives: one cannot simply divide the new expression by the derivative of the inside, like one multiplies the new expression by the derivative of the inside when employing the chain rule. This creates havoc a la the quotient rule. But one can perform an integral in the spirit of the differentiation rules (3.6)–(3.10) if the integrand already has a factor of the derivative of the inside function as a factor. To see what we mean by this perhaps abstract discussion, let us specifically examine the following rules.

Theorem 5.1.5 — The Five Basic Integrals ***Plus*****.** The following integration formulas hold:

$$\int (u(x))^n u'(x)\,dx = \frac{(u(x))^{n+1}}{n+1} + C, \qquad \text{for } n \neq -1, \tag{5.7}$$

$$\int \sin(u(x))\,u'(x)\,dx = -\cos(u(x)) + C, \tag{5.8}$$

$$\int \cos(u(x))\,u'(x)\,dx = \sin(u(x)) + C, \tag{5.9}$$

$$\int e^{u(x)} u'(x)\,dx = e^{u(x)} + C, \tag{5.10}$$

$$\int \frac{u'(x)}{u(x)}\,dx = \ln|u(x)| + C. \tag{5.11}$$

These formulas follow immediately from the differentiation formulas given by equations (3.6)–(3.10), as, in each case, the integrand on the left-hand side is the derivative of the right-hand side. Again, it should be stressed: we are doing nothing more than working the regular derivative rules backwards.

■ **Example 5.10** Compute the integral

$$\int 9x^2 \sin\left(3x^3\right)\,dx.$$

Here, we use equation (5.8). The inside function is $u(x) = 3x^3$, which has a derivative $9x^2$ that is clearly present as a factor within the integrand. Therefore, we obtain

$$\int 9x^2 \sin\left(3x^3\right)\,dx = -\cos\left(3x^3\right) + C.$$

By differentiating the right-hand side, we can indeed confirm that we obtain the integrand on the left. ■

■ **Example 5.11** Compute the integral

$$\int \tan(x)\,dx.$$

This is interesting, because, following the basic derivative rules, we do not, a priori, recall a function whose derivative is the tangent function. We recognize, however, that the tangent function may be written as

$$\tan(x) = \frac{\sin(x)}{\cos(x)},$$

and rule (5.11) may be employed. We obtain

$$\int \tan(x)\,dx = -\ln|\cos(x)| + C = \ln|\sec(x)| + C.$$

The final equality follows since $-\ln|u| = \ln|u^{-1}|$. ■

5.1.4 One-dimensional Particle Motion: Constant Acceleration

These basic techniques of integration can be used to uncover a fundamental equation in physics: the position, as a function of time, of a particle that moves with constant acceleration.

Theorem 5.1.6 Let $x(t)$ represent the position along the x-axis of a particle at time t. If the particle's acceleration is a constant a, i.e., $x''(t) = a$, then

$$x(t) = x_0 + v_0 t + \frac{1}{2}at^2, \tag{5.12}$$

where x_0 and v_0 are constants that represent the *initial position* and *initial velocity* of the particle, respectively.

Proof. We will use the basic definitions of velocity and acceleration; to wit, the velocity $v(t)$ is the time derivative of the position ($v(t) = x'(t)$) and the acceleration $a(t)$ is the time derivative of the velocity ($a(t) = v'(t) = x''(t)$).

To begin the analysis, we first write down our underlying constant-acceleration premise:

$$a(t) = a.$$

Since the acceleration is the derivative of velocity, the velocity must be one of the antiderivatives of the acceleration, so that

$$v(t) = \int a(t)\,dt = \int a\,dt = at + C.$$

In order to give the constant of integration C a practical interpretation, we note that the velocity is initially equal to $v(0) = C$. By defining a new parameter v_0 to represent the initial velocity, we may now write

$$v(t) = v_0 + at. \tag{5.13}$$

This equation is used in conjunction with equation (5.12) to solve many problems.

Now that we have an equation for the velocity of the particle, we may integrate it, yet again, to recover the position:

$$x(t) = \int v(t)\,dt = \int (v_0 + at)\,dt = v_0 t + \frac{1}{2}at^2 + D,$$

where D is the new constant of integration. Again, evaluating this expression at the initial time, we find $x(0) = D$. We may therefore, again, introduce a new parameter x_0 to represent the particle's initial position, thereby recovering equation (5.12).

■

Exercises

For Exercises 1–30, compute the indicated indefinite integral.

1. $\int (x^4 + 5x^2 + 17)\, dx.$
2. $\int \left(x + 1 + \frac{1}{x} + \frac{1}{x^2}\right) dx.$
3. $\int \left(\sqrt{x} + \frac{1}{\sqrt{x}}\right) dx.$
4. $\int \pi\, dx.$
5. $\int (x^8 + \sin(3x) + e^{8x})\, dx.$
6. $\int (e^x + e^{-x})\, dx.$
7. $\int \cot(x)\, dx.$
8. $\int (8\cos(3x) + 12\sin(4x))\, dx.$
9. $\int (1+x)^{99}\, dx.$
10. $\int \left(7 + \frac{7}{x}\right) dx.$
11. $\int \sin^2(x)\cos(x)\, dx.$
12. $\int x^7 \left(8x^8 + 17\right)^{88} dx.$
13. $\int \sin(3x+5)\, dx.$
14. $\int 2x\sin\left(x^2\right) dx.$
15. $\int x\sin\left(x^2\right) dx.$
16. $\int x^2\cos\left(x^3\right) dx.$
17. $\int xe^{-x^2}\, dx.$
18. $\int \frac{15x^4 + 16x^3 + 15x^2 + 7}{3x^5 + 4x^4 + 5x^3 + 7x + 8}\, dx.$
19. $\int \cos(3x)e^{\sin(3x)}\, dx.$
20. $\int \left(x^7 + \frac{1}{x^7}\right) dx.$
21. $\int \frac{\cos(x)}{\sqrt{\sin(x)}}\, dx.$
22. $\int \frac{x\cos(x^2)}{\sin(x^2)}\, dx.$
23. $\int \frac{3}{x}\, dx.$
24. $\int \cos(\sin(x))\cos(x)\, dx.$
25. $\int (9\sin(3x) + 8\cos(2x))\, dx.$
26. $\int \frac{e^x}{e^x + 7}\, dx.$
27. $\int \frac{\cos(x)}{\sin(x) + \pi}\, dx.$
28. $\int x^2 e^{x^3}\, dx.$
29. $\int \sin^9(x)\cos(x)\, dx.$
30. $\int \frac{\sec^2(x)}{\tan(x)}\, dx.$

Problems

31. A man jumps off the ground with an initial upward velocity of 5 m/s. What is the maximum height of his vertical jump if (a) he is on Earth, where acceleration due to gravity is $g = -9.8$ m/s^2? (b) he is on Mars, where acceleration due to gravity is $g = -3.7$ m/s^2? (c) he is on the Moon, where acceleration due to gravity is $g = -1.6$ m/s^2? What is his total “air time” (i.e.,

duration of the jump) in each case?

32. A penny is dropped from the roof of the Empire State Building, at a height of 381 m. Assuming constant, downward gravitational acceleration of 9.8 m/s^2, how long will it take before the penny hits the ground? What will its velocity be the moment just before impact? You may neglect air resistance.
33. At time $t = 1$, a particle is at rest at the origin. Assuming that its acceleration is given by $a(t) = 1/t^2$, for $t \geq 1$, determine its velocity and position as a function of time. What happens to its velocity as $t \to \infty$? What happens to its position as $t \to \infty$?
34. The marginal cost $MC(q)$ and marginal revenue $MR(q)$ functions of a particular commodity are shown as a function of the quantity q of goods produced in Figure 5.3

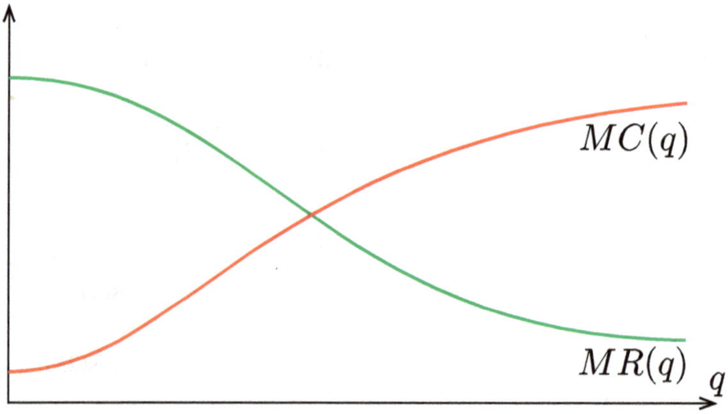

Figure 5.3: The marginal revenue and marginal cost as a function of the quantity q of goods produced; Problem 34.

(a) Draw a graph of the revenue function $R(q)$ as accurately as possible (increasing/decreasing, concavity, etc.) Is it possible to determine the constant of integration? Why or why not? What is the constant of integration and what does it represent?

(b) Draw a graph of the cost function $C(q)$ as accurately as possible (increasing/decreasing, concavity, etc.) Is it possible to determine the constant of integration? Why or why not? What is the constant of integration and what does it represent?

(c) On the graph, indicate the optimal quantity q^* of goods that should be produced in order to maximize profit.

5.2 Summation Notation

In this section, we introduce an important shorthand that will permeate much of our future discussion. We will soon be interested in *summing* things, large quantities of things, and, as such, we will require a shorthand to facilitate the process. We first lay out the notation, reveal a few simple summation formulas, and then discuss the topic of geometric series.

5.2.1 Summation Notation

The main notation is one called *summation notation*, or *sigma notation*.

Definition 5.2.1 — Summation Notation. Suppose we have n numbers, $a_1, \ldots, a_n$, ordered by an index $i = 1, \ldots, n$. Then we say that *the sum as i goes from* 1 *to n of* a_i, written

$$\sum_{i=1}^{n} a_i,$$

is equal to the sum of those n numbers, i.e.,

$$\sum_{i=1}^{n} a_i = a_1 + a_2 + \cdots + a_n. \tag{5.14}$$

When reading this notation, the symbol Σ is called *the summation sign*, or *sigma*, the symbol i is referred to as the *index*, or as the *index of summation*, and the terms a_i are collectively referred to as the *summand*. The summation itself is often referred to by mathematicians as a *series*.

(R) The symbol Σ is the capital Greek letter sigma, which is why the notation is sometimes alternatively referred to as sigma notation.

(R) The index i is used for convenience. Since one is "summing over i," it does not matter what name we give this index, i.e., the series

$$\sum_{i=1}^{n} a_i \quad \text{and} \quad \sum_{j=1}^{n} a_j$$

both represent the same sum.

■ **Example 5.12** Consider the following examples:

$$\begin{aligned}
\sum_{i=1}^{5} i &= 1+2+3+4+5, \\
\sum_{k=1}^{5} k^2 &= 1+4+9+16+25, \\
\sum_{i=1}^{5} \frac{1}{i} &= 1+\frac{1}{2}+\frac{1}{3}+\frac{1}{4}+\frac{1}{5}, \\
\sum_{j=1}^{5} \frac{1}{j^2} &= 1+\frac{1}{4}+\frac{1}{9}+\frac{1}{16}+\frac{1}{25}, \\
\sum_{k=1}^{5} 1 &= 1+1+1+1+1.
\end{aligned}$$

The name of the index and its starting value are both found underneath the summation sign Σ. The terminating value of the index is located above the summation sign. For instance, $\sum_{i=1}^{5}$ means that the index is i, and that the index initially starts with the value $i = 1$, and that the summation stops when $i = 5$. For each integer value of the index between the starting and ending values, one takes the expression to the right of the summation sign and inputs that value of the index into that expression. For example, in the first summation, we are given

$$\sum_{i=1}^{5} i.$$

The ith term of the sum is $a_i = i$. Thus, $a_1 = 1$, $a_2 = 2$, $a_3 = 3$, $a_4 = 4$, and $a_5 = 5$. Since we are summing with a starting and ending value of 1 and 5, respectively, this summation represents the sum $1+2+3+4+5$.

The other four examples follow in a similar pattern. Notice that the name of the index is really immaterial, as the value of the sum, in the end, no longer depends on the name of the index. ■

The terms of a sum do not always have to be given by a clearcut formula. For instance, we may define a_i to be the ith digit of the number π. Then, for example, we can write

$$\sum_{i=0}^{5} a_i = 3+1+4+1+5+9.$$

Here, there is no simple formula that generates the ith term of the sum as a simple function of i.

Theorem 5.2.1 — Summation Formulas. The following summation formulas hold:

$$\sum_{i=1}^{n} 1 = n, \tag{5.15}$$

$$\sum_{i=1}^{n} i = \frac{n(n+1)}{2}, \tag{5.16}$$

$$\sum_{i=1}^{n} i^2 = \frac{n(n+1)(2n+1)}{6}, \tag{5.17}$$

$$\sum_{i=1}^{n} i^3 = \left(\frac{n(n+1)}{2}\right)^2, \tag{5.18}$$

$$\sum_{i=1}^{n} i^4 = \frac{n(n+1)(2n+1)(3n^2+3n-1)}{30}. \tag{5.19}$$

(R) The formulas on the right-hand side of equations (5.15)–(5.19) are referred to as *closed-form solutions* for the summations on the left-hand sides. Here, "closed form" simply means that the value can be evaluated in a few algebraic steps, as opposed to the *open-form* expressions on the left-hand sides, which require as many algebraic steps as there are terms in the series.

Proof. Equation (5.15) is simple enough to understand. The summation represents

$$\sum_{i=1}^{n} 1 = \underbrace{1+1+\cdots+1+1}_{n \text{ times}}.$$

Of course, if we sum the number one together n times, we obtain exactly n.

Equation (5.16) is interesting. To compute this sum, let us write out the terms of the sum *twice*, in opposite order. First, let us define

$$S = \sum_{i=1}^{n} i.$$

That is, S is a variable that represents the total sum. Then

$$\begin{aligned} S &= 1+2+3+\cdots+(n-2)+(n-1)+n, \\ S &= n+(n-1)+(n-2)+\cdots+3+2+1. \end{aligned}$$

If we add this sum with itself, in the fashion

$$\begin{aligned} 1+n &= n+1, \\ 2+(n-1) &= n+1, \\ 3+(n-2) &= n+1, \end{aligned}$$

and so forth, we find that

$$2S = \underbrace{(n+1)+(n+1)+\cdots+(n+1)+(n+1)}_{n \text{ times}}.$$

And, therefore, we have

$$S = \frac{n(n+1)}{2},$$

which proves our result.

Though the use of judicious trickery, the remaining summation formulas can also be shown. ■

■ **Example 5.13** Compute the sum of the first 100 integers, i.e., determine $1+2+\cdots+99+100$.

To do this, we simply compute

$$\sum_{i=1}^{100} i = \frac{100(101)}{2} = 5050.$$

We thus summed one hundred integers using only a few computations. ■

■ **Example 5.14** Compute the sum of the first 100 integers squared, i.e., determine $1^2+2^2+\cdots+99^2+100^2$.

Again, we use the appropriate formula:

$$\sum_{i=1}^{100} i^2 = \frac{100(101)(201)}{6} = 338,350.$$

■

5.2.2 Infinite Series

We will also be interested in series that contain not a finite, but an infinite number of terms. It is all too possible that one can add an infinity of nonzero terms and reach only a finite amount. To be precise with what we mean by an infinite summation, we introduce the following definition.

Definition 5.2.2 An *infinite series*, by definition, is assigned the value

$$\sum_{i=1}^{\infty} a_i = \lim_{n\to\infty}\left(\sum_{i=1}^{n} a_i\right), \tag{5.20}$$

if this limit exists. In the case that the limit exists, we say that the infinite series *converges* to that value. Otherwise, we say that the infinite series *diverges*.

Thus, an *infinite sum* is simply assigned the value that the finite sums approach, if one keeps adding more and more terms to the series ad infinitum.

To see how the sum of an infinite number of numbers can add up to a finite amount, consider the following example.

■ **Example 5.15** Determine the value of

$$\sum_{i=1}^{\infty} \frac{1}{2^i} = \frac{1}{2} + \frac{1}{4} + \frac{1}{8} + \frac{1}{16} + \frac{1}{32} + \frac{1}{64} + \cdots.$$

In order to compute this, we begin with a unit square. We then cut the square in half, thereby obtaining two equal sections, each with area 1/2, as shown in Figure 5.4. In one of these halves, we again halve it, to obtain two equal sections, each with area 1/4. We now have three sections, with areas 1/2, 1/4, and 1/4. We repeat this process, until we recognize that the total square can be partitioned into an infinite number of sections, with areas 1/2, 1/4, 1/8, 1/16, 1/32, and so on. We

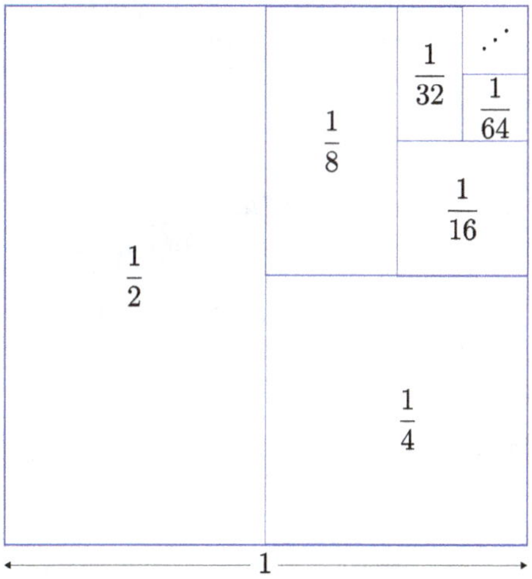

Figure 5.4: The sum $\sum_{i=1}^{\infty} \frac{1}{2^i} = \frac{1}{2} + \frac{1}{4} + \frac{1}{8} + \cdots$.

conclude that

$$\sum_{i=1}^{\infty} \frac{1}{2^i} = 1.$$

This example demonstrates, geometrically, how the sum of an infinite number of numbers can add up to something finite. The idea is, the size of those things vanishes fast enough so that the total sum reaches only a finite limit. ■

5.2.3 Geometric Series

It is difficult to come by actual closed-form formulas for summations. One noteworthy exception that has found its use in a variety of applications is for the case of *geometric* series. The series is *geometric* in the sense that each term is obtained by multiplying the previous term by the same constant. Geometric growth is to arithmetic growth as exponential functions are to linear functions.

Definition 5.2.3 A *finite geometric series* is a series of the form

$$\sum_{i=0}^{n-1} ar^i = a + ar + ar^2 + \cdots + ar^{n-1}. \tag{5.21}$$

The parameter a is called the *initial value*, and the parameter r is called the *growth factor*.

(R) The standard way of writing a finite geometric series is by summing from $i = 0$ to $i = (n-1)$. This represents the first n terms of the series. ($i = 0$ is the first term; $i = 1$ is the second term, and so forth.) In practice, a geometric series may not always present itself in that clean form; one may have to apply several algebraic steps—for example, adding or subtracting some terms or pulling out a common multiple—in order to manipulate a given expression into the standard form.

Theorem 5.2.2 — Finite Geometric Series. The value of a finite geometric series may be written in closed form via the formula

$$\sum_{i=0}^{n-1} ar^i = a\left(\frac{r^n - 1}{r - 1}\right). \tag{5.22}$$

Proof. Let us denote the finite geometric series by the symbol S. Next, consider

$$\begin{aligned} S &= a + ar + ar^2 + \cdots + ar^{n-2} + ar^{n-1}, \\ rS &= ar + ar^2 + ar^3 + \cdots + ar^{n-1} + ar^n. \end{aligned}$$

Subtracting, we find

$$\begin{aligned} rS - S &= ar^n - ar^{n-1} + ar^{n-1} - ar^{n-2} + ar^{n-2} + \cdots - ar^2 + ar^2 - ar + ar - a \\ &= ar^n - a. \end{aligned}$$

Therfore,

$$(r-1)S = ar^n - a,$$

and the result follows. ■

■ Example 5.16 Legend has it that a mathematician invented the game of chess for an ancient emperor (Figure 5.5). The emperor was so pleased with the game, he asked the mathematician to name his reward. The mathematician replied,

"Place a single grain of your rice upon
the first noble square of this great chessboard;
on the second square place two grains of rice;
on the third place four, and then next place eight.

> Continue in this same manner until
> you have placed on every square a few rice,
> each square double the last until you reach
> the very last square; then my fee is paid."

The emperor blushed; he would have been willing to pay a handsome reward, and here the mathematician only asked for a few grains of rice! But how much rice did the mathematician really ask for?

Figure 5.5: A chessboard.

To solve this, we must set up a geometric series. Let us consider the following progression:

square	number of grains	cumulative amount of rice
1	1	1
2	2	3
3	4	7
4	8	15
5	16	31
6	32	63
7	64	127
8	128	255
n	2^{n-1}	$2^n - 1$

On the first square lies a single grain of rice; at this point one total grain of rice has been paid. On the second square lies two grains of rice; cumulatively, three grains now sit upon the board. On the third square there are four grains, and so forth.

Notice that, in general, there are 2^{n-1} grains of rice on the nth square of the board. Cumulatively, there is a total of $2^n - 1$ grains on the first n squares. (Also, notice that on each new square, more rice is added than has been added cumulatively on all of the preceding squares.) We may conclude that on the 64th square, there are 2^{63} grains of rice, and that, on the entire board, there will be

$$2^{64} - 1 = 18,446,744,073,709,551,615$$

grains of rice. Considering that a grain of rice has a mass of approximately 0.02 grams, this represents approximately 3.7×10^{14} kg of rice, or about 544 times the worldwide rice consumption in 2009.

We may alternatively express the total number of grains of rice on the board in the form of a geometric series:

$$\sum_{i=0}^{63} 2^i = 1 + 2 + 4 + 8 + 16 + 32 + \cdots + 2^{63} = \frac{2^{64} - 1}{2 - 1} = 2^{64} - 1,$$

thereby obtaining the same answer. When assigning the index i for the purpose of the summation notation, we began with $i = 0$ for the first square, simply for convenience. ■

XXXX

A wonderful consequence of Theorem 5.2.2 is the following:

Theorem 5.2.3 — Infinite Geometric Series. The value of an *infinite geometric series* is given by

$$\sum_{i=0}^{\infty} ar^i = \frac{a}{1 - r}, \tag{5.23}$$

whenever $|r| < 1$. If $|r| \geq 1$, then the series diverges.

Proof. According to the definition of an infinite series, which is given by equation (5.20), we have

$$\sum_{i=0}^{\infty} ar^i = \lim_{n \to \infty} \left(\sum_{i=0}^{n-1} ar^i \right).$$

We already have a formula, however, for the finite geometric series, as given by equation (5.22). Using this, we obtain

$$\sum_{i=0}^{\infty} ar^i = \lim_{n \to \infty} \left(a \frac{r^n - 1}{r - 1} \right).$$

However, if $|r| < 1$, then we know that

$$\lim_{n \to \infty} r^n = 0.$$

Thus, for the case $|r| < 1$, we may conclude that

$$\sum_{i=0}^{\infty} ar^i = \lim_{n \to \infty} \left(a \frac{r^n - 1}{r - 1} \right) = \frac{-a}{r - 1} = \frac{a}{1 - r}.$$

This verifies equation (5.23). ■

■ **Example 5.17** Consider again the series

$$\sum_{i=1}^{\infty} \frac{1}{2^i} = \frac{1}{2} + \frac{1}{4} + \frac{1}{8} + \cdots.$$

This is equivalent to

$$\sum_{i=1}^{\infty} \frac{1}{2^i} = \left(\sum_{i=0}^{\infty} \frac{1}{2^i}\right) - 1 = \frac{1}{1-1/2} - 1 = 1.$$

This verifies the result of Example 5.15. ■

5.2.4 Applications to Banking

In this paragraph, we will use what we know about geometric series to determine a formula that tells us what monthly mortgage payment is required to pay off a home mortgage loan in a fixed amount of time. The main result is stated in the following theorem.

Theorem 5.2.4 Given a home mortgage loan with principal balance P_0 and a monthly interest growth factor of

$$r = 1 + \frac{\text{annual interest rate}}{12},$$

the monthly payment A required to pay off the loan in N months is given by

$$A = P_0 r^N \left(\frac{r-1}{r^N - 1}\right). \tag{5.24}$$

Proof. In order to prove equation (5.24), we will first derive a formula that tells us the balance of a home mortgage loan after n months. To achieve this, we first recognize that two things happen to the loan each month:

1. interest is added to the previous balance of the loan; and
2. a monthly payment is made in order to pay down the outstanding balance of the loan.

These events occur in the specified order: the bank charges interest *first*, and then credits the account with the monthly payment.

In order to charge interest, the previous balance is multiplied by the monthly growth factor r. In order to credit a monthly payment to the account, the amount of the payment A is subtracted from the new balance.

Let P_n be the balance after the nth payment. The initial balance of the loan is P_0. After the first month, interest is charged, bringing the total amount owed to $P_0 r$. Then, a monthly payment is credited to the account, bringing the new balance after the first month to

$$P_1 = P_0 r - A.$$

This process repeats each month. Thus, after the second month, we first add interest, bringing the new balance to $P_1 r = P_0 r^2 - Ar$. Then, we credit the account with a payment of A, bringing the total balance after the second payment to

$$P_2 = P_0 r^2 - Ar - A.$$

If we proceed another month, we obtain

$$P_3 = P_2 r - A = P_0 r^3 - Ar^2 - Ar - A.$$

It does not take long before a pattern emerges. In fact, one recognizes that after n payments, the balance will be

$$P_n = P_0 r^n - Ar^{n-1} - Ar^{n-2} - \cdots - Ar - A.$$

In summation notation, this is equivalent to

$$P_n = P_0 r^n - A \sum_{i=0}^{n-1} r^i.$$

Using equation (5.22), we may write this in closed form as

$$P_n = P_0 r^n - A\left(\frac{r^n - 1}{r - 1}\right). \tag{5.25}$$

This simple formula tells us exactly how much will be due on a home mortgage, with principal balance P_0, monthly interest factor r, and monthly payment A, after n months.

Now, suppose we want the balance to be paid off in exactly N months. Therefore, we desire $P_N = 0$. To enforce this condition, we set equation (5.25) to zero after the Nth month, thereby obtaining

$$P_0 r^N - A\left(\frac{r^N - 1}{r - 1}\right) = 0.$$

Solving for the monthly payment A, we recover the formula (5.24). ■

■ **Example 5.18** A couple purchases a silicon-valley condo financed by a home mortgage with principal balance \$480,000. The mortgage is a 30-year fixed mortgage with an annual interest rate of 4.5%. Determine the monthly payment of the loan.

First, we compute the monthly interest growth factor. We obtain

$$r = 1 + \frac{0.045}{12} = 1.00375.$$

Next, since the mortgage is a 30-year loan, the total number of months is $N = 360$. The monthly payment on the loan is therefore given by

$$A = \$480{,}000(1.00375)^{360}\left(\frac{1.00375 - 1}{1.00375^{360} - 1}\right) = \$2{,}432.09.$$

R For fun, create a spreadsheet that confirms this number. To start, enter 480,000 in the first row of the spreadsheet, i.e., in position A1. In the second row, enter "=A1*1.00375-2432.09." Then copy this formula all the way down to row 361, in order to generate the amount owed after each month. The value in A361 should equal $0.00.

Thus, the couple's monthly mortgage payment will be $2,432.09.

R The interest paid to the bank on a home mortgage is tax deductible, i.e., the total amount one pays in interest is subtracted from the gross income before income taxes are applied. For the first month's payment in this example, $1,800 constitutes interest that is paid to the bank. Therefore, the tax liability of the couple will be decreased by $1,800, just for that month's payment. In this example, the couple will pay a total of $21,441.59 in interest over the course of the first twelve payments. Assuming those twelve payments occur during the same calendar year, the couple will receive a home-mortgage tax deduction of $21,441.59 for the first year of their mortgage. Supposing, for example, the couple is in the 25% tax bracket, this means that they will pay $5,360.40 *less* in federal income tax than they would have paid without the mortgage. (They will also save on their state income tax.) Essentially, this is equivalent to $446.70 of the monthly mortgage payments, on average, for the first year, being paid for by the federal government. As time goes on, a larger and larger fraction of the monthly payment will go toward the principal, rather than toward interest, and, therefore, the tax deduction will be less and less every year.

If we wished to figure out the balance of the mortgage after ten years, we simply use formula (5.25). We obtain

$$P_{120} = \$480,000(1.00375)^{120} - \$2,432.09\left(\frac{1.00375^{120} - 1}{1.00375 - 1}\right) = \$384,429.21.$$

Similarly, it can be shown that the amount due on the loan after twenty years will be $P_{240} = \$234,670.47$. One can see how paying off a mortgage accelerates: almost half of the principal is paid off in the last ten years. ∎

Exercises

For Exercises 1–10, determine the value of the given sum.

1. $\sum_{i=1}^{10} (3i + 4i^2)$.
2. $\sum_{i=1}^{25} (i^3 - i)$.
3. $1 + 16 + 81 + 256 + \cdots + 10,000$.
4. $8 + 27 + 64 + 125 + \cdots + 1,000$.
5. $\sum_{i=1}^{N} (1+i)^2$.
6. $\sum_{i=1}^{9} \cos(i\pi)$.
7. $\sum_{i=0}^{10} \sin\left(\frac{i\pi}{2}\right)$.
8. $\sum_{i=0}^{100} 3$.
9. $\sum_{i=1}^{100} \frac{2^i}{3^{i-1}}$.

10. $\displaystyle\sum_{i=1}^{10} \frac{3(-1)^i}{2^i}$.

For Exercises 11–20, determine a formula for the ith term of the indicated sum. Express each sum using sigma notation.

11. $1+2+3+\cdots+100$.

12. $1+\frac{1}{2}+\frac{1}{3}+\cdots+\frac{1}{100}$.

13. $1+\frac{1}{4}+\frac{1}{9}+\cdots+\frac{1}{400}$.

14. $1-\frac{1}{2}+\frac{1}{3}-\frac{1}{4}+\frac{1}{5}-\cdots+\frac{1}{99}$.

15. $\frac{1}{2}+\frac{1}{4}+\frac{1}{6}+\frac{1}{8}+\cdots+\frac{1}{20}$.

16. $1+\frac{1}{3}+\frac{1}{5}+\frac{1}{7}+\frac{1}{9}+\cdots+\frac{1}{29}$.

17. $1+\frac{1}{2}+\frac{1}{4}+\frac{1}{8}+\frac{1}{16}+\frac{1}{32}$.

18. $\frac{1}{2}+\frac{3}{4}+\frac{9}{8}+\frac{27}{16}$.

19. $1-\frac{1}{2}+\frac{1}{6}-\frac{1}{24}+\frac{1}{120}$.

20. $1+\frac{1}{4}+\frac{1}{16}+\frac{1}{64}+\frac{1}{256}$.

Problems

21. Compute the value of the infinite sum

$$1+\frac{2013}{2014}+\frac{2013^2}{2014^2}+\frac{2013^3}{2014^3}+\cdots.$$

22. Consider the number $0.\overline{999} = 0.999\cdots$ (i.e., zero point nine repeating), formed with an infinite number of nines following the decimal point.
    (a) Write $0.\overline{999}$ as an infinite geometric series.
    (b) What is the value of its sum?

23. Compute the total amount of interest paid to the bank over the course of the 30-year mortgage described in Example 5.18. (*Hint*: There is an easy way to do this.)

24. A patient in a hospital is given a 200 mg pill of a certain drug every six hours. The half-life of the pill in the body is two hours, meaning that it takes two hours for 50% of the current amount of the drug to be metabolized by the body.
    (a) What fraction of the original amount of drug in the body is present after six hours, assuming the patient does not take any additional pills during those six hours?
    (b) Suppose the patient takes her initial pill, resulting in an amount of $A_1 = 200$ mg of the drug in her body. Let A_n represent how much of the drug is present in her body after she takes the nth pill. Determine a closed-form formula for A_n.
    (c) Determine $\lim_{n\to\infty} A_n$. Describe what this result means in practical terms, i.e., using everyday, ordinary language.

25. Suppose that a tax cut is currently being considered in the United States. Economists are predicting that people will spend 80% of the tax cut and save the remaining 20%. The money that is spent constitutes additional income for others; assume that these others will spend 80% of their additional income

and save the remaining 20%, and so on, and so on. Assuming this process continues indefinitely, how many dollars does the national economy gain per \$1 spent by the government in the form of a tax cut? This number is an example of a *fiscal multiplier*.

26. In this problem, we will construct a geometric proof for the summation formula given by equation (5.18).
    (a) Draw a square of side-length
$$1+2+\cdots+n=\frac{n(n+1)}{2}.$$
    (b) In the lower left-hand corner, draw a square of side-length 1. In the lower left-hand corner of the large square, again, draw a square of side-length $(1+2)$. Then a square of side-length $(1+2+3)$, and so forth. This is done for the case $n=5$ in Figure 5.6.

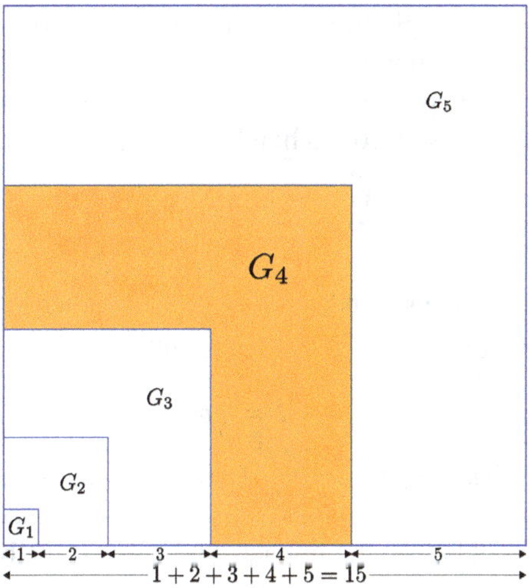

Figure 5.6: Gnomons for $\sum_{i=1}^{5} i^3$; Problem 26.

    (c) The ith *gnomon*, denoted G_i, is the difference between the ith square and the $(i-1)$th square. The gnomon G_4 is shaded in Figure 5.6. Show that, in general, the area of the ith gnomon is given by $\text{area}(G_i)=i^3$.
    (d) Explain why the area of the large square is given by both
$$\left(\sum_{i=1}^{n} i\right)^2 \quad \text{and} \quad \sum_{i=1}^{n} i^3.$$
    Conclude that the summation formula (5.18) is valid.

27. Consider a home mortgage with monthly interest growth factor r and principal P_0.

(a) Show that if a monthly payment of A is made on the balance, the loan will be paid off in N months, where N is given by the formula

$$N = \frac{-\ln\left(1 - \frac{P_0(r-1)}{A}\right)}{\ln r}. \tag{5.26}$$

(b) What happens if $A = P_0(r-1)$? What is the significance of this number? What if $A < P_0(r-1)$? Explain in practical terms why this is so.

(c) Suppose that the monthly payment required to pay off the mortgage in 30 years is A_0, and that, instead, a monthly payment of $A > A_0$ is applied to the mortgage. Show that the total amount saved S when implementing this strategy is given by

$$S = 360A_0 + \frac{A\ln\left(1 - \frac{P_0(r-1)}{A}\right)}{\ln r}.$$

(d) Suppose the couple from Example 5.18 were to pay an extra \$100 per month on their mortgage payment. How much earlier would they have their mortgage paid off? What is the total dollar amount they will save?

28. When a ball is released from a height h, it has *potential energy* V given by the formula

$$V = mgh,$$

where m is the mass of the ball and g is the acceleration due to gravity. At the moment it hits the ground, all of its potential energy has been transferred into kinetic energy T, which is given by the formula

$$T = \frac{1}{2}mv^2.$$

(a) The *coefficient of restitution* C between the ball and the floor is the ratio of the ball's speed the moment after the impact to the ball's speed the moment before the impact. Show that if the ball is released from height h, after it bounces one time, it will reach a height of C^2h after its first bounce.

(b) Suppose a ball with a coefficient of restitution equal to $C = \sqrt{0.8}$ is initially released from a height of 10 m. Show that the height h_n the ball reaches after its nth bounce is given by

$$h_n = 10(0.8)^n.$$

(c) Show that the total vertical distance D_n that the ball travels before bouncing off the ground the nth time is given by

$$D_n = 90 - 100(0.8)^n.$$

(d) Show that the time t_n it takes between the ball's nth bounce and $(n+1)$st bounce is

$$t_n = \sqrt{\frac{8h_n}{g}} = \sqrt{\frac{80}{9.8}}(0.8)^{n/2}.$$

Chapter 6

The Definite Integral

6.1 Velocity, Area, and Distance

In this section, we take our first look at the relation between area, velocity, and distance. The observations we make herein will give us an intuitive understanding of the fundamental theorem of calculus and will also help motivate our formulation of the definite integral.

6.1.1 A Look at Constant-velocity Motion

We have previously explored the relation between position and velocity: the velocity of a particle is the derivative of its position with respect to time. Next, we will look at a particle's velocity, and try to use this information to determine its position. We will begin with a simple example that will have profound consequences.

■ **Example 6.1** A car is traveling along a highway at a constant speed of 75 mph. What is the total distance that the car traversed between 3:00 and 5:00?

But, of course, the car traveled a total distance of 150 miles during these two hours! It travels 75 miles per hour; so, in two hours, it travels twice as far as it would in one. What can be the importance of such a simple exercise? The real intellectual achievement of this example, however, is not in finding the answer, but in a spark of insight that will come when one *graphs* the answer.

Let us examine now Figure 6.1, in which the velocity of the car is graphed as a function of time. Since the car's velocity is constant, $v(t) = 75$, the graph is simply a horizontal line. But the founders of calculus went a step further and asked an interesting question: what is the the total *area* that lies beneath the graph of the velocity, above the t-axis, and between the endpoints at 3:00 and 5:00? This is the area of a rectangle. The width of this rectangle is 2 hours, and the height of this rectangle is 75 mph. Notice that both the width and height of the rectangle are treated as dimensionally meaningful quantities (i.e., they each have their own

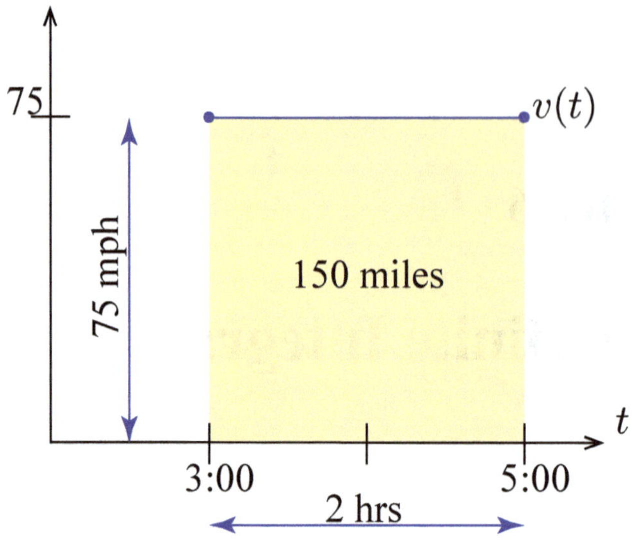

Figure 6.1: The velocity of a car as a function of time; Example 6.1.

units). When we multiply the two together, we obtain

$$\text{area} = (2 \text{ hours}) \times \left(75\ \frac{\text{miles}}{\text{hour}}\right) = 150 \text{ miles}.$$

This is quite an astonishing thing: the *area* under the graph of the velocity turns out to equal the total distance that was traveled. At least, it does so when the velocity is *constant*. There can be no hope that a similar result holds if the velocity is a crazy, non-constant function of time; or can there? ∎

6.1.2 Non-constant Velocities

Next, let us look at the case of a particle moving with a non-constant velocity.

∎ **Example 6.2** Let us now consider a car with a non-constant velocity, which is graphed in Figure 6.2. The car may be driving slower or faster depending on *when* we ask what it's speed is. The question before us is: does the area beneath the graph of the car's velocity still represent the total distance driven by the car?

As astonishing as it may seem, the answer is a resounding *yes*: the shaded area in Figure 6.2 *exactly* measures how far the car traveled between 3:00 and 5:00. In fact, upon painstaking examination, it can be shown that the depicted area is approximately 138 miles. We may conclude that the car, whose velocity is logged by the graph in Figure 6.2, traveled approximately 138 miles between 3:00 and 5:00. As a side note, its *average* speed must have been 69 mph. Note that to obtain

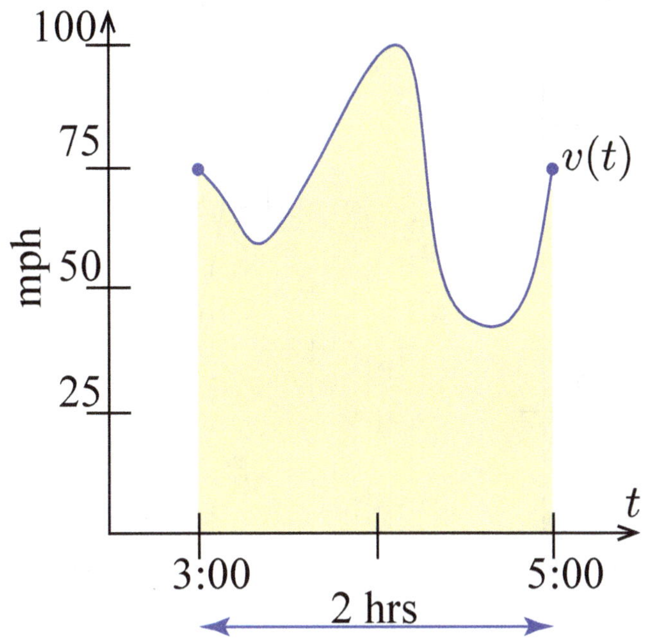

Figure 6.2: The velocity of a car as a function of time; Example 6.2.

the average speed, you divide the area (which represents the total distance traveled) by the total time.

To see *why* this is true—i.e., *why* the area under the curve represents the total distance traveled—we only have to recognize that the overall interval may be broken up into a large number of smaller intervals. In fact, the interval between 3:00 and 5:00 may be divided so finely that on each subinterval the car's velocity appears as though it is constant. One such refinement is shown in Figure 6.3, in which the interval between 3:00 and 5:00 is divided into 32 subintervals, each of approximately 3 minutes and 45 seconds. Over each subinterval, the velocity is much *closer* to being a constant than it was when considered over the full two hours. Now imagine what the picture would look like if we again divided *each* of these 32 subintervals into 225 subintervals, each having a width of one single second. During one second of driving, the car's velocity should look very close to constant. If it does not, that single second can be divided into, for example, 1,000 milliseconds. Over any given millisecond, a car's velocity should certainly be extremely close to a constant.

R Let's do a quick reality check to test our assumption that a car's velocity is approximately constant over any given millisecond. The fastest Formula One race cars have been clocked doing 0–60 mph in about 1.6 seconds. That represents a change in velocity of 37.5 mph/s (miles per hour per

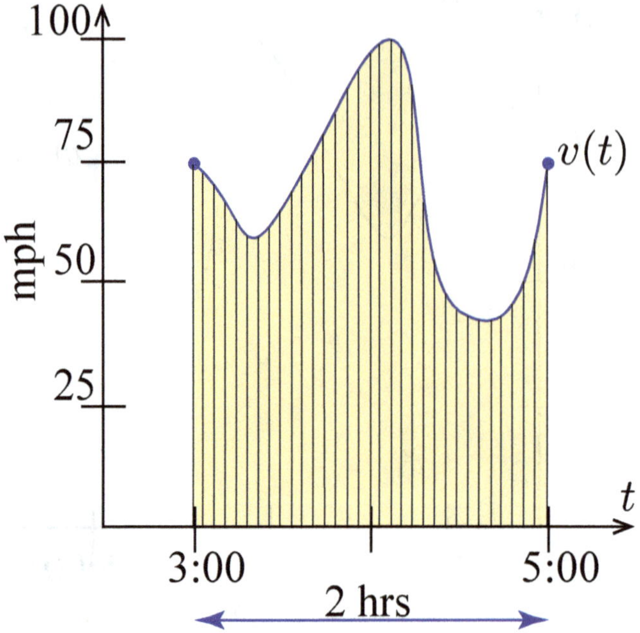

Figure 6.3: The velocity of a car as a function of time; Example 6.2.

second), or an average acceleration of 0.0375 mph/ms (miles per hour per millisecond). Thus, even the strongest accelerating race cars will only change their velocity by a fraction of a mph in any given millisecond. The assumption seems sound.

Once the macroscopic interval from 3:00 to 5:00 is subdivided finely enough, so that the velocity *appears* approximately constant over each individual subinterval, we can apply our previous result from Subsection 6.1.1: over each of these tiny microscopic subintervals, the velocity is approximately constant, and, therefore, the distance traveled in that very short amount of time should equal the area under the curve between the beginning and ending of that given subinterval. But the total area is just the sum of all these tiny areas, and the total distance is just the sum of the tiny distances the car drove over each millisecond. Therefore, the total area should represent the total distance traveled by the car, and the result is valid.

■

Our goal for the rest of the chapter is to take these concepts, which we just introduced on an intuitive level, and refine them into the development of integral calculus. What we just saw, in the preceding example, is actually an intuitive statement of the fundamental theorem of calculus: the area under the curve of the graph of a function represents the net change in that function's antiderivative.

Before returning to this idea, we will more carefully lay out a limit-based definition of area and, subsequently, a careful definition of the definite integral.

6.1.3 Negative Velocities

Velocity is a measure of both *speed* and *direction.* Speed, by itself, only tells how fast a particle is moving, without any indication of direction. This distinction will be more clear during the study of multivariable and vector calculus, in which particles are allowed to move around in three-dimensions. Even in our present context of one-dimensional motion, however, there is a distinction: positive velocity indicates that the particle's x-position is increasing, whereas negative velocity indicates that the particle's x-position is decreasing. If the particle is moving on a horizontal line, positive velocity indicates motion to the right, whereas negative velocity indicates motion to the left. If the particle is moving up and down, such as under the influence of gravity, positive velocity indicates upward motion, whereas negative velocity indicates downward motion. The *speed* of the particle is the absolute value (or *magnitude*) of its velocity, with no indication of the direction.

Therefore, when measuring the *area* between the graph of a particle's velocity and the t-axis, one wishes to distinguish the area above the t-axis and the area below the t-axis. The area *above* the t-axis is *added* to the particle's position, whereas the area *below* the t-axis is *subtracted* from the particle's position.

Problems

1. A car accelerates linearly from 0 mph to 60 mph in six seconds.
   (a) Draw a graph of the car's speed as a function of time for those six seconds.
   (b) How far did the car travel during those six seconds? Explain your reasoning.
   (c) If the car instead accelerated from 0 mph to 60 mph in three seconds, how much distance would it have traveled during those three seconds?
2. A man standing at a fountain begins to jog away from the fountain, and his velocity (ft/s) is shown as a function of time (s) in Figure 6.4. Suppose that the positive x-axis is pointing away from the fountain.
   (a) What was the man's position (i) after 9 seconds? (ii) after 15 seconds? (iii) after 21 seconds?
   (b) When was he running (i) the fastest? (ii) the slowest?
   (c) When was his acceleration the greatest?
   (d) What was the total *distance* he ran (i.e., count distance ran in both directions)?
   (e) What was his net *displacement* (i.e., only count the distance between his starting and ending point)?
3. During a road trip, a woman records the speed of her car every five minutes, as shown in Table 6.1.

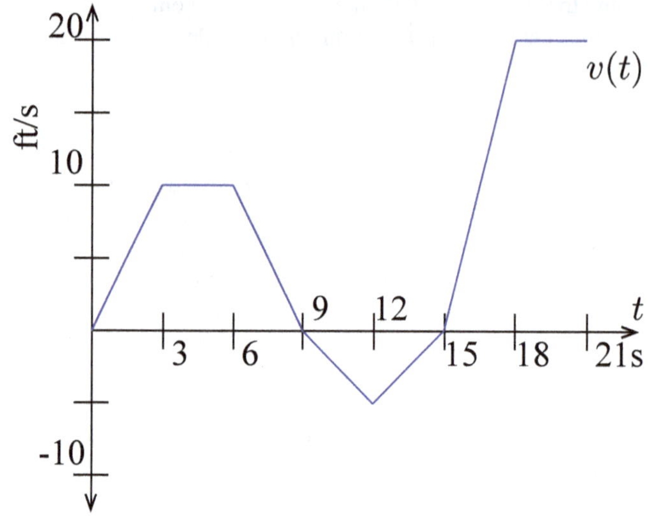

Figure 6.4: A man's velocity as a function of time; Problem 2.

time	5:00	5:05	5:10	5:15	5:20	5:25	5:30
speed (mph)	65	70	72	68	68	71	74

Table 6.1: The car's velocity for Problem 3.

Estimate how far the woman traveled during the thirty minutes. State any assumptions you might make and clearly explain your approach; there may be several ways to estimate this distance.

4. If the acceleration acting on a particle is constant, then the particle's velocity can be modeled as

$$v(t) = v_0 + at,$$

where v_0 is its initial velocity and a is its constant acceleration.

   (a) Plot $v = v(s)$. Use formulas from geometry to determine the area under the curve between $s = 0$ and $s = t$, for some endpoint t. (Count area below the s-axis as negative.

   (b) Explain what your answer from part (a) represents in practical terms, i.e., in terms of the particle.

   (c) Use your answer from part (a) to show that the particle's position at time t is given by

$$x(t) = x_0 + v_0 t + \frac{1}{2}at^2.$$

5. Suppose that a particle's velocity is given by $v(t) = \sqrt{400 - t^2}\ m/s$, for $t \in [0, 20]$. Determine how far it travels during these 20 seconds.

6. Suppose that a car is traveling at a speed of w_0 ft/s, and that its maximum braking acceleration is 22 ft/s^2.
   (a) Suppose that a driver has an average reaction time of 2.5 s, meaning that it takes 2.5 seconds to recognize a danger and apply the brakes. Show that the car's velocity $w(t)$ as measured in ft/s may be represented by the piecewise linear function

$$w(t) = \begin{cases} w_0 & \text{for } t < 2.5 \\ w_0 - 22(t - 2.5) & \text{for } t \geq 2.5 \end{cases}$$

   (b) Draw a graph of $w(t)$ and determine the amount of time it takes for the car to come to a stop *after* the brakes have been applied.
   (c) Determine the total stopping distance D of the car, measured in feet, as a function of its initial velocity w_0, as measured in ft/s.
   (d) Using the conversion 1 mph = 22/15 ft/s, express the total stopping distance D, measured in feet, as a function of the car's initial velocity v, measured in mph.
   (e) Given that the average car length is 16 ft, how many car lengths does it take to come to a stop if a car is traveling at 30 mph? at 60 mph? (Include the distance traveled during the reaction time.)

6.2 Riemann Sums

In our previous section, we saw how the area under the graph of the velocity of a particle represents the distance traveled by the particle. Of course, knowing the area of a region can be useful in its own regard; for instance, how can we compute the area of a circle? In addition to these two simple examples, the theory we will unfold will have a diverse array of applications: computing volume, the amount of energy required to drain a swimming pool, the total hydrostatic force on a dam, computations of probability, and present and future values of income streams. For the moment, however, we will concentrate on the question of how one can compute the area under a graph.

6.2.1 Introduction

When we wish to find *the area under a graph*, we mean the area bounded between the graph of a function, the horizontal axis, and two vertical lines. For instance, consider the area under the graph of $f(x)$ between the points $x = a$ and $x = b$, as shown in Figure 6.5.

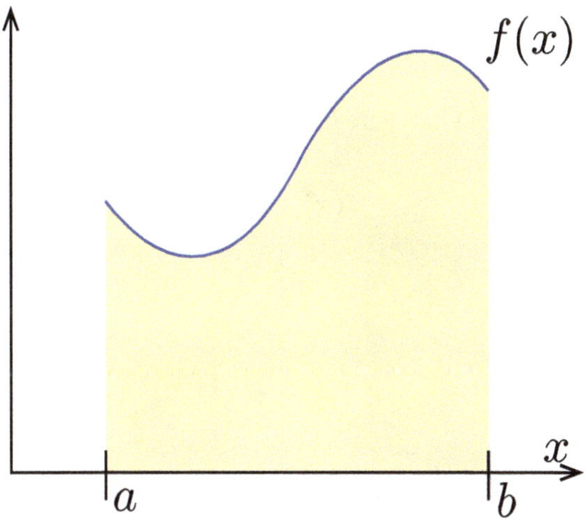

Figure 6.5: Area under a graph.

Our goal is to determine a systematic method for computing areas such as these. To achieve this, we will use an estimation called a *Riemann sum*. A Riemann sum with N subintervals basically divides the interval $a \leq x \leq b$ into N equal subintervals, and approximates the function as a constant on each subinterval. One therefore reduces the actual function with a series of rectangles. For example, we may approximate the total area under the graph of the function shown in Figure 6.5 using two or four subintervals, as shown in Figure 6.6 (left) and (right). Instead of computing the actual area, we compute the area of the two red rectangles as an approximation (left) or four red rectangles as an even better approximation (right).

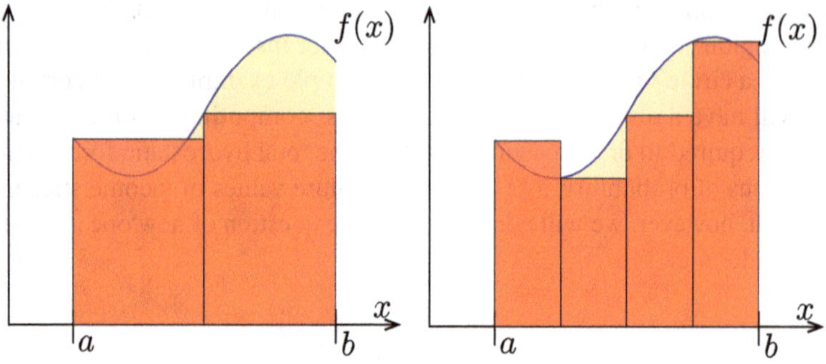

Figure 6.6: Area under a graph approximated with a Riemann sum with two (left) or four (right) subintervals.

The height of each rectangle only needs to be *some* output of the function f that is achieved on the given subinterval. Here, we simply evaluated the function f at the leftmost point on each subinterval of the domain.

Similarly, we might use eight or sixteen subintervals to approximate the area under the graph, as shown in Figure 6.7.

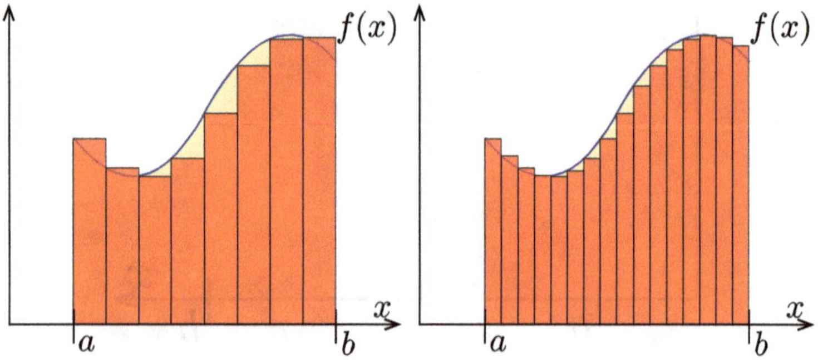

Figure 6.7: Area under a graph approximated with a Riemann sum with eight (left) and sixteen (right) subintervals.

The more subintervals we use, i.e., the more finely we partition the interval $a \leq x \leq b$, the better our approximation to the actual area we will obtain. It therefore makes sense to *define* area as a limit of this process. In other words, the true area is what we active in the limit as the number of subintervals tends to infinity.

6.2.2 Setting Up: Defining the Subintervals

In order to formalize these notions, we will require some notation. To start, we will examine the interval $[a,b]$ of the domain over which we wish to compute the area of some function f.

Definition 6.2.1 Given a closed interval $[a,b]$ and a number N of desired subintervals, we say that the *width* of each subinterval is Δx, which is given by the formula

$$\Delta x = \frac{b-a}{N}, \tag{6.1}$$

and that the *partition points* $\{x_0, x_1, \ldots, x_N\}$ are the points that separate the individual subintervals. The partition points are, specifically, the points located at

$$x_i = a + i\Delta x, \tag{6.2}$$

for $i = 0, \ldots, N$.

First, notice that the width of each subinterval, as given by equation (6.1), is simply the total length of the interval $[a,b]$ divided by the number of subintervals. For example, the interval $[a,b]$ is partitioned into eight separate subintervals in Figure 6.8. Each subinterval has width

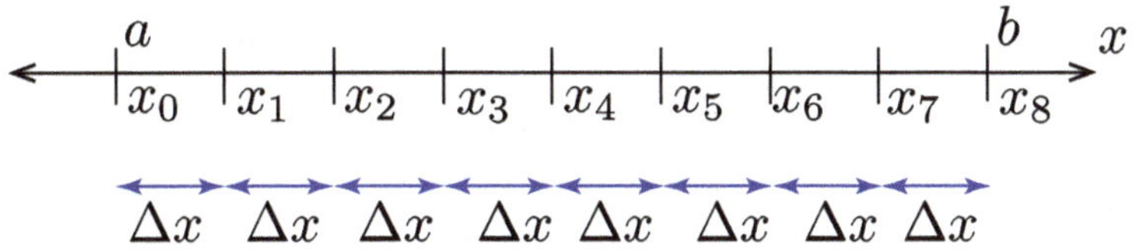

Figure 6.8: The interval $[a,b]$ partitioned into eight equal subintervals.

$$\Delta x = \frac{b-a}{N} = \frac{b-a}{8}.$$

Also, the partition points $\{x_0, \ldots, x_N\}$ are located at

$$\begin{aligned} x_0 &= a, \\ x_1 &= a + \Delta x, \\ x_2 &= a + 2\Delta x, \\ x_3 &= a + 3\Delta x, \\ &\vdots \\ x_8 &= a + 8\Delta x = b. \end{aligned}$$

Notice that the last partition point x_N is always equal to the right endpoint b, since

$$x_N = a + N\Delta x = a + N\left(\frac{b-a}{N}\right) = a + (b-a) = b.$$

Further, notice that there are always one more partition points than the number of subintervals. For example, the eight subintervals shown in Figure 6.8 are partitioned using the *nine* partition points x_0, x_1, x_2, x_3, x_4, x_5, x_6, x_7, and x_8.

Finally, notice the way in which the subintervals fall between the partition points:

$$\begin{aligned}
\text{the first subinterval} &= [x_0, x_1],\\
\text{the second subinterval} &= [x_1, x_2],\\
\text{the third subinterval} &= [x_2, x_3],\\
&\;\vdots\\
\text{the eighth subinterval} &= [x_7, x_8].
\end{aligned}$$

By observing the pattern, we have the following result.

Proposition 6.2.1 When a closed interval $[a,b]$ is divided into N subintervals, using a set of partition points $\{x_0, \ldots, x_N\}$, the ith subinterval is located at

$$\text{the } i\text{th subinterval} = [x_{i-1}, x_i], \tag{6.3}$$

for $i = 1, \ldots, N$.

This proposition tells us where each of the subintervals are located. For example, the first subinterval is located at $[x_0, x_1]$, the second subinterval is located at $[x_1, x_2]$, and so forth, exactly as we observed in our example.

Finally, we will require a set of points at which we can evaluate the function f in order to compute the height of our rectangles.

Definition 6.2.2 Given a closed interval $[a,b]$ that is divided into N subintervals, the ith *test point* x_i^* is a selected point from the ith subinterval, i.e.,

$$x_i^* \in [x_{i-1}, x_i], \tag{6.4}$$

for $i = 1, \ldots, N$ (i.e., the point x_i^* is a point that satisfies the inequalities $x_{i-1} \leq x_i^* \leq x_i$).

There is no unique method for selecting which test points to use to approximate the area under the graph of a function; in fact, one may even use a random assortment of points, just so long as that each test point is a point on the corresponding subinterval.

6.2.3 General Riemann Sums

The idea behind the Riemann sum is to approximate the area under a curve. As we have discussed, this is accomplished by first dividing the interval $[a,b]$, over which the area is to be computed, into a number of subintervals. Next, we approximate the value of the function at a given test point located on each subinterval. In this manner, one may easily compute the area of each of these rectangles. In particular, we have the following outcome.

Definition 6.2.3 Given a function f, a closed interval $[a,b]$ on its domain, and a natural number N, we may partition the given interval as specified in equations (6.1) and (6.2) into N subintervals. We say that a *Riemann sum* S_n of the function f over the domain $[a,b]$ with N subintervals is a sum of the form

$$S_n = \sum_{i=1}^{N} f(x_i^*)\Delta x, \tag{6.5}$$

where $x_i^* \in [x_{i-1}, x_i]$, i.e., x_i^* is a test point located on the ith subinterval.

To illustrate this definition, examine Figure 6.9, which depicts a Riemann sum of a function over an interval $[a,b]$ using $N = 4$ subintervals. On each of the four

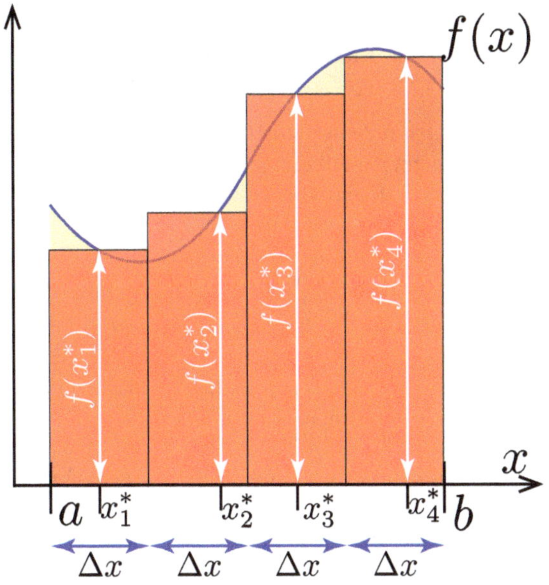

Figure 6.9: A general Riemann sum with four subintervals.

subintervals, a point is selected; these points are labeled x_1^*, x_2^*, x_3^*, and x_4^*. The function f is then evaluated at each of these points; the outputs are the heights of the illustrated rectangles. Meanwhile, each rectangle has a width given by equation (6.1). Therefore, the Riemann sum is equal to the sum of the areas of these four rectangles:

$$S_4 = \underbrace{f(x_1^*)}_{\text{height}} \underbrace{\Delta x}_{\text{width}} + \underbrace{f(x_2^*)}_{\text{height}} \underbrace{\Delta x}_{\text{width}} + \underbrace{f(x_3^*)}_{\text{height}} \underbrace{\Delta x}_{\text{width}} + \underbrace{f(x_4^*)}_{\text{height}} \underbrace{\Delta x}_{\text{width}}.$$

6.2.4 Right- and Left-hand Riemann Sums

There are various *types* of Riemann sums that will be of interest to us. By *type*, we simply mean a method for determining *which* test points to use. To the general rubric of a Riemann sum will remain, only the particular test point x_i^* for each subinterval will be determined based on a specific method. In particular, we will introduce the right- and left-hand Riemann sums, meaning that the test point of each subinterval will either be the left endpoint or the right endpoint of that particular subinterval, respectively.

Definition 6.2.4 The *right-hand Riemann sum* R_N of a function f with N subintervals from a to b, i.e., over the closed interval $[a,b]$, is the Riemann sum obtained by selecting $x_i^* = x_i$, i.e., each test point is selected to be the right endpoint of the given subinterval. The right-hand Riemann sum is given by

$$R_N = \sum_{i=1}^{N} f(x_i)\Delta x. \tag{6.6}$$

Similarly, the *left-hand Riemann sum* L_N of a function f with N subintervals from a to b is the Riemann sum obtained by selecting $x_i^* = x_{i-1}$, i.e., each test point is selected to be the left endpoint of the given subinterval. The left-hand Riemann sum is given by

$$L_N = \sum_{i=1}^{N} f(x_{i-1})\Delta x. \tag{6.7}$$

A left-hand Riemann sum of a function with $N = 4$ subintervals is illustrated in Figure 6.10. The value of this Riemann sum will be

$$L_4 = f(x_0)\Delta x + f(x_1)\Delta x + f(x_2)\Delta x + f(x_3)\Delta x. \tag{6.8}$$

Thus, we see how we are using the left endpoints of the four subintervals as test points. Evaluating the function f at these left endpoints, we obtain the height of the graph at the left endpoints. Using these as our rectangle heights, we obtain our Riemann sum.

Similarly, a right-hand Riemann sum of a function with $N = 4$ subintervals is illustrated in Figure 6.11. Here, the Riemann sum has a value equal to

$$R_4 = f(x_1)\Delta x + f(x_2)\Delta x + f(x_3)\Delta x + f(x_4)\Delta x. \tag{6.9}$$

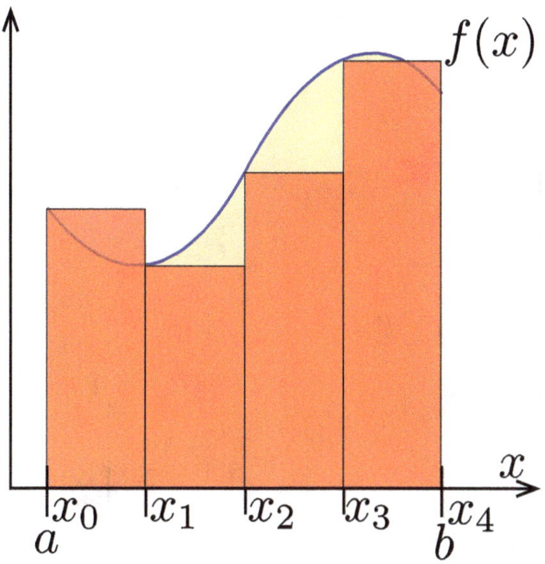

Figure 6.10: A left-hand Riemann sum with $N = 4$ subintervals.

Instead of using the left endpoints for each rectangle, we evaluate the function at the right endpoints.

When comparing the left-hand and right-hand Riemann sums with $N = 4$ subintervals, as given by equations (6.8) and (6.9), we observe that three of the four terms are the same in both sums. These terms that are identical correspond to the inner partition points x_1, x_2, and x_3. These three inner partition points are used for both Riemann sums; additionally, the left-hand Riemann sum uses the endpoint $x_0 = a$ whereas the right-hand Riemann sum uses the endpoint $x_N = b$. By means of example, we have therefore arrived at the following theorem.

Theorem 6.2.2 Let $\{x_0, \ldots, x_N\}$ be a partition of the closed interval $[a, b]$ into N subintervals, and suppose L_N and R_N are the left-hand and right-hand Riemann sums of a function f over the interval $[a, b]$ with respect to this partition. Then the difference between these two Riemann sums is given by

$$R_N - L_N = (f(b) - f(a))\Delta x. \tag{6.10}$$

Proof. The right-hand and left-hand Riemann sums are given by equations (6.6) and (6.7), respectively. Therefore, the difference between the two may be computed

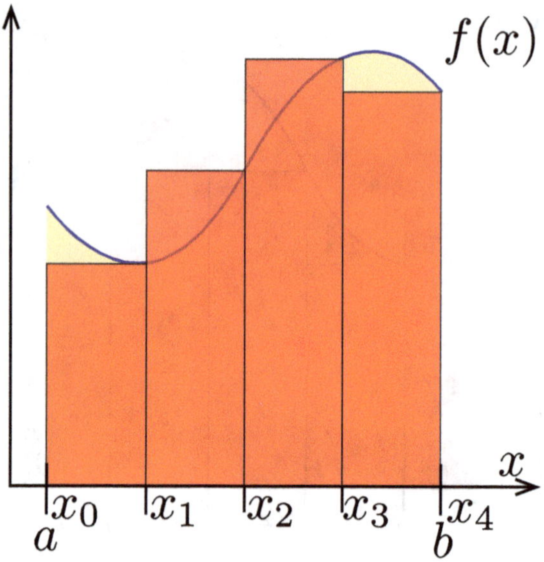

Figure 6.11: A right-hand Riemann sum with $N = 4$ subintervals.

as

$$\begin{aligned} R_N - L_N &= \sum_{i=1}^{N} f(x_i)\Delta x - \sum_{i=1}^{N} f(x_{i-1})\Delta x \\ &= \left[f(x_1)\Delta x + f(x_2)\Delta x + \cdots + f(x_{N-1})\Delta x + f(x_N)\Delta x\right] \\ &\quad - \left[f(x_0)\Delta x + f(x_1)\Delta x + \cdots + f(x_{N-2})\Delta x + f(x_{N-1})\Delta x\right] \\ &= f(x_N)\Delta x - f(x_0)\Delta x \\ &= (f(b) - f(a))\,\Delta x. \end{aligned}$$

This completes the proof. ■

■ **Example 6.3** Compute the left-hand and right-hand Riemann sum of the function $f(x) = \sin(x)$ over the interval $[0, \pi/2]$ using $N = 4$ subintervals. Draw a sketch of both Riemann sums.

For both cases, the width of each subinterval is given by

$$\Delta x = \frac{b-a}{4} = \frac{\pi/2}{4} = \frac{\pi}{8}.$$

Since $a = 0$ and $b = \pi/2$, the five partition points are given by

$$x_0 = 0, \quad x_1 = \frac{\pi}{8}, \quad x_2 = \frac{\pi}{4}, \quad x_3 = \frac{3\pi}{8}, \quad \text{and} \quad x_4 = \frac{\pi}{2}.$$

The left-hand Riemann sum is obtained by evaluating the function f at the left endpoints of each subinterval, i.e., at the points x_0, x_1, x_2, and x_3. For the left-hand Riemann sum, we obtain

$$\begin{aligned} L_4 &= \Delta x\left[f(x_0)+f(x_1)+f(x_2)+f(x_3)\right] \\ &= \frac{\pi}{8}\left[0+\sin\left(\frac{\pi}{8}\right)+\sin\left(\frac{\pi}{4}\right)+\sin\left(\frac{3\pi}{8}\right)\right] \\ &\approx 0.7907662602. \end{aligned}$$

The left-hand Riemann sum is pictured in Figure 6.12. Notice that for each of

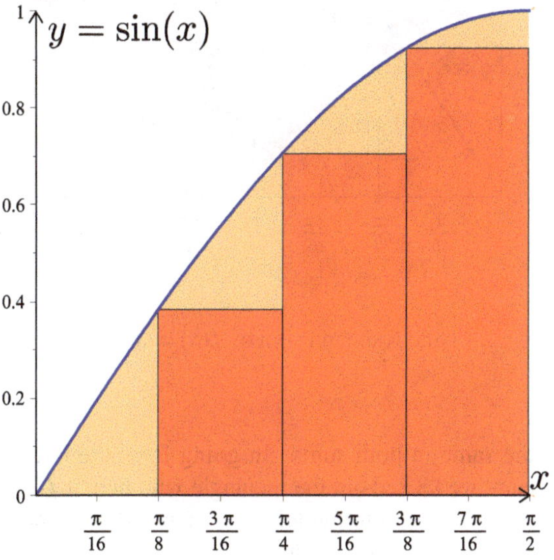

Figure 6.12: The left-hand Riemann sum of $f(x) = \sin(x)$ over $[0, \pi/2]$ with $N = 4$ subintervals.

the four "slots," the height of the corresponding rectangle is determined by the height of the function at the left-most endpoint of the corresponding slot. Since the function starts off with zero height, the first rectangle has zero area and does not contribute to the sum.

Similarly, in order to compute the right-hand Riemann sum, we use the test points x_1, x_2, x_3, and x_4. We therefore obtain

$$\begin{aligned} R_4 &= \Delta x\left[f(x_1)+f(x_2)+f(x_3)+f(x_4)\right] \\ &= \frac{\pi}{8}\left[\sin\left(\frac{\pi}{8}\right)+\sin\left(\frac{\pi}{4}\right)+\sin\left(\frac{3\pi}{8}\right)+\sin\left(\frac{\pi}{2}\right)\right] \\ &\approx 1.183465342. \end{aligned}$$

The right-hand Riemann sum is pictured in Figure 6.13. Notice that three of the

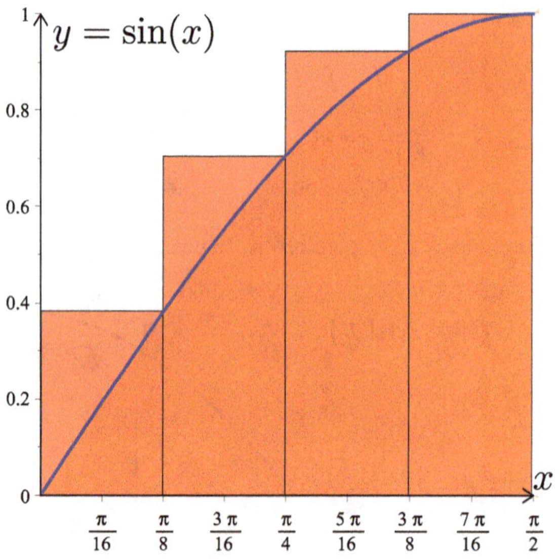

Figure 6.13: The right-hand Riemann sum of $f(x) = \sin(x)$ over $[0, \pi/2]$ with $N = 4$ subintervals.

rectangles are the same in both sums. In going from the left-hand to the right-hand Riemann sum, we kicked out the rectangle with height $f(x_0) = f(a)$, slid the remaining rectangles one slot to the left, and then added a rectangle with height $f(x_N) = f(b)$. This is a visual representation of formula (6.10), which predicts that

$$R_4 - L_4 = \left[\sin\left(\frac{\pi}{2}\right) - \sin(0)\right]\frac{\pi}{8} = \frac{\pi}{8} \approx 0.3926990818.$$

We can confirm this by subtracting the value of the left-hand Riemann sum from the value of the right-hand Riemann sum, as we computed above.

One additional observation: In this example, the actual value of the area must lie somewhere between the values of the left-hand and right-hand Riemann sums. One can see this by examining the graphs shown in Figures 6.12 and 6.13. For the left-hand Riemann sum, the rectangles always lie below the curve. (At least, in this example.) Similarly, for the right-hand Riemann sum, the rectangles always lie above the curve. (Again, for this example.) We may conclude that the left-hand Riemann sum is an *underestimate* of the true area, and the right-hand Riemann sum is an *overestimate* of the true area, i.e.,

$$LH_4 \leq \text{area} \leq RH_4 \quad \text{(for this example)}.$$

Therefore, we may conclude that the actual area has a value somewhere between 0.79 and 1.18. ∎

The previous example hints at the following theorem.

Theorem 6.2.3 If a function f is increasing on an interval $[a,b]$, then any left-hand Riemann sum of f over $[a,b]$ will be an underestimate of the true area and any right-hand Riemann sum of f over $[a,b]$ will be an overestimate of the true area, i.e.,

$$L_N \leq \text{area} \leq R_N.$$

Similarly, if a function f is decreasing on an interval $[a,b]$, then any left-hand Riemann sum of f over $[a,b]$ will be an overestimate of the true area and any right-hand Riemann sum of f over $[a,b]$ will be an underestimate of the true area, i.e.,

$$R_N \leq \text{area} \leq L_N.$$

A generic increasing function is shown along with its left-hand Riemann sum with eight subintervals in Figure 6.14. By studying the figure, one sees why the

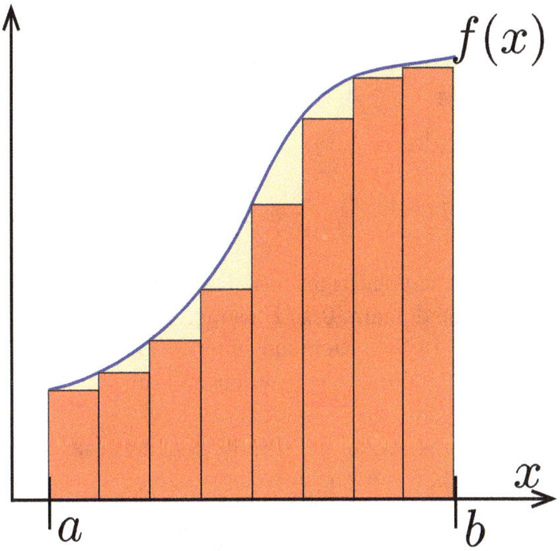

Figure 6.14: A generic increasing function and its left-hand Riemann sum with $N = 8$ subintervals.

previous theorem works. Since the function f is an increasing function, the left-end point over any subinterval must be the smallest output value of the function over that subinterval. That is, we have $f(x_{i-1}) \leq f(x)$, for all $x \in [x_{i-1}, x_i]$. Therefore, constructing a rectangle whose height is equal to the value of f at the left endpoint produces a rectangle that lies entirely below the graph of f. This sort of reasoning holds for the similar cases: a right-hand Riemann sum of an increasing function is an overestimate; a left-hand and right-hand Riemann sum of a decreasing function is an overestimate and an underestimate, respectively.

Exercises

For Exercises 1–10, compute the left-hand and right-hand Riemann sums for the indicated function over the given domain using the given number of subintervals. Verify formula (6.10) for each exercise.

1. $f(x) = 2x, \ [0,3], \ N = 6.$
2. $f(x) = x^2, \ [0,1], \ N = 5.$
3. $f(x) = x^2, \ [0,1], \ N = 10.$
4. $f(x) = x^3, \ [0,1], \ N = 10.$
5. $f(x) = e^x, \ [0,2], \ N = 4.$
6. $f(x) = \sin(x), \ [0,\pi], \ N = 4.$
7. $f(x) = \sin(x), \ \left[0, \frac{\pi}{2}\right], \ N = 3.$
8. $f(x) = \ln(x), \ [1,4], \ N = 6.$
9. $f(x) = \frac{1}{x}, \ [1,5], \ N = 4.$
10. $f(x) = \cos^2(x), \ [0,\pi], \ N = 6.$

For Exercises 11–15, determine the number of subintervals that are needed so that the the difference between the right-hand and left-hand Riemann sums is less than 0.01, i.e., determine N so that $|R_N - L_N| < 0.01$.

11. $f(x) = x^2, \ [0,6].$
12. $f(x) = e^x, \ [0,1].$
13. $f(x) = \sin(x), \ \left[0, \frac{\pi}{2}\right].$
14. $f(x) = \frac{1}{x}, \ [1,5].$
15. $f(x) = \ln(x), \ [1,e].$

Problems

16. (a) Determine the left- and right-hand Riemann sums for the function $f(x) = \sin(x)$ over the domain $[0,\pi/2]$ using 3 subintervals.
    (b) Draw a graph of both Riemann sums.
    (c) You are told that for a certain number of subintervals N, $L_N = 0.98823$ and $\Delta x = 0.02344$. Determine R_N.
17. During a nine-months period in 2013, the popular instant-messaging service WhatsApp was adding new customers at a rate of

$$r(t) = 17(2)^{t/9}$$

million customers per month, t months into this period. If WhatsApp had 225 million users at the start of this nine-month period, determine an underestimate and overestimate for the number of users it had at the end of the period, using a left- and right-hand Riemann sum with nine subintervals.

18. Some data for the July 04, 2006 launch of space shuttle Discovery on mission STS 121 to the International Space Station are assembled in Table 6.2. Certain events are listed, such as solid rocket booster (SRB) cutoff and jettison, negative return to launch site (RTLS) (i.e., the last possible moment for an RTLS abort), and main engine cutoff (MECO).
    Fill in the accompanying worksheet (Table 6.3). Some terms are defined below. ΔD represents the total change in the flight path distance function

time (min:sec)	time (s)	altitude (mi)	downrange (mi)	speed (mph)	event
0:00	0	0	0	0	liftoff
0:50	50	6	2	750	
1:47	107	22	18	2600	
2:04	124	29	24	2880	SRB jettison
2:35	155	39	49	3200	
3:30	210	53	100	4000	
4:08	248	61	150	5000	negative RTLS
4:39	279	65	200	-	
5:15	315	-	255	7000	
6:10	370	-	-	10000	
7:00	420	75	516	-	
7:30	450	-	-	14400	
8:30	510	70	-	17500	MECO

Table 6.2: telemetry data for STS 121: July 04, 2006

over the course of the interval. $D(t_{final})$ represents your approximation to $D(t)$ at the final time of the interval (the right endpoint). Use the trapezoid rule to compute each ΔD (i.e., average the speed at both endpoints and multiply by time). A useful conversion: 1 g = 22 mph/s.

- Rate of Ascent (ROA), [mph]: time derivative of altitude
- groundspeed (ground), [mph]: time derivative of downrange distance
- acceleration (acc), [mph/s] and [g's]: time derivative of space shuttle's speed
- flight path distance function [mi]: $D(t)$ the total distance traveled between launch and time t.

time	avg ROA [mph]	avg ground [mph]	avg acc [mph/s]	avg acc [g's]	ΔD [mi]	$D(t_{final})$ [mi]
0:00 - 0:50						
0:50 - 1:47						
1:47 - 2:04						
2:04 - 2:35						
2:35 - 3:30						
3:30 - 4:08						
4:08 - 4:39						
4:39 - 5:15						
5:15 - 6:10						
6:10 - 7:00						
7:00 - 7:30						
7:30 - 8:30						

Table 6.3: worksheet for Problem 16

6.3 The Definite Integral

We previously saw how Riemann sums could be constructed in order to approximate the area under the graph of a function. As one uses more and more subintervals, i.e., as one divides the interval $[a,b]$ into a finer and finer partition, this approximation becomes more and more accurate. In the limit, we obtain the true area under the graph. With this in mind, we next define the definite integral of a function.

6.3.1 Definite Integrals

Definition 6.3.1 The *definite integral* of a function f over an interval $[a,b]$ is defined as the limit

$$\int_a^b f(x)\,dx = \lim_{N\to\infty}\left(\sum_{i=1}^{N} f(x_i^*)\Delta x\right), \tag{6.11}$$

whenever that limit exists. Here, the object in the parentheses is a Riemann sum for the function f over the interval $[a,b]$. The numbers a and b are called the *limits of integration*.

(R) The definite integral of a function over an interval is a number, i.e., it no longer depends on the independent variable. This is why, when we write

$$\int_a^b f(x)\,dx,$$

we say that we *integrate over* x from a to b. When we integrate over x, we obtain a value that no longer depends on x, as this value represents something that happens over the entire range of x-values from $x = a$ to $x = b$.

(R) The definite integral treats area *below* the horizontal axis as a *negative* quantity. This follows since one uses the function outputs $f(x_i^*)$ in the Riemann sum, and not their absolute values. This also ties in with our discussion from Section 6.1: negative velocity *subtracts* from ones position; it does not add. Thus, the integral of a velocity function will output the *net* change in the particle's position over the given interval of time.

It is important to recognize that the limit in equation (6.11) is not a limit in the typical sense. In theory, one should be able to come up with the formula for a Riemann sum with N subintervals, as a function of N, and then take the limit as N tends to infinity. In practice, however, this is only achievable in a few simple cases, which we will outline below. In general, though, we should think of the definite integral as a process: one makes a number of subsequent estimates, using Riemann sums, and observes the behavior of those estimates in the limit as the number of subintervals used tends toward infinity.

The difference between the definite integral and the Riemann sum, in both practice and notation, is illustrated in Figure 6.15. The integral is obtained in the limit as the number of subintervals for a given Riemann sum goes to infinity. The integral then replaces the sigma sign; the function f is evaluated at *every point* instead of at a discrete spectrum of points; and the width of the subinterval becomes infinitesimally small instead of discrete.

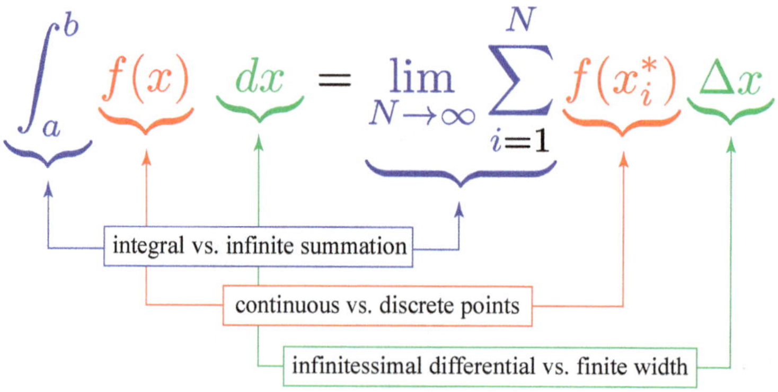

Figure 6.15: An anatomy of the integral notation.

Note that one cannot use the notation $\sum_{i=1}^{\infty}$ in the definition of the integral; one must take the *limit* $\lim_{N\to\infty} S_N$, where S_N is the Riemann sum with N subintervals. This follows since, for different N-values, the terms comprising the Riemann sum change; that is, absent a formula, the Riemann sum needs to be recomputed for every value of N, at least theoretically.

■ **Example 6.4** Compute the integral

$$\int_1^5 x\,dx.$$

To start, let us divide the closed interval $[1,5]$ into N subintervals, each of width

$$\Delta x = \frac{5-1}{N} = \frac{4}{N}.$$

The ith partition point is then located at

$$x_i = 1 + i\Delta x = 1 + \frac{4i}{N},$$

for $i = 0,\ldots,N$. Using a right-hand Riemann sum, for convenience, we obtain

$$R_N = \sum_{i=1}^{N} f(x_i)\Delta x = \sum_{i=1}^{N} x_i\Delta x = \sum_{i=1}^{N}(1+i\Delta x)\Delta x = \sum_{i=1}^{N}\left(\Delta x + i\Delta x^2\right).$$

Using equations (5.15) and (5.16), we may write this summation in closed form as

$$R_N = N\Delta x + \frac{N(N+1)\Delta x^2}{2}. \tag{6.12}$$

Replacing Δx, this is equivalent to

$$R_N = 4 + \frac{8N(N+1)}{N^2}.$$

In the limit, we obtain

$$\int_1^5 x\,dx = \lim_{N\to\infty} R_N = \lim_{N\to\infty}\left(4 + \frac{8N(N+1)}{N^2}\right) = 12.$$

In this example, the equation (6.12) gives us a specific, closed-form expression for the right-hand Riemann sum using N subintervals. (Such an expression is not always attainable.) If one were to actually sit down and compute a right-hand Riemann sum using, let's say, 100 intervals, then one would obtain, according to the formula, $R_{100} = 12.08$. Using 1,000 subintervals, one would similarly obtain $R_{1000} = 12.008$, and so on. *In the limit* as the number of subintervals approaches infinity, we arrive at the exact value for the definite integral, which is precisely 12. We know this, because we have a simple expression for the Riemann sum R_N as a function of the number of subintervals, and we may therefore compute the limit as $N \to \infty$ of this expression to see what happens. ■

6.3.2 Properties of Definite Integrals

Definite integrals enjoy several key properties, which are often useful in performing computations or simplifications.

Theorem 6.3.1 — Integral Properties. Given two functions f and g and constants a, b, c, α, and β, the following properties hold.

$$\int_a^b (\alpha f(x) + \beta g(x))\,dx = \alpha\int_a^b f(x)\,dx + \beta\int_a^b g(x)\,dx, \tag{6.13}$$

$$\int_a^b f(x)\,dx = -\int_b^a f(x)\,dx, \tag{6.14}$$

$$\int_a^b f(x)\,dx = \int_a^c f(x)\,dx + \int_c^b f(x)\,dx. \tag{6.15}$$

Property (6.13) is the *linearity* property of integrals—it allows us to pull constants out of the integral and break an integral up into individual terms. This property corresponds to the linearity property for derivatives that allowed us to differentiate, for example, $f(x) = 8x^4 + 13x^3 + 12x^2 + 19x + 7$, by applying the derivative rules for x^4, x^3, x^2, x, and 1.

Property (6.13) can also be understood graphically. The function $f+g$ is simply the function whose height is equal to the sum of the heights of functions f and g.

It makes sense therefore that the integral of the sum is the sum of the integrals, as shown in Figure 6.16. Similarly, if we were to multiply a function f by a constant α, then we obtain a function whose output, at every point, is a factor of α greater than it was before. Therefore, the area under the graph of the function should *also* increase by a factor of α.

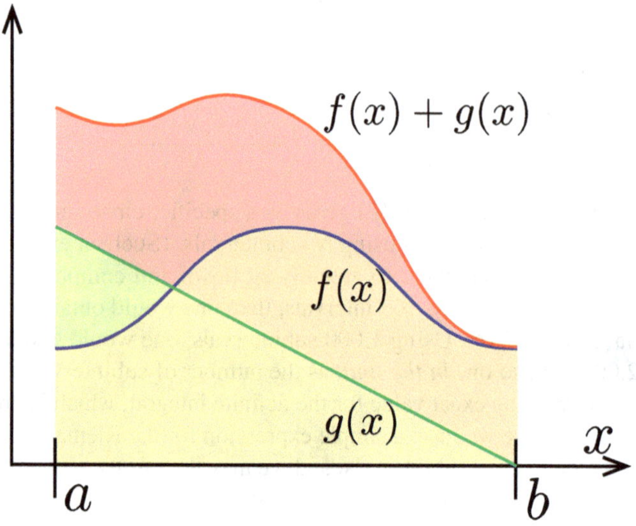

Figure 6.16: Linearity of the definite integral: $\int_a^b (f(x)+g(x))\ dx = \int_a^b f(x)\ dx + \int_a^b g(x)\ dx$.

Property (6.14) is an algebraic property that tells us that switching the order of the limits of integration a and b comes at the cost of an additional minus sign. This fact is a result of the inner workings of Riemann sums; in fact, it comes directly from equation (6.1) that defines the subinterval width Δx. If one were to switch the ordering of a and b, then the width Δx would switch in sign. For example, if we do a Riemann sum from 1 to 5 with N subintervals, we obtain $\Delta x = (5-1)/N$; on the other hand, if we do a Riemann sum *from* 5 *to* 1 with N subintervals, we instead obtain $\Delta x = (1-5)/N$, a quantity with the opposite sign as before.

Property 6.14 also extends meaning to an integral in which the lower limit of integration is *greater* than the upper limit of integration. Though this is sometimes useful in algebraic manipulation of integrals, it is preferred, in practice, to write a definite integral so that the lower limit of integration is *less* than the upper limit of integration.

Finally, property 6.15 must be understood in two separate cases: if c lies inside the interval from a to b and if c lies outside of the interval from a to b. For the first case, consider Figure 6.17. Here, it should be clear that the area under the graph between a and b is just the sum of the areas under the graph from a to c and from c

to b.

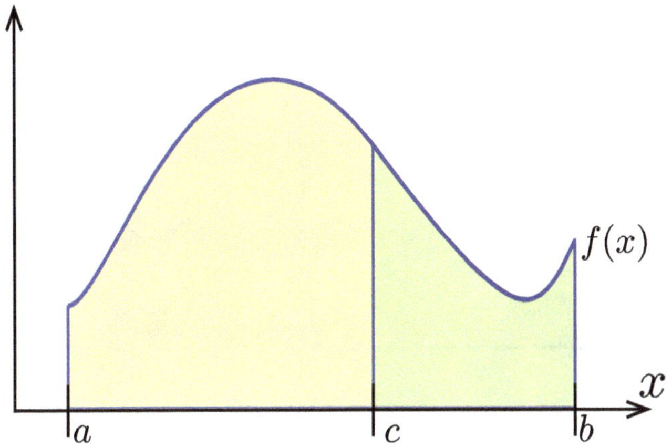

Figure 6.17: Additive property of the integral—Case 1: $c \in (a,b)$.

The second case is if $c \notin [a,b]$. To be concrete, let us suppose that $c > b$, though a similar argument holds if $c < a$. This situation is shown in Figure 6.18. In this case, property (6.15) still holds. The idea in this case is that the integral from a to c contains *more* area than just the integral from a to b; when we add the integral from c to b, however, we are *subtracting* the area between b and c. (When we integrate from c to b, we are integrating this portion of the domain *backwards*, and we thus pickup a minus sign pursuant to property (6.14).)

To see what we mean more closely, consider that

$$\int_a^c f(x)\,dx = \int_a^b f(x)\,dx + \int_b^c f(x)\,dx.$$

This follows since the point b *is* between the points a and c. It just says that the total area pictured in Figure 6.18, i.e., the total area between a and c, is just the sum of the area between a and b and the area between b and c. From the preceding equation, however, we may obtain

$$\int_a^b f(x)\,dx = \int_a^c f(x)\,dx - \int_b^c f(x)\,dx.$$

Using property (6.14), we may next rewrite this as

$$\int_a^b f(x)\,dx = \int_a^c f(x)\,dx + \int_c^b f(x)\,dx.$$

We have now obtained equation (6.15) for the case in which $c > b$. Essentially, it is because of the convention (6.14) that allows equation (6.15) to work for the general case.

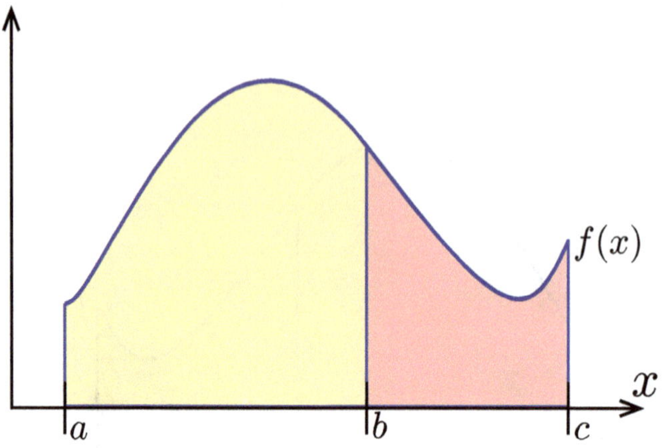

Figure 6.18: Additive property of the integral—Case 2: $c \notin [a,b]$.

6.3.3 Symmetric Integrals of Even and Odd Functions

We begin by recalling the definitions of even and odd functions.

Definition 6.3.2 A function f is an *even function* if

$$f(-x) = f(x), \tag{6.16}$$

for all x.

A function f is an *odd function* if

$$f(-x) = -f(x), \tag{6.17}$$

for all x.

(R) A given function does not have to be either even or odd; i.e., there are plenty of functions that are neither even nor odd. The properties (6.16) and (6.17) are properties that some functions may have; they are a type of *symmetry*. Thus, in general, a function can be even, odd, or neither.

An even function is a function that is equal to its own horizontal reflection. An example of an even function and an odd function is shown in Figure 6.19.

Notice that the even function is symmetric about the y-axis: the function evaluated at -3, for example, yields the same result as the function evaluated at $+3$. The odd function, on the other hand, has the opposite property: the function evaluated at -3 yields the opposite number, in sign, as the function evaluated at $+3$.

Following these symmetry properties, we have the following special properties of definite integrals for even and odd functions.

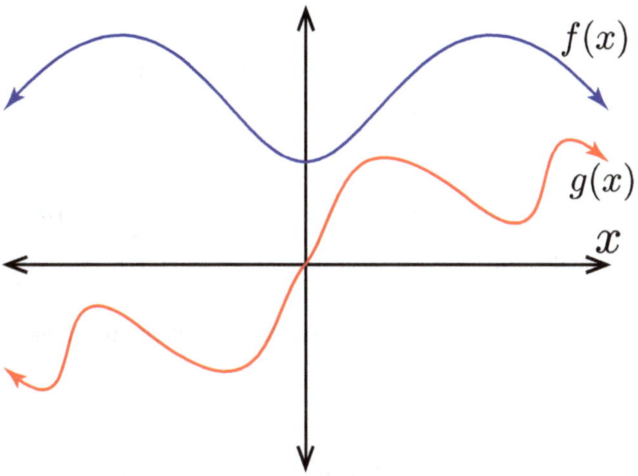

Figure 6.19: An even function f and an odd function g.

Theorem 6.3.2 If f is an even function and $a > 0$, then

$$\int_{-a}^{a} f(x)\,dx = 2\int_{0}^{a} f(x)\,dx. \tag{6.18}$$

If f is an odd function and $a > 0$, then

$$\int_{-a}^{a} f(x)dx = 0. \tag{6.19}$$

Exercises

For Exercises 1–5, use the summation formulas (5.15)–(5.19) and the definition of the definite integral to prove the given formula.

1. $\displaystyle\int_{a}^{b} dx = b - a.$
2. $\displaystyle\int_{a}^{b} x\,dx = \frac{b^2 - a^2}{2}.$
3. $\displaystyle\int_{a}^{b} x^2\,dx = \frac{b^3 - a^3}{3}.$
4. $\displaystyle\int_{a}^{b} x^3\,dx = \frac{b^4 - a^4}{4}.$
5. $\displaystyle\int_{-a}^{a} x\,dx = 0$, for $a > 0$.

For Exercises 6–10, suppose that f is an even function, g is an odd function, $\int_0^3 f(x)\,dx = 5$, $\int_0^3 g(x)\,dx = 7$, and $\int_0^5 f(x)\,dx = 9$. Determine the value of the indicated integral.

6. $\displaystyle\int_{0}^{3} (1 + 2f(x))\,dx.$
7. $\displaystyle\int_{-5}^{-3} f(x)\,dx.$
8. $\displaystyle\int_{-3}^{3} f(x)g(x)\,dx.$
9. $\displaystyle\int_{0}^{3} g(-x)\,dx.$
10. $\displaystyle\int_{-5}^{5} (f(x) + g(x))\,dx.$

Problems

11. Use symmetry to determine the value of the integral $\int_{-\pi}^{\pi} \sin(x)\,dx$.
12. Use geometry and the interpretation of the definite integral as an area to evaluate $\int_a^b x\,dx$.
13. A particle traveling along the x-axis has a velocity $v(t) = \sqrt{100 - t^2}$ m/s.
    (a) Describe the meaning of the definite integral $\int_0^{10} \sqrt{100 - t^2}\,dt$ in practical terms.
    (b) Determine the distance the particle traveled during the interval $0 \le t \le 10$.
14. Suppose that $v(t)$ is the velocity of a particle moving along the x-axis, so that $x'(t) = v(t)$.
    (a) Explain the meaning of the integral $\int_a^b v(t)\,dt$ in practical terms.
    (b) Explain the meaning of the integral $\int_a^b |v(t)|\,dt$ in practical terms.
    (c) Explain the meaning of the quantity $\left|\int_a^b v(t)\,dt\right|$ in practical terms.
15. Suppose that a car has a velocity $v(t)$ (mph), for a period of time $[0, T]$, and that its fuel economy η (mpg) is given as a function of its velocity, i.e., $\eta = \eta(v)$.
    (a) Describe the meaning of the definite integral $\int_0^T v(t)\,dt$ in practical terms.
    (b) Describe the meaning of the definite integral $\int_0^T \frac{v(t)}{\eta(v(t))}\,dt$ in practical terms.
    (c) Suppose that a car accelerates at a constant rate from a speed of 30 mph to a speed of 60 mph over a half-hour period, and that various values for its fuel economy as a function of its speed are given in the table below.

v (mph)	30	35	40	45	50	55	60
$\eta(v)$ (mpg)	40	39	38	35	33	32	29

    Approximate the amount of fuel this car consumed over the half-hour period.
16. An oil spill on the ocean has left a circularly symmetric oil patch with a density $\rho(r)$ (kg/m^2) at a distance of r from the source of the spill.
    (a) Since the density is constant everywhere along concentric circles centered at the spill's source, draw a thin, circular annulus with radius r_i, and explain why the amount of oil contained in this annulus can be approximated by $2\pi r\rho(r)\Delta r$.
    (b) Explain the meaning of the integral $\int_0^{R_{\max}} 2\pi r\rho(r)\,dr$ in practical terms.

17. A NASA satellite is collecting energy with its solar array and storing it in its battery at the rate $R(t)$ Watts, t hours after the deployment of the solar array.

(a) Using dimensional analysis, write down the conversion between Watt-hours and kiloJoules. Report the conversion in the form: 1 W·h = ____ kJ.

Now suppose you are given the following telemetry data:

t [hrs]	0	1.5	3	4.5	6
$R(t)$ [W]	17.2	17.6	17.7	18.1	18.3

(b) Compute the Left- and Right-Hand Riemann Sum approximations for $\int_0^6 R(t)\,dt$.

(c) Explain what this integral represent in practical terms.

(d) Which Riemann Sum is an underestimate and which is an overestimate to the exact value of the integral? Explain your reasoning and state any assumptions you used to answer this question.

(e) Suppose the exact value of the integral asked for in part (d) is 107 kJ. Does this fit the lower and upper bounds that were previously determined? If not, which assumption must have been wrong? Give a practical physical explanation.

(f) In order to perform a certain maneuver, the satellite requires 200 kJ to be stored in its battery. If the exact value of the integral is $\int_0^6 R(t)\,dt = 107$ kJ, does the satellite have enough energy stored in its battery at time $t = 6$ to perform that maneuver? (The answer is open ended. What condition is necessary for there to be enough energy in the battery at time $t = 6$?)

6.4 The Fundamental Theorem of Calculus

In Section 6.1, we made a heuristic argument that the integral of the velocity with respect to time should simply equal the net change in a particle's position. In this section, we uncover a powerful and closely related tool for computing definite integrals.

.4.1 The Fundamental Theorem

We begin by stating the fundamental theorem, which shows us how we may easily compute a definite integral of a function that possess a known antiderivative.

Theorem 6.4.1 — The Fundamental Theorem of Calculus. Given a function f with an antiderivative F defined on the domain $[a,b]$, we have

$$\int_a^b f(x)\,dx = F(b) - F(a). \tag{6.20}$$

Proof. We begin with the general Riemann sum S_N of the function f from a to b, which is given by equations (6.5) as

$$S_N = \sum_{i=1}^{N} f(x_i^*)\Delta x,$$

where $\Delta x = (b-a)/N$, $x_i = a + i\Delta x$, for $i = 0, \ldots, N$, and $x_i^* \in [x_{i-1}, x_i]$, as usual.

The key observation that powers the fundamental theorem is the realization that one doesn't actually need to know which value of x_i^* one is using; one only *really* needs to know the value of the output $f(x_i^*)$ for *some* point x_i^* located within the ith subinterval. At first, that may seem like a silly thing to say—after all, how can one compute the value of $f(x_i^*)$ *without* knowing where x_i^* actually is? That is where the supposition, that the function f has an antiderivative, and an old friend come in.

Recall the mean-value theorem (as stated in Theorem 2.3.3 on page 146). In particular, let us apply the mean-value theorem to the antiderivative F over the ith subinterval $[x_{i-1}, x_i]$. In this situation, the mean-value theorem guarantees that there exists a point $x_i^* \in [x_{i-1}, x_i]$ such that

$$F'(x_i^*) = \frac{F(x_i) - F(x_{i-1})}{\Delta x}. \tag{6.21}$$

The derivative F', however, is simply equal to the function f. For each subinterval, let us use the point x_i^*, for $i = 1, \ldots, N$, that is guaranteed by the mean-value theorem as the test point for the Riemann sum. Even though we don't actually know where the point x_i^* is, for each of the i, we do know that it is somewhere within the given subinterval. Proceeding in this fashion, our Riemann sum simplifies as follows

$$\begin{aligned} S_n N &= \sum_{i=1}^{N} f(x_i^*)\Delta x \\ &= \sum_{i=1}^{N} \left[\frac{F(x_i) - F(x_{i-1})}{\Delta x} \right] \Delta x \\ &= \sum_{i=1}^{N} [F(x_i) - F(x_{i-1})]. \end{aligned}$$

Thus, the Riemann sum becomes a telescoping series:

$$\begin{aligned} S_N &= [F(x_1) - F(x_0)] + ([F(x_2) - F(x_1)] + [F(x_3) - F(x_2)] + \cdots \\ &\quad + [F(x_{N-2}) - F(x_{N-3})] + [F(x_{N-1}) - F(x_{N-2})] + [F(x_N) - F(x_{N-1})]. \end{aligned}$$

Each term cancels except for two:

$$S_N = F(x_N) - F(x_0).$$

But recall that $x_0 = a$ and $x_N = b$; moreover, these two relations are *independent* of the number of subintervals used. Therefore, by using the mean-value theorem to generate the sample points, our Riemann sum with N subintervals becomes

$$S_N = F(b) - F(a).$$

This simplifies this way *only* for this very special type of Riemann sum. Notice now that this result no longer depends on N: it only depends on the antiderivative F and the endpoints a and b. Therefore, when we take the limit, we obtain the same:

$$\int_a^b f(x)\, dx = \lim_{N\to\infty} [F(b) - F(a)] = F(b) - F(a).$$

Since it does not matter what type of Riemann sum we use in the definition of the definite integral (all Riemann sums must approach the same exact value for the area in the limit as $N \to \infty$), we conclude that this is the true value of the definite integral. This completes the proof. ■

6.4.2 Using the Fundamental Theorem

When an antiderivative is available, it is always preferred to use the fundamental theorem of calculus, as given by equation (6.20), to compute the value of a definite integral. The sheer simplicity of finding the change in the antiderivative between the points a and b, by simply computing its difference, yields a tremendous advantage over the laborious tedium of determining limits of Riemann sums. The fundamental theorem also adds a degree of insight into the meaning of the definite integral as the net change in the function's antiderivative. When the function is the velocity of a particle, the integral over time tells us the net change in the particle's position. When the function is the acceleration, the integral tells us the net change in the particle's velocity.

R In practice, the right-hand side of equation (6.20) is sometimes expressed using the notation

$$F(x)|_{x=a}^{x=b} = F(b) - F(a).$$

The left-hand side of this equation is just a shorthand that says: take the function $F(x)$, evaluate it at $x = a$ and $x = b$, and find the difference.

R Now, in the wake of the fundamental theorem, we may understand the difference between definite and indefinite integrals. An indefinite integral represents the entire family of antiderivatives, whereas the definite integral one computes the difference in one of the antiderivatives between a and b. The indefinite integral still depends on the independent variable (e.g., x), whereas the definite integral is an integral over a definite domain, and so one obtains only a number.

■ **Example 6.5** Compute the area of the trapezoid bounded by the x-axis, the function $f(x) = x$, and the lines $x = 1$ and $x = 3$, as shown in Figure 6.20.

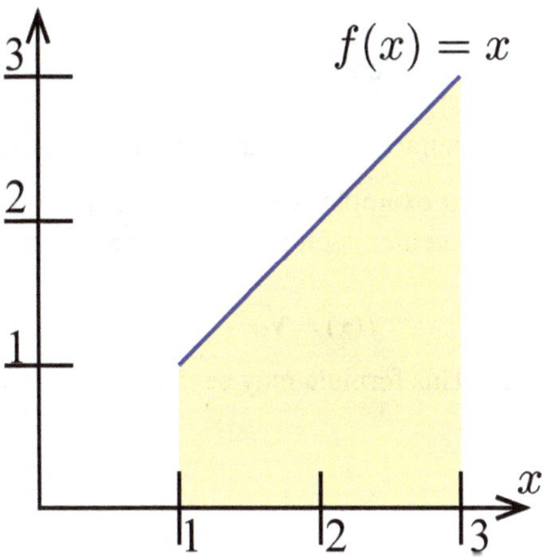

Figure 6.20: Area under the curve $f(x) = x$.

It's not hard to see that the answer should be 4. We can use this example, however, as a check on our theory. Let us compute the definite integral

$$\int_1^3 x\,dx,$$

as this answer should equal the area of the trapezoid shown in Figure 6.20. The antiderivative of x is simply $x^2/2$; therefore,

$$\int_1^3 x\,dx = \frac{x^2}{2}\bigg|_{x=1}^{x=3} = \frac{3^2 - 1^2}{2} = 4,$$

as it should be. ■

Example 6.6 Compute the area bounded by the graph of $f(x) = \sin(x)$, the x-axis, and the lines $x = 0$ and $x = \pi$.

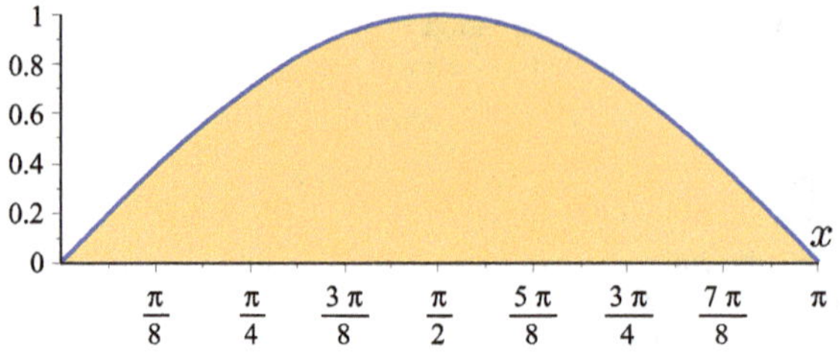

Figure 6.21: Area under the curve $f(x) = \sin x$.

To achieve this, we simply recall that $\int \sin(x)\, dx = -\cos(x) + C$ and we compute the integral

$$\int_0^{\pi} \sin(x)\, dx = -\cos(x)|_{x=0}^{x=\pi} = -\cos(\pi) + \cos(0) = 2.$$

Thus, the area under one "hump" of the graph of the sine wave is equal to two. ∎

Example 6.7 In our next example, we will derive the formula for the area of a circle of radius a. If we center such a circle at the origin, its top half may be described by the function

$$f(x) = \sqrt{a^2 - x^2},$$

as shown in Figure 6.22. (This formula may be derived by using the equation for a circle: $x^2 + y^2 = a^2$.)

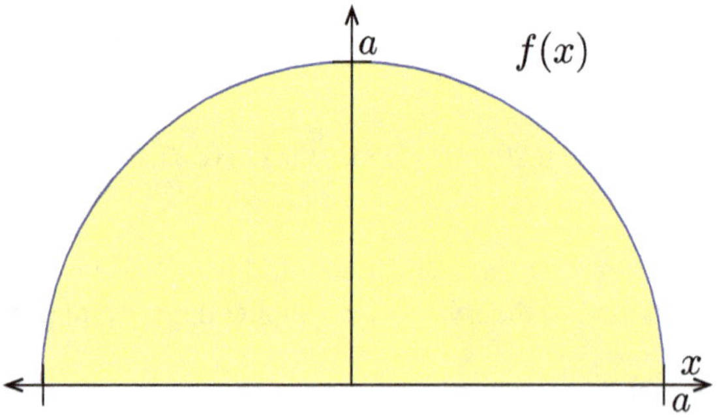

Figure 6.22: Area under the curve $f(x) = \sqrt{a^2 - x^2}$.

To help in this pursuit, we will rely on the antiderivative

$$\int \sqrt{a^2 - x^2}\, dx = \frac{x}{2}\sqrt{a^2 - x^2} + \frac{a^2}{2} \sin^{-1}\left(\frac{x}{a}\right).$$

(We will learn how this is computed later on; for now, let us take this formula as a given.) Therefore,

$$\begin{aligned}\int_{-a}^{a}\sqrt{a^2-x^2}\,dx &= \frac{x}{2}\sqrt{a^2-x^2}+\frac{a^2}{2}\sin^{-1}\left(\frac{x}{a}\right)\Bigg|_{x=-a}^{x=a}\\ &= \left(\frac{a}{2}\sqrt{a^2-a^2}+\frac{a^2}{2}\sin^{-1}(1)\right)\\ &\quad -\left(\frac{-a}{2}\sqrt{a^2-(-a)^2}+\frac{a^2}{2}\sin^{-1}(-1)\right)\\ &= \frac{a^2}{2}\left(\sin^{-1}(1)-\sin^{-1}(-1)\right)=\frac{\pi a^2}{2}.\end{aligned}$$

Since this integral only represents *half* of the area of the full circle (since we integrated from the x-axis to the function), the area of the full circle is therefore equal to πa^2. We have thus proven an important formula from geometry. ■

6.4.3 The Average Value of a Function

As a final application for this section, we introduce the concept of the average value of a function.

Definition 6.4.1 Given a function f defined on an interval $[a,b]$, we say that the *average value of f* over $[a,b]$, denoted $\overline{f}$, is the number determined by

$$\overline{f}=\frac{1}{b-a}\int_a^b f(x)\,dx. \tag{6.22}$$

There are three ways that one may understand the motivation for this definition. The first is by analogy with taking the average of a set of numbers. Suppose, for example, that we would like to compute the average value of the numbers 2, 3, 5, and 6. This is achieved by summing and dividing:

$$\text{average}=\frac{2+3+5+6}{4}=4.$$

If we represent these data as a bar graph, however, we gain a new understanding of the meaning of the average value. This is shown in Figure 6.23 (left). Here, the sum $2+3+5+6$ can be interpreted as the area under the graph, and the total number of values, 4, can be viewed as the width of the overall interval (one unit equals one number in our data set). Thus, the average value can be interpreted graphically as the area divided by the total width, thereby motivating the definition found in equation (6.22) for the case of a continuous function.

The second way of thinking about equation (6.22) is depicted in Figure 6.23 (right). Here, the average value $\overline{f}$ is selected so that the total area under the average value, between $x=a$ and $x=b$ is the same as the area under the graph of the function. Thus, we define $\overline{f}$ so that the area of under the graph of f, which is given

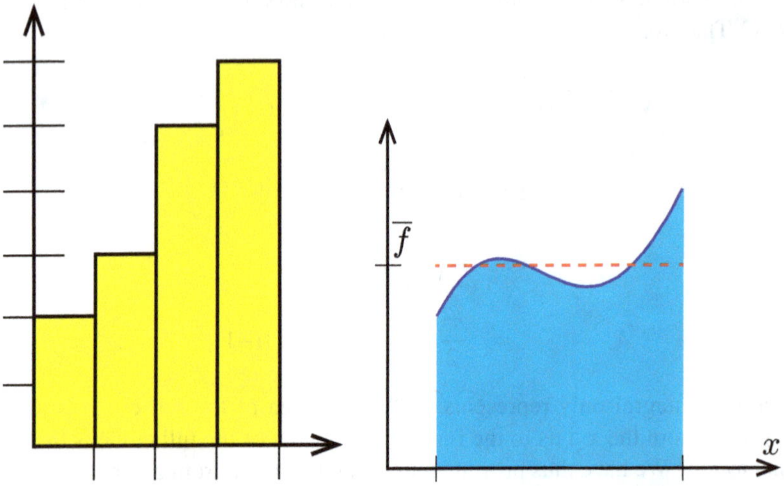

Figure 6.23: The average value of the numbers 2, 3, 5, and 6 (left) and the average value $\overline{f}$ of a function f (right).

by the value of the definite integral, is equal to the area of the rectangle with height $\overline{f}$ and width $(b-a)$, so that

$$\overline{f} \cdot (b-a) = \int_a^b f(x)\,dx.$$

By dividing by the length $(b-a)$, we recover equation (6.22).

The third and final way we may understand the average value of a function returns us to our discussion on average and instantaneous rates of change. Suppose that the function f has an antiderivative F, so that $F'(x) = f(x)$. In this way, we say that the function f represents the instantaneous rate of change of the function F. Therefore, it makes sense that we should define the average value of the function f as the average rate of change of its antiderivative F:

$$\overline{f} = \frac{F(b) - F(a)}{b-a}.$$

This formula, however, reverts to our definition (6.22) via the fundamental theorem of calculus. Thus, as seen through the lense of the fundamental theorem, the right-hand side of equation (6.22) is just the average rate of change of any antiderivative of the function f, when an antiderivative actually exists.

Exercises

For Exercises 1–10, use the fundamental theorem of calculus to compute the given integral. Draw a graph of the integrand and shade an area equal to the value of the given integral.

1. $\int_0^a x\,dx.$
2. $\int_0^5 (4x^3+3x^2+2x+7).$
3. $\int_0^1 e^x\,dx.$
4. $\int_0^1 e^{-0.5x}\,dx.$
5. $\int_0^{\pi/6} \sin(x)\,dx.$
6. $\int_0^4 \sqrt{x}\,dx.$
7. $\int_0^2 x^2\,dx.$
8. $\int_1^e \frac{1}{x}\,dx.$
9. $\int_0^{\pi/4} \sec^2(x)\,dx.$
10. $\int_0^{\pi/4} \tan(x)\,dx.$

For Exercises 11–20, compute the average value of the indicated function on the given interval. Use the fundamental theorem of calculus to perform any integrals.

11. $f(x)=e^x,\ [0,1].$
12. $f(x)=\sin(x),\ [0,2\pi].$
13. $f(x)=x^2,\ [0,3].$
14. $f(x)=x,\ [0,10].$
15. $f(x)=\frac{1}{x},\ [1,2].$
16. $f(x)=\sin(x),\ \left[0,\frac{\pi}{2}\right].$
17. $f(x)=\sqrt{x},\ [0,9].$
18. $f(x)=\sqrt{3x+7},\ [0,2].$
19. $f(x)=2^x,\ [0,1].$
20. $f(x)=7+3\cos(2\pi x),\ [0,1].$

Problems

21. Determine the average height of the function $f(x)=\sqrt{a^2-x^2}$ over the interval $[0,a]$.
22. Suppose that a particle traveling along the x-axis has a constant acceleration a.
    (a) Use integration to show that its velocity is given by $v(t)=v_0+at$, where v_0 is its initial velocity.
    (b) Use integration to show that its position is given by $x(t)=x_0+v_0t+\frac{1}{2}at^2$, where x_0 is its initial position.
23. Suppose that a particle is initially located at $x(0)=5$, and its velocity is given by $v(t)=\sin(t)$. Determine the particle's location at time $t=3\pi/2$.
24. Suppose that a particle is initially located at $x(1)=3$, and its velocity is given by $v(t)=1/t$, for $t\geq 1$. Determine the particle's location at time $t=10$.
25. As the radius of a circle increases, its area A changes at a rate

$$\frac{dA}{dr}=2\pi r.$$

This is since a small increase in radius Δr should result in a corresponding increase in area ΔA according to

$$\Delta A \approx 2\pi r \Delta r,$$

i.e., the change in area is approximately the circumference times the change in radius. Use this fact to determine the total area of a circle of radius a.

26. As the radius of a sphere increases, its volume V changes at a rate

$$\frac{dV}{dr} = 4\pi r^2.$$

This is since a small increase in radius Δr should result in a corresponding increase in volume ΔV according to

$$\Delta V \approx 4\pi r^2 \Delta r,$$

i.e., the change in volume is approximately equal to the surface area times the change in radius. Use this fact to determine the total volume of a sphere of radius a.

27. A *catenary* is the shape that a hanging cable or wire makes when it is only supported at its two endpoints. The equation of a catenary is given by the hyperbolic cosine function[1]. Suppose that the cables of the Golden Gate Bridge, shown in Figure 6.24, are modeled by the transformed catenary equation

$$h(x) = 75\cosh\left(\frac{x}{361}\right),$$

for $x \in [-640, 640]$, as shown in Figure 6.25. Height is measured above sea level. Determine the average height of the cable.

28. Suppose that demand for apples from a particular orchard fluctuates seasonally, so that consumers wish to purchase apples at the rate

$$A(t) = 36 + 4\cos(4\pi t)$$

bushels of apples per day on the tth day of the year. If each acre of the orchard can produce 300 bushels of apples per year, how many acres of apple orchard would be required to meet the consumers' total annual demand?

29. In Problem 17 in Section 6.2, we saw that the instant-messaging service WhatsApp grew at a rate of

$$r(t) = 17(2)^{t/9}$$

million customers per month, for $t \in [0, 9]$, during a nine-month period in 2013.

(a) Express the total number of new customers during this nine-month period as a definite integral.

---

[1] Recall that $\cosh(x) = \dfrac{e^x - e^{-x}}{2}$.

Figure 6.24: The Golden Gate Bridge at night.

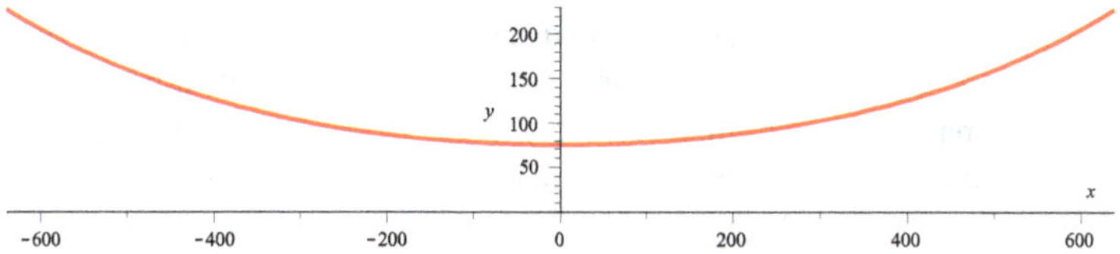

Figure 6.25: The height of the cable on the Golden Gate Bridge.

(b) Evaluate the definite integral you obtained in part (a).
(c) If there were 225 million users at the beginning of this period, how many users were there at the end of the nine months?

30. Suppose that the marginal cost associated with producing a quantity q of a certain commodity is given by

$$MC(q) = 7 + \frac{100}{\sqrt{q}} \text{ dollars.}$$

(a) If the fixed cost for the company is \$12,000, determine how much it will cost to produce the first 500 units of the commodity.
(b) What is the average cost per item for the first 500 units?
(c) How much will it cost to produce the first 1,000 units of the commodity?
(d) What is the average cost per item for the first 1,000 units?
(e) Determine the average cost $A(q)$ per unit of producing the first q units.

31. Suppose that the marginal revenue for a particular product decreases exponentially, so that

$$MR(q) = ae^{-rq},$$

for some choice of the positive parameters $a, r > 0$. Furthermore, suppose that the marginal cost is constant, $MC(q) = c$.
    (a) Describe the meaning of the parameters a and c in practical terms.
    (b) Determine a formula for the optimal quantity q^* that maximizes profit.
    (c) Use the fundamental theorem of calculus to determine the profit function $\Pi(q)$.
32. *Bernoulli's equation* in physics can be used to show that the time T required to drain a right-circular cylindrical tank is given by

$$T = \frac{A}{A_e\sqrt{2g}} \int_0^H \frac{dh}{\sqrt{h}},$$

where A is the cross-sectional area of the cylinder, A_e is the cross-sectional area of the exit nozzle (at the bottom of the tank), g is acceleration due to gravity, and H is the initial height of the water level in the tank. Use the fundamental theorem of calculus to integrate the preceding equation, thereby completing the derivation of the formula for the total draining time.
33. In physics, an *impulsive force* is a large-magnitude force that acts on an object for a short duration in time, such as when one hits a baseball with a bat or a nail with a hammer. In such situations, instead of modeling a usually complicated forcing function, one instead defines the *impulse* J of the force as

$$J = \int_{t_1}^{t_2} F(t)\, dt.$$

(Typically, $\Delta t = t_2 - t_1$ is of a very short duration.)
    (a) Explain why the equation $J = F_{\text{avg}}\Delta t$ holds, where F_{avg} is the average value of the force over the interval $[t_1, t_2]$.
    (b) The *impulse–momentum theorem* states that the impulse acting on a particle is equal to its total change in momentum, i.e., $J = mv_2 - mv_1$. Using Newton's second law for a particle with constant mass, $F = ma$, and the fundamental theorem of calculus, explain why the impulse–momentum theorem is true.

6.5 Representing Antiderivatives with Integrals

In this section, we examine an alternate form of the fundamental theorem of calculus, which deals with variable limits of integration.

6.5.1 Integrals with Variable Endpoints

We begin by discussing functions defined with the form

$$F(x) = \int_a^x f(t)\,dt, \tag{6.23}$$

i.e., given a function f, we define a new function F by the relation (6.23). Our first goal is to simply understand what such a construction would look like. In our next paragraph, we will see that this new function F is actually the particular antiderivative for f that satisfies the initial condition $F(a) = 0$. (This is actually an alternate statement of the fundamental theorem of calculus.)

The first common point of confusion in understanding equation (6.23) is that the integrand depends on the variable t instead of the variable x. The reason for using two distinct variables—a variable of integration t and a variable endpoint x—is the separate roles that those variables play.

First, notice that the integral in equation (6.23) is a *definite* integral. When computing definite integrals, the result no longer depends on the variable in question. That is, the following quantities are each the same:

$$\int_a^b f(x)\,dx, \ \int_a^b f(s)\,ds, \ \int_a^b f(t)\,dt, \text{ and } \int_a^b f(w)\,dw.$$

Each of these represents the area under the graph of the function f between the values a and b of the independent variable: the result does not care how we label the horizontal axis, and is the same regardless of whether we call the independent variable x, s, t, w, or "smiley face." When we compute a definite integral, *we integrate over the independent variable*, resulting in a quantity, which no longer depends on the independent variable. For this reason, the variable of integration is sometimes called a *dummy variable*—it is just a placeholder used to compute the integral and, once computed, the result no longer depends on this variable. An index of summation is another instance of a dummy variable, as the particular variable used as an index does not affect the result: $\sum_{i=1}^n a_i = \sum_{j=1}^n a_j$.

In equation (6.23), however, the end result *does* depend on a variable, as this equation is being used to define a function of x. Writing this integral in the form

$$\int_a^x f(x)\,dx \qquad \text{(poor notation)}$$

is therefor to be avoided, as it confuses the functional dependance on x (i.e., x as a variable upper limit of integration) with the role of the variable of integration, which is a dummy variable used to achieve a certain end.

(R) When expressing integrals using variable endpoints, do not commingle variables, i.e., always use a different variable of integration than the variable describing the endpoints, as the result should no longer depend on the variable of integration.

The integral of equation (6.23) is represented by the shaded area in Figure 6.26. The function F, defined by that equation, therefore has the following graphical

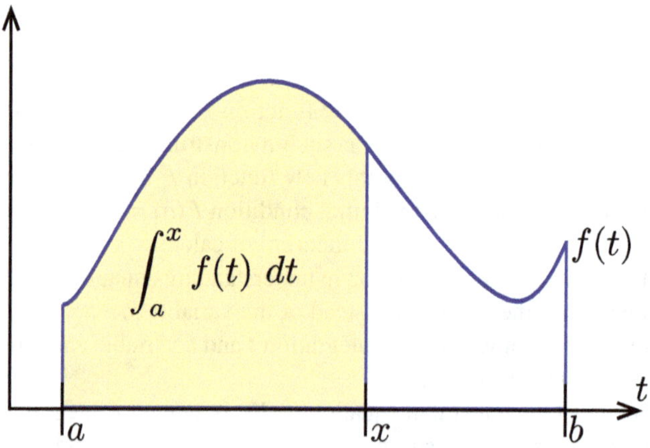

Figure 6.26: Illustration of the quantity $\int_a^x f(t)\,dt$.

interpretation: it accumulates area under the graph of the function f, as the endpoint x sweeps from left to right. That is, $F(a)$ starts off at zero, since there is zero area under the curve between $t = a$ and $t = a$. Then, as x increases, $F(x)$ returns the cumulative area from the point $t = a$ up to the point $t = x$.

6.5.2 The Fundamental Theorem: Part 2

In this paragraph, we explore a way of describing particular antiderivatives of a given function.

Theorem 6.5.1 — Fundamental Theorem of Calculus (Part 2). Let f be a continuous function defined on a closed interval $[a,b]$. The function F defined by

$$F(x) = \int_a^x f(t)\,dt \tag{6.24}$$

is an antiderivative of f on the domain $[a,b]$, i.e., $F'(x) = f(x)$; moreover, F is the particular antiderivative that satisfies the condition $F(a) = 0$.

(R) An alternate statement of Theorem 6.5.1 is given by the single equation

$$\frac{d}{dx}\left[\int_a^x f(t)\,dt\right] = f(x). \tag{6.25}$$

(R) There is a simple proof of Theorem 6.5.1 in the special case in which f has a known antiderivative G. From the fundamental theorem of calculus (Theorem 6.4.1), the function F defined by equation (6.24) can be expressed as

$$F(x) = \int_a^x f(t)\,dt = G(x) - G(a).$$

Since $G(a)$ is a constant and since $G(x)$ is an antiderivative for $f(x)$, it follows that $F(x)$ must also be an antiderivative for $f(x)$. Moreover, $F(a) = G(a) - G(a) = 0$. Theorem 6.5.1 is more general than this, as it makes no requirement on first obtaining an antiderivative for the function f. In fact, functions of this form can be used to *define* the antiderivative, unambiguously, even when the antiderivative cannot otherwise be expressed using elementary functions.

Next, we prove Theorem 6.5.1.

Proof. The derivative of the function F is defined by

$$F'(x) = \lim_{h\to 0} \frac{F(x+h) - F(x)}{h}.$$

Substituting the expression (6.24) for F, this is equivalent to

$$\begin{aligned} F'(x) &= \lim_{h\to 0} \frac{1}{h}\left(\int_a^{x+h} f(t)\,dt - \int_a^x f(t)\,dt\right) \\ &= \lim_{h\to 0}\left(\frac{1}{h}\int_x^{x+h} f(t)\,dt\right). \end{aligned}$$

In the limit as $h \to 0$, the interval $[x, x+h]$ becomes infinitely small. Therefore, we can approximate this integral using a left-hand Riemann sum with a single subinterval:

$$\int_x^{x+h} f(t)\,dt \approx f(x)h.$$

This approximation is illustrated in Figure 6.27. Since this approximation becomes exact in the limit as $h \to 0$, we may conclude that

$$F'(x) = \lim_{h\to 0}\left(\frac{1}{h}\int_x^{x+h} f(t)\,dt\right) = f(x).$$

This completes the proof. ■

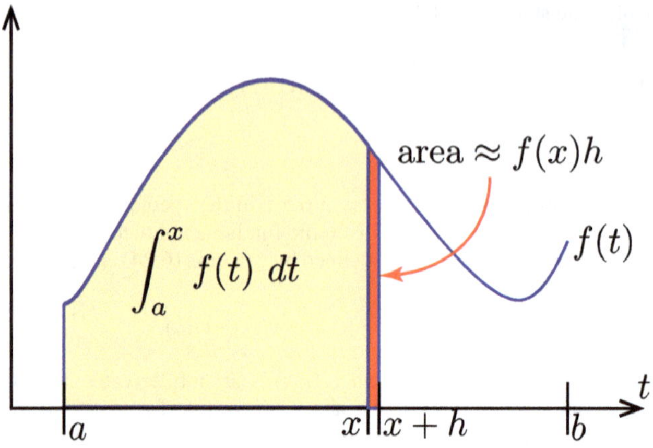

Figure 6.27: Illustration of the quantity $\int_x^{x+h} f(t)\,dt$.

■ **Example 6.8** Compute the derivative of the function defined by

$$\mathrm{Si}(x) = \int_0^x \frac{\sin(t)}{t}\,dt. \tag{6.26}$$

This function is known as the *sine integral function*. A graph of the sine integral function is shown in Figure 6.28.

From Theorem 6.5.1, we may write this derivative as

$$\mathrm{Si}'(x) = \frac{\sin(x)}{x}.$$

Notice that, like the sine function, the derivative $\mathrm{Si}'(x)$ alternates sign every π units, which means that the sine integral function alternates from increasing to decreasing every π units. Moreover, the rate of change decays due to the factor of $1/x$, causing the sine integral curve's oscillations to damp out as we move away from the origin. The factor $1/x$ also changes sign as x passes through zero, causing the sine integral function to be an odd function. These conclusions obtained from the derivative confirm the shape of the graph we observe in Figure 6.28. ■

6.5.3 Acceleration, Velocity, and Position

The advantage of specifying an antiderivative in the form of equation (6.24) is that it uniquely selects the particular antiderivative such that $F(a) = 0$. Thus, it acts as an accumulator of change starting at the point $x = a$. This makes handling a nonzero initial condition easy: one simply adds an initial value to this integral. In particular, this technique provides an elegant way to represent a particle's velocity when given its acceleration, as outlined in the following theorem.

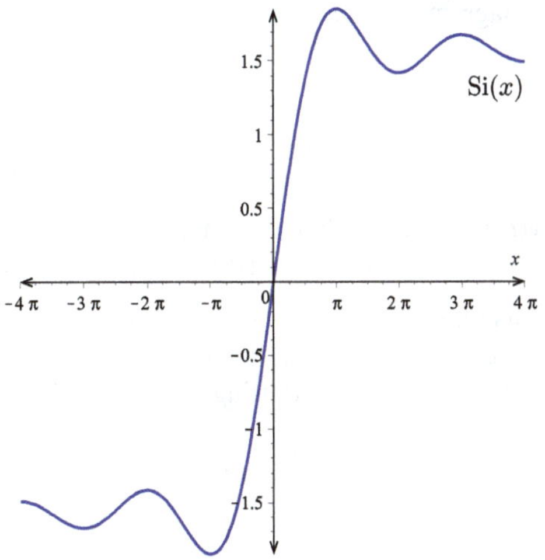

Figure 6.28: The sine integral function Si(x).

Theorem 6.5.2 Let $a(t)$ represent the (continuous) acceleration of a particle traveling along the x-axis as a function of time t. Further, suppose the particle's initial position and velocity are given by x_0 and v_0, respectively, specified at a starting time $t = a$. Then the particle's velocity may be expressed as

$$v(t) = v_0 + \int_a^t a(s)\, ds. \tag{6.27}$$

Once computed, the velocity can be used to represent the particle's position by the relation

$$x(t) = x_0 + \int_a^t v(s)\, ds. \tag{6.28}$$

(R) Computing the velocity and position of a particle using equations (6.27) and (6.28), as opposed to finding an antiderivative and then using the appropriate initial conditions to solve for the constant of integration C, has several advantages. First, the velocity may be expressed concisely using a single equation involving an integral of the acceleration (and, similarly, the position as a single equation involving the integral of the velocity). Second, we do not need to perform any algebra to solve for the constant of integration C, as this is done automatically by correctly evaluating the *definite* integrals as stated in these equations.

Equations (6.27) and (6.28) have a simple interpretation—starting value plus the net change—as diagramed here:

$$v(t) = \underbrace{v_0}_{\text{initial velocity}} + \underbrace{\int_a^t a(s)\,ds}_{\text{net change in velocity}}.$$

Here, the velocity $v(t)$ is efficaciously represented as an initial velocity (at time a) plus the net change in velocity (accumulated from time a to time t).

■ **Example 6.9** Let us return to the example of constant-acceleration particle motion. Assuming a constant acceleration a, initial position x_0, and initial velocity v_0, the particle's velocity may be expressed using equation (6.27) as follows:

$$v(t) = v_0 + \int_0^t a\,ds = v_0 + at.$$

Similarly, the position is given by

$$x(t) = x_0 + \int_0^t (v_0 + as)\,ds = x_0 + v_0 t + \frac{1}{2}at^2.$$

Thus, we have recovered the usual expressions for constant-acceleration motion with two short computations. ■

■ **Example 6.10** Consider a particle with initial position $x_0 = 4$ and initial velocity $v_0 = 3$. If its acceleration is given by

$$a(t) = t^2 - 1,$$

determine its position and velocity as a function of time. What is the position and velocity of the particle at time $t = 2$?

First, we compute the particle's velocity using equation (6.27):

$$v(t) = 3 + \int_0^t (s^2 - 1)\,ds = 3 + \frac{t^3}{3} - t.$$

Notice that the integrand in the preceding equation is simply the function $a(s)$, i.e., we rewrite the acceleration as a function of the variable of integration s for the purpose of computing this definite integral. Evaluating the antiderivative between $s = 0$ and $s = t$ yields the result. Next, we compute the particle's position using equation (6.28):

$$x(t) = 4 + \int_0^t \left(3 + \frac{s^3}{3} - s\right) ds = 4 + 3t + \frac{t^4}{12} - \frac{t^2}{2}.$$

Finally, we simply evaluate these expressions at $t = 2$ to find

$$x(2) = \frac{28}{3} \quad \text{and} \quad v(2) = \frac{11}{3}.$$

This completes the example. ■

(R) It is important not to get too comfortable with the lower limit of integration evaluating to zero: such is not always the case. There are plenty of examples for which the lower limit of integration comes into play.

■ **Example 6.11** Suppose the velocity of a particle is given by $v(t) = \sin(\pi t)$. If its initial position is $x_0 = 2$, determine its position as a function of time.

Here, we utilize equation (6.28), obtaining

$$x(t) = 2 + \int_0^t \sin(\pi s)\, ds = 2 - \left(\frac{\cos(\pi s)}{\pi}\right)\Bigg|_{s=0}^{s=t} = 2 - \frac{\cos(\pi t)}{\pi} + \frac{1}{\pi}.$$

We observe how this sort of computation automatically yields the correct initial condition:

$$x(0) = 2 - \frac{\cos(0)}{\pi} + \frac{1}{\pi} = 2 - \frac{1}{\pi} + \frac{1}{\pi} = 2.$$

The lower limit of integration comes into play in order to enforce this. ■

Exercises

For Exercises 1–5, determine the velocity and position of a particle given its acceleration and initial conditions.

1. $a(t) = 3t$, $v(0) = 2$, $x(0) = 0$.
2. $a(t) = \sin(2\pi t)$, $v(0) = 1$, $x(0) = 0$.
3. $a(t) = 3e^{-0.5t}$, $v(0) = 5$, $x(0) = 0$.
4. $a(t) = 1/t^2$, $v(1) = 0$, $x(1) = 0$.
5. $a(t) = -t(t-1)$, $v(0) = -1$, $x(0) = 1$.

For Exercises 6–10, compute the derivative of the indicated function.

6. $f(x) = \int_0^x s\sin(3s)\, ds$.
7. $f(x) = \int_1^x \ln|t|\, dt$.
8. $f(x) = \int_1^{\sqrt{x}} \frac{\cos(u)}{u}\, du$.
9. $f(x) = \int_x^{x^2} \csc(w)\, dw$.
10. $f(x) = \int_x^{x^3} \frac{\sin(u)}{u}\, du$.

Problems

11. A Boeing 747 accelerates down a runway at the rate 3 m/s^2.
    (a) Determine its position and velocity as a function of time.
    (b) If the aircraft's takeoff speed is 80 m/s, determine the length of runway that the aircraft used before it became airborne.
12. The velocity of a particle (measured in m/s) that is traveling along the x-axis is given by the *error function*:

$$v(t) = \operatorname{erf}(t) = \frac{2}{\sqrt{\pi}} \int_0^t e^{-x^2}\, dx$$

where $t \geq 0$ is time (measured in s). Some values of erf(t) are listed in the table below.

t	0.2	0.4	0.6	0.8	1.0
$\text{erf}(t)$	0.222703	0.428392	0.603856	0.742101	0.842701

(a) What is the initial velocity $v(0)$ of the particle? Include units.

(b) Let $I = \int_0^1 v(t)\,dt$. What is the physical meaning of I? Include units.

(c) Approximate the integral $\int_0^1 v(t)\,dt$ using a right-hand Riemann sum. State the number of subintervals that were used. Use only information about $\text{erf}(t)$ provided in the table above. Keep six digits in your answer and include units.

(d) Let L_{57} and R_{57} be the left- and right-hand Riemann sum approximations to $\int_0^1 v(t)\,dt$, with 57 subintervals, respectively. Is the following inequality necassarily true? Explain your reasoning.

$$L_{57} \le \int_0^1 v(t)\,dt \le R_{57}$$

(e) Find the acceleration of the particle $a(t) = v'(t)$.

(f) Find the particle's average acceleration over the interval $[0.4, 1.0]$.

13. Suppose that the function $P(t)$ models the productivity of employees of a certain company t hours after the start of a shift.

(a) Describe the meaning of the function

$$A(t) = \frac{1}{t}\int_0^t P(s)\,ds$$

in practical terms.

(b) Compute the derivative $A'(t)$ in terms of the functions A and P.

(c) Show that the second derivative may be expressed as

$$A''(t) = \frac{1}{t}\left[P'(t) + \frac{1}{t}(A - P)\right].$$

(d) Show that a necessary condition for the function A to have a critical point at time $t = t^*$ is $P(t^*) = A(t^*)$. Explain why this condition makes sense using practical terms.

(e) Use the second-derivative test to show that a critical point t^* of A is a local maximum only if $P'(t^*) < 0$.

Chapter 7

Techniques of Integration

7.1 Substitution Method

Our first advanced integration technique is a formalism of the integration rules given by equations (5.7)–(5.11). This technique is essentially a systematic way of undoing the chain rule. We first present the general rule for constructing antiderivatives, and then specialize to the application of the rule for definite integrals.

7.1.1 Substitution Method for Antiderivatives

We begin by first recalling the chain rule, which states

$$\frac{d}{dx}\left[f(u(x))\right] = f'(u(x))u'(x).$$

The substitution method is simply a reverse of the chain rule. In particular, we have the following.

Theorem 7.1.1 — Substitution Method. Given a function f and a differentiable function u, such that f has an antiderivative F, the following equality holds

$$\int f(u(x))u'(x)\,dx = F(u(x)) + C. \tag{7.1}$$

Proof. We can verify this result by differentiating the right-hand side of equation (7.1), using the chain rule. We obtain

$$\frac{d}{dx}[F(u(x))] = F'(u(x))u'(x).$$

Since this is the integrand on the left-hand side, the formula is verified. One could, alternatively, integrate the preceding equation, which is given by the chain rule, whilst noting that

$$\int \frac{d}{dx}[F(u(x))]\,dx = F(u(x)) + C.$$

■

(R) The factor $u'(x)\,dx$ in the integral on the left-hand side of equation (7.1) is often interpreted as simply "du," so that

$$du = u'(x)\,dx.$$

Furthermore, it is permissible to use a substitution as long as $u'(x)\,dx$ appears *up to a constant multiple*. Constant multiples are no barrier to the technique, as they may pass freely through the integral sign.

■ **Example 7.1** Compute

$$\int e^{x^2} x\,dx.$$

First, we identify the inside function as $u(x) = x^2$. We may then compute its differential, obtaining

$$du = 2x\,dx.$$

Up to a constant, this differential is contained within the integrand; we have thus identified

$$\int \underbrace{e^{x^2}}_{e^u} \underbrace{x\,dx}_{du/2}.$$

We may therefore "substitute" x^2 for u and $x\,dx$ for $du/2$, obtaining instead the integral

$$\frac{1}{2}\int e^u du = \frac{e^u}{2} + C.$$

Substituting the x^2 back in for u, we have our final answer

$$\int e^{x^2} x\,dx = \frac{e^{x^2}}{2} + C.$$

By differentiating the right-hand side, we can easily confirm that this result is correct. ■

■ **Example 7.2** Compute the value of the integral

$$\int x^3\sqrt{1+x^4}\,dx.$$

Here, we identify the inside function as $u(x) = 1 + x^4$. Its differential is

$$du = 4x^3\,dx.$$

This differential, up to a constant multiple, is sitting inside the integrand. We may therefore write

$$\int x^3\sqrt{1+x^4}\,dx = \int \sqrt{1+x^4}\,\frac{4}{4}x^3\,dx = \frac{1}{4}\int \underbrace{\sqrt{1+x^4}}_{\sqrt{u}}\,\underbrace{4x^3\,dx}_{du}$$

Thus,

$$\int x^3\sqrt{1+x^4}\,dx = \frac{1}{4}\int \sqrt{u}\,du = \frac{1}{6}u^{3/2} + C = \frac{1}{6}\sqrt{(1+x^4)^3} + C.$$

By differentiating the right-hand side—a very instructive practice—one sees exactly how the substitution rule is an undoing of the chain rule. ■

■ **Example 7.3** Consider the integral

$$\int \tan(x)\,dx = \int \frac{\sin(x)}{\cos(x)}\,dx.$$

Here, we may let $u(x) = \cos(x)$. Then the differential will be

$$du = -\sin(x)\,dx.$$

Thus,

$$\int \frac{\sin(x)}{\cos(x)}\,dx = -\int \frac{1}{u}\,du = -\ln|u| + C = \ln|\sec(x)| + C.$$

The last equality follows since $-\ln x = \ln x^{-1}$. ■

■ **Example 7.4** Next, we consider a famous example that requires a certain stroke of insight. Suppose we wish to determine the integral

$$\int \sec(x)\,dx.$$

The trick here is to multiply and divide by a factor of $\sec(x) + \tan(x)$, thereby obtaining

$$\int \sec(x)\,dx = \int \sec(x)\left(\frac{\sec(x)+\tan(x)}{\sec(x)+\tan(x)}\right)dx = \int \frac{\sec^2(x)+\sec(x)\tan(x)}{\sec(x)+\tan(x)}\,dx.$$

One may wonder why anyone would wish to complicate matters with such a satanic sprinkling of trigonometry; reservations quenched, however, when one recalls that

$$\frac{d}{dx}[\tan(x)] = \sec^2(x) \qquad \text{and} \qquad \frac{d}{dx}[\sec(x)] = \sec(x)\tan(x).$$

Identifying $u(x) = \sec(x) + \tan(x)$ and $du = \big(\sec(x)\tan(x) + \sec^2(x)\big)\,dx$, we immediately recover

$$\int \sec(x)\,dx = \int \frac{du}{u} = \ln|u| + C = \ln|\sec(x) + \tan(x)| + C.$$

We have therefore derived the useful integration formula

$$\int \sec(x)\,dx = \ln|\sec(x) + \tan(x)| + C. \tag{7.2}$$

■

7.1.2 Substitution Method for Definite Integrals

Theorem 7.1.1 can be extended to the case of definite integrals. Though the result is practically immediate, there is a subtlety in how to handle the limits of integration. Namely, the limits of integration must be appropriately transformed when transforming the integral from one over the x domain into one over the u domain. To make this concrete, we consider now the following theorem.

Theorem 7.1.2 — Substitution Method for Definite Integrals. Given a function f and a differentiable function u, such that f has an antiderivative F, the following equality holds

$$\int_a^b f(u(x))u'(x)\,dx = \int_{u(a)}^{u(b)} f(u)\,du = F(u(b)) - F(u(a)). \tag{7.3}$$

The substitution method for definite integrals is substantially similar to the one for computing antiderivatives, only one must also take care to exchange the limits of integration for the appropriate ones pursuant to the new domain. That is, if the variable x is integrated from a to b, then the variable u should be integrated between $u(a)$ and $u(b)$, the corresponding domain points.

■ **Example 7.5** Compute the value of the definite integral

$$\int_0^{\pi/2} e^{\sin(x)}\cos(x)\,dx.$$

Here, we identify the inside function as $u(x) = \sin(x)$, so that $du = \cos(x)\,dx$. Furthermore, we compute the new limits of integration as

$$u(0) = \sin(0) = 0 \qquad \text{and} \qquad u(\pi/2) = \sin(\pi/2) = 1.$$

Whereas the x-domain is the interval $[0,\pi/2]$, the u-domain is the interval $[0,1]$. We therefore obtain

$$\int_0^{\pi/2} e^{\sin(x)}\cos(x)\,dx = \int_0^1 e^u\,du = e^u\Big|_{u=0}^{u=1} = e^1 - e^0 = e - 1.$$

Notice that without transforming the limits of integration, we would have obtained the incorrect answer. ■

■ **Example 7.6** Compute the value of the definite integral

$$\int_0^1 x\sqrt{1-x^2}\,dx.$$

Again, we first identify the inside function as $u = 1 - x^2$, with its differential $du = -2x\,dx$. Computing the new limits of integration, we obtain

$$u(0) = 1 \qquad \text{and} \qquad u(1) = 0.$$

Therefore,

$$\int_0^1 x\sqrt{1-x^2}\,dx = \frac{-1}{2}\int_1^0 \sqrt{u}\,du = \frac{1}{2}\int_0^1 \sqrt{u}\,du = \left.\frac{u^{3/2}}{3}\right|_{u=0}^{u=1} = \frac{1}{3}.$$

Notice we used the integral property (6.14) to achieve the second equality in the preceding equation chain. Thus, the negative sign in the du worked in conjunction with swapping the limits of integration in order to achieve the correct result. ▪

▪ **Example 7.7** Compute the value of the definite integral

$$\int_0^{\pi/3} \tan(x)\,dx.$$

As before, we use the substitution $u = \cos(x)$ and $du = -\sin(x)\,dx$. Converting the limits of integration, we find

$$u(0) = \cos(0) = 1 \qquad \text{and} \qquad u(\pi/3) = \cos(\pi/3) = 0.5.$$

Thus,

$$\int_0^{\pi/3} \tan(x)\,dx = -\int_1^{0.5} \frac{du}{u} = \int_{0.5}^1 \frac{du}{u} = \ln|u|\Big|_{u=0.5}^{u=1} = \ln(1) - \ln(0.5) = \ln(2).$$

Here, we used the fact that $-\ln(0.5) = \ln(2)$. ▪

7.1.3 Using an Integral Table

The substitution method can also be applied in conjunction with a *table of integrals*, such as the one given in Appendix D. A table of integrals is essentially a listing of known antiderivatives that people have figured out in the past. Oftentimes, these antiderivatives are not used directly, but are applied using the substitution method.

▪ **Example 7.8** Compute the value of the integral

$$\int_0^{\pi} e^{-x}\sin(2x)\,dx.$$

Here, we wish to use the integration formula (D.40), with the choice of parameters $a = -1$ and $b = 2$. In this case, the substitution of the independent variable is trivial: $u = x$, $du = dx$. With the aid of this integration formula, we therefore obtain

$$\begin{aligned}
\int_0^{\pi} e^{-x}\sin(2x)\,dx &= \left.\frac{e^{-x}}{1+2^2}\left(-\sin(2x) - 2\cos(2x)\right)\right|_{x=0}^{x=\pi} \\
&= \frac{-e^{-\pi}}{5}\left(\sin(2\pi) + 2\cos(2\pi)\right) + \frac{1}{5}\left(\sin(0) + 2\cos(0)\right) \\
&= \frac{2}{5} - \frac{2e^{-\pi}}{5} = \frac{2}{5}\left(1 - e^{-\pi}\right) \approx 0.38271.
\end{aligned}$$

This completes the exercise. ▪

Example 7.9 Compute the value of the integral

$$\int_0^1 \frac{e^x}{\sqrt{9-e^{2x}}}\,dx.$$

For this integral, we can use equation (D.13), identifying $u = e^x$ and $du = e^x\,dx$. (Note that $u^2 = e^{2x}$.) We also identify the parameter a in this integration formula with the value $a = 3$. The new limits of integration are given by

$$u(0) = e^0 = 1 \qquad \text{and} \qquad u(1) = e^1 = e.$$

Therefore,

$$\begin{aligned}\int_0^1 \frac{e^x}{\sqrt{9-e^{2x}}}\,dx &= \int_1^e \frac{du}{\sqrt{9-u^2}} \\ &= \sin^{-1}\left(\frac{u}{3}\right)\Bigg|_{u=1}^{u=e} = \sin^{-1}\left(\frac{e}{3}\right) - \sin^{-1}\left(\frac{1}{3}\right) \approx 0.79412.\end{aligned}$$

The second line comes from the integration formula (D.13). ∎

Example 7.10 Determine the value of the definite integral

$$\int_0^{\pi/6} \frac{\cos x}{1-\sin^2 x}\,dx.$$

Here, we identify $u = \sin(x)$ and $du = \cos(x)\,dx$, and then utilize formula (D.11). The new limits of integration are

$$u(0) = 0 \qquad \text{and} \qquad u(\pi/6) = 0.5.$$

Therefore,

$$\int_0^{\pi/6} \frac{\cos x}{1-\sin^2 x}\,dx = \int_0^{0.5} \frac{du}{1-u^2} = \frac{1}{2}\ln\left|\frac{1+u}{1-u}\right|\Bigg|_{u=0}^{u=0.5} = \frac{\ln(3)}{2}.$$

∎

For each of the examples in this section, notice that our choice of the inside function u is usually a part of a larger function. Moreover, in all cases, the derivative of the function $u = u(x)$ *must*, up to a multiplicative constant, be sitting directly in the integrand. Otherwise there is no du that will make the substitution work.

Exercises

For Exercises 1–10, use the substitution method to evaluate the given integral.

1. $\int x\sin(x^2)\,dx.$
2. $\int \sin(\sin(x))\cos(x)\,dx.$
3. $\int \frac{x^2}{1+x^3}\,dx.$
4. $\int \sin(x)\cos(x).$

5. $\int e^x\sqrt{1+e^x}\,dx.$

6. $\int \tan(x)\sec^2(x)\,dx.$

7. $\int \sin^2(x)\cos(x)\,dx.$

8. $\int \cos^3(x)\,dx.$

9. $\int \frac{\ln|x|}{x}\,dx.$

10. $\int \frac{\ln(x^2)}{x}\,dx.$

For Exercises 11–20, use the substitution method to evaluate the given definite integral.

11. $\int_0^{\pi} x\sin(x^2)\,dx.$

12. $\int_0^{\pi/4} \sec^2(x)e^{\tan(x)}\,dx.$

13. $\int_0^{\pi/2} \sin(x)\cos(x)\,dx.$

14. $\int_0^{\pi/2} \cos(x)\sqrt{1-\sin(x)}\,dx.$

15. $\int_0^1 \frac{3x^2+2}{x^3+2x+1}\,dx.$

16. $\int_0^3 \frac{x}{\sqrt{1+x^2}}\,dx.$

17. $\int_{\pi/4}^{\pi/3} \cot(x)\,dx.$

18. $\int_{\pi/6}^{\pi/3} \frac{\sec^2(x)}{\tan(x)}\,dx.$

19. $\int_1^e \frac{\ln|x|}{x}\,dx.$

20. $\int_1^e \frac{(\ln|x|)^2}{x}\,dx.$

Problems

21. Suppose that a 100 g sample of the radioisotope silicon-31 is prepared and that t minutes later the sample is decaying at a rate of

$$R(t) = 0.44e^{-0.0044t}\text{ g/min.}$$

Determine how much of the sample is remaining after 100 minutes.

22. Data for the functions f and g are shown in Table 7.1.

x	0	1	2	3	4	5
$f(x)$	2	3	11	19	21	27
$g(x)$	3	2	4	7	6	5

Table 7.1: Data from function f; Problem 22

(a) Determine the value of $\int_0^5 f'(g(x))g'(x)\,dx.$

(b) Determine the value of $\int_0^1 g'(f(x))f'(x)\,dx.$

23. In this problem, we will study *the rocket equation*. In particular, suppose we have a rocket with a velocity v, such that the cross-sectional area of the exhaust nozzle is A_e, and the velocity of the exhaust gas relative to the rocket is v_e, as shown in Figure 7.1.

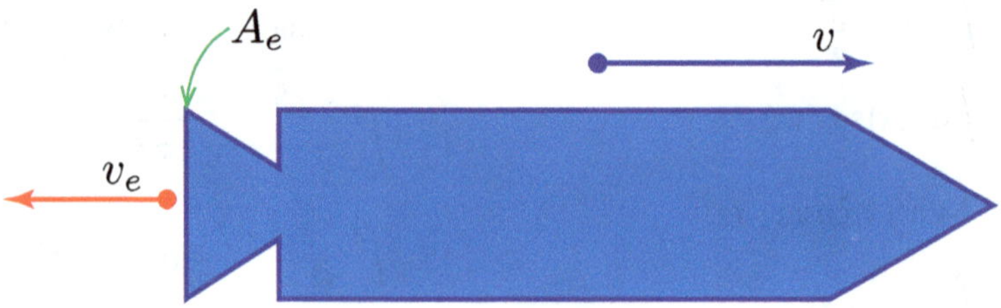

Figure 7.1: Schematic of a rocket.

Assuming the rocket is in free space, in the absence of gravity or pressure forces, the *momentum equation*[1] can be used to derive the equation

$$m\frac{dv}{dt} = -v_e\frac{dm}{dt}.$$

(Note that $m = m(t)$ and $v = v(t)$, i.e., both the rocket's mass and velocity are changing in time; v_e is a constant.)

(a) Explain why

$$\int_{t_0}^{t_1} v'(t)\,dt = -v_e\int_{t_0}^{t_1}\frac{m'(t)}{m}\,dt.$$

(b) Use the substitution method to integrate the preceding equation, thereby deriving *the rocket equation*

$$\Delta v = v_e \ln\left(\frac{m_0}{m_1}\right), \tag{7.4}$$

where $\Delta v = v(t_1) - v(t_0)$, $m_0 = m(t_0)$, and $m_1 = m(t_1)$.

(c) Suppose that a rocket with an exhaust velocity of $v_e = 4{,}400$ m/s is designed to accelerate from rest to escape velocity, which is approximately 11,200 m/s (i.e., approximately 25,000 mph). What fraction of its launch mass must be fuel?

(d) Suppose a Δv of 23.2 km/s is required to send astronauts to Mars, consisting of $\Delta v_1 = 7.9$ km/s launch to low-Earth orbit, $\Delta v_2 = 6.6$ km/s low-Earth orbit to transfer orbit, $\Delta v_3 = 7.2$ km/s injection into Martian orbit, and $\Delta v_4 = 1.5$ km/s landing on Mars. Further, suppose that the return trip costs $\Delta v = 15.3$ km/s (the cost of going from low-Earth orbit to landing is achieved by utilizing pressure forces during atmospheric reentry and a parachute). Assuming an exhaust velocity of

[1] The momentum equation for fluid mechanics is

$$\iiint_\Omega \rho\mathbf{f}\,dV - \iint_{\partial\Omega} p\,d\mathbf{S} = \iiint_\Omega \frac{\partial(\rho\mathbf{v})}{\partial t}\,dV + \iint_{\partial\Omega} \rho\mathbf{v}\mathbf{v}\cdot d\mathbf{S}.$$

This is the general form of Newton's second law.

$v_e = 4.0$ km/s, how much fuel would be required to launch a 5,000 kg capsule from Earth and land it on Mars? What if we also want to bring the capsule back?

24. The gravitational force that a massive disk of radius a exerts on a point P that is a distance d away is given by the formula

$$F(d) = 2\pi G\sigma \int_0^a \frac{x}{x^2+d^2}\,dx,$$

where G is the gravitational constant and σ is the disk's constant mass density (mass per unit area). The system is shown in Figure 7.2. Evaluate

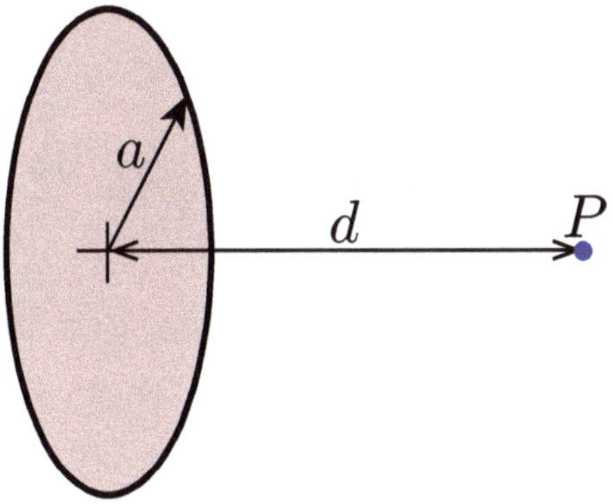

Figure 7.2: Gravitational field of a disk; Problem 24.

this integral to determine the strength of the gravitational field as a function of the distance between the point P and the center of the disk.

25. In physics, a *conservative force field* is a force that depends on a position variable x and that is the exact (negative) derivative of a function U called the *potential energy*, so that

$$F(x) = -U'(x).$$

(a) Newton's second law of motion for a particle with constant mass states that

$$F = mx''(t),$$

or, alternatively, $F = mv'(t)$, where $x(t)$ is a particle's position at time t, m is the mass of the particle, and $v(t) = x'(t)$ is the particle's velocity. If we assume that the velocity can be expressed as a function of position if the forcing function is a conservative force field, explain why the chain rule can allow us to write

$$mvv'(x) = -U'(x),$$

where $v'(x)$ is the derivative of the velocity *with respect to the position.*

(b) Integrate both sides of the preceding equation with respect to x from $x = x_1$ to $x = x_2$. Assume that $v(x_1) = v_1$ and $v(x_2) = v_2$. Derive the equation

$$\frac{1}{2}mv_2^2 + U(x_2) = \frac{1}{2}mv_1^2 + U(x_1).$$

This relation is known as the *law of conservation of energy*. The term $\frac{1}{2}mv^2$ is the *kinetic energy* of the particle.

We note that there are many examples of conservative force fields in physics, such as the restoring force of a spring, a constant gravitational field, Newton's law of gravitation, electric fields, and so forth.

7.2 Integration by Parts

As the substitution method essentially "undoes" the chain rule, in this section we shall learn a technique that acts as counterpart to the product rule for differentiation, and this technique is called *integration by parts*. This technique will enable us to compute integrals of *products* of simpler functions, in the same way that the product rule allowed us to compute derivatives of products of functions.

7.2.1 Integration by Parts for Antiderivatives

We begin by considering the case of indefinite integrals. As before, we will use a notation to ease some of the computational complexity of the problem. When employing the substitution method, we identified an inner function u and its differential du. Here, we will add to this ensemble with the introduction of a second function v and its differential dv; thus there are four key constituents to any integration-by-parts problem. The mechanics of the technique are contained in the following theorem.

Theorem 7.2.1 — Integration by Parts. Consider a differentiable function u and an exact differential dv. Then

$$\int u\,dv = uv - \int v\,du. \tag{7.5}$$

The preceding equation is well worth memorizing. This at-first seemingly unwieldy concoction of symbolism turns to to be, quite to the contrary, a welcomed friend who comes in aid; it is a condensation of a complex idea into a simple form. It should be noted that u and v themselves are to be regarded as functions of a common independent variable, usually x, the same as their differentials $du = u'(x)dx$ and $dv = v'(x)dx$, which also are capped with a dx. Taking for granted the usage of this notation for a moment, even if it is still somewhat opaque, we first prove the integration-by-parts formula; we will then turn to examples that will illuminate any present obscurity of what is meant here.

Proof. Suppose that u and v are both differentiable functions of a single independent variable x. Then, the product rule tells us that

$$\frac{d}{dx}[u(x)v(x)] = u(x)v'(x) + v(x)u'(x).$$

Rearranging, we may write

$$u(x)v'(x) = \frac{d}{dx}[u(x)v(x)] - v(x)u'(x).$$

Integrating this formula with respect to x, we obtain

$$\begin{aligned}\int u(x)v'(x)\,dx &= \int \frac{d}{dx}[u(x)v(x)]\,dx - \int v(x)u'(x)\,dx \\ &= u(x)v(x) - \int v(x)u'(x)\,dx.\end{aligned}$$

Identifying $du = u'(x)\,dx$ and $dv = v'(x)\,dx$, we achieve our result. ■

The proof sheds some light on the use of our notation:

$$\int \underbrace{u(x)}_{u}\underbrace{v'(x)\,dx}_{dv} = \underbrace{u(x)v(x)}_{uv} - \int \underbrace{v(x)}_{v}\underbrace{u'(x)\,dx}_{du}. \tag{7.6}$$

In this way, the *u*-*v*-*du*-*dv* notation acts as a shorthand.

(R) Essentially, integration by parts allows us to trade an integral of $u\,dv$ for an integral of $v\,du$. Thus, we start off with a product of two functions, and we differentiate one of them and integrate the other, in hopes of arriving at a simpler integral.

The function u is therefore typically chosen as whatever part of the integrand that would become simpler if differentiated. The differential dv is usually selected as a part of the integrand that has a known antiderivative, by which the function v may be computed.

■ **Example 7.11** Compute the integral

$$\int xe^x\,dx.$$

Here, we choose $u = x$ and $dv = e^x\,dx$. Differentiating u and integrating dv, we have

$$du = dx \qquad \text{and} \qquad v = e^x.$$

We thus arrive at

$$\int \underbrace{x}_{u}\underbrace{e^x\,dx}_{dv} = \underbrace{x}_{u}\underbrace{e^x}_{v} - \int \underbrace{e^x}_{v}\underbrace{dx}_{du}.$$

Therefore,

$$\int xe^x\,dx = xe^x - e^x + C.$$

Notice that had we instead selected $u = e^x$ and $dv = x$, we would have created a more complicated integral for ourselves, as the integral of xe^x would have been replaced with an integral of x^2e^x, the opposite way as the way toward simplification.
■

■ **Example 7.12** Use integration by parts to determine a formula for the integral

$$\int \ln|x|dx.$$

At first glance, this does not appear as an integration-by-parts problem. Integration by parts turns out, nevertheless, to aid in this computation, as we may set $u = \ln|x|\,dx$ and $dv = dx$, so that

$$du = \frac{1}{x}\,dx \qquad \text{and} \qquad v = x.$$

We therefore obtain

$$\int \underbrace{\ln|x|}_{u} \underbrace{dx}_{dv} = \underbrace{\ln|x|}_{u} \underbrace{x}_{v} - \int \underbrace{x}_{v} \underbrace{\frac{1}{x}dx}_{du}.$$

We have therefore recovered the integration formula

$$\int \ln|x| \, dx = x\ln|x| - x + C, \tag{7.7}$$

which agrees with equation (D.17). ∎

Example 7.13 Use integration by parts to determine

$$\int \sin^2(\theta) \, d\theta.$$

Here, we may set $u = \sin(\theta)$ and $dv = \sin(\theta) \, d\theta$. Differentiating u and integrating v, we have

$$du = \cos(\theta) \, d\theta \qquad \text{and} \qquad v = -\cos(\theta),$$

respectively. We thus obtain

$$\begin{aligned}
\int \sin^2(\theta) &= \underbrace{-\sin(\theta)\cos(\theta)}_{uv} - \int \underbrace{-\cos(\theta)}_{v}\underbrace{\cos(\theta) \, d\theta}_{du} \\
&= -\sin(\theta)\cos(\theta) + \int \cos^2(\theta) \, d\theta \\
&= -\sin(\theta)\cos(\theta) + \int \left(1 - \sin^2(\theta)\right) \, d\theta \\
&= -\sin(\theta)\cos(\theta) + \theta - \int \sin^2(\theta) \, d\theta.
\end{aligned}$$

Therefore,

$$\int \sin^2(\theta) = \frac{\theta - \sin(\theta)\cos(\theta)}{2} + C.$$

This expression is equivalent to equation (D.23), as is apparent when one considers the double-angle identity $\sin(2x) = 2\sin(x)\cos(x)$. ∎

Sometimes a function may require several applications of integration by parts before an answer is obtained, as we will observe in our next example.

Example 7.14 Use integration by parts to compute the integral

$$\int e^x \sin(x) \, dx.$$

To set this up, let us choose $u = e^x$ and $dv = \sin(x) \, dx$. (We encourage the reader to try this the other way, setting u to $\sin(x)$ and dv to $e^x \, dx$; the following procedure will still work.) We thus have

$$du = e^x \, dx \qquad \text{and} \qquad v = -\cos(x).$$

Applying integration by parts, we therefore obtain

$$\int \underbrace{e^x}_{u} \underbrace{\sin(x)\,dx}_{dv} = \underbrace{-e^x\cos(x)}_{uv} - \int \underbrace{-\cos(x)}_{v}\underbrace{e^x\,dx}_{du}.$$

Sans annotation, this reads

$$\int e^x \sin(x)\,dx = -e^x\cos(x) + \int e^x\cos(x)\,dx. \tag{7.8}$$

Was our experiment a success? At first, it seems as though we are no better off than before: certainly the integral of $e^x\cos(x)$ is just as confounding and elusive as the integral of $e^x\sin(x)$. The preceding one is still a valid equation, since we correctly applied integration by parts, though it at first doesn't seem to have helped us. As the process of discovery is often one of trial and experiment, let us brave a *second* application of integration by parts—applied to the integral on the right-hand side—just to see where it takes us.

To apply integration by parts, a second time, to the integral on the right-hand side of equation (7.8), we need to reset our choice of u and dv. For the new integral—$\int e^x\cos(x)\,dx$—we may select $u = e^x$ and $dv = \cos(x)\,dx$. By differentiating and integrating this selection, we find

$$du = e^x\,dx \qquad \text{and} \qquad v = \sin(x).$$

Therefore,

$$\int \underbrace{e^x}_{u} \underbrace{\cos(x)}_{dv} = \underbrace{e^x\sin(x)}_{uv} - \int \underbrace{\sin(x)}_{v}\underbrace{e^x\,dx}_{du}.$$

Substituting this result into equation (7.8), we obtain

$$\int e^x\sin(x)\,dx = -e^x\cos(x) + e^x\sin(x) - \int e^x\sin(x)\,dx.$$

At first it seems our ending is despair; but then, behold, what if we *add* our unknown integral $e^x\sin(x)\,dx$ to both sides of the equation? We obtain

$$2\int e^x\sin(x)\,dx = e^x(\sin(x) - \cos(x)).$$

Our pursuit was not fruitless after all, for we have reached our desired end:

$$\int e^x\sin(x)\,dx = \frac{e^x(\sin(x)-\cos(x))}{2} + C.$$

This is really a rather amazing result, obtained by implementing a few simple steps. Using integration by parts, we have thus obtained an integration formula that we probably would have never divined by insight alone. ∎

7.2.2 Integration by Parts for Definite Integrals

When generalizing integration by parts to definite integrals, we must be slightly more proper with our use of notation in order to clearly represent which variable the limits of integration correspond to. Therefore, we write our formula more in lines with equation (7.6) than with the shorthand offered by equation (7.5). The result is stated as follows.

Theorem 7.2.2 — Integration by Parts for Definite Integrals. Consider a differentiable function u and an exact differential dv. Then

$$\int_a^b u(x)\, v'(x)dx = u(x)v(x)\Big|_{x=a}^{x=b} - \int_a^b v(x)u'(x)\, dx. \tag{7.9}$$

This really is just plain old integration by parts; the only difference, however, is that we are now evaluating the result between the appropriate limits of integration.

■ **Example 7.15** Determine the value of

$$\int_0^{\pi/2} x\cos(x)\, dx.$$

As usual, we identify $u = x$ and $dv = \cos(x)\, dx$. Therefore,

$$du = dx \qquad \text{and} \qquad v = \sin(x).$$

We thus have

$$\begin{aligned}\int_0^{\pi/2} \underbrace{x}_{u(x)}\, \underbrace{\cos(x)}_{v'(x)}\, dx &= \underbrace{x\sin(x)}_{u(x)v(x)}\Big|_{x=0}^{x=\pi/2} - \int_0^{\pi/2} \underbrace{\sin(x)}_{v(x)}dx \\ &= (x\sin(x)+\cos(x))\Big|_{x=0}^{x=\pi/2} = \frac{\pi}{2} - 1,\end{aligned}$$

and our result is obtained. ■

7.2.3 Table Method

In this paragraph, we will explore a rare way to systematize integration by parts for multiple repetitions. We begin with a theorem, which will lay out the theoretical underpinnings of the method, and then dispense the technique.

Theorem 7.2.3 Suppose that f eventually differentiates to zero, and that k is the smallest integer such that $f^{(k)}(x) = 0$. Then

$$\int f(x)g(x)\, dx = fg^{\wedge} - f'g^{\wedge\wedge} + f''g^{\wedge\wedge\wedge} - \cdots + (-1)^{k-1}f^{(k-1)}g^{\wedge k}, \tag{7.10}$$

where $g^{\wedge}$ is an antiderivative of g; $g^{\wedge\wedge}$ is an antiderivative of $g^{\wedge}$, and so forth, until $g^{\wedge k}$ is the kth antiderivative of g.

(R) The "g-wedge" notation is not standard; we will only use it momentarily as we have a need to represent multiple antiderivative of the function g, and as no common notation is preapproved for such a use.

Proof. Theorem 7.2.3 is simply the result of a repeated application of integration by parts. To begin, let us identify

$$u = f, \qquad dv = g\,dx, \qquad du = f'\,dx, \qquad v = g^{\wedge}.$$

Applying integration by parts, we therefore obtain

$$\int fg\,dx = fg^{\wedge} - \int f'g^{\wedge}dx.$$

We are closer to having solved this integral, as the function f eventually differentiates to zero. Applying integration by parts again, we have

$$u = f', \qquad dv = g^{\wedge}\,dx, \qquad du = f''\,dx \qquad v = g^{\wedge\wedge}.$$

Thus,

$$\int fg\,dx = fg^{\wedge} - f'g^{\wedge\wedge} + \int f''g^{\wedge\wedge}\,dx.$$

Continuing in this manner, we eventually arrive at

$$\int fg\,dx = fg^{\wedge} - f'g^{\wedge\wedge} + \cdots + (-1)^{k-1}f^{(k-1)}g^{\wedge k} + (-1)^k \int f^k g^{\wedge k}\,dx.$$

But $f^k = 0$, so $\int f^k g^{\wedge k}\,dx = C$, and the result is proved. ■

The Table Method

We may apply Theorem 7.2.3 to create a systematic method for efficaciously computing multiple applications of integration by parts. To begin the table method, let us suppose that we have the conditions of Theorem 7.2.3, i.e., we are given a function f that eventually differentiates to zero, and that k is the smallest integer such that $f^{(k)}(x) = 0$. Further, we wish to calculate

$$\int f(x)g(x)\,dx.$$

To achieve this, we begin by writing down a table with three columns, labeled S for *sign*, D for *derivative*, and I for *integral*. In the second row, we start with the entries "+", $f(x)$, and $g(x)$, as shown below.

S	D	I
+	f	g

Next, we complete the table as follows. First, the entries in the S-column alternate sign: +, −, +, −, and so on. Each entry in the D-column is the derivative of the corresponding entry that precedes it; similarly, each entry in the I-column is an antiderivative of the corresponding entry that precedes it. We continue this until the k-th row, in which f^k is identified with zero. We then have something like this:

S	D	I
+	f	g
−	f'	$g^{\wedge}$
+	f''	$g^{\wedge\wedge}$
⋮	⋮	⋮
⋮	$f^{(k-1)}$	$g^{\wedge(k-1)}$
⋮	0	$g^{\wedge k}$

Next, we draw a series of lines through these entries, in the manner shown in Figure 7.3. Each line contains one entry from each of the three columns. By

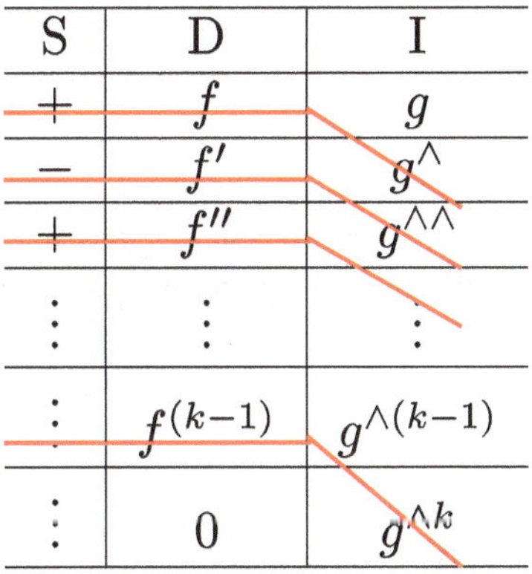

Figure 7.3: Schematic of the table method.

multiplying everything connected by each line, and adding the results together, we recover the formula expressed in equation (7.10).

■ **Example 7.16** Determine the value of the integral

$$\int x^4 \sin(2x)\, dx.$$

To do this, we first set up the table method:

S	D	I
+	x^4	$\sin(2x)$

The first entry in the D-column must be something that will eventually differentiate to zero; the factor x^4 does the trick here. Next, we simply fill out the rest of the table until we reach a zero in the D-column:

S	D	I
+	x^4	$\sin(2x)$
−	$4x^3$	$-\frac{1}{2}\cos(2x)$
+	$12x^2$	$-\frac{1}{4}\sin(2x)$
−	$24x$	$\frac{1}{8}\cos(2x)$
+	24	$\frac{1}{16}\sin(2x)$
−	0	$-\frac{1}{32}\cos(2x)$

We may now read the answer straight out:

$$\begin{aligned}\int x^4 \sin(2x)\,dx &= -\frac{1}{2}x^4\cos(2x)+x^3\sin(2x)+\frac{3}{2}x^2\cos(2x) \\ &\quad -\frac{3}{2}x\sin(2x)-\frac{3}{4}\cos(2x)+C.\end{aligned}$$

By differentiating the right-hand side, using the product rule on four of the terms, we recover the integrand $x^4\sin(2x)$. The table method has therefore allowed us to apply four separate instances of integration by parts in relatively short order, by simply making a table and differentiating and integrating the base functions a few times. ∎

Exercises

For Exercises 1–10, use integration by parts to compute the indicated integral.

1. $\int x\sin(x)\,dx.$
2. $\int x\ln x\,dx.$
3. $\int \cos^2(x)\,dx.$
4. $\int x\sec^2(x)\,dx.$
5. $\int xe^{5x}.$
6. $\int \sqrt{x^2+x^3}\,dx.$
7. $\int x^2\sin(x)\,dx.$
8. $\int x^2e^{3x}\,dx.$
9. $\int e^x\cos(x).$
10. $\int x^3e^{x^2}\,dx.$

For Exercises 11–20, use integration by parts to determine the value of the definite integral.

11. $\int_0^{\pi/2} x\sin(2x)\,dx.$
12. $\int_0^1 xe^{-x}\,dx.$

13. $\int_0^{\pi} \sin^2(x)\, dx.$

14. $\int_0^1 x^5 \ln(x)\, dx.$

15. $\int_1^e \ln(x)\, dx.$

16. $\int_0^{\sqrt[3]{\pi}} x^5 \sin(x^3)\, dx.$

17. $\int_0^{\pi/2} e^x \sin(2x)\, dx.$

18. $\int_0^{\pi/2} (1-x^2)\sin(x)\, dx.$

19. $\int_0^1 x^3 e^{-x^2}\, dx.$

20. $\int_0^{\pi/6} \cos(x)\ln(\sec(x)+\tan(x))\, dx.$

Problems

21. Consider two functions f and g. Some values for function f appear in Table 7.2, and a graph of function g is shown in Figure 7.4. The function f also has the property that its derivative f' is an even function.

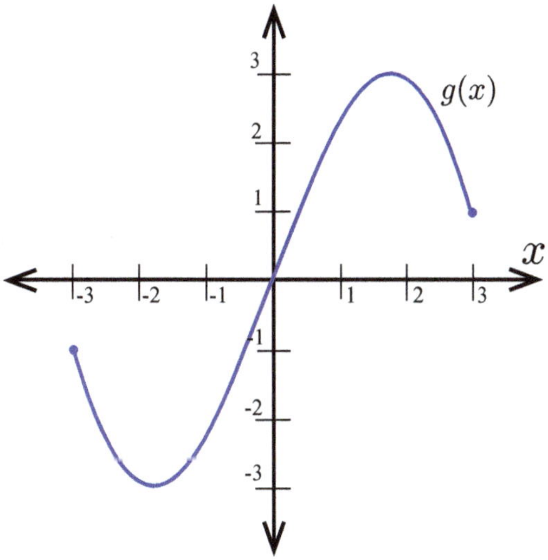

Figure 7.4: The function g; Problem 21.

x	-3	-2	-1	0	1	2	3
$f(x)$	8	9	11	16	21	23	24

Table 7.2: Data from function f; Problem 21

(a) Assuming $f'(x)$ is even, explain why $\int_{-3}^{3} g(x)f'(x)\, dx = 0$.

(b) Using Integration by Parts, find the exact value of $\int_{-3}^{3} f(x)g'(x)\, dx$.

(c) Using u-substitution, find the exact value for $\int_{-3}^{3} f'(g(x))g'(x)\, dx$.

7.3 Partial Fractions

Partial fractions is an algebraic technique that can be used to simplify certain rational functions into a sum of simpler rational functions. Its use in integration derives from the fact that those several simpler rational functions are usually more amenable to basic integration techniques and therefore easier to work with.

7.3.1 Overview: Integration of Rational Functions

In this section, we will concern ourselves with integrals of the form

$$\int \frac{P(x)}{Q(x)}\,dx,$$

where the functions P and Q are polynomials. The partial-fraction decomposition allows us to rewrite integrals of this form as a sum of simpler integrals, based on the factorization of the denominator polynomial Q. In particular, suppose that the function Q has polynomial factors f and g, such that $Q(x) = f(x)g(x)$. Then the partial-fraction decomposition of the function $1/Q$ may look something like this:

$$\frac{1}{f(x)g(x)} = \frac{A}{f(x)} + \frac{B}{g(x)},$$

for a certain choice of constants A and B. Here, each term on the right-hand side is a *partial fraction* of the left-hand side. The whole of the right-hand side is known as the *partial-fraction decomposition* of the left-hand side. (The partial-fraction decomposition may, in general, consist of more than two terms.) The reason the technique works is that both sides of this equation share the same common denominator. The right-hand side is therefore equivalent to

$$\frac{Ag(x)+Bf(x)}{f(x)g(x)}.$$

By comparing this expression to the left-hand side, we may deduce the values of the constants A and B that will make the decomposition work. In particular, we compare corresponding powers in the numerator of both sides of the equation one at a time. Thus, the general procedure may be outlined as follows:

1. Determine the correct form for the partial-fraction decomposition. (We will discuss how this is done for a number of different cases throughout the rest of the section.) Each partial fraction in this form has an unknown constant coefficient, which must be solved for.
2. Combine each of the partial fractions into one expression by finding the common denominator. The common denominator must equal the denominator of the original expression.
3. Compare the numerators from the original rational function and from the resulting rational function from step 2. In particular, equate the coefficients of the corresponding powers of these expressions.

4. Use the equalities obtained from the comparisons in step 3 to solve for the unknown constant coefficients.
5. Replace the constant coefficients in the general form from step 1 with the values obtained in step 4. The result is the partial-fraction decomposition of the original function.

7.3.2 Simple Linear Factors

We begin by considering rational functions with denominators that can be factored using linear binomials.

Proposition 7.3.1 Suppose the function P is a linear function or a constant and that the quadratic function Q may be factored as $Q(x) = (x-a)(x-b)$. Then the partial-fraction decomposition of P/Q may be written in the form

$$\frac{P(x)}{(x-a)(x-b)} = \frac{A}{x-a} + \frac{B}{x-b},$$

for a certain choice of the constants A and B.

■ **Example 7.17** Compute the integral

$$\int \frac{x}{x^2+x-12}\,dx.$$

As partial fractions is an algebraic technique used to simplify the integrand, we begin by factoring the denominator in the integrand as

$$\frac{x}{x^2+x-12} = \frac{x}{(x-3)(x+4)}.$$

Since the denominator can be factored, we may proceed with computing its partial-fractions decomposition. To achieve this, we assume that, for a certain choice of constants A and B, this fraction may be written as a sum

$$\frac{x}{(x-3)(x+4)} = \frac{A}{x-3} + \frac{B}{x+4}.$$

The rationale behind this assumption is that the common denominator of the two terms on the right-hand side is simply the denominator on the left-hand side. Our task is to determine the values of A and B that will make this expression true. To do this, we go back the other way: we simplify the right-hand side by finding a common denominator:

$$\frac{A}{x-3} + \frac{B}{x+4} = \frac{A(x+4)}{(x-3)(x+4)} + \frac{B(x-3)}{(x-3)(x+4)} = \frac{(A+B)x+(4A-3B)}{x^2+x-12}.$$

Equating this back to the original expression, we may restate the problem as follows: determine A and B so that the equality

$$\frac{x}{x^2+x-12} = \frac{(A+B)x+(4A-3B)}{x^2+x-12}$$

holds. This is easily solved by comparing the coefficient of each power of x in the numerator individually. In particular, we require that

$$A+B=1 \qquad \text{and} \qquad 4A-3B=0.$$

A little algebra shows that these equations are satisfied for the choice

$$A=\frac{3}{7} \qquad \text{and} \qquad B=\frac{4}{7}.$$

Thus, we have proven the partial-fraction decomposition

$$\frac{x}{(x-3)(x+4)}=\frac{3/7}{x-3}+\frac{4/7}{x+4}.$$

Our original integral is therefore solved by writing

$$\int \frac{x}{(x-3)(x+4)}\,dx=\int \frac{3/7}{x-3}+\frac{4/7}{x+4}\,dx=\frac{3}{7}\ln|x-3|+\frac{4}{7}\ln|x+4|+C.$$

■

Proposition 7.3.1 may be generalized to the case of multiple linear factors as follows.

Proposition 7.3.2 Suppose that the nth degree polynomial Q has n distinct linear factors, so that

$$Q(x)=\prod_{i=1}^{n}(x-a_i)=(x-a_1)(x-a_2)\cdots(x-a_n),$$

where each of the constants $\{a_1,\ldots,a_n\}$ is unique (i.e., the polynomial Q has n real, distinct roots, none of which are repeated roots). Further suppose that the polynomial P has a degree no greater than $(n-1)$. Then the partial-fraction decomposition of P/Q may be written in the form

$$\frac{P(x)}{Q(x)}=\sum_{i=1}^{n}\frac{A_i}{x-a_i}=\frac{A_1}{x-a_1}+\cdots+\frac{A_n}{x-a_n},$$

for a certain choice of the constants $\{A_1,\ldots,A_n\}$.

(R) The symbol $\prod$ (capital pi) is used for *product notation*. It is used similarly to $\sum$ in summation notation, except individual terms are multiplied together instead of added. For example, the factorial may be expressed in product notation as

$$n!=\prod_{i=1}^{n} i=1\cdot 2\cdots n.$$

■ **Example 7.18** Compute the integral

$$\int \frac{x^2+x+4}{x^3+2x^2-5x-6}\,dx.$$

We begin by factoring the denominator:

$$\frac{x^2+x+4}{x^3+2x^2-5x-6} = \frac{x^2+x+4}{(x+1)(x-2)(x+3)}.$$

Given this factorization, we assume that this rational function ha the partial-fraction decomposition

$$\frac{x^2+x+4}{x^3+2x^2-5x-6} = \frac{A}{x+1} + \frac{B}{x-2} + \frac{C}{x+3}.$$

Upon combining these three terms with a common denominator, we find that the right-hand side is equivalent to

$$\frac{A(x-2)(x+3)+B(x+1)(x+3)+C(x+1)(x-2)}{x^3+2x^2-5x-6},$$

which, in turn, is equivalent to

$$\frac{(A+B+C)x^2+(A+4B-C)x+(-6A+3B-2C)}{x^3+2x^2-5x-6}.$$

Since we want this expression to equal our original function, we may equate the corresponding coefficients of each power of x, obtaining the system of equations

$$\begin{aligned} A+B+C &= 1, \\ A+4B-C &= 1, \\ -6A+3B-2C &= 4. \end{aligned}$$

The solution of this system is

$$A = \frac{-2}{3}, \qquad B = \frac{2}{3}, \qquad C = 1.$$

Therefore, the partial-fraction decomposition is

$$\frac{x^2+x+4}{x^3+2x^2-5x-6} = \frac{1}{x+3} + \frac{2/3}{x-2} - \frac{2/3}{x+1},$$

and our original integral yields

$$\begin{aligned} \frac{x^2+x+4}{x^3+2x^2-5x-6} &= \ln|x+3| + \frac{2}{3}\ln|x-2| - \frac{2}{3}\ln x+1+C \\ &= \ln\left|(x+3)\left(\frac{x-2}{x+1}\right)^{2/3}\right| + C. \end{aligned}$$

■

7.3.3 Repeated Linear Factors

So far, we have seen how to construct a partial-fraction decomposition when the denominator has real, distinct (non-repeated) roots. Next, we turn to the case in which the denominator possesses repeated roots.

Proposition 7.3.3 Suppose that the nth degree polynomial Q can be factored as

$$Q(x) = \prod_{i=1}^{m}(x-a_i)^{r_i} = (x-a_1)^{r_1}\cdots(x-a_m)^{r_m},$$

where $r_1+\cdots+r_m = n$, and suppose that the polynomial P is of order no greater than $(n-1)$. Then the partial-fraction decomposition of P/Q is given by

$$\frac{P(x)}{Q(x)} = \sum_{i=1}^{m}\frac{R_i(x-a_i)}{(x-a_i)^{r_i}},$$

where R_i represents a polynomial of degree $r_i - 1$, with unknown constant coefficients.

The meaning of this result is easiest understood through examples. One such example is as follows:

$$\frac{1}{(x-2)^2(x-3)^3} = \frac{A(x-2)+B}{(x-2)^2} + \frac{C(x-3)^2+D(x-3)+E}{(x-3)^3}.$$

This illustrates the correct form of the partial-fraction decomposition. It should be apparent that the task of solving for the unknown coefficients easily becomes laborious. As a simpler example, we consider the following.

Example 7.19 Compute the integral

$$\int \frac{1}{(x-2)^2x^3}\,dx.$$

Following Proposition 7.3.3, the partial-fraction decomposition should take the form

$$\frac{1}{(x-2)^2x^3} = \frac{A(x-2)+B}{(x-2)^2} + \frac{Cx^2+Dx+E}{x^3}.$$

The right-hand side is equivalent to

$$\frac{x^3\left[A(x-2)+B\right]+(x-2)^2\left(Cx^2+Dx+E\right)}{x^3(x-2)^2},$$

which, in turn, is equivalent to

$$\frac{(A+C)x^4+(-2A+B+D-4C)x^3+(E-4D+4C)x^2+(-4E+4D)x+4E}{x^3(x-2)^2}.$$

Comparing with our original, we require that the x^4, x^3, x^2, and x coefficients in the numerator should each equal zero, and the constant term $4E$ should equal 1.

We therefore obtain the system of equations

$$\begin{aligned} A+C &= 0, \\ -2A+B+D-4C &= 0, \\ E-4D+4C &= 0, \\ -4E+4D &= 0, \\ 4E &= 1. \end{aligned}$$

We may solve these one at a time, obtaining

$$A=\frac{-3}{16}, \qquad B=\frac{1}{8}, \qquad C=\frac{3}{16}, \qquad D=\frac{1}{4}, \qquad E=\frac{1}{4}.$$

Our original expression is therefore equivalent to

$$\begin{aligned} \frac{1}{(x-2)^2x^3} &= \frac{-3/16(x-2)+1/8}{(x-2)^2}+\frac{3x^2/16+x/4+1/4}{x^3} \\ &= \frac{-3}{16}\frac{1}{(x-2)}+\frac{1}{8}\frac{1}{(x-2)^2}+\frac{3}{16x}+\frac{1}{4x^2}+\frac{1}{4x^3}. \end{aligned}$$

Our result follows:

$$\int \frac{1}{(x-2)^2x^3}\,dx=\frac{-3}{16}\ln|x-2|-\frac{1}{8(x-2)}+\frac{3}{16}\ln|x|-\frac{1}{4x}-\frac{1}{8x^2}+C.$$

■

7.3.4 Irreducible Quadratic Factors

Not all polynomials, however, can be factored into a product of linear binomials. The *fundamental theorem of algebra* guarantees that any nth degree polynomial has precisely n roots, when counted according to their multiplicities, and when including imaginary roots. Moreover, imaginary roots must occur in complex-conjugate pairs. Thus, if $x=a+bi$ is a complex root of a certain polynomial, then $x=a-bi$ must also be a root to the same polynomial. A pair of complex-conjugate roots $x=a\pm bi$ therefore create the following irreducible quadratic factor:

$$[x-(a+bi)]\,[x-(a-bi)]=x^2-2ax+a^2+b^2.$$

This is called an irreducible quadratic as it cannot be factored into a product of two linear binomials of the form $(x-a)(x-b)$ without using imaginary numbers. When such irreducible quadratic factors appear, we can complete the square to obtain

$$x^2-2ax+a^2+b^2=(x-a)^2+b^2.$$

This is an important observation to hang onto for later.

Proposition 7.3.4 Suppose that the nth degree polynomial Q has a non-repeated irreducible quadratic factor (ax^2+bx+c) (i.e., the discriminant b^2-4ac is less than zero, resulting in imaginary roots). Then the partial fraction decomposition of the rational function $P(x)/Q(x)$ inherits a term of the form

$$\frac{Ax+B}{ax^2+bx+c},$$

where A and B represent unknown constant coefficients.

■ **Example 7.20** Determine a formula for the integral

$$\int \frac{x^2+11x+7}{x^3+4x-5x^2-20}\,dx.$$

First, observe that the denominator can be factored as

$$\frac{x^2+11x+7}{x^3+4x-5x^2-20} = \frac{x^2+11x+7}{(x-5)(x^2+4)}.$$

The partial-fraction decomposition therefore takes the form

$$\frac{x^2+11x+7}{(x-5)(x^2+4)} = \frac{Ax+B}{x^2+4} + \frac{C}{x-5}.$$

We continue by computing the common denominator:

$$\begin{aligned} \frac{Ax+B}{x^2+4} + \frac{C}{x-5} &= \frac{(Ax+B)(x-5)+C(x^2+4)}{x^3+4x-5x^2-20} \\ &= \frac{(A+C)x^2+(B-5A)x+(4C-5B)}{x^3+4x-5x^2-20}. \end{aligned}$$

By comparing the corresponding coefficients of each power in the numerator of this equation and our original, we obtain the following conditions on the unknown constants A, B, and C:

$$\begin{aligned} A+C &= 1 \\ B-5A &= 11 \\ 4C-5B &= 7. \end{aligned}$$

These equations may be solved to obtain $A=-2$, $B=1$, and $C=3$. We have therefore shown the partial-fraction decomposition

$$\frac{x^2+11x+7}{x^3+4x-5x^2-20} = \frac{-2x+1}{x^2+4} + \frac{3}{x-5} = \frac{-2x}{x^2+4} + \frac{1}{x^2+4} + \frac{3}{x-5}.$$

The partial-fraction decomposition is split into three separate terms for ease of integration. In particular, the first term can be integrated using the substitution

$u = x^2 + 4$. The second term can be integrating using formula (D.9), and the final term can be integrated with the substitution $u = x - 5$. In the end, we obtain

$$\int \frac{x^2 + 11x + 7}{x^3 + 4x - 5x^2 - 20}\,dx = -\ln(x^2 + 4) + \frac{1}{2}\tan^{-1}\left(\frac{x}{2}\right) + 3\ln|x-5| + C.$$

This may alternatively be simplified to obtain

$$\int \frac{x^2 + 11x + 7}{x^3 + 4x - 5x^2 - 20}\,dx = \ln\left|\frac{(x-5)^3}{x^2+4}\right| + \frac{1}{2}\tan^{-1}\left(\frac{x}{2}\right) + C.$$

■

■ **Example 7.21** Determine a formula for the integral

$$\int \frac{x^2 + 4x + 7}{x^3 - 8x^2 + 29x - 52}\,dx.$$

First, observe that the denominator may be factored as follows:

$$\frac{x^2 + 4x + 7}{x^3 - 8x^2 + 29x - 52} = \frac{x^2 + 4x + 7}{(x-4)(x^2 - 4x + 13)}.$$

Proceeding as usual, we introduce a partial-fraction decomposition in the form

$$\begin{aligned}
\frac{x^2 + 4x + 7}{(x-4)(x^2 - 4x + 13)} &= \frac{Ax + B}{x^2 - 4x + 13} + \frac{C}{x-4} \\
&= \frac{(Ax+B)(x-4) + C(x^2 - 4x + 13)}{x^3 - 8x^2 + 29x - 52} \\
&= \frac{(A+C)x^2 + (B - 4A - 4C)x + (13C - 4B)}{x^3 - 8x^2 + 29x - 52}.
\end{aligned}$$

Comparing the corresponding coefficients of equal powers, we obtain the system of equations

$$\begin{aligned}
A + C &= 1 \\
B - 4A - 4C &= 4 \\
13C - 4B &= 7.
\end{aligned}$$

The combination of coefficients that solves this system of equations is $A = -2$, $B = 8$, and $C = 3$. We have therefore shown the partial-fraction decomposition

$$\frac{x^2 + 4x + 7}{x^3 - 8x^2 + 29x - 52} = \frac{-2x + 8}{x^2 - 4x + 13} + \frac{3}{x-4} = \frac{-2x + 8}{(x-2)^2 + 9} + \frac{3}{x-4}.$$

The last equality is obtained by completing the square in the denominator of the first term. Using the substitution $u = x - 2$, the first term transforms into

$$\frac{-2x + 8}{(x-2)^2 + 9} = \frac{-2u}{u^2 + 9} + \frac{4}{u^2 + 9},$$

which can be integrated using a natural logarithm and the formula (D.9). All together, we obtain

$$\int \frac{x^2+4x+7}{x^3-8x^2+29x-52}\,dx = -\ln\left[(x-2)^2+9\right] + \frac{4}{3}\tan^{-1}\left(\frac{x-2}{3}\right) + 3\ln|x-4| + C,$$

which is the desired result. ■

For completeness, we will also mention how to perform a partial-fraction decomposition for *repeated* irreducible quadratic factors, as outlined in the following proposition.

Proposition 7.3.5 Suppose that the nth degree polynomial Q has an irreducible quadratic factor (ax^2+bx+c) (i.e., the discriminant b^2-4ac is less than zero, resulting in imaginary roots) that is repeated r times, i.e., $(ax^2+bx+c)^r$ is a factor of Q. Then the partial fraction decomposition of the rational function $P(x)/Q(x)$ inherits a term of the form

$$\frac{(A_1x+B_1)(ax^2+bx+c)^{r-1}+\cdots+(A_rx+B_r)}{(ax^2+bx+c)^r},$$

where A_i and B_i represent unknown constant coefficients.

■ **Example 7.22** Compute a formula for the integral

$$\int \frac{x^3+81}{x(x^2+9)^2}\,dx.$$

In this example, the denominator has a repeated irreducible quadratic factor $(x^2+9)^2$. The form of the partial-fraction decomposition is therefore given as follows:

$$\frac{x^3+81}{x(x^2+9)^2} = \frac{(Ax+B)(x^2+9)+(Cx+D)}{(x^2+9)^2} + \frac{E}{x}.$$

Notice that this is equivalent to the following:

$$\frac{x^3+81}{x(x^2+9)^2} = \frac{(Ax+B)}{x^2+9} + \frac{Cx+D}{(x^2+9)^2} + \frac{E}{x}.$$

As usual, we proceed by rewriting our expression using a common denominator:

$$\frac{x^3+81}{x(x^2+9)^2} = \frac{(A+E)x^4+Bx^3+(9A+C+18E)x^2+(9B+D)x+81E}{x(x^2+9)^2}.$$

Comparing the corresponding powers of this expression and the original, we obtain a system of equations for the unknown coefficients:

$$\begin{aligned} A+E &= 0 \\ B &= 1 \\ 9A+C+18E &= 0 \\ 9B+D &= 0 \\ 81E &= 81. \end{aligned}$$

The solution is given by $A = -1$, $B = 1$, $C = -9$, $D = -9$, $E = 1$. We have therefore shown the partial-fraction decomposition

$$\frac{x^3+81}{x(x^2+9)^2} = \frac{-x+1}{x^2+9} + \frac{-9x-9}{(x^2+9)^2} + \frac{1}{x}.$$

Upon breaking this into five terms, using substitutions where appropriate, and relying on formulas (D.9) and (D.10), we obtain the end result

$$\int \frac{x^3+81}{x(x^2+9)^2}\,dx = -\frac{1}{2}\ln(x^2+9) + \frac{1}{6}\tan^{-1}\left(\frac{x}{3}\right) + \frac{-x+9}{2(x^2+9)} + \ln|x| + C.$$

■

Exercises

For Exercises 1–20, use partial fractions to perform the given integration.

1. $\displaystyle\int \frac{1}{x^2-2x}\,dx.$
2. $\displaystyle\int \frac{x+7}{x^2-2x}\,dx.$
3. $\displaystyle\int \frac{1}{x^2-1}\,dx.$
4. $\displaystyle\int \frac{3x+7}{x^2-1}\,dx.$
5. $\displaystyle\int \frac{1}{x^2-15x+56}\,dx.$
6. $\displaystyle\int \frac{x}{x^2-15x+56}\,dx.$
7. $\displaystyle\int \frac{x+1}{x^2-5x+6}\,dx.$
8. $\displaystyle\int \frac{x^2+9}{x^3-9x}\,dx.$
9. $\displaystyle\int \frac{x^3+7}{x^4-9x^2}\,dx.$
10. $\displaystyle\int \frac{1}{x^3-2x^2}\,dx.$
11. $\displaystyle\int \frac{x}{(x-2)(x-3)^2}\,dx.$
12. $\displaystyle\int \frac{x^2+21}{(x+2)(x-3)^2}\,dx.$
13. $\displaystyle\int \frac{x^3+7}{x^4-x^3}\,dx.$
14. $\displaystyle\int \frac{x^2}{x^2-9}\,dx.$
15. $\displaystyle\int \frac{x^2}{x^2+9}\,dx.$
16. $\displaystyle\int \frac{x+5}{x^3+25x}\,dx.$
17. $\displaystyle\int \frac{x^2-32}{x^3-8x^2+32x}\,dx.$
18. $\displaystyle\int \frac{x^2-100}{x^4+8x^3+20x^2}\,dx.$
19. $\displaystyle\int \frac{1}{x(x^2+4)^2}\,dx.$
20. $\displaystyle\int \frac{1}{x(x^2+8x+20)^2}\,dx.$

Problems

21. In chemistry, the *law of mass action* states that the time T to form a quantity x of a chemical product is given by the integral

$$T = k\int_0^x \frac{ds}{(a-s)(b-s)},$$

where k is a positive constant and where the parameters a and b represent the initial amounts of the reactants. Use partial fractions to integrate this, thereby obtaining T as an explicit function of x.

22. *Logistic growth* occurs when a population P satisfies the differential equation

$$\frac{dP}{dt} = -kP\left(1 - \frac{P}{M}\right),$$

for certain constants k and M. For small population sizes (i.e., for $P << M$), growth is similar to exponential growth. But as the population becomes large, the population size levels off at M, known as the *carrying capacity* for the given system. Such a differential equation is solved by separating the variables, a procedure that results in the equation

$$\frac{dP}{P(1-P/M)} = kdt.$$

(a) Use partial fractions to integrate the left-hand side, i.e., show that

$$\int \frac{dP}{P(1-P/M)} = \ln\left|\frac{P}{M-P}\right| + C.$$

(b) Combining this with the right-hand side, we obtain the relationship

$$\ln\left|\frac{P}{M-P}\right| = kt + C.$$

Solve this relationship to obtain $P(t)$.

(c) Show that $P(t)$ can be cast into an equivalent form

$$P(t) = \frac{M}{1+Ae^{-kt}},$$

for some constant A.

7.4 Trigonometric Substitution

Next, we learn a technique that is used when dealing with integrals that contain certain quadratic functions. The technique is inspired by right triangles and is useful for dealing with integrals related to the quantity $a^2 - x^2$, $a^2 + x^2$, or $x^2 - a^2$.

7.4.1 Basic Substitutions

When encountering quantities such as $\sqrt{a^2 - x^2}$, $\sqrt{a^2 + x^2}$, or $\sqrt{x^2 - a^2}$, for some positive constant a, the idea is to make a substitution based on trigonometric identities and the Pythagorean theorem. The main suggested substitutions are outlined in Table 7.3, along with their domain of validity and the associated differential.

quantity	substitution	domain	differential
$\sqrt{a^2 - x^2}$	$x = a\sin(\theta)$	$\theta \in [-\pi/2, \pi/2]$	$dx = a\cos(\theta)\, d\theta$
$\sqrt{a^2 + x^2}$	$x = a\tan(\theta)$	$\theta \in [-\pi/2, \pi/2]$	$dx = a\sec^2(\theta)\, d\theta$
$\sqrt{x^2 - a^2}$	$x = a\sec(\theta)$	$\theta \in [0, \pi]$	$dx = a\sec(\theta)\tan(\theta)\, d\theta$

Table 7.3: Trigonometric Substitutions

The rationale for these substitutions is illustrated in Figure 7.5. For each case, the quantity under the square root is identified with a leg of a certain right triangle. In this way, these quantities may be replaced by an appropriate trigonometric function.

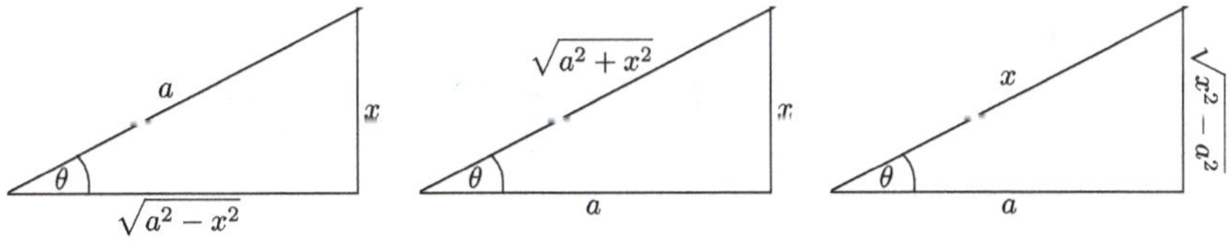

Figure 7.5: Illustration of the basic trigonometric substitutions.

■ **Example 7.23** Compute a formula for the integral

$$\int \frac{dx}{\sqrt{a^2 - x^2}}.$$

From Table 7.3, we choose the substitution

$$x = a\sin(\theta).$$

Its corresponding differential is given by

$$dx = a\cos(\theta)\, d\theta.$$

Accordingly, the quantity $\sqrt{a^2 - x^2}$ is equivalent to

$$\sqrt{a^2 - x^2} = \sqrt{a^2 - a^2\sin^2(\theta)} = a\sqrt{1 - \sin^2(\theta)} = a\cos(\theta).$$

Our substitution therefore yields

$$\int \frac{dx}{\sqrt{a^2 - x^2}} = \int \frac{a\cos(\theta)\, d\theta}{a\cos(\theta)} = \int d\theta = \theta + C.$$

Substituting back in for x, we obtain our final result:

$$\int \frac{dx}{\sqrt{a^2 - a^2}} = \sin^{-1}\left(\frac{x}{a}\right) + C.$$

This reproduces the integral (D.13). ■

■ **Example 7.24** Consider an underground storage tank in the shape of a horizontal right-circular cylinder, as shown in Figure 7.6, where R represents the radius of the circular cross sections and L represents the length of the tank. For a given tank, determine the volume V of liquid as a function of the water level h.

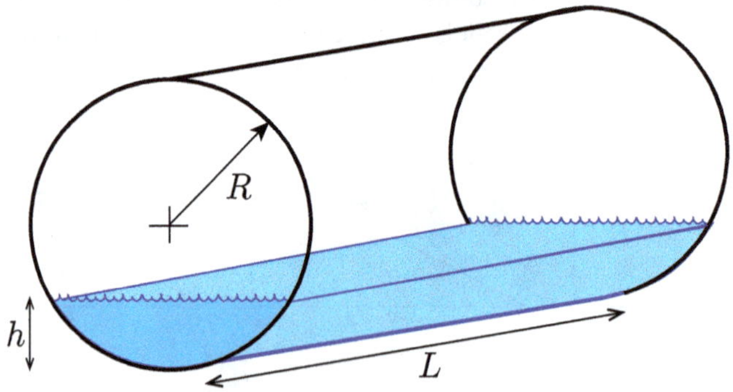

Figure 7.6: A partially filled horizontal tank; Example 7.24.

Essentially, we must determine the area of a partially filled circle, as the volume will equal the area of this partial circle times the tank's length L. Our result will be in terms of the parameters R and L, which must be treated as constants.

Our first task is to define a variable of integration. We choose a vertical axis with $x = 0$ corresponding to the center of the tank, as shown in Figure 7.7.

The idea is to determine the length of the cross section as a function of the variable of integration x. This can be achieved by considering the right triangle

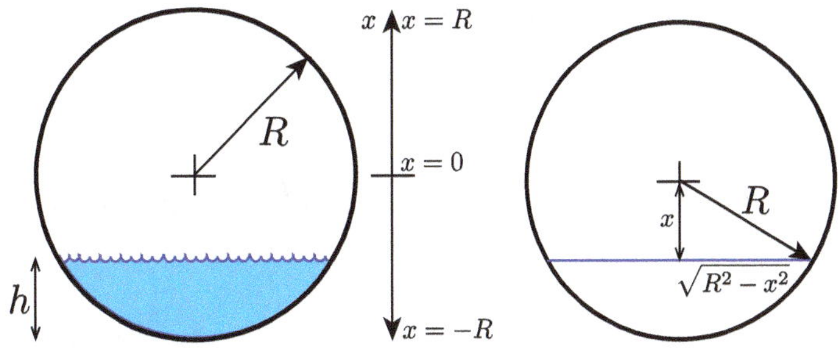

Figure 7.7: Vertical axis (left) and horizontal cross section (right) for the tank in Example 7.24.

shown in Figure 7.7 (right). At position x, as measured from the circle's center, the cross sectional width is given by $2\sqrt{R^2 - x^2}$. Notice that this width is independent of the sign of x; the same formula therefore holds if the cross section is above or below the centerline.

The volume of water in the tank, when the water level is h, is therefore given by the integral

$$V(h) = 2L \int_{-R}^{h-R} \sqrt{R^2 - x^2}\, dx.$$

The limits of integration are given by the endpoints between which the water lies; the bottom of the tank corresponds to $x = -R$, and the x-position of the water level is $x = h - R$, which is h units higher than the bottom.

To proceed further, we must next determine a formula for the integral

$$\int \sqrt{R^2 - x^2}\, dx.$$

Inspired from Table 7.3, we consider the substitution

$$x = R\sin(\theta).$$

Using this substitution, the integrand becomes

$$\sqrt{R^2 - x^2} = \sqrt{R^2 - R^2 \sin^2(\theta)} = R\cos(\theta).$$

Similarly, the differential is given by $dx = R\cos(\theta)\, d\theta$. Our integral is therefore converted into the form

$$\int \sqrt{R^2 - x^2}\, dx = R^2 \int \cos^2(\theta)\, d\theta.$$

Using the identity (B.9), this is equivalent to

$$\begin{aligned} R^2 \int \frac{1 + \cos(2\theta)}{2}\, d\theta &= \frac{R^2\theta}{2} + \frac{R^2 \sin(2\theta)}{4} + C \\ &= \frac{R^2}{2}\left[\theta + \sin(\theta)\cos(\theta)\right] + C. \end{aligned}$$

The second equality follows from identity (B.7). Substituting back in for x, we obtain

$$\int \sqrt{R^2 - x^2}\, dx = \frac{R^2}{2}\left[\sin^{-1}\left(\frac{x}{R}\right) + \frac{x}{R}\cos\left(\sin^{-1}\left(\frac{x}{R}\right)\right)\right] + C.$$

Since our trigonometric substitution was based off the first triangle in Figure 7.5, we realize that $\sin^{-1}(x/R) = \theta$, and that $\cos(\theta) = \sqrt{R^2 - x^2}/R$. We therefore obtain a final simplification:

$$\int \sqrt{R^2 - x^2}\, dx = \frac{R^2}{2}\sin^{-1}\left(\frac{x}{R}\right) + \frac{x\sqrt{R^2 - x^2}}{2} + C.$$

This recovers the integration formula (D.14). Finally, we must consider the limits of integration. The volume is therefore given by the formula

$$\begin{aligned} V(h) &= 2L\int_{-R}^{h-R} \sqrt{R^2 - x^2}\, dx \\ &= LR^2 \sin^{-1}\left(\frac{x}{R}\right) + Lx\sqrt{R^2 - x^2}\Big|_{x=-R}^{x=h-R} \\ &= \frac{LR^2\pi}{2} + LR^2 \sin^{-1}\left(\frac{h-R}{R}\right) + L(h-R)\sqrt{2hR - h^2}. \end{aligned}$$

One may easily confirm that $V(0) = 0$, $V(R) = \pi R^2 L/2$, and $V(2R) = \pi R^2 L$, as expected.

We may take this answer one step further by *nondimensionalizing* this formula. By dividing the result by $\pi R^2 L$, we obtain the volume of water as a fraction of the total capacity of the tank. Also, by introducing a new variable $x = h/(2R)$, our response turns into a function of the water level as a fraction of the tank's diameter $2R$. This nondimensionalized formula is given by

$$f(x) = \frac{1}{2} + \frac{1}{\pi}\sin^{-1}(2x - 1) + \frac{4x - 2}{\pi}\sqrt{x - x^2}. \tag{7.11}$$

Equation (7.11) is graphed in Figure 7.8.

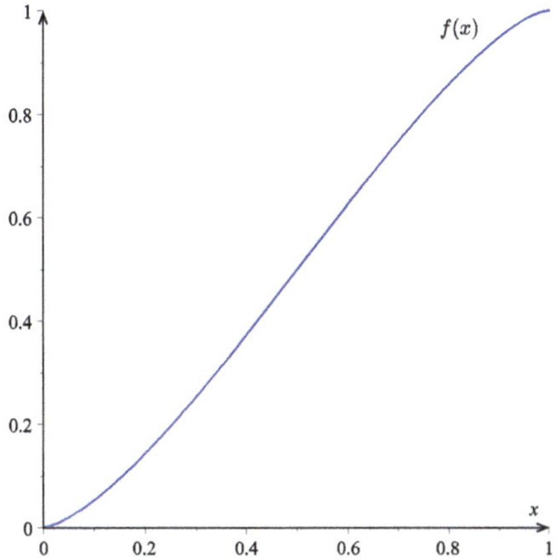

Figure 7.8: Fraction of volume f as a function of the non-dimensional water level x; Example 7.24.

Notice that when $f(0) = 0$, $f(0.5) = 0.5$, and $f(1) = 1$, indicating that when the water level is at 0%, 50%, and 100%, then the tank is 0%, 50%, and 100% filled (volume-wise), respectively.

This formula is also useful if one wished to correctly label a dipstick for a sidewise cylindrical tank, revealing the quantity of volume remaining in the tank. Equation (7.11) cannot be inverted analytically; though a computer has no difficulty solving equations like $f(x) = 0.2$ for x. The approximate water levels corresponding to 10% increments of the tanks volume are given in Table 7.4.

volume	water level
0.1	0.1564755869
0.2	0.2540690836
0.3	0.3401542451
0.4	0.4211319031
0.5	0.5
0.6	0.5788680969
0.7	0.6598457549
0.8	0.7459309164
0.9	0.8435244131
1.0	1.0

Table 7.4: Fraction of tank's volume and the corresponding non-dimensional water level; Example 7.24.

These data can be used to construct a scale revealing the fraction of volume of liquid in the tank, as shown in Figure 7.9.

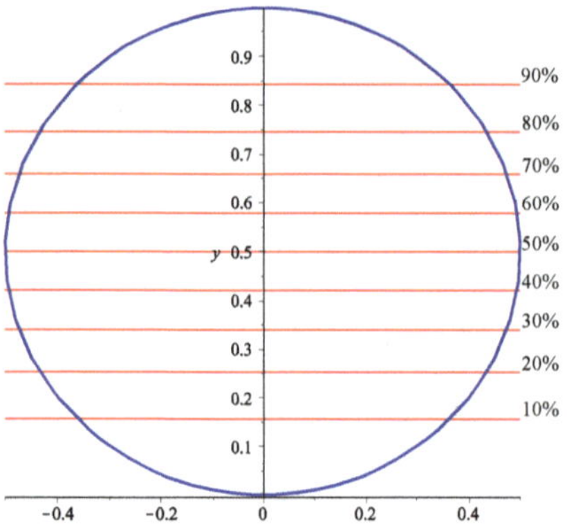

Figure 7.9: Horizontal lines represent the fraction of volume in the tank in 10% increments; Example 7.24.

■

Exercises

For Exercises 1–10, use trigonometric substitution to derive a formula for the indicated integral.

1. $\int x\sqrt{a^2 - x^2}\, dx.$

2. $\int \frac{dx}{a^2 + x^2}.$

3. $\int \frac{x}{\sqrt{x^2 - a^2}}\, dx.$

4. $\int x\sqrt{x^2 - a^2}\, dx.$

5. $\int \frac{dx}{\sqrt{x^2 + a^2}}.$

6. $\int \frac{dx}{\sqrt{x^2 - a^2}}.$

7. $\int \frac{x}{\sqrt{a^2 - x^2}}\, dx.$

8. $\int \frac{x}{a^2 - x^2}\, dx.$

9. $\int \frac{x}{(a^2 - x^2)^{3/2}}\, dx.$

10. $\int x\sqrt{a^2 + x^2}\, dx.$

Problems

11. Use a trigonometric substitution to show that the area of a sector of a circle is given by $\frac{1}{2}r^2\theta$. *Hint*: Divide the sector into two regions, as shown in Figure 7.10.

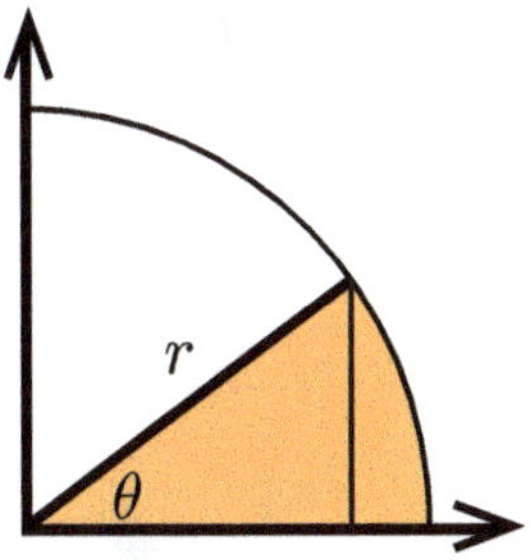

Figure 7.10: A sector of a circle; Problem 11.

12. A 12" diameter pizza is to be divided into three equal portions by making two parallel cuts. How far from the edge of the pizza should each cut be in order to achieve this?
13. The *sound intensity* I at point P, located at a distance d from the center of a sound bar of length $2a$, is given by the integral

$$I = \int_{-a}^{a} \frac{k}{x^2 + d^2}\, dx,$$

for some positive constant $k > 0$. This arrangement is shown in Figure 7.11. Use a trigonometric substitution to show that

$$I(d) = \frac{2k}{d} \tan^{-1}\left(\frac{a}{d}\right).$$

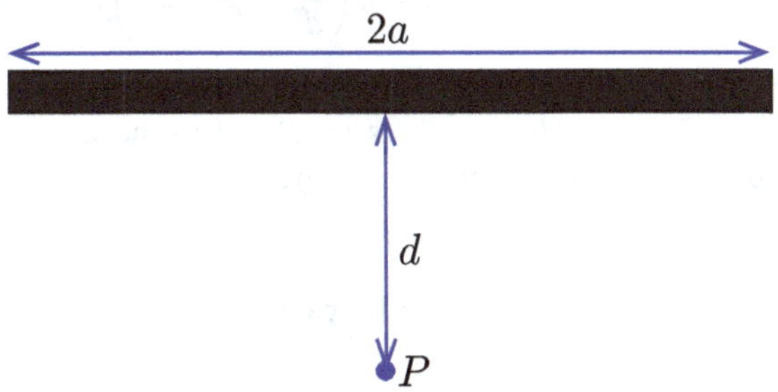

Figure 7.11: A sound bar; Problem 13.

14. A satellite travels along a linear path near a planet P, as shown in Figure 7.12. The parameter d represents the shortest distance between the planet and the satellite's path, and the satellite travels a total distance a, as shown in the figure. The total *work*[2] done by the satellite against the planet's gravitational

[2]In physics, *work* measures the quantity of energy required to move an object along its path. It equals the force applied times the distance traveled.

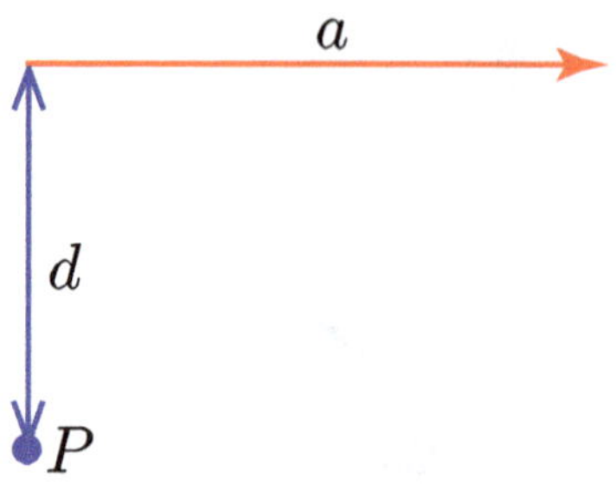

Figure 7.12: A satellite passing by a planet; Problem 14.

field is given by the integral

$$W = GMm \int_0^a \frac{x}{(x^2+d^2)^{3/2}}\,dx,$$

where G represents Newton's gravitational constant, and the parameters M and m represent the mass of the planet and the mass of the satellite, respectively.

(a) Use the substitution $u = x^2 + d^2$ to compute an explicit formula for the total work.

(b) Repeat the calculation from part (a), only instead using a trigonometric substitution.

(c) Show that both results are equivalent to the formula

$$W = GMm\left(\frac{1}{d} - \frac{1}{\sqrt{a^2+d^2}}\right).$$

Observe that this expression equals the change in the satellite's potential energy $U = -GMm/r$. (This is related to the *work-kinetic energy theorem* and *conservation of energy*.)

15. In Exercise 5, you showed that

$$\int \frac{dx}{\sqrt{x^2+a^2}} = \ln\left|x + \sqrt{a^2+x^2}\right| + C.$$

(a) Use the *hyperbolic trigonometric substitution* $x = a\sinh(\theta)$ to derive an alternate formula for this integral.

(b) By evaluating each antiderivative at $x = 0$, demonstrate that the following identity must hold:

$$\ln\left|x + \sqrt{a^2+x^2}\right| = \sinh^{-1}\left(\frac{x}{a}\right) + \ln(a).$$

16. Determine the area enclosed by the ellipse

$$\frac{x^2}{a^2} + \frac{y^2}{b^2} = 1.$$

7.5 Improper Integrals

In this section, we will learn how to handle integrals involving infinity. This can happen several different ways: one may have an infinite domain or a domain in which the function possesses a vertical asymptote. In these cases, we say that the definite integral is improper.

Definition 7.5.1 Given a function f, we say that the definite integral

$$I = \int_a^b f(x)\, dx,$$

where $a, b \in \{\mathbb{R}, \pm\infty\}$, is *improper* if either:

1. at least one of the limits of integration a and b is equal to $\pm\infty$; or
2. the function f has a vertical asymptote at some point c contained in the interval bounded by the limits of integration, i.e., $c \in [a,b]$.

We call these Type I and Type II, respectively.

An integral that is not improper is said to be *proper*.

Thus, an improper integral is an integral that involves infinity in some way. For a Type I improper integral, the entanglement with infinity is a direct consequence of the domain of integration, as one is literally dealing with an integral over a domain with infinite length. For a Type II improper integral, the "infinity" comes in not due to the size of the domain, but due to the size of the outputs of the function, which is not bounded on the given domain.

The solution for treating both types of improper integrals relies on taking limits; for both cases, we first construct a *proper* definite integral over a variable domain, and then allow the bounds of the domain of integration to slowly expand up to the point that creates the problem. We then look at the behavior of these proper integrals *in the limit* as the domain approaches the domain over which the integral is improper. We will carefully examine what is meant by this in the upcoming paragraphs.

7.5.1 Integrals over Infinite Domains

We first study how to integrate a function over an infinite domain.

Definition 7.5.2 The value assigned to the various Type I improper integrals are as follows:

$$\int_a^\infty f(x)\,dx = \lim_{T\to\infty}\left(\int_a^T f(x)\,dx\right), \tag{7.12}$$

$$\int_{-\infty}^a f(x)\,dx = \lim_{T\to-\infty}\left(\int_T^a f(x)\,dx\right), \tag{7.13}$$

$$\int_{-\infty}^\infty f(x)\,dx = \int_{-\infty}^0 f(x)\,dx + \int_0^\infty f(x)\,dx. \tag{7.14}$$

For each case, if the limit on the right-hand side exists, we say that the given integral *converges*. If the limit does not exist, the integral *diverges*.

Given a Type I improper integral over an infinite domain, we literally compute the integral over a finite, but variable, domain, and then take the limit as the domain is extended to infinity (in the appropriate way). For the third case, when one wishes to compute an integral over all of $\mathbb{R}$, one breaks the integral into a composite of two half domains $(-\infty,0]$ and $[0,\infty)$. In this case, *each* of the two integrals on the right-hand side of equation (7.14) must converge separately in order for the *whole* integral over $(-\infty,\infty)$ to converge as well. By definition. (In advanced mathematics, for particular applications, there are other ways to define an integral over $\mathbb{R}$, such as by taking the *symmetric* limit $\lim_{T\to\infty}\int_{-T}^{T} f(x)\,dx$. We will only use equation (7.14) for our purposes here, however.)

The meaning of the definition (7.12) is illustrated in Figure 7.13. In order to

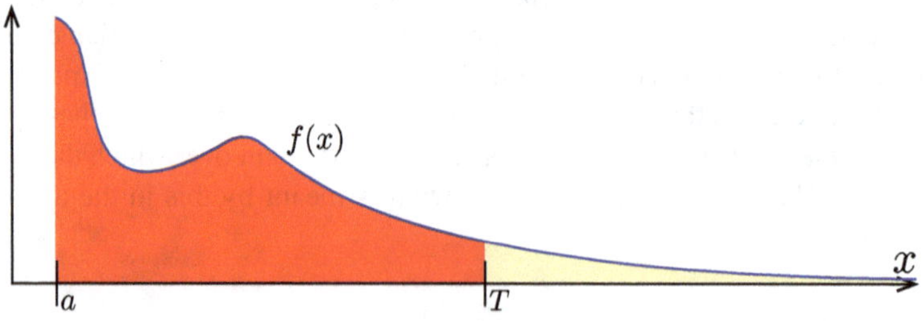

Figure 7.13: Illustration of an integral over an infinite domain.

integrate from $x = a$ to infinity, we instead perform an integration from $x = a$ to $x = T$. This captures a definitive amount of area, and this amount of area will depend on the upper bound T. We may then see what happens as we raise the upper bound T to infinity. If the limit exists, the total area under the graph from $x = a$ to infinity is only a finite amount of area, despite its infinite length. This can only happen if $\lim_{x\to 0} f(x) = 0$, though this condition turns out not to be sufficient. That is, not only must the function f go to zero as $x\to\infty$, but the function f must go to zero *fast enough*, so that it only captures a finite amount of area under its curve.

■ **Example 7.25** Compute the value of the improper integral

$$\int_0^\infty e^{-x}\,dx.$$

To do this, we implement equation (7.12), to obtain

$$\begin{aligned}\int_0^\infty e^{-x} &= \lim_{T\to\infty}\left(\int_0^T e^{-x}\,dx\right)\\ &= \lim_{T\to\infty}\left(-e^{-x}\Big|_{x=0}^{x=T}\right)\\ &= \lim_{T\to\infty}\left(1-e^{-T}\right)=1.\end{aligned}$$

We can observe two things from this calculation: the first is that the total area under the graph of $f(x)=e^{-x}$ from $x=0$ to $x=T$ is given by $A(T)=1-e^{-T}$. For example, the area between $x=0$ and $x=10$ is

$$A(10)=1-e^{-10}\approx 0.9999546,$$

and so forth. Since e^{-T} exponentially decays to zero as $T\to\infty$, the total area under the graph from $x=0$ to infinity is simply equal to 1. We therefore say that

$$\int_0^\infty e^{-x}\,dx=1,$$

and that this integral converges. ■

■ **Example 7.26** Compute the value of the improper integral

$$\int_1^\infty \frac{dx}{x},$$

if it converges. Again, we begin by applying the definition given by equation (7.12):

$$\begin{aligned}\int_1^\infty \frac{dx}{x} &= \lim_{T\to\infty}\left(\int_1^T \frac{dx}{x}\right)\\ &= \lim_{T\to\infty}\left(\ln T\Big|_{x=1}^{x=T}\right)\\ &= \lim_{T\to\infty}\ln T=\infty.\end{aligned}$$

Since the logarithmic function increases without bound, this limit does not exist, and therefore the integral diverges. Even though the function $f(x)=1/x$ goes to zero, it does not go to zero fast enough, as there is an infinite amount of area under its graph between $x=1$ and infinity. ■

7.5.2 Integrals over Vertical Asymptotes

We apply a similar technique if a function possesses a vertical asymptote on the domain of integration: we relax the interval of integration down to the problem point by means of taking the limit.

Definition 7.5.3 Suppose that the function f has a vertical asymptote at the point $x = c$, and that $c \in (a,b)$, for some $a < b$. Then the value assigned to the various Type II improper integrals is as follows:

$$\int_c^b f(x)\,dx = \lim_{T \to c^+} \left(\int_T^b f(x)\,dx \right), \tag{7.15}$$

$$\int_a^c f(x)\,dx = \lim_{T \to c^-} \left(\int_a^T f(x)\,dx \right), \tag{7.16}$$

$$\int_a^b f(x)\,dx = \int_a^c f(x)\,dx + \int_c^b f(x)\,dx. \tag{7.17}$$

For each case, if the integral on the right-hand side exists, we say that the given integral *converges*. If the limit does not exist, the integral *diverges*.

If the function f has a vertical asymptote at the point $c \in (a,b)$, and if we wish to compute a definite integral using the point c as one of the endpoints, either on the left or the right of the interval of integration, we instead compute the definite integral over a variable domain that does not contain the point c, and then take the limit as this variable domain is expanded to the desired domain.

The meaning of equation (7.15), for example, is illustrated in Figure 7.14. It

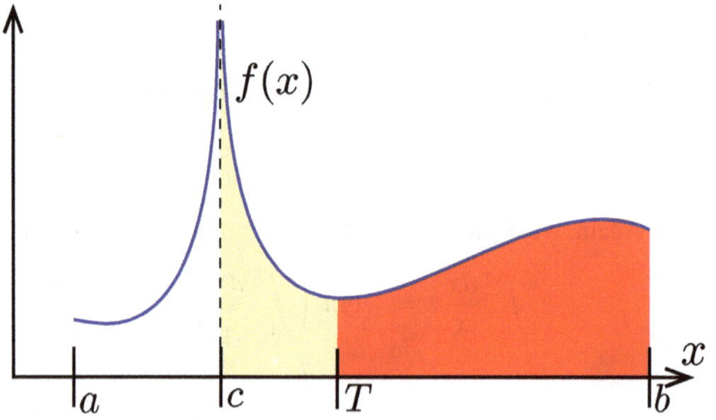

Figure 7.14: Illustration of an integral up to a vertical asymptote.

is problematic to simply integrate from $x = c$ to $x = b$, since there is a vertical asymptote at the point $x = c$, thus the height of the graph is unbounded as one approaches the point $x = c$. This is the sideways image of the Type I improper integral in which the width goes to infinity, whereas here the height is going to infinity. Instead of integrating from $x = c$ to $x = b$, we instead integrate from $x = T$ to $x = b$, for $T > c$, and then take the limit as T relaxes toward c from the right.

If one wishes to integrate from $x = a$ to $x = b$—an interval in which f contains a vertical asymptote—one breaks the domain into two separate domains a la equation (7.17). If either of these separate integrals diverges, we say that the whole integral diverges.

Example 7.27 Compute

$$\int_0^1 \frac{dx}{\sqrt{x}}.$$

The integrand has a vertical asymptote at $x = 0$, therefore we proceed using equation (7.15), by writing

$$\begin{aligned}\int_0^1 \frac{dx}{\sqrt{x}} &= \lim_{T\to 0^+}\left(\int_T^1 \frac{dx}{\sqrt{x}}\right)\\ &= \lim_{T\to 0^+}\left(2\sqrt{x}\Big|_{x=T}^{x=1}\right)\\ &= \lim_{T\to 0^+}\left(2-2\sqrt{T}\right) = 2.\end{aligned}$$

The area under $f(x) = 1/\sqrt{x}$ between $x = T$ and $x = 1$ is equal to $2(1-\sqrt{T})$. As we relax T down to zero, from the right, we therefore obtain the total area under f from $x = 0$ to $x = 1$, and that area is equal to exactly 2. ■

Example 7.28 Compute the value of the integral

$$\int_{-1}^1 \frac{dx}{x}. \tag{7.18}$$

The function $f(x) = 1/x$ is plotted over the domain $[-1,1]$ in Figure 7.15. It is

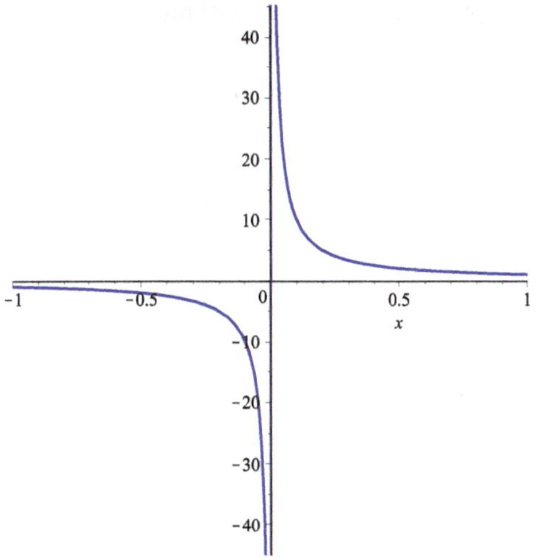

Figure 7.15: Graph of $f(x) = 1/x$ between $x = -1$ and $x = 1$.

tempting to assuming that the definite integral (7.18) is equal to zero, as the area between $x = 0$ and $x = 1$ *should* cancel out exactly with the area between $x = -1$ and $x = 0$. The function f is, after all, an odd function, so any symmetric integral *should* yield zero. This, however, is not the case here; equation (7.17) *defines* the meaning of this integral as

$$\int_{-1}^{1} \frac{dx}{x} = \int_{-1}^{0} \frac{dx}{x} + \int_{0}^{1} \frac{dx}{x}.$$

As *the greater contains the less*[3], if either of these separate integrals diverge, the whole integral must diverge by definition. Let us therefore begin by examining

$$\int_{0}^{1} \frac{dx}{x}.$$

Applying equation (7.15), we obtain

$$\begin{aligned} \int_{0}^{1} \frac{dx}{x} &= \lim_{T\to 0^+} \left(\int_{T}^{1} \frac{dx}{x} \right) \\ &= \lim_{T\to 0^+} \left(\ln(x) \Big|_{x=T}^{x=1} \right) \\ &= \lim_{T\to 0^+} (-\ln(T)) = \infty. \end{aligned}$$

Since $\lim_{T\to 0^+} \ln(T) = -\infty$ (the function $f(x) = \ln x$ has a vertical asymptote at $x = 0$), we conclude that our integral diverges. Since this integral from $x = 0$ to $x = 1$ is contained within the integral (7.18), we conclude that the latter diverges as well. ■

■ **Example 7.29** Compute the value of the definite integral

$$\int_{0}^{1} \ln(x)\,dx.$$

The graph of $f(x) = \ln(x)$ over $[0, 1]$ is shown in Figure 7.16. We can integrate this with the aid of equation (7.7). We proceed as usual:

$$\begin{aligned} \int_{0}^{1} \ln(x)\,dx &= \lim_{T\to 0^+} \left(\int_{T}^{1} \ln(x)\,dx \right) \\ &= \lim_{T\to 0^+} \left(x\ln(x) - x \Big|_{x=T}^{x=1} \right) \\ &= \lim_{T\to 0^+} (-1 - (T\ln(T) - T)) \\ &= \lim_{T\to 0^+} (T - 1 - T\ln(T)). \end{aligned}$$

The obstacle now is to compute the limit of $T\ln(T)$ as $T \to 0^+$. We can rewrite this limit as

$$\lim_{T\to 0^+} T\ln(T) = \lim_{T\to 0^+} \frac{\ln(T)}{1/T}.$$

[3] This expression is actually a codified maxim of jurisprudence: California Civil Code § 3536.

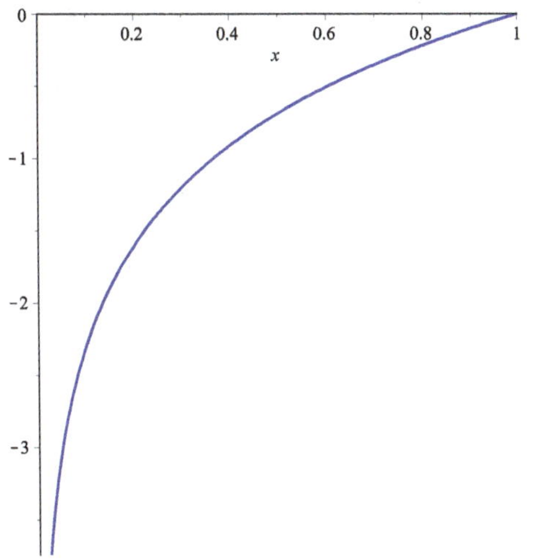

Figure 7.16: Graph of $f(x) = \ln(x)$ between $x = 0$ and $x = 1$.

We recognize this as an *indeterminate form*, and we may therefore apply L'Hospital's rule to obtain

$$\lim_{T\to 0^+} \frac{\ln(T)}{1/T} = \lim_{T\to 0^+} \frac{1/T}{-1/T^2} = \lim_{T\to 0^+} -T = 0.$$

We conclude that

$$\int_0^1 \ln(x)\, dx = -1.$$

This answer makes sense, as the graph of $f(x) = \ln(x)$ is the reflection of the graph of e^x about the line $y = x$, and this area should therefore equal the negative of the area bounded by e^x from $-\infty$ to $x = 0$. ■

7.5.3 The p-test

A common class of functions that we will be interested is the class of functions of the form $f(x) = 1/x^p$. These functions tend to zero as $x \to \infty$, and they tend to infinity as $x \to 0$. It turns out that there is a simple way of spotting whether a given function from this class converges or diverges, as laid out in the following theorem.

Theorem 7.5.1 — The p-test. The Type I improper integral

$$\int_1^\infty \frac{dx}{x^p} \quad \begin{cases} \text{converges for } p > 1, \text{ and} \\ \text{diverges for } p \leq 1. \end{cases} \tag{7.19}$$

The Type II improper integral

$$\int_0^1 \frac{dx}{x^p} \quad \begin{cases} \text{converges for } p < 1, \text{ and} \\ \text{diverges for } p \geq 1. \end{cases} \tag{7.20}$$

As we saw in Examples 7.26 and 7.28, the integral of the function $f(x) = 1/x$ diverges *both* on the intervals $[0,1]$ and $[1,\infty)$. This confirms that the case $p = 1$ should be result in divergence for both integrals (7.19) and (7.20). This is reasonable, as $\int dx/x = \ln(x) + C$, and the logarithmic function goes to $\pm\infty$ as x goes to both zero and infinity. For the other cases, let us consider the following proof.

Proof. Suppose $p \neq 1$, and consider the integral

$$\int_1^\infty \frac{dx}{x}.$$

Using the definition from equation (7.12), we may compute

$$\begin{aligned} \int_1^\infty \frac{dx}{x^p} &= \lim_{T\to\infty} \left(\int_1^T \frac{dx}{x^p} \right) \\ &= \lim_{T\to\infty} \left(\left. \frac{x^{1-p}}{(1-p)} \right|_{x=1}^{x=T} \right) \\ &= \lim_{T\to\infty} \left(\frac{T^{1-p} - 1}{(1-p)} \right). \end{aligned}$$

If $p > 1$, the quantity $(1-p)$ will be negative, and hence

$$\lim_{T\to\infty} T^{1-p} = 0.$$

On the other hand, if $p < 1$, the quantity $(1-p)$ becomes positive, and therefore

$$\lim_{T\to\infty} T^{1-p} = \infty,$$

and the integral will diverge. Adding to this the result of Example 7.26, all of the cases for equation (7.19) have now been recovered.

Next, let us consider (again for the case $p \neq 1$) the Type II improper integral

$$\int_0^1 \frac{dx}{x^p}.$$

Applying equation (7.15), we obtain

$$\begin{aligned} \int_0^1 \frac{dx}{x^p} &= \lim_{T\to 0^+} \left(\int_T^1 \frac{dx}{x^p} \right) \\ &= \lim_{T\to 0^+} \left(\left. \frac{x^{1-p}}{(1-p)} \right|_{x=T}^{x=1} \right) \\ &= \lim_{T\to\infty} \left(\frac{1 - T^{1-p}}{(1-p)} \right). \end{aligned}$$

If $p > 1$, the quantity $(1 - p)$ will again be negative, resulting in

$$\lim_{T\to 0^+} T^{1-p} = \infty.$$

(Essentially, the negative exponent will create a vertical asymptote at $T = 0$, and at $T = 0$, one is "dividing by zero.")

On the other hand, if $p < 1$, the quantity $(1 = p)$ will be positive, and thus

$$\lim_{T\to 0^+} T^{1-p} = 0,$$

since we are essentially computing a positive power of zero. Adding to this the result of Example 7.28, we have thus recovered all of the cases for equation (7.20). This completes the proof. ■

7.5.4 The Comparison Test

The comparison test offers us a way to determine whether a particular improper integral converges or diverges, without actually computing the value of the integral.

Theorem 7.5.2 — The Comparison Test. Let $a \in \mathbb{R} \cup \{-\infty\}$ and $b \in \mathbb{R} \cup \{+\infty\}$, and suppose that $0 \le f(x) \le g(x)$, for all $x \in [a,b]$. The following statements hold:

1. If $\int_a^b g(x)\,dx$ converges, then $\int_a^b f(x)\,dx$ also converges.
2. If $\int_a^b f(x)\,dx$ diverges, then $\int_a^b g(x)\,dx$ also diverges.

The comparison test basically says that *if the larger function converges, so too must the smaller* and that *if the smaller function diverges, so too must the larger.*

(R) Theorem 7.5.2 yields *no information* if either $\int_a^b g(x)\,dx$ diverges or if $\int_a^b f(x)\,dx$ converges.

Thus, if the integral of the larger function g diverges, then the integral of f may still be either finite or infinite. On the other hand, if the integral of the smaller function f converges, then the integral of g may still be either finite or infinite.

The comparison test often works in conjunction with the p-test to show that a given integral converges or diverges.

■ **Example 7.30** Use the comparison test to determine whether or not the integral

$$\int_1^\infty \frac{5 + 2\sin\left(8\sqrt{x} + 17\right)}{x^2 + 2x + 7}\,dx$$

converges or diverges. The integrand is graphed on the interval $[1,10]$ in Figure 7.17.

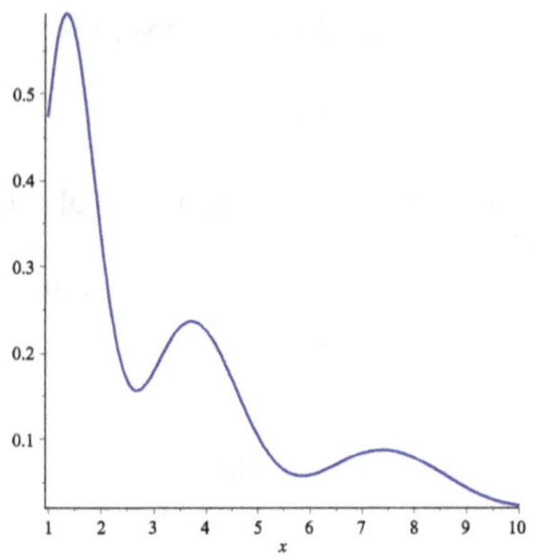

Figure 7.17: Function from Example 7.30.

Clearly, finding an antiderivative for such a function is an insurmountable task. We can, however, say something on the issue of convergence. We start with a guess. By examining the integrand, the x^2 term in the denominator seems to be the dominating factor as $x \to \infty$. This is because the numerator only fluctuates between 3 and 7, and because the x^2 in the denominator towers the linear terms $2x+7$ as x becomes large. We can say that the integral should behave like $1/x^2$ as x becomes large, and, therefore, it should converges.

To *show* that it converges, however, we must compare it with a larger function that does converge—and, ideally, this larger function will be a multiple of $1/x^2$.

We begin by comparing the numerator with a function that is larger. Since $-1 \leq \sin(x) \leq 1$ for all x, we can conclude that

$$5+2\sin\left(8\sqrt{x}+7\right) \leq 7,$$

for all x. Since the act of reciprocation reverses inequalities, we wish to compare the denominator with a function that is smaller, but that still retains the x^2 characteristic. To do this, we simply observe that

$$x^2+2x+7 > x^2.$$

Therefore,

$$\frac{1}{x^2+2x+7} < \frac{1}{x^2}.$$

Combining the previous result, we obtain

$$\frac{5+2\sin\left(8\sqrt{x}+17\right)}{x^2+2x+7} < \frac{7}{x^2}.$$

Since

$$\int_1^{\infty} \frac{7}{x^2}\, dx$$

converges, by the p-test, we therefore conclude, by the comparison test, that the integral

$$\int_1^\infty \frac{5+2\sin\left(8\sqrt{x}+17\right)}{x^2+2x+7}\,dx$$

also converges. ∎

Exercises

For Exercises 1–20, classify the given integral as either a Type I or Type II improper integral; if it converges, determine its value, otherwise show that it diverges.

1. $\displaystyle\int_1^\infty \frac{dx}{x^2}$.
2. $\displaystyle\int_1^\infty \frac{dx}{x^3}$.
3. $\displaystyle\int_0^1 \frac{dx}{\sqrt{x}}$.
4. $\displaystyle\int_0^{\pi/2} \tan(x)\,dx$.
5. $\displaystyle\int_0^\infty xe^{-x}\,dx$.
6. $\displaystyle\int_0^\infty xe^{-x^2}\,dx$.
7. $\displaystyle\int_0^\infty x^2e^{-x^3}\,dx$.
8. $\displaystyle\int_0^\infty x^3e^{-x^2}\,dx$.
9. $\displaystyle\int_0^\infty e^{-x}\sin(x)\,dx$.
10. $\displaystyle\int_0^3 \frac{x^2-9}{x-3}\,dx$.
11. $\displaystyle\int_0^1 \frac{dx}{\sqrt[3]{x}}$.
12. $\displaystyle\int_0^1 x\ln(x)\,dx$.
13. $\displaystyle\int_0^3 \frac{x}{x^2-9}\,dx$.
14. $\displaystyle\int_0^3 \frac{dx}{\sqrt{9-x^2}}$.
15. $\displaystyle\int_0^{\pi/2} \frac{\sin(x)}{\sqrt{\cos(x)}}\,dx$.
16. $\displaystyle\int_{-\infty}^0 e^x\,dx$.
17. $\displaystyle\int_0^\infty \frac{dx}{x^5}$.
18. $\displaystyle\int_{\pi/4}^{\pi/2} \frac{\sec^2(x)}{\tan^2(x)}\,dx$.
19. $\displaystyle\int_0^{e^{-1}} \frac{dx}{x(\ln x)^2}$.
20. $\displaystyle\int_0^\infty \frac{e^{\tan^{-1}(x)}}{1+x^2}\,dx$.

For Exercises 21–30, use the comparison test to determine whether the given improper integral converges or diverges.

21. $\displaystyle\int_2^\infty \frac{\sin^2(x)}{x^3}\,dx$.
22. $\displaystyle\int_1^\infty \frac{dx}{x+e^x}$.
23. $\displaystyle\int_1^\infty \frac{dx}{1+x^3}$.
24. $\displaystyle\int_1^\infty \frac{dx}{1+x}$.
25. $\displaystyle\int_2^\infty \frac{x(2+\sin(3x))}{x^2+e^{-x}}\,dx$.
26. $\displaystyle\int_0^2 \frac{x^2+7}{\sqrt{x^3+x}}\,dx$.
27. $\displaystyle\int_0^1 \frac{x^4+x^2}{x^2+\sqrt{x^3}}\,dx$.
28. $\displaystyle\int_0^1 \frac{8+\sin(1/x)}{x+\sqrt{x}}\,dx$.

29. $\int_0^1 \frac{x+\sqrt{x}}{x^2+x}\,dx.$

30. $\int_0^1 \frac{dx}{\sin(x)}.$

(*Hint* for Exercise 30: compare $\sin(x)$ with its tangent line.)

Problems

31. Consider the graph of $f(x)$, $g(x)$, $h(x)$, $1/x$, and $1/x^2$, as shown in Figure 7.18. You may assume that

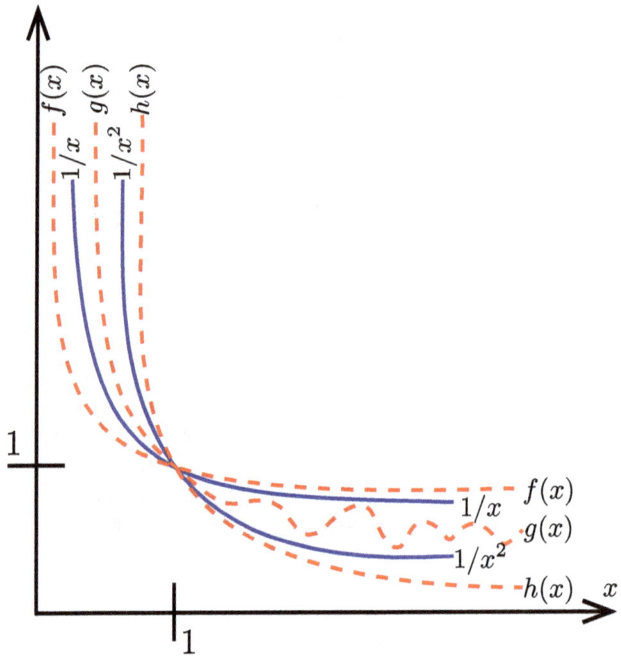

Figure 7.18: Graph of the functions f, g, h, $1/x$, and $1/x^2$.

$$h(x) < \frac{1}{x^2} < g(x) < \frac{1}{x} < f(x),$$

for all $x > 1$, and that

$$f(x) < \frac{1}{x} < g(x) < \frac{1}{x^2} < h(x),$$

for all $0 < x < 1$. Determine the convergence or divergence of each integral whenever possible. If this is not possible, explain why.

(a) $\int_1^\infty f(x)\,dx.$

(b) $\int_1^\infty g(x)\,dx.$

(c) $\int_1^\infty h(x)\,dx.$

(d) $\int_0^1 f(x)\,dx.$

(e) $\int_0^1 g(x)\,dx.$

(f) $\int_0^1 h(x)\,dx.$

32. A power surge can be modeled using the function

$$p(t) = ate^{-rt}. \quad (7.21)$$

The function $p(t)$ represents the power (energy per unit time) at time t, for given positive parameters $a, r > 0$. An example function of this form is shown in Figure 7.19.

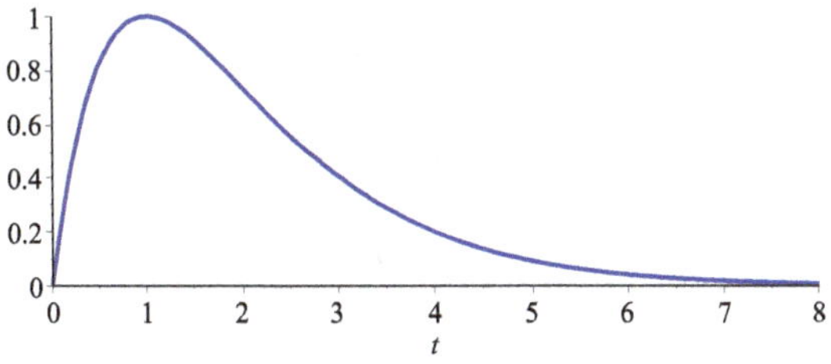

Figure 7.19: Model of a power surge; Problem 32.

(a) Determine the time at which peak power occurs. What is the peak power level? Your answer may be in terms of the parameters a and r.
(b) Explain what the quantity

$$\int_0^\infty ate^{-rt}\,dt \quad (7.22)$$

represents in practical terms.
(c) Evaluate the integral (7.22). Your answer may be in terms of the parameters a and r.

33. Consider a planet (or star) with mass M and radius R, both assumed to be constants. Let G be Newton's gravitational constant ($G = 6.67384 \times 10^{-11}\mathrm{m}^3\mathrm{kg}^{-1}\mathrm{s}^{-2}$). If an object on the surface of a planet is given an initial upward velocity of v_0, then its initial kinetic energy is

$$K_0 = \frac{v_0^2}{2}.$$

As the object continues its ascent, its kinetic energy is taken away by the gravitational field of the earth. The amount of kinetic energy taken away from the time of launch, when the object is on the surface of the planet $r = R$, and the time when the object reaches a distance H away from the center of the planet, $H > R$, is given by

$$\text{Kinetic Energy taken away} = \int_R^H \frac{GM}{r^2}\, dr.$$

If the object has *exactly* the necessary amount of energy to escape the gravitational field of the planet entirely and reach infinity, then the following must be true:

$$\frac{v_0^2}{2} = \int_R^\infty \frac{GM}{r^2}\, dr.$$

This special velocity v_0 for which this is true is known as the planet's *escape velocity.*

(a) Show that the escape velocity of a planet is given by:

$$v_0 = \sqrt{\frac{2GM}{R}}$$

(b) For earth, $R = 6.37 \times 10^6$ m and $M = 5.98 \times 10^{24}$ kg. Calculate the earth's escape velocity. Using the conversion 1 m/s = 2.24 mph, what is the earth's escape velocity in mph?

34. A result of Einstein's general theory of relativity tells that the amount of time, T, that the surrounding universe ages as a beam of light falls into a black hole with radius a, from an arbitrary starting position $r = b > a$, is given by:

$$T = \int_a^b \frac{r}{r-a}\, dr$$

By using the substitution $u = r - a$, transform this integral into an equivalent one, and then show that the integral diverges.

35. An important integral in probability theory is

$$I = \int_{-\infty}^{\infty} e^{-x^2}\, dx.$$

Using techniques from multivariable calculus, one can show that

$$I^2 = 2\pi \int_0^\infty e^{-r^2}\, r\, dr.$$

Compute the value of this integral and conclude that

$$\int_{-\infty}^{\infty} e^{-x^2}\, dx = \sqrt{\pi}. \qquad (7.23)$$

36. Consider the *Gaussian distribution*, which is the function defined by

$$p(x) = \frac{1}{\sqrt{2\pi}} e^{\frac{-(x-\mu)^2}{2}}, \qquad (7.24)$$

where μ is a constant (the mean value of the distribution). The standard deviation of this distribution is 1. Derive the following results. You may use equation (7.23).

(a) Using the change of variables

$$u = \frac{x-\mu}{\sqrt{2}},$$

show that

$$\int_{-\infty}^{\infty} p(x)\,dx = 1. \tag{7.25}$$

(b) Using the result from part (a) with integration by parts, show that

$$\int_{-\infty}^{\infty} xp(x)\,dx = \mu. \tag{7.26}$$

This value is called the *mean*. We will understand this in more detail in Section 8.6.

37. In physics, a *blackbody* is an object that does not reflect any light, so that it appears completely black. Blackbodies nevertheless have been shown to radiate *thermal* energy at various wavelengths. Blackbodies at a given absolute temperature have a quality called *radiancy*, which is a function of the wavelength of its thermal radiation. The *radiancy* $R(\lambda)$, measured in $\mathrm{W/m^2/\mu m}$, of radiation with wavelength λ μm, is defined so that the total intensity of thermal radiation, dI, measured in $\mathrm{W/m^2}$, due to all radiation whose wavelengths lie between λ and $\lambda + d\lambda$, is given by

$$dI = R(\lambda)d\lambda.$$

The radiancy of a blackbody is given by the formula

$$R(\lambda) = \frac{a}{\lambda^5\left(e^{b/\lambda}-1\right)}\ \frac{\mathrm{W}}{\mathrm{m^2\mu m}},$$

where a and b are physical constants.

(a) What does the quantity

$$I(\lambda) = \int_0^{\lambda} \frac{a}{x^5\left(e^{b/x}-1\right)}\,dx$$

represent physically?

(b) Show that

$$R'(\lambda) = \frac{a}{\lambda^5\left(e^{b/\lambda}-1\right)}\left[\frac{be^{b/\lambda}}{\lambda^2(e^{b/\lambda}-1)} - \frac{5}{\lambda}\right]\frac{\mathrm{W}}{\mathrm{m^2\mu m^2}}.$$

(c) Show that this implies $R'(\lambda) = 0$ if and only if

$$b - 5\lambda + 5\lambda e^{-b/\lambda} = 0.$$

(d) Consider a blackbody with temperature 300K, $a = 3.7417 \times 10^8$ W$(\mu\text{m})^4/\text{m}^2$and $b = 47.946\ \mu$m. Using these numbers and with the aid of a computer or graphing calculator, determine the value of λ^* such that $R'(\lambda^*) = 0$. Using a computer, graph $R(\lambda)$ on the interval $\lambda \in [0, 20]$.

(e) Using a computer or graphing calculator, compute the value of $I(\lambda^*)$ defined by the definite integral listed above.

(f) Using a computer or graphing calculator, compute the limit

$$\lim_{\lambda\to\infty} I(\lambda) = \int_0^\infty R(\lambda) d\lambda.$$

What does this number represent physically?

(g) What does the function

$$f(\lambda) = \frac{\displaystyle\int_0^\lambda R(x)\, dx}{\displaystyle\int_0^\infty R(x)\, dx}$$

correspond to physically?

38. **[The Gamma Function]** The *Gamma function* $\Gamma(z)$ is defined by the relation

$$\Gamma(z) = \int_0^\infty t^{z-1} e^{-t} dt. \tag{7.27}$$

A plot of the Gamma function is shown in Figure 7.20.

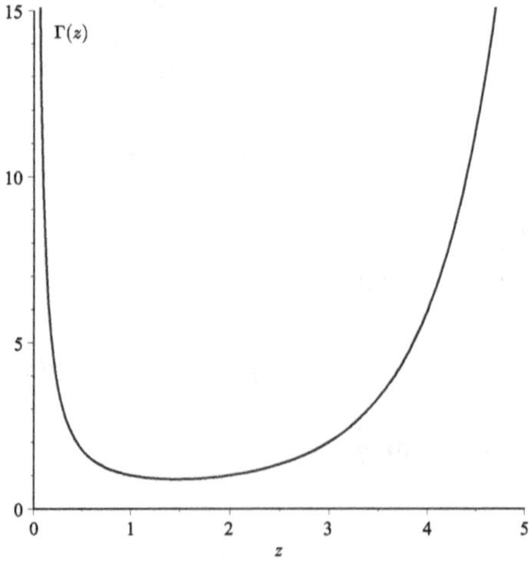

Figure 7.20: The Gamma function $\Gamma(z)$.

(a) Compute $\Gamma(1)$, $\Gamma(2)$, and $\Gamma(3)$.

(b) Use integration by parts to show that

$$\Gamma(z+1) = z\Gamma(z). \tag{7.28}$$

(c) Use the result from part (a) with equation (7.28) to determine a formula for $\Gamma(n+1)$, where $n \in \mathbb{N}$ (i.e., where n is a natural number).
Thus, the Gamma function is an extension of the factorial that is valid for general positive numbers, not just positive integers. It is also for this reason that, by convention, mathematicians define the factorial of zero as $0! = 1$.

(d) It is possible to show that $\Gamma(1/2) = \sqrt{\pi}$. Using this fact and the relation (7.28), compute $\Gamma(3/2)$, $\Gamma(5/2)$, and $\Gamma(7/2)$.

Chapter 8

Applications of Integration

8.1 Arc Length and Area

In this section, we explore applications of integration to computing the length of a path and the area between two curves.

8.1.1 Arc Length of a Graph

The arc length of the graph of a smooth function may be computed as follows.

Proposition 8.1.1 Let f be a differentiable function on the interval $[a,b]$. Then the *arc length* $\mathcal{L}$ of the graph of f between $x = a$ and $x = b$ is

$$\mathcal{L} = \int_a^b \sqrt{1 + f'(x)^2}\, dx. \tag{8.1}$$

Proof. To show that this is true, we must simply determine the differential amount ds by which the arc length changes corresponding to a differential change dx in the independent variable x. Since the function f is differentiable, it will appear as an (approximate) straight line if we zoom in close enough to any given point on the graph. The derivative f' connects an infinitesimal change in x to an infinitesimal change in y according to the relation

$$dy = f'(x)\, dx.$$

The associated change in arc length ds is therefore given by the Pythagorean theorem as

$$ds = \sqrt{dx^2 + (f'(x)dx)^2} = \sqrt{1 + f'(x)^2}\, dx,$$

as shown in Figure 8.1. Upon integrating the arc-length differential ds over the corresponding domain of x, we obtain the total arc length of the curve. This proves the result. ■

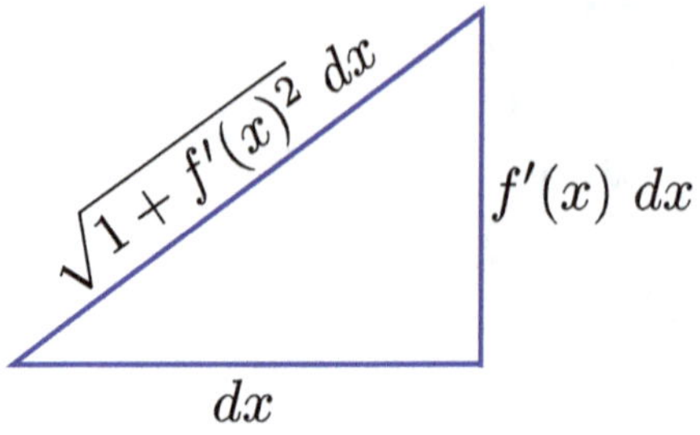

Figure 8.1: Schematic of a differential arc length.

Example 8.1 Compute the circumference of a circle of radius a.

To simplify matters, let us focus on the portion of the circle in the first quadrant. Here, the (quarter) circle is defined by the function

$$f(x) = \sqrt{a^2 - x^2},$$

for $x \in [0, a]$. The derivative of f is obtained using the chain rule as

$$f'(x) = \frac{-x}{\sqrt{a^2 - x^2}}.$$

The arc length of the circle is therefore

$$\mathcal{L} = 4\int_0^a \sqrt{1 + \frac{x^2}{a^2 - x^2}}\ dx = 4\int_0^a \frac{a}{\sqrt{a^2 - x^2}}\ dx.$$

Using the trigonometric substitution $x = a\sin(\theta)$, this integral transforms into

$$\mathcal{L} = 4a\int_0^{\pi/2} d\theta.$$

(We have $dx = a\cos(\theta)\ d\theta$ and $\sqrt{a^2 - x^2} = a\cos(\theta)$; also, the endpoints $x = 0$ and $x = a$ correspond to $\theta = 0$ and $\theta = \pi/2$, respectively.) This integral may easily be performed to yield the arc length of a circle with radius a as

$$\mathcal{L} = 2\pi a,$$

thereby recovering the well known formula. ■

Example 8.2 Determine the arc length of the hyperbolic cosine function over the interval $[-1, 1]$.

Here, we will be considering the function $f(x) = \cosh(x)$ and its derivative $f'(x) = \sinh(x)$. The arc length is therefore given as

$$\mathcal{L} = \int_{-1}^{1} \sqrt{1 + \sinh^2(x)}\ dx = \int_{-1}^{1} \cosh(x)\ dx.$$

(Recall the hyperbolic trigonometric identity $\cosh^2(x) - \sinh^2(x) = 1$.) This easily integrates to yield

$$\mathcal{L} = \sinh(1) - \sinh(-1) = 2\sinh(1).$$

The final equality holds since the hyperbolic sine is an odd function. ■

8.1.2 Arc Length of Parametric Curves

So far, we have seen how to compute the arc length of the graph of a function. Some planar curves, however, are not described by functions, but by a set of parametric equations. Computing the arc lengths of such curves is as simple as integrating the speed.

Proposition 8.1.2 Let $x = x(t)$ and $y = y(t)$ parameterize a curve γ on the interval $t \in [a,b]$. Then the arc length $\mathcal{L}$ of the curve γ is given by

$$\mathcal{L} = \int_a^b \sqrt{x'(t)^2 + y'(t)^2}\, dt. \tag{8.2}$$

(R) Recall from equation (4.22) that the quantity

$$v(t) = \sqrt{x'(t)^2 + y'(t)^2}$$

represents the speed at which we trace the curve, as a function of time t. Thus, by integrating the speed, we obtain the total distance traveled. This essentially reduces a two-dimensional problem into a one-dimensional one.

As it turns out, Proposition (8.1.1) is but a special case of Proposition (8.1.2), as we shall see in our next example.

■ **Example 8.3** Use Proposition (8.1.2) to compute the arc length of the graph of a function $f(x)$ over the interval $[a,b]$.

The secret here is that we can always parameterize the graph using the independent variable x. Identifying x with t, to avoid confusion, we may write

$$x(t) = t \qquad \text{and} \qquad y(t) = f(t).$$

Applying equation (8.2), we obtain

$$\mathcal{L} = \int_a^b \sqrt{1 + f'(t)^2}\, dt,$$

which is clearly identical to equation (8.1). Thus, the arc length of a graph is really just a special case of the arc length of a set of parametric equations. ■

■ **Example 8.4** Consider the parameterized circle

$$x(t) = R\cos(\omega t) \qquad \text{and} \qquad y(t) = R\sin(\omega t).$$

These parametric equations represent a circle of radius R being traced with a constant angular speed ω, resulting in a period of

$$T = \frac{2\pi}{\omega}.$$

The x- and y-velocities are given by

$$x'(t) = -R\omega\sin(\omega t) \qquad \text{and} \qquad y'(t) = R\omega\cos(\omega t).$$

The particle's speed is obtained using the Pythagorean theorem:

$$v(t) = \sqrt{x'(t)^2 + y'(t)^2} = R\omega.$$

Integrating over a single cycle, we obtain

$$\mathcal{L} = \int_0^{2\pi/\omega} \omega R\, dt = 2\pi R,$$

which again recovers the circumference of a circle. ■

■ **Example 8.5** A particle is launched from a cannon at an angle θ above the horizontal. Its x- and y-velocities are given by

$$\begin{aligned} x'(t) &= v_0\cos(\theta) \\ y'(t) &= v_0\sin(\theta) - gt, \end{aligned}$$

where g is the (constant) downward acceleration due to gravity and v_0 represents the projectile's initial speed. (The initial angle θ is also a constant.) Determine the total length of the path the projectile travels before it lands.

First, let us compute the total time during which the particle remains airborne. Assuming that the particle is both launched and lands at ground level, $y = 0$, we may integrate the y-velocity to obtain

$$y(t) = v_0\sin(\theta)t - \frac{g}{2}t^2.$$

This represents the projectile's height off the ground as a function of time. The roots of this quadratic function are at the launch time $t = 0$ and at the landing time

$$T = \frac{2v_0\sin(\theta)}{g}.$$

The projectile's speed is given by

$$v(t) = \sqrt{v_0^2\cos^2(\theta) + (v_0\sin(\theta) - gt)^2}.$$

We will not further expand the expression under the root, as it is already in its complete-the-square form.

The total distance of the particle's path is obtained from integrating the particle's speed over its time of flight:

$$\mathcal{L} = \int_0^{2v_0 \sin(\theta)/g} \sqrt{v_0^2 \cos^2(\theta) + (v_0 \sin(\theta) - gt)^2}\, dt.$$

As a matter of style, let us pull a factor of v_0 from the integrand. (This, admittedly, was only something I thought of having already worked the problem once and seen that it leads to a simplification later on.) We obtain

$$\mathcal{L} = v_0 \int_0^{2v_0 \sin(\theta)/g} \sqrt{\cos^2(\theta) + \left(\sin(\theta) - \frac{gt}{v_0}\right)^2}\, dt.$$

Next, let us consider the substitution

$$u = \sin(\theta) - \frac{gt}{v_0}.$$

Its differential is given by

$$du = \frac{-g}{v_0}\, dt,$$

and the limits of integration transform into their new manifestations:

$$u(0) = \sin(\theta) \qquad \text{and} \qquad u(2v_0 \sin(\theta)/g) = -\sin(\theta).$$

The integral is therefore equivalent to

$$\mathcal{L} = \frac{v_0^2}{g} \int_{-\sin(\theta)}^{\sin(\theta)} \sqrt{\cos^2(\theta) + u^2}\, du.$$

Notice that we used the minus sign inherited from substituting dt for $-v_0\, du/g$ in order to reverse the order of the limits of integration. We may next utilize formula (D.16) to obtain

$$\mathcal{L} = \frac{v_0^2}{2g} \left(u\sqrt{\cos^2(\theta) + u^2} + \cos^2(\theta) \ln\left| u + \sqrt{\cos^2(\theta) + u^2} \right| \right) \Bigg|_{u=-\sin(\theta)}^{u=\sin(\theta)}$$

Evaluating this antiderivative at the upper and lower limits of integration, we obtain

$$\mathcal{L} = \frac{v_0^2 \sin(\theta)}{g} + \frac{v_0^2 \cos^2(\theta)}{2g} \ln\left(\frac{1 + \sin(\theta)}{1 - \sin(\theta)}\right). \tag{8.3}$$

This formula gives the total arc length of the projectile's parabolic trajectory in terms of the problem parameters v_0, g, and θ. ■

8.1.3 Area Between Two Curves

In Chapter 5, we discussed how one can use definite integrals and the fundamental theorem of calculus to determine the area under the graph of a function and above the horizontal axis. This result may easily be modified to determine the area bounded *between* two curves.

Proposition 8.1.3 Consider two continuous functions f and g with the property that $g(x) \leq f(x)$ for $x \in [a,b]$. Then the area A bounded by the functions f and g and the vertical lines $x = a$ and $x = b$ is obtained by integrating the difference of the two functions:

$$A = \int_a^b \left[f(x) - g(x)\right]\, dx. \tag{8.4}$$

Proof. To see why equation (8.4), consider the Riemann sum depicted in Figure 8.2. The area bounded by the graphs of functions f and g between $x = a$ and $x = b$

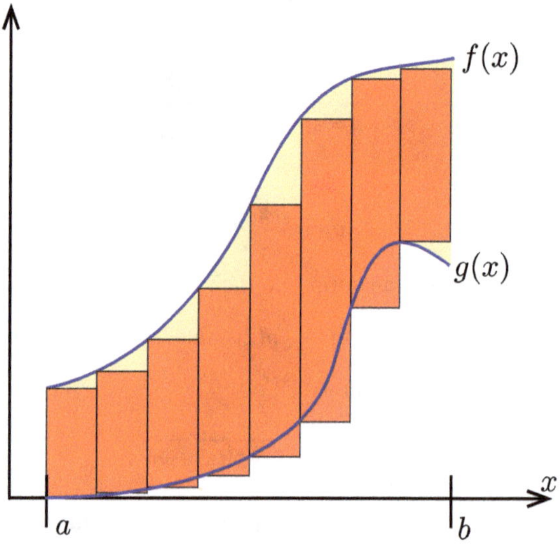

Figure 8.2: Riemann sum illustrating the area between two curves.

can clearly be approximated by a set of rectangles similar to the ones depicted in the figure. The height of the rectangle at x_i is given simply by the difference $f(x_i) - g(x_i)$. Therefore, the Riemann-sum approximation for the area between the two curves is given by

$$\sum_{i=1}^{N} \left[f(x_i) - g(x_i)\right] \Delta x.$$

In the limit as $N \to \infty$, this sum approaches the exact area, which is therefore given by the definite integral

$$A = \int_a^b \left[f(x) - g(x)\right]\, dx.$$

This completes our proof. ■

▪ **Example 8.6** Compute the area bounded between the curves $y = x^2$ and $y = \sqrt{x}$, as shown in Figure 8.3.

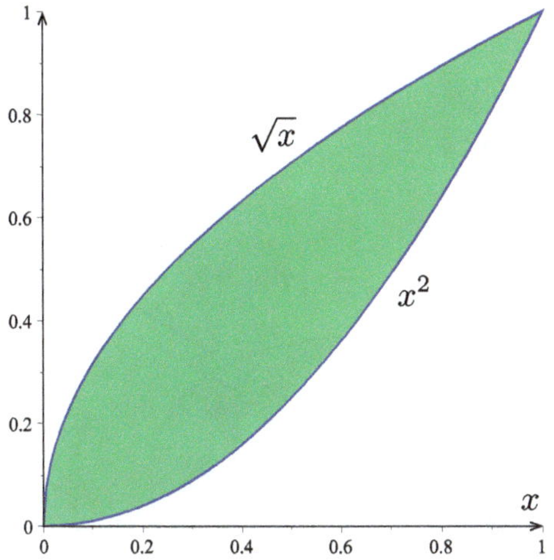

Figure 8.3: Area bounded between $y = x^2$ and $y = \sqrt{x}$.

To accomplish this, we only need to compute the integral

$$A = \int_0^1 \left(\sqrt{x} - x^2\right)\, dx.$$

Using the fundamental theorem of calculus, we obtain

$$A = \left(\frac{2x^{3/2}}{3} - \frac{x^3}{3}\right)\Bigg|_{x=0}^{x=1} = \frac{1}{3}.$$

This is a rather astonishing result; despite the slender appearance of the leaf in Figure 8.3, the shaded area occupies 1/3 of the total area in the unit square $[0,1] \times [0,1]$. ∎

8.1.4 Area in Polar Coordinates

We can also use integration to determine the area enclosed within a polar plot, as shown in our next proposition. (Recall that we introduced polar functions in Section 4.5.)

Proposition 8.1.4 Given a continuous polar function $f(\theta)$ defined on an interval $[a,b]$, the area enclosed between the rays $\theta = a$ and $\theta = b$ and the curve $r = f(\theta)$ is given by

$$A = \frac{1}{2}\int_a^b f(\theta)^2\, d\theta. \tag{8.5}$$

Proof. To prove the proposition, we will consider a differential angular sector bounded by the rays $\theta = \theta_i$ and $\theta = \theta_i + \Delta\theta$, as shown in Figure 8.4.

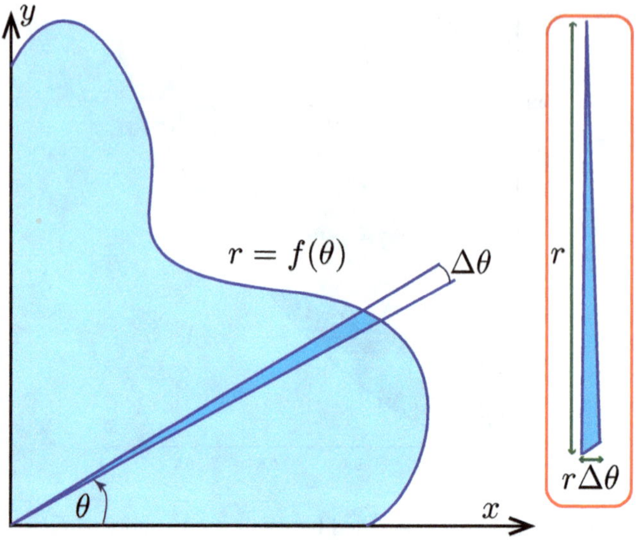

Figure 8.4: Differential area in a differential angular sector.

The graph lies a distance $r = f(\theta)$ from the origin, at angular location θ. Approximating $f(\theta)$ as a constant over the interval $[\theta_i, \theta_i + \Delta\theta]$ (in the spirit of a left-hand Riemann sum), our differential sector becomes an arc of a circle. The arc length along the edge of such a circle is given by $r\Delta\theta = f(\theta_i)\Delta\theta$. Thinking of this shape as an isosceles triangle with height $r = f(\theta_i)$ and width $r\Delta\theta = f(\theta_i)\Delta\theta$, the Riemann sum approximation for the total area bounded by the polar function $r = f(\theta)$ is given by

$$A \approx \sum_{i=1}^{N} \frac{1}{2} f(\theta_i)^2 \Delta\theta.$$

As $N \to \infty$, this becomes exact, resulting in the definite integral given by equation (8.5). ■

■ **Example 8.7** Compute the area of one petal of the graph of the polar function $f(\theta) = \cos(2\theta)$, as shown in Figure 8.5.

We first identify the bounds of integration as $\theta = -\pi/4$ and $\theta = +\pi/4$. (These are two consecutive zeros of the function $f(\theta) = \cos(2\theta)$.) The area of a single petal can thus be obtained from equation (8.5) as

$$A = \frac{1}{2} \int_{-\pi/4}^{\pi/4} \cos^2(2\theta)\, d\theta.$$

Using the fundamental theorem of calculus and formula (D.24), we obtain

$$A = \frac{1}{4} \left(\theta + \frac{\sin(4\theta)}{4} \right) \Bigg|_{\theta=-\pi/4}^{\theta=\pi/4} = \frac{\pi}{8}.$$

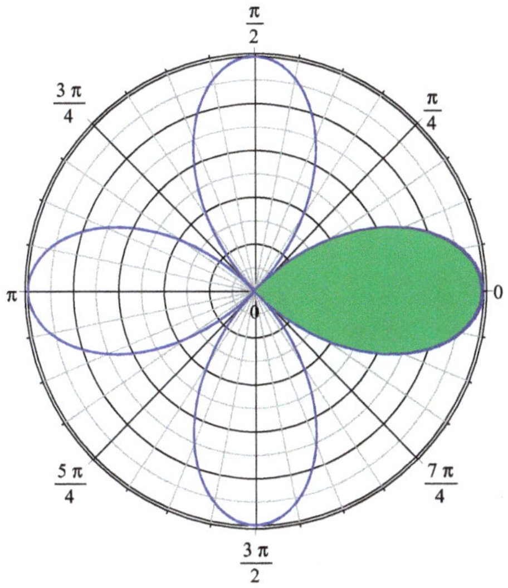

Figure 8.5: Graph of the polar function $f(\theta) = \cos(2\theta)$.

The area of a single petal is therefore $\pi/8$. Consequently, the area of all four petals is $\pi/2$. ■

Exercises

For Exercises 1–10, determine the arc length of the indicated curve.

1. $f(x) = x^2$, $0 \le x \le 2$.
2. $f(x) = x^{3/2}$, $0 \le x \le 2$.
3. $f(x) = \ln|\sec(x)|$, $0 \le x \le \pi/4$.
4. $f(x) = \sqrt{1-x^2}$, $0 \le x \le 1$.
5. ★ $f(x) = 2\sqrt{x}$, $0 \le x \le 1$.
6. $x(t) = a + bt$,
   $y(t) = c + dt$; $0 \le t \le 1$.
7. $x(t) = t^2$,
   $y(t) = t^3$; $0 \le t \le 1$.
8. $x(t) = 3t^4$,
   $y(t) = t^6$; $0 \le t \le 1$.
9. $x(t) = \dfrac{\sin(t)}{\sqrt{2}} + \dfrac{\sin^2(t)}{\sqrt{8}}$,
   $y(t) = \dfrac{\sin(t)}{\sqrt{2}} - \dfrac{\sin^2(t)}{\sqrt{8}}$;
   $\dfrac{-\pi}{2} \le t \le \dfrac{\pi}{2}$.
10. $x(t) = \sin(t) + \sin^2(t)$,
    $y(t) = \sin(t) - \sin^2(t)$; $-\pi/2 < t < \pi/2$.

For Exercises 11–15, determine the area bounded by the indicated polar function.

11. $f(\theta) = 2 + 4\cos(\theta)$; $\theta \in \left[\dfrac{2\pi}{3}, \dfrac{4\pi}{3}\right]$.
12. $f(\theta) = \theta$; $\theta \in [0, \pi]$.
13. $f(\theta) = e^{-\theta}$; $\theta \in [0, \ln(2)]$.
14. $f(\theta) = \sin(\theta)$; $\theta \in \left[\dfrac{-\pi}{2}, \dfrac{\pi}{2}\right]$.
15. $f(\theta) = \sqrt{\theta}$; $\theta \in [0, 4]$.

For Exercises 16–20, determine the area between the graphs of functions f and g, on the indicated domain.

16. $f(x) = \sin(x)$; $g(x) = \cos(x)$; $x \in [\pi/4, 5\pi/4]$.
17. $f(x) = \cos(x)$; $g(x) = 1 - \dfrac{x^2}{2}$; $x \in [0, \pi/2]$.
18. $f(x) = x^2$; $g(x) = x^3$; $x \in [0, 1]$.
19. $f(x) = \sin(x)$; $g(x) = x$; $x \in [0, \pi/2]$.
20. $f(x) = \cosh(x)$; $g(x) = \sinh(x)$; $x \in [0, \infty)$.

Problems

21. Determine the area between the curves $y = x^3$ and $y = \sqrt[3]{x}$. Then determine a general formula for the area between $y = x^n$ and $y = \sqrt[n]{x}$, for $n = 2, 3, \ldots$.
22. Determine the area between the curves

$$f(x) = x + \frac{1}{x^2} \qquad \text{and} \qquad g(x) = x - \frac{1}{x^2}$$

for $x > 1$.
23. Consider the curve defined by the parametric equations

$$\begin{aligned} x(t) &= \ln(t) \\ y(t) &= 2\sqrt{t}, \end{aligned}$$

for $t \in [3, 8]$.
   (a) Write an expression for the arc length of this curve. Show that it is equivalent to the following

$$\mathcal{L} = \int_3^8 \frac{dt}{\sqrt{t+1}} + \int_3^8 \frac{dt}{t\sqrt{t+1}}.$$

   (b) Compute the value of the first integral.
   (c) For the second integral, use the substitution $\sqrt{t+1} = \tanh(u)$. Evaluate this integral.
   (d) Use your results from (b) and (c) to write an exact expression for the arc length of the given curve.

8.2 Solids of Revolution

In this section, we use the theory of integration to analyze volumes of certain shapes. Some form of symmetry will be required for each of these volume problems, as we have not as of yet developed the tools of multivariable calculus. In this way, the volume will be renderable by a single integral. We begin with a classic example—solids of revolution.

8.2.1 Disk Method

We begin by examining what is meant by a solid of revolution.

Definition 8.2.1 A *solid of revolution* is the three-dimensional shape obtained by revolving a two-dimensional about an axis.

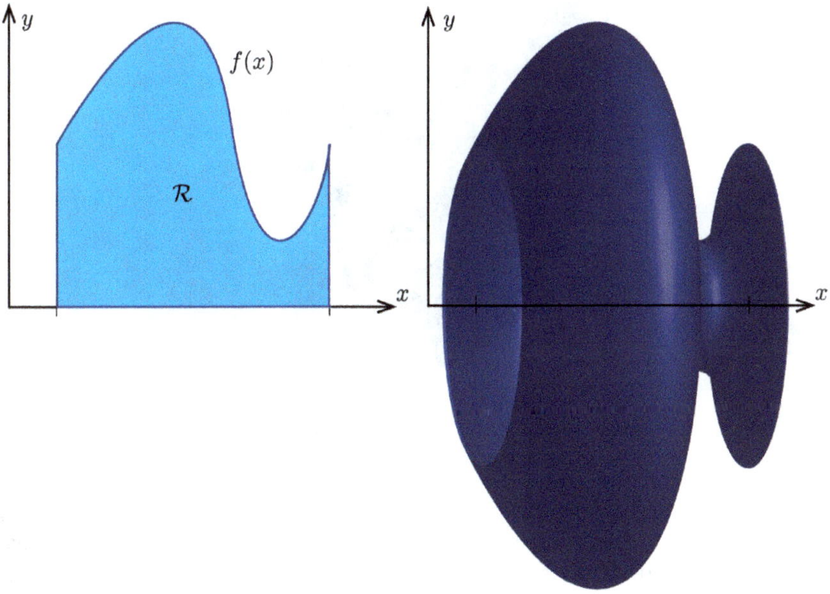

Figure 8.6: The region $\mathcal{R}$ revolved about the x-axis.

To illustrate this, consider the region $\mathcal{R}$ bounded by the graph of the function $f(x)$ and the lines $y = 0$, $x = a$, and $x = b$. The solid of revolution obtained by revolving region $\mathcal{R}$ about the x-axis is shown in Figure 8.6.

In order to calculate the volume of a solid of revolution, we apply the following theorem.

Theorem 8.2.1 — The Disk Method. Let $\mathcal{R}$ be the region bounded by the graph of a nonnegative function f, the x-axis, and the lines $x = a$ and $x = b$. The volume of the solid of revolution obtained by revolving region $\mathcal{R}$ about the x-axis is given by

$$V = \pi \int_a^b f(x)^2 \, dx. \tag{8.6}$$

Proof. To prove this, let us approximate the volume using a Riemann sum. Par-

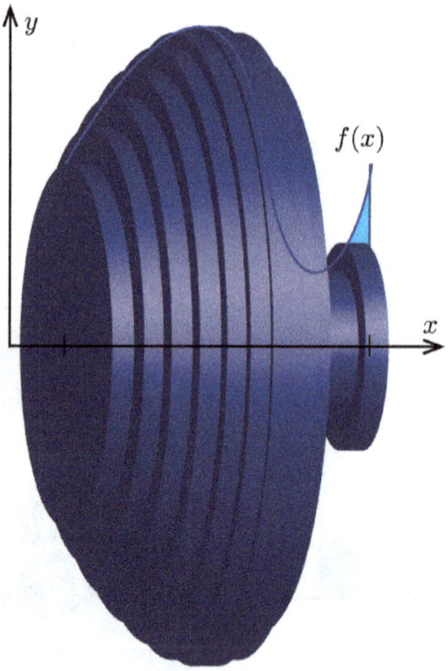

Figure 8.7: The region $\mathcal{R}$ revolved about the x-axis.

titioning the interval $[a,b]$ into N subintervals, we may approximate the solid of revolution by a stack of cylindrical disks, as shown in Figure 8.7. The ith disk has a circular face with radius $f(x_i)$. (Recall that $f(x)$ is the distance between a point on the graph of f and the x-axis.) Therefore, the area of the face of the ith disk is given by

$$A_i = \pi f(x_i)^2.$$

Since each disk has a width of Δx, the volume of the ith disk is given by

$$\Delta V_i = \pi f(x_i)^2 \Delta x.$$

Therefore, the Riemann-sum approximation to the volume is

$$V = \sum_{i=1}^{N} \pi f(x_i)^2 \Delta x.$$

Taking the limit as $N \to \infty$, this Riemann sum turns into the definite integral given by equation (8.6), which proves our result. ■

■ **Example 8.8** Compute the volume of a sphere of radius a.

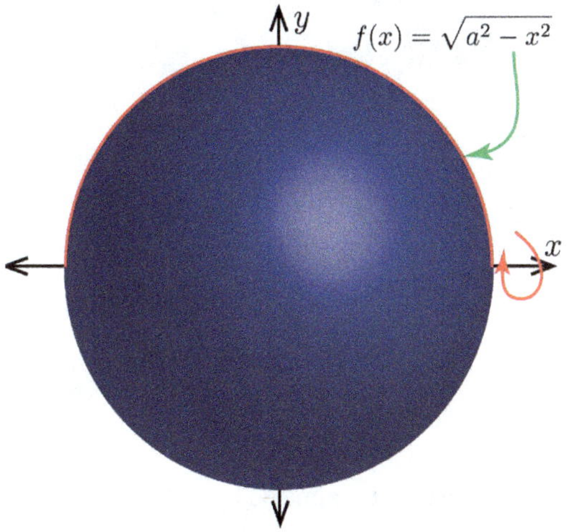

Figure 8.8: Graph of the function $f(x) = \sqrt{a^2 - x^2}$ revolved about the x-axis.

A sphere of radius a is obtained by revolving the region bounded by the graph of the function $f(x) = \sqrt{a^2 - x^2}$ and the x-axis about the x-axis, as shown in Figure 8.8. We may approximate this sphere using a stack of N disks, as shown in Figure 8.9. The disk starting at position x_i has radius $f(x_i)$, as shown in the Figure. Therefore, the area of the face of the ith disk is given by

$$A_i = \pi(a^2 - x_i^2),$$

with a corresponding volume of

$$\Delta V_i = \pi(a^2 - x_i^2)\Delta x.$$

Summing the volume of these N disks and taking the limit as $N \to \infty$, we once again obtain the integral of equation (8.6). For this function, that integral yields

$$\begin{aligned} V &= \pi \int_{-a}^{a} (a^2 - x^2)\, dx \\ &= \pi \left(a^2 x - \frac{x^3}{3} \right) \Bigg|_{x=-a}^{x=a} \\ &= \pi \left[\left(a^3 - \frac{a^3}{3} \right) - \left(-a^3 + \frac{a^3}{3} \right) \right] = \frac{4}{3}\pi a^3. \end{aligned}$$

This proves the famous result. ■

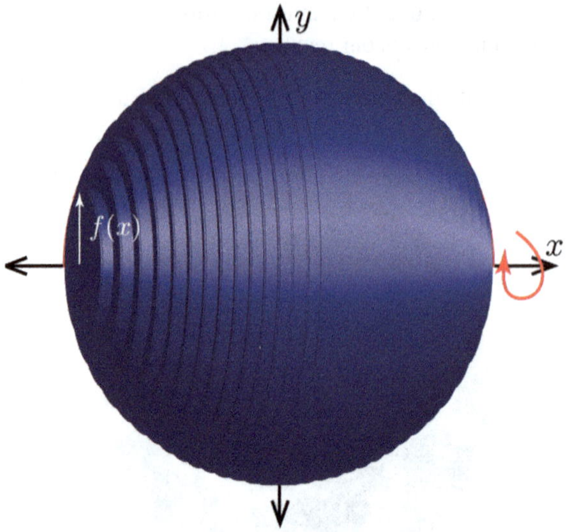

Figure 8.9: A sphere approximated by a stack of circular disks.

(R) It is a good habit to visualize the disks that approximate the given volume, as opposed to blindly relying on equation (8.6). The approximating disks will always lead to the correct formula.

■ **Example 8.9** Determine the volume of a right-circular cone with base radius r and height h.

A right-circular cone may be viewed as a solid of revolution. Consider the region bounded by the function $f(x) = rx/h$ and the lines $y = 0$ and $x = h$, as shown in Figure 8.10. Upon revolving this region about the x-axis, we obtain our cone.

The volume of this cone is therefore given by

$$V = \pi \int_0^h \frac{r^2 x^2}{h^2}\, dx = \frac{\pi r^2 x^3}{3h^2}\bigg|_{x=0}^{x=h} = \frac{\pi r^2 h}{3}.$$

Another well known formula is thus recovered! ■

8.2.2 Washer Method

When a solid of revolution contains a hollow cavity, we utilize *washers* instead of disks to compute its volume.

Theorem 8.2.2 — The Washer Method. Consider the region $\mathcal{R}$ bounded by the curves $y = f(x)$ and $y = g(x)$ and the lines $x = a$ and $x = b$, where $0 \le g(x) \le f(x)$, for all $x \in [a, b]$. Then the solid of revolution obtained by revolving region $\mathcal{R}$

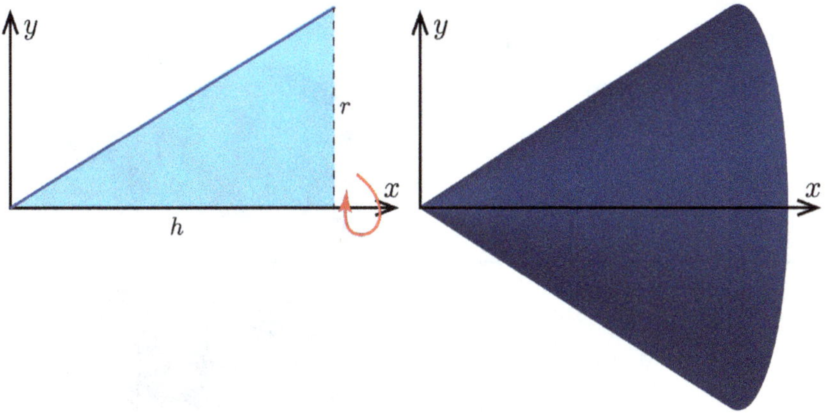

Figure 8.10: The linear function $f(x) = rx/h$ revolved about the x-axis to form a cone.

about the x-axis has a volume given by

$$V = \pi \int_a^b \left[f(x)^2 - g(x)^2 \right] \, dx. \tag{8.7}$$

Proof. First, let us consider the illustration shown in Figure 8.11. Here, the region bounded by the functions f and g between $x = a$ and $x = b$ is revolved about the x-axis, forming a hollow shell. Each cross section has the shape of a circular annulus, with inner radius $r_1 = g(x)$ and outer radius $r_2 = f(x)$. The approximate volume is shown alongside a single washer in Figure 8.12.

The area of a circular annulus (i.e., a circle with a smaller circle removed from its center) is given by

$$A = \pi(r_2^2 - r_1^2).$$

(This is literally the area of the larger circle minus the area of the smaller circle.) Thus, the volume of the ith washer, located at position x_i, is given by

$$V_i = \pi \left[f(x_i)^2 - g(x_i)^2 \right] \Delta x.$$

The total volume may be approximated by summing the volumes of these individual washers:

$$V = \sum_{i=1}^{N} \pi \left[f(x_i)^2 - g(x_i)^2 \right] \Delta x.$$

In the limit as $N \to \infty$, this approximation becomes exact, and the above Riemann sum converges to the value of the definite integral (8.7). This completes the proof. ∎

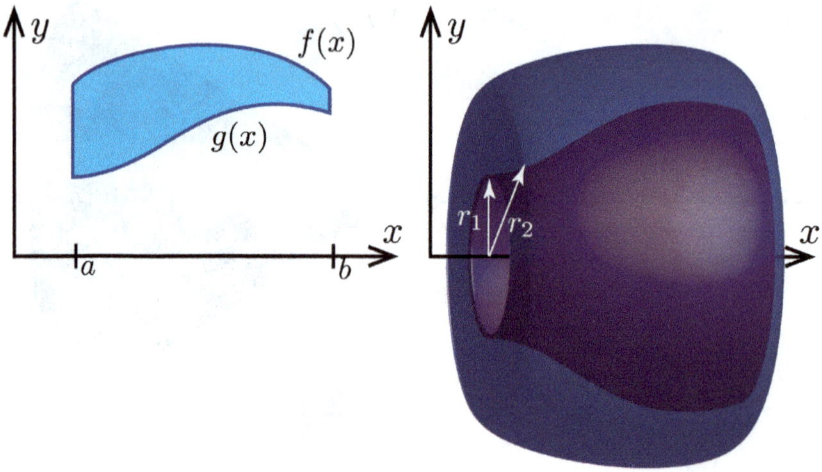

Figure 8.11: The region bounded by the graphs of the functions f and g and the vertical lines $x = a$ and $x = b$, revolved about the x-axis.

■ **Example 8.10** The cup of a goblet is to be constructed by revolving the region $\mathcal{R}$ bounded by the curves $y = 2x^{1/4}$, $y = 2(x-1)^{1/4}$, $x = 0$, and $x = 4$ about the x-axis, as shown in Figure 8.13. Determine the volume of glass required to construct such an object.

We will consider the intervals $[0,1]$ and $[1,4]$ separately. On the domain $[0,1]$, the solid of revolution has no cavity, and therefore its volume is given by

$$V_1 = \pi \int_0^1 4\sqrt{x}\,dx = \frac{8\pi}{3}.$$

On the second portion of the domain $[1,4]$, the solid of revolution has a cavity. We therefore use the washer method to obtain the volume of the second part:

$$V_2 = 4\pi \int_1^4 \left(\sqrt{x} - \sqrt{x-1}\right)\,dx = \frac{8\pi}{3}\left(x^{3/2} - (x-1)^{3/2}\right)\Big|_{x=1}^{x=4} = \frac{8\pi}{3}\left(7 - \sqrt{27}\right).$$

The total volume of this cup is therefore equal to

$$V = \frac{8\pi}{3}\left(8 - \sqrt{27}\right) \approx 23.48945853496103.$$

■

8.2.3 Surface Area

In addition to computing the volume of a solid of revolution, we may also compute its surface area.

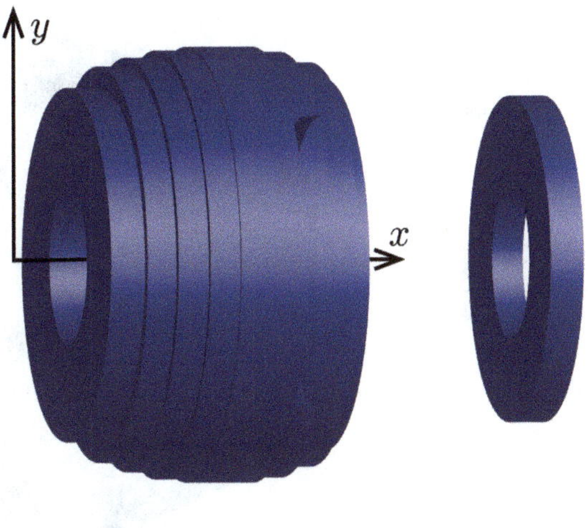

Figure 8.12: Approximation of a solid of revolution using washers.

Theorem 8.2.3 — Surface Area. Let $\mathcal{R}$ be the region bounded by the graph of a nonnegative function f, the x-axis, and the lines $x = a$ and $x = b$. The surface area of the solid of revolution obtained by revolving region $\mathcal{R}$ about the x-axis is given by

$$S = 2\pi \int_a^b f(x)\sqrt{1 + f'(x)^2}\, dx. \quad (8.8)$$

(R) The surface area given in equation (8.8) consists only of the *sides* of the volume of revolution. The surface area of the two end caps are simply given by

$$\pi f(a)^2 \qquad \text{and} \qquad \pi f(b)^2,$$

as these end caps are circles with radii $f(a)$ and $f(b)$.

Proof. As we previously seen, the circular disk located at x_i has radius $f(x_i)$. Its circumference is therefore

$$2\pi f(x_i).$$

To approximate the surface area of this disk, it does not suffice to simply multiply this circumference by the interval width Δx, as this does not take into account the slant of the graph of $f(x)$. Instead, we multiply the circumference about the x-axis by the differential arc length $\Delta s = \sqrt{1 + f'(x)^2}\Delta x$. (Compare with equation (8.1).) Thus, the side area of the given disk is equal to

$$2\pi f(x_i)\sqrt{1 + f(x_i)^2}\Delta x.$$

Upon summing over each disk and then taking the limit as $N \to \infty$, we recover the integral in equation (8.8). ■

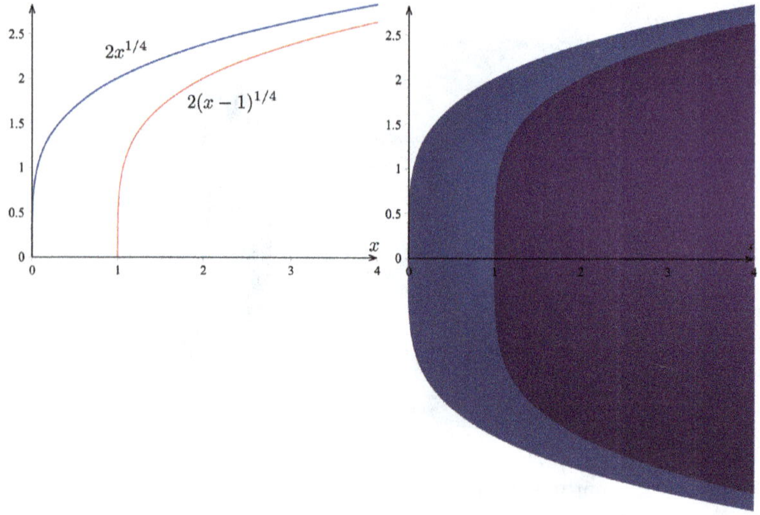

Figure 8.13: The solid of revolution from Example 8.10.

■ **Example 8.11** Determine the surface area of a sphere of radius a.

In example 8.8, we saw that a sphere of radius a is formed by revolving the area under the graph of the function $f(x) = \sqrt{a^2 - x^2}$ about the x-axis. To apply equation (8.8), we need to compute the derivative of this function, which works out to be

$$f'(x) = \frac{-x}{\sqrt{a^2 - x^2}}.$$

Now,

$$\sqrt{1 + f'(x)^2} = \sqrt{1 + \frac{x^2}{a^2 - x^2}} = \frac{a}{\sqrt{a^2 - x^2}}.$$

Therefore, the surface area given by equation (8.8) simplifies to

$$S = 2\pi a \int_{-a}^{a} dx = 4\pi a^2.$$

Thus, the surface area of a sphere of radius a is given by $4\pi a^2$. ■

■ **Example 8.12** Determine the surface area of a right-circular cone with base radius r and height h.

As we saw in example (8.9), such a cone is obtained by revolving the line segment $f(x) = rx/h$ between $x = 0$ and $x = h$ about the x-axis. The derivative is $f'(x) = r/h$. Therefore, the surface area is given by

$$S = 2\pi \int_0^h \frac{rx}{h} \sqrt{1 + \frac{r^2}{h^2}}\, dx = \frac{2\pi r}{h^2} \sqrt{h^2 + r^2} \int_0^h x\, dx = \pi r \sqrt{h^2 + r^2}.$$

Thus, the surface area equals π times the base radius times the slant height of the cone, another familiar formula from basic geometry. ∎

Exercises

For Exercises 1–10, determine the (a) volume and (b) surface area of the solid of revolution formed by revolving the area enclosed by the given curves about the indicated axis.

1. $y = x^2$, $y = 0$, $x = 0$, and $x = 1$; revolved about the x-axis.
2. $y = \sqrt{x}$, $y = 0$, $x = 0$, and $x = 1$; revolved about the x-axis.
3. $y = x^3$, $y = 0$, $x = 0$, and $x = 1$; revolved about the x-axis .
4. $y = e^x$, $y = 0$, $x = 0$, and $x = 1$; revolved about the x-axis.
5. $y = \sin(x)$, $y = 0$, $x = 0$, and $x = \pi$; revolved about the x-axis.
6. $y = \sinh(x)$, $y = 0$, $x = 0$, and $x = \ln(2)$; revolved about the x-axis.
7. $y = \cosh(x)$, $y = 0$, $x = -1$, and $x = 1$; revolved about the x-axis.
8. $y = \frac{1}{36}\left(2\sqrt{3} - x\right)^3$, $y = 0$, $x = 0$, and $x = 2\sqrt{3}$; revolved about the x-axis.
9. $y = x - x^3/3$, $y = 0$, $x = 0$, and $x = 1$; revolved about the x-axis.
10. $y = \tan(x)$, $y = 0$, $x = 0$, and $x = \pi/4$; revolved about the x-axis.

For Exercises 11–20, determine the volume of the solid of revolution formed by revolving the area enclosed by the given curves about the indicated axis.

11. $y = x^2$, $y = \sqrt{x}$; revolved about the x-axis.
12. $y = x^3$, $y = \sqrt[3]{x}$; revolved about the x-axis.
13. $y = 2$, $y = 2 + \sin(x)$, $x = 0$, $x = \pi$; revolved about the x-axis.
14. $y = 2 + \sin(x)$, $y = 1 + \cos(x)$, $x = 0$, $x = 3\pi/2$; revolved about the x-axis.
15. $y = e^{-x}$, $y = x^2$, $x = 0$, $x = 0.7$; revolved about the x-axis.
16. $y = x^2$, $y = x^3$; revolved about the line $y = -1$.
17. $y = x^2$, $y = x^3$; revolved about the line $y = 1$.
18. $y = \sin(x)$, $y = 1 + \sin(x)$, $x = 0$, $x = 10\pi$; revolved about the line $y = 5$.
19. $y = x^2$, $y = x^3$; revolved about the y-axis.
20. $y = \sqrt{x}$, $y = 0$, $x = 0$, $x = 4$; revolved about the y-axis.

Problems

21. *Gabriel's horn* is the solid of revolution obtained by revolving the region bounded by the graph of $f(x) = 1/x$ and the lines $y = 0$ and $x = 1$ about the x-axis. This region extends infinitely in the positive x-direction. Gabriel's horn is shown in Figure 8.14.
    (a) Calculate the volume of Gabriel's horn.

Figure 8.14: Gabriel's Horn.

(b) Set up an (improper) definite integral for the surface area. Show that this integral diverges.

Gabriel's horn therefore has the interesting property that it contains only a finite amount of volume, but has an infinite surface area. It would make for a very inefficient paper cup, as it would require an infinite amount of paper to construct, but would only be able to contain a finite amount of water.

22. A cylindrical hole with radius b is drilled through a solid sphere with radius a, as shown in Figure 8.15. Determine the volume that remains.

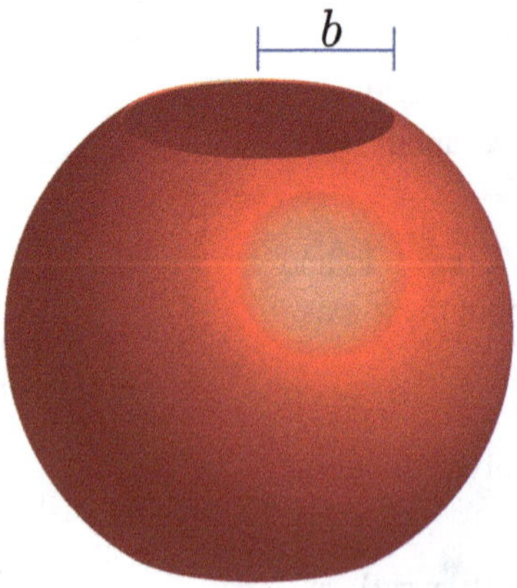

Figure 8.15: Sphere with a cylindrical hole.

23. Determine the volume of the cap of a sphere of radius a, as shown in Figure 8.16. The cap consists of everything that is within a vertical distance h to the top of the sphere.
24. A wine barrel with height h and maximum radius R is modeled by revolving the region bounded by $y = R - ax^2$, $y = 0$, $x = -h/2$, and $x = h/2$ about the x-axis.
    (a) Determine the volume of such a barrel.

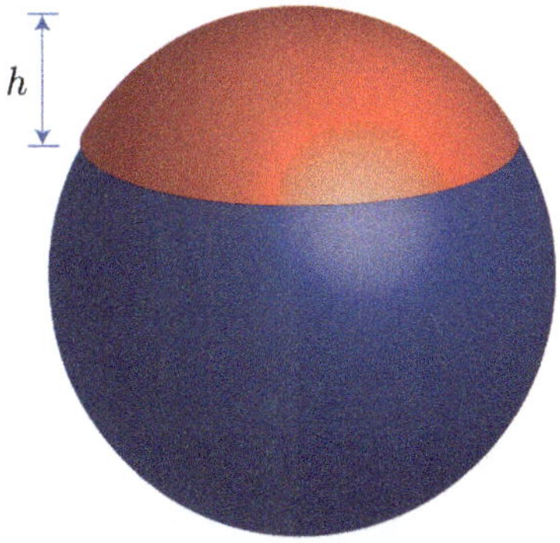

Figure 8.16: End cap of a sphere.

(b) If a particular barrel has a height of $h = 1$, maximum radius $R = 1/2$, and parameter $a = 1/2$, determine its volume and surface area.

25. A *torus* is constructed by revolving a circle of radius a about an axis that is located at a distance b from its center, as shown in Figure 8.17. Determine the volume of this torus.

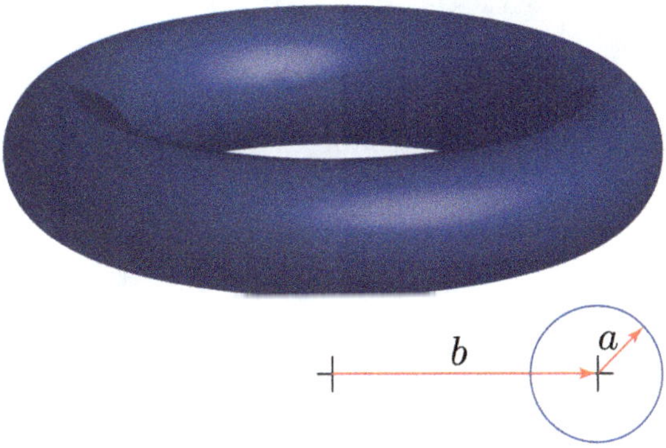

Figure 8.17: Picture of a torus.

26. Consider the region $\mathcal{R}$ bounded by the curves $y = f(x)$, $y = g(x)$, $x = a$, and $x = b$, where $0 \leq g(x) \leq f(x)$. Let V represent the volume obtained by revolving region $\mathcal{R}$ about the x-axis, and let V' represent the volume obtained by revolving region $\mathcal{R}$ about the line $y = -k$, for $k > 0$. Show that

$$V' = V + 2\pi kA,$$

where A is the area of region $\mathcal{R}$.

8.3 Volumes of Symmetric Regions

In the previous section, we explored specific examples of computing volumes of solids of revolution using both the disk and washer methods. In this paragraph, we will generalize these concepts to a broader picture.

8.3.1 Axis and Cross Sections

Since we are presently limited to a single integral, we can only consider volumes that possess some form of symmetry. Solids of revolution, for example, are symmetric about their axis of revolution, which duals as an axis of symmetry. One such solid of revolution is illustrated in Figure 8.18, concurrently with one of the approximating disks. More generally—that is, without considering that these disks

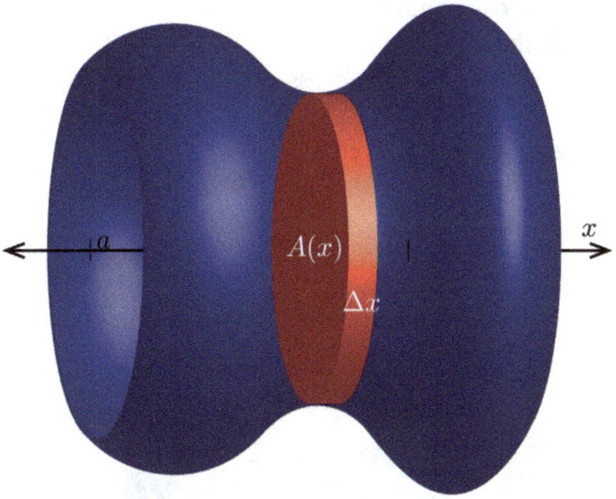

Figure 8.18: Volume as an integral of cross-sectional areas.

have a circular face—the volume of one "slice" of the solid of revolution is given by

$$V_i = A(x_i)\,\Delta x,$$

where the function $A(x)$ represents the area of the planar cross section, perpendicular to the axis of integration, at position x. For the disk method, $A(x) = \pi f(x)^2$; for the washer method, $A(x) = \pi\left[f(x)^2 - g(x)^2\right]$. The total approximate volume when using N slices is therefore given by

$$V \approx \sum_{i=1}^{N} A(x_i)\Delta x.$$

In the limit as $N \to \infty$, this converges to the exact volume, which is therefore given by the definite integral

$$V = \int_a^b A(x)\,dx.$$

This is the generalized form of the disk and washer methods: an integral of the cross-sectional areas yields the volume.

Proposition 8.3.1 — Strategy for Computing Volume. The following strategy can be used for computing the volume V of a three-dimensional solid E.

1. Identify the axis of integration. Select the axis that has the simplest perpendicular cross sections (e.g., circles, squares, etc.).
2. Introduce a variable x that determines position along the axis of integration. Define the zero point $x = 0$ and the positive direction.
3. Determine the x-values at which the volume begins ($x = a$) and ends ($x = b$).
4. Visualize the cross sections perpendicular to the axis of integration. Determine their shape.
5. Determine the area A of the cross-sectional slices as a function of the variable x.

The volume of the solid is then given by

$$V = \int_a^b A(x)\,dx. \tag{8.9}$$

This is analogous to a sliced loaf of bread, as shown in Figure 8.19. Determine

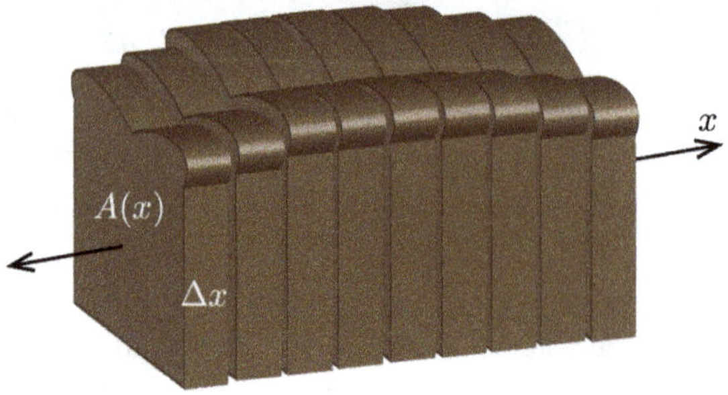

Figure 8.19: Volume as an integral of cross-sectional areas; loaf-of-bread illustration.

the area of each slice of bread individually, multiplying by the width of that slice. (This approximates the volume of the ith slice of bread.) Add the approximate volumes of all the slices of bread together, and the result yields the approximate volume of the full loaf of bread. This approximation becomes exact in the limit as the bread is sliced infinitely thin. Thus, the total volume of the loaf is given by equation (8.9). Recall that this follows from the definition of the definite integral:

$$\lim_{N\to\infty}\left(\sum_{i=1}^{N} A(x_i)\,\Delta x\right) = \int_a^b A(x)\,dx.$$

The summation represents the approximate volume (using N total slices) by adding the individual cross-sectional areas multiplied by the corresponding slice's thickness. In the limit, we obtain the exact volume.

■ **Example 8.13** Determine the volume of a pyramid with height h and with base side length s, as shown in Figure 8.20.

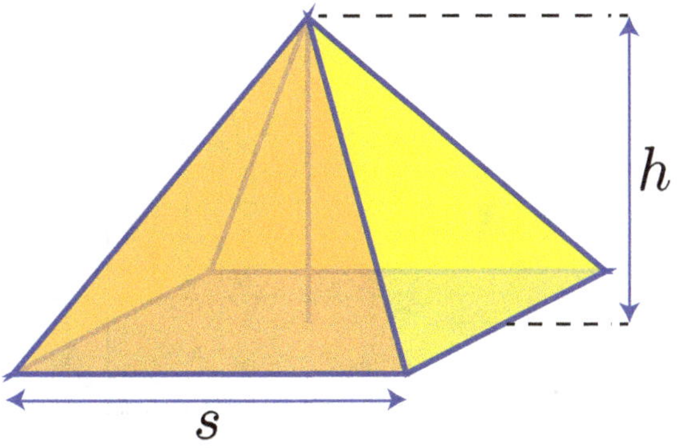

Figure 8.20: A pyramid.

We select the vertical axis as the axis of integration, since the horizontal cross sections are easily understood shapes (in this example, squares). In particular, we define the variable x to measure the vertical distance to the top of the pyramid. The trick is to consider the right triangle formed by the central vertical axis of the pyramid, and the bisector of one of its triangular faces, as shown in Figure 8.21.

Using similar triangles, we may express the length y as

$$y = \frac{sx}{2h}.$$

Therefore, the side length of the square cross sectional slice at vertical position x is given by sx/h, as shown in Figure 8.21.

The area of the horizontal cross-sectional slices is therefore given by the formula

$$A(x) = \frac{s^2x^2}{h^2}.$$

The total volume of the pyramid may therefore be derived using equation (8.9), which yields

$$V = \frac{s^2}{h^2}\int_0^h x^2\,dx = \frac{s^2h}{3}.$$

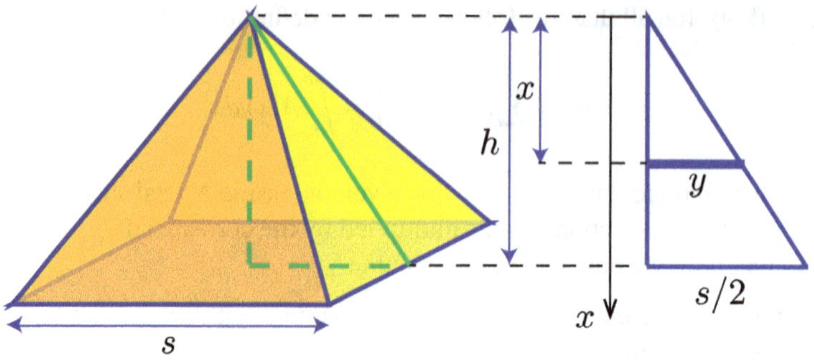

Figure 8.21: A pyramid with a vertical right-triangular cross section removed.

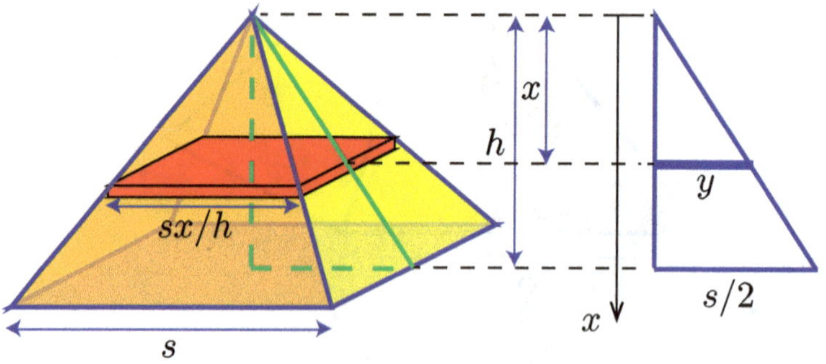

Figure 8.22: A pyramid with a horizontal square cross-sectional slice.

The volume of a pyramid is thus equal to one third the area of its base times its height. ■

8.3.2 Cylindrical Shells

In addition to general cross sections, there are additional methods that are often useful in determining volumes of certain solid objects. These additional methods fall under the general guise of "shells." The main idea here is that a solid may be constructed using a sequence of similar shells, each with a common geometric shape, similar to the idea of Russian nesting dolls. Each shell contains a slightly smaller version of itself, which, again, contains a smaller version, and so on, until we reach the center of the figure. The volume is therefore approximated by summing together the products of the surface area and width of each shell.

Our first example of a shell method is *cylindrical shells*, as outlined in the following.

Theorem 8.3.2 — Method of Cylindrical Shells. Consider a nonnegative function f defined on an interval $[a,b]$, where $a \geq 0$, and the region $\mathcal{R}$ bounded by the graph of f and the lines $y = 0$, $x = a$, and $x = b$. The volume V of the solid of revolution obtained by revolving region $\mathcal{R}$ about the y-axis is equal to

$$V = 2\pi \int_a^b xf(x)\, dx. \tag{8.10}$$

Proof. An example solid of revolution about the y-axis, as described in Theorem 8.3.2, is shown in Figure 8.23. In theory, one may apply the washer method to

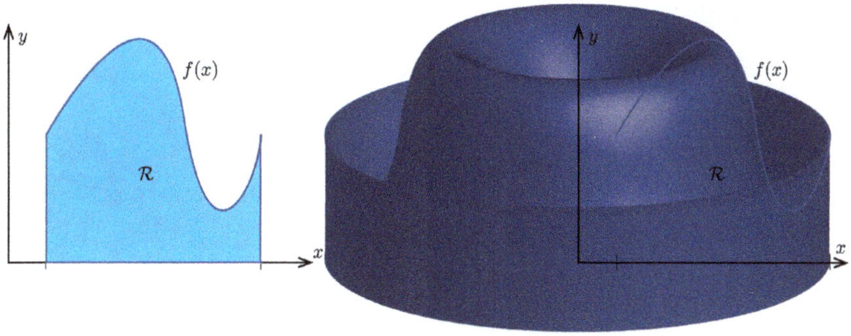

Figure 8.23: A solid of revolution about the y-axis.

obtain this volume. In practice, however, the washer method may be labored with difficulties. In particular, to apply the washer method, one needs to use the y-axis as the axis of integration and determine the x-position for points on the graph as a function of y. Sometimes this is easily achievable. In general, however, there can be a problem with the invertibility of the function f. Even if the function f is locally invertible, there is often no simple formula for its (local) inverse.

The cylindrical-shell method circumvents this issue by utilizing an approach more suited to the problem at hand—by considering a nested sequence of cylindrical shells. Each shell may be visualized itself as the solid of revolution obtained by taking one of the Riemann-sum rectangles for the integral of the function f and revolving it about the y-axis, as shown in Figure 8.24.

Each of these approximating shells has a lateral surface area equal to the surface area of a cylinder: circumference of the (circular) base times height. Thus, the lateral surface area of the ith approximating shell located at position x_i is given by

$$A_i = 2\pi x_i f(x_i).$$

Here, x_i is the radius (given as the distance between the Riemann-sum rectangle and the y-axis) and $f(x_i)$ represents the height. The volume of the ith cylindrical

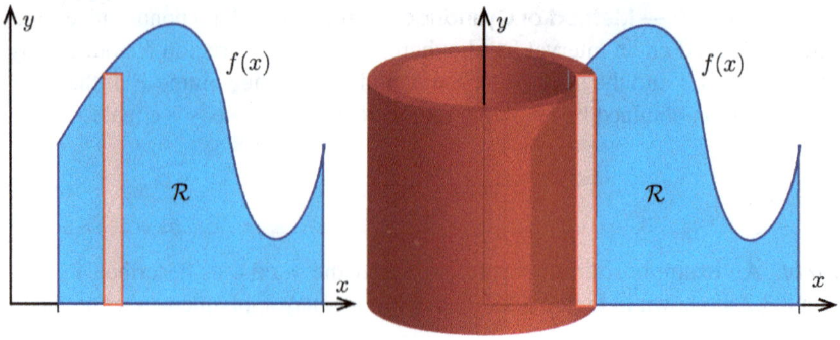

Figure 8.24: A solid of revolution about the y-axis.

shell is therefore

$$\Delta V_i = 2\pi x_i f(x_i)\Delta x.$$

(The width of the shell's wall is Δx, so we obtain volume by multiplying its lateral area by its width.) The total volume of the given solid of revolution may therefore be approximated by the Riemann sum

$$V \approx \sum_{i=1}^{N} 2\pi x_i f(x_i)\Delta x.$$

This sum becomes exact in the limit as $N \to \infty$. In the said limit, however, the summation converges to the definite integral given by equation (8.10), and the theorem is proven. ■

▪ **Example 8.14** Consider the region $\mathcal{R}$ bounded by the curves $y = x^3 - x^6$, $y = 0$, $x = 0$, and $x = 1$. Determine the volume of the solid of revolution obtained by revolving the region $\mathcal{R}$ about the y-axis, as shown in Figure 8.25.

First, let us consider the inadequacies of the washer method for this particular application. A single washer (i.e., a single horizontal slice) at vertical location y_i will have an inner and outer radius given by the x-values that solve the equation $y_i = f(x)$, i.e., $y_i = x^3 - x^6$, as shown in Figure 8.26.

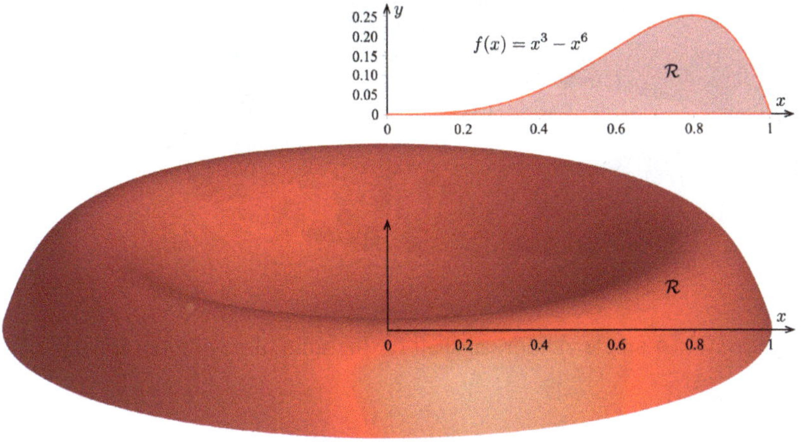

Figure 8.25: The solid of revolution formed by revolving the area under the graph of $f(x) = x^3 - x^6$ about the y-axis; Example 8.14.

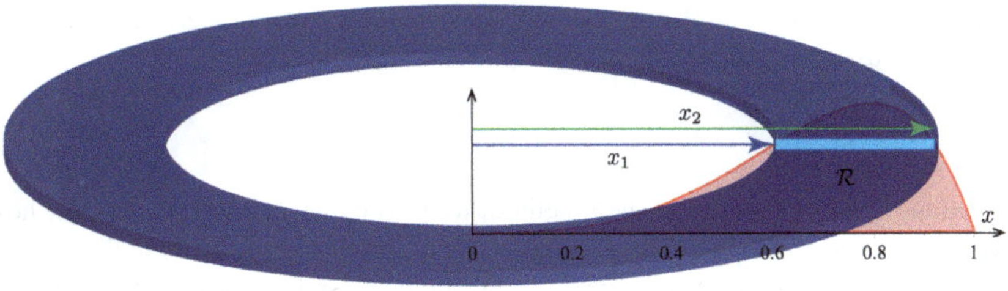

Figure 8.26: A single washer for the solid of revolution; Example 8.14.

Solving this equation for x as a function of y quickly becomes unwieldy. In this example it is possible to use the quadratic formula to solve for the two x-values as a function of y (this is not always the case), and the washer method would result in the integral

$$V = \frac{\pi}{\sqrt[3]{4}} \int_0^{0.25} \left[\left(1 + \sqrt{1-4y}\right)^{2/3} - \left(1 - \sqrt{1-4y}\right)^{2/3} \right] dy.$$

This integral, however, cannot be evaluated using elementary functions.

Let us compare this approach with the shell method. Instead of slices (which take the shape of the washers shown in Figure 8.26), we use cylindrical shells, as shown in Figure 8.27. For the shell method, the variable of integration is

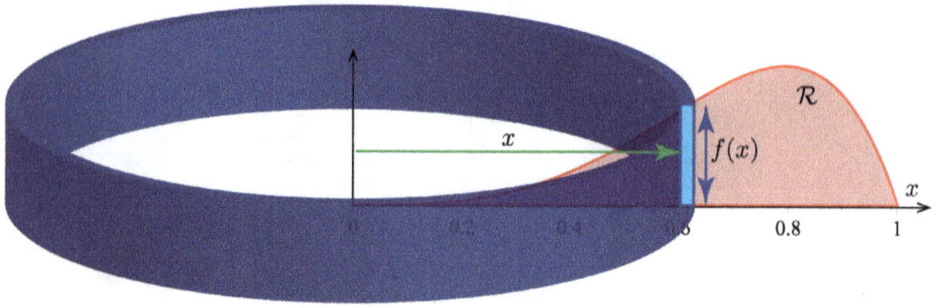

Figure 8.27: A single cylindrical shell for the solid of revolution; Example 8.14.

x (measuring the radial distance to the axis of revolution, i.e., the axis of each cylindrical shell). Each shell has radius x, height $f(x)$, surface area $2\pi x f(x)$, and volume $2\pi x f(x)\Delta x$. The volume of the solid of revolution therefore equals

$$V = 2\pi \int_0^1 \left(x^4 - x^7\right) dx = \frac{3\pi}{20}.$$

Clearly, this method is to be preferred. (Rest assured, the integral obtained using the washer method, though complicated, does yield the same numerical result when evaluated using a computer.) ■

8.3.3 Circular Shells

Continuing our theme of the shell method, we next consider we next consider how to compute the area of a circle of radius a using concentric circular annuli. Despite its limited application at present, this example is far from a merely academic exercise. In our next section, we will use this setup to compute the mass of distributions with radial symmetry.

Theorem 8.3.3 The area A of a circle with radius a may be computed using the integral

$$A = 2\pi \int_0^a r\, dr. \tag{8.11}$$

Proof. To see why this theorem holds, let us divide the circle of radius a into N circular strips, as shown in Figure 8.28. These concentric circular annuli play a role analogous to the rectangles of a Riemann sum. Since the radial range $[0,a]$ is divided into N subintervals, each has a width

$$\Delta r = \frac{a}{N},$$

j and the ith circular strip exists on the radial interval $[r_{i-1}, r_i]$, where $r_i = i\Delta r$, for $i \in \{0, 1, \ldots, N\}$.

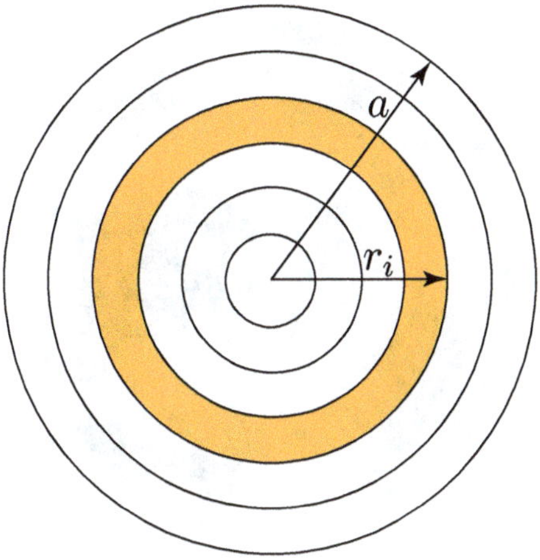

Figure 8.28: A Riemann sum consisting of circular annuli.

We may approximate the area of the ith strip by taking its outer circumference and multiplying by its width, thereby obtaining

$$A_i = 2\pi r_i \Delta r.$$

The total area of this circle is therefore approximated by

$$A \approx \sum_{i=1}^{N} 2\pi r_i \Delta r.$$

This approximation becomes exact in the limit as $N \to \infty$. Further, this summation approaches the integral given in equation (8.11) in this same limit. The result holds. ■

8.3.4 Spherical Shells

Returning to three dimensions, the same concept applied in Theorem 8.3.3 may be applied to find the volume of a sphere.

Theorem 8.3.4 The volume V of a sphere with radius a may be computed using the integral

$$V = 4\pi \int_0^a a^2 \, da. \tag{8.12}$$

Proof. The proof of Theorem 8.3.4 proceeds in an analogous fashion as the proof of Theorem 8.3.4. The spherical radius r (measuring distance to the origin in

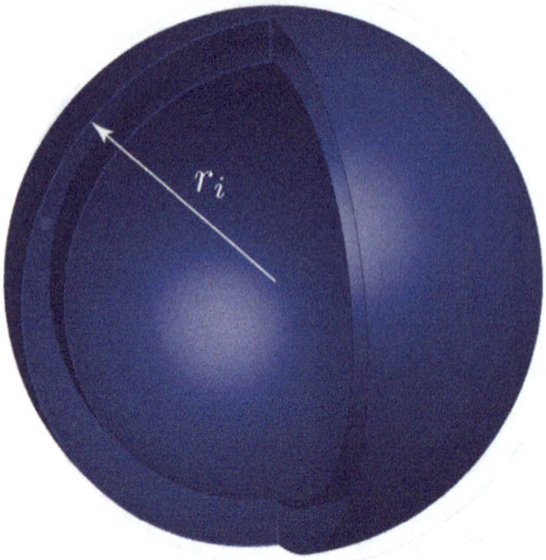

Figure 8.29: A Riemann sum consisting of concentric spherical shells, shown with an angular slice removed to reveal the hollow inside cavity.

three-dimensional space) ranges over the interval $[0, a]$. If we divide this interval into subintervals, we obtain spherical shells, such as the one shown in Figure 8.29.

Since the surface area of a sphere of radius r is $4\pi r^2$, the approximate volume of the ith spherical shell, located on the radial interval $[r_{i-1}, r_i]$ is

$$V_i = 4\pi r_i^2 \Delta r.$$

The approximate volume of the overall sphere is

$$V \approx \sum_{i=1}^{N} 4\pi r_i^2 \Delta r.$$

In taking the limit as $N \to \infty$, the exact volume is obtained, and we recover the formula from equation (8.12). ■

Exercises

For Exercises 1–10, draw a picture of the indicated solid and a representative cross-sectional slice. Compute the volume of the given solid.

1. The solid bounded by $x = 0$ and $x = 2$, such that the cross sections perpendicular to the x-axis are squares with side length x^2.

2. The solid bounded by $x = 0$ and $x = \pi$, such that the cross sections perpendicular to the x-axis are circles with radius $\sqrt{\sin(x)}$.

3. The solid bounded by $x = 0$ and $x = \pi$, such that the cross sections perpendicular to the x-axis are circles with radius $\sin(x)$.

4. A conical frustum with height h, base radius R, and top radius r, where $0 < r < R$. (This is a por-

tion of a right-circular cone.)

5. The solid bounded by $y = 0$ and $y = h$, such that the cross sections perpendicular to the y-axis are equilateral triangles with side length y.

6. The solid bounded by $y = 0$ and $y = h$, such that the cross sections perpendicular to the y-axis are semicircles with side radius y.

7. The solid with a base bounded by $y = e^{-x}$, $y = 0$, $x = 0$, and $x = 1$, such that the cross sections perpendicular to the x-axis are squares.

8. The solid with a base bounded by $y = 1/x$, $y = 0$, $x = 1$, and $x = 10$, such that the cross sections perpendicular to the x-axis are squares.

9. The solid with a base bounded by $x = 0$, $y = x^2$, and $y = 1$, such that the cross sections perpendicular to the y-axis are semicircles.

10. The solid with a base bounded by $x = 0$, $y = x^2$, and $y = h$, such that the cross sections perpendicular to the y-axis are equilateral triangles.

For Exercises 11–20, sketch the indicated solid and a representative shell. Then use the shell method to determine the volume of the indicated solid.

11. The solid obtained by revolving the region bounded by $y = 0$, $y = x^2$, and $x = 2$ about the y-axis.

12. The solid obtained by revolving the region bounded by $y = 0$ and $y = x(1-x)$ about the y-axis.

13. The solid obtained by revolving the region bounded by $y = 0$, $y = 2 + \sin(x)$, $x = 0$, and $x = 2\pi$ about the y-axis.

14. The solid obtained by revolving the region bounded by $y = 0$, $y = x$, and $x = 1$ about the y-axis.

15. The solid obtained by revolving the region bounded by $y = 0$, $y = 2 - x^4 + 6x^3 - 11x^2 + 6x$, and $x = 3$ about the y-axis.

16. The solid obtained by revolving the region bounded by $y = x$, $y = 2$, and $x = 0$ about the x-axis.

17. The solid obtained by revolving the region bounded by $y = x^2$, $y = 4$, and $x = 0$ about the x-axis.

18. The solid obtained by revolving the region bounded by $y = \ln(x)$, $y = 1$, $x = 1$ about the x-axis.

19. The solid obtained by revolving the region bounded by $y = \sqrt{x}$, $y = 2$, and $x = 0$ about the x-axis.

20. The solid obtained by revolving the region bounded by $y = x - x^3$, $y = -2$, $x = 0$, and $x = 1$ about the y-axis.

Problems

21. Use circular shells to show that the area of a circular annulus, with inner radius r_1 and outer radius r_2, is given by

$$A = \pi\left(r_2^2 - r_1^2\right).$$

22. Determine the area of a square with side length a using the formula for the

perimeter of a square and concentric square strips, similar to the method used in Theorem 8.3.3.

23. Determine the volume of a right-circular cylinder with base radius a and height h using the formula for the surface area of a cylinder and concentric cylindrical shells, similar to the method used in Theorem 8.3.4.
24. A hemispherical swimming pool of radius R has water level h.
    (a) Write down an integral which represents the total volume $V(h)$ of water contained in this swimming pool as a function of the water level h.
    (b) Evaluate this integral, obtaining an explicit formula for $V(h)$.
    (c) A certain hemispherical swimming pool with radius $R = 10$ feet is being filled with water at a rate of $0.5\ \text{ft}^3/\text{s}$. What is the instantaneous rate of change of the water level h at the moment when the water level is $h = 4$ feet?
25. In this problem, we will be estimating the volume of an axisymmetric container of milk, as shown in Figure 8.30. The circumference, $C(H)$, of the

Figure 8.30: A container of milk.

jug at height H, both measured in inches, is given at various points by:

H (inches)	1	2	3	4	5	6	7	8	9
$C(H)$ (inches)	16	16	16	16	15	12	8	5.5	5.25

The top surface of the milk reaches a height of 9 inches, and the jug is symmetric about its central axis. The *exact* amount of milk contained in the jug is $115\ \text{inch}^3$, i.e. one half gallon.

(a) Write a general Riemann sum, in terms of the function $C(H)$, which represents an approximation to the total volume of milk in the jug.

(b) Write down a definite integral, in terms of the function $C(H)$, which represents the exact volume contained in the jug.
(c) Approximate the volume of milk in the jug using a right-hand Riemann sum and the data provided in the table.

26. Consider two ceramic mugs, as shown in Figure 8.31(a) and (b). Suppose

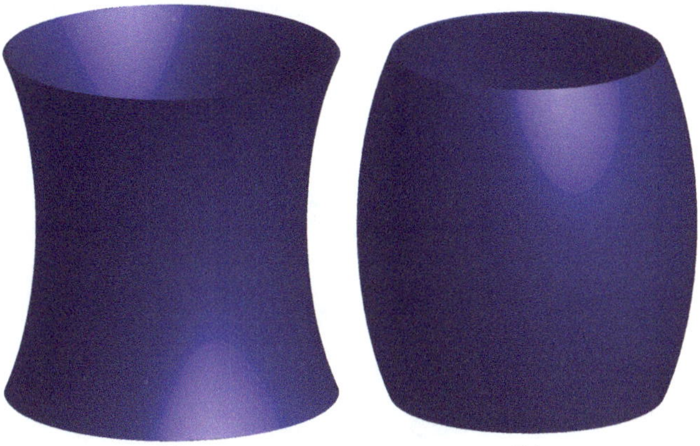

(a) (b)

Figure 8.31: Two mugs.

that liquid is poured into these mugs at a constant rate. Draw a sketch of the volume V as a function of time t. For each, determine whether the function $V(t)$ should be concave up or concave down?

27. Consider a container that has a horizontal cross-sectional area $A(x)$ at height x. Let $x = x(t)$ represent the water level in the container as a function of time, and let $V = V(t)$ represent the volume of liquid in the container as a function of time.
(a) Explain why

$$V'(t) = A(x(t))x'(t).$$

(b) Explain why, if the container starts being filled from empty, the function V may be expressed as

$$V(t) = \int_0^{x(t)} A(s)\, ds.$$

(c) Use the result from part (b) and the fundamental theorem of calculus to derive the formula in part (a).

8.4 Mass, Center of Mass, and the Moments of Inertia

In addition to calculating the volume of geometric shapes, the theory of integration can also be applied to determine the mass, center of mass, and moments of inertia of three-dimensional solids.

8.4.1 Mass of an Object

In this paragraph, we consider how to determine the mass of an object with a nonhomogeneous mass distribution, i.e., an object with spatially varying density.

Definition 8.4.1 *Density* is defined as mass per unit space. In particular, we define:

1. the *linear density* λ of a one-dimensional rod or ring is the amount of mass per unit length;
2. the *planar density* σ of a two-dimensional plate or lamina is the amount of mass per unit area;
3. the *volumetric density* ρ of a three-dimensional object is the amount of mass per unit volume.

(R) A more precise definition of the volumetric density of an object at a point is given as follows. Let p represent a point in three-dimensional space. Let S_ε represent a sphere of radius ε centered at the point p, and let M_ε represent the amount of mass of the solid object contained within the sphere S_ε. Then the *volumetric density* $\rho(p)$ of the solid at point p is defined by

$$\rho(p) = \lim_{\varepsilon \to 0} \frac{3M_\varepsilon}{4\pi\varepsilon^3}. \tag{8.13}$$

A similar definition may be appropriately constructed for the case of linear and planar densities.

(R) The Greek letters λ, σ, and ρ are conventionally used to distinguish between linear, planar, and volumetric densities, respectively. We will use this notation here.

We will begin by analyzing the mass of one-dimensional rods.

Mass of one-dimensional rods

Theorem 8.4.1 Consider a one-dimensional rod aligned with an axis with position variable x that lies between the points $x = a$ and $x = b$. If the nonnegative function $\lambda(x)$ represents the linear density of the rod as a function of the position x, then the total mass of the rod is given by

$$M = \int_a^b \lambda(x)\, dx. \tag{8.14}$$

Proof. To prove this theorem, we begin by slicing the interval $[a, b]$ into N equal subintervals of width Δx, as shown in Figure 8.32.

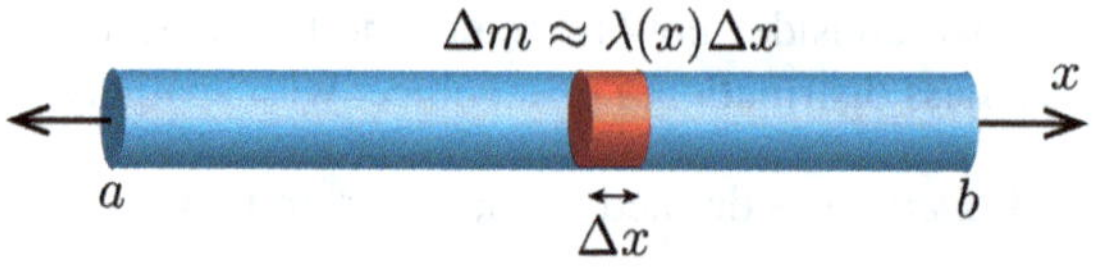

Figure 8.32: A one-dimensional rod with linear density $\lambda(x)$.

Since the density function $\lambda(x)$ describes the mass per unit length, the *mass* of the ith slice, located on the interval $[x_{i-1}, x_i]$ (using standard Riemann-sum notation) is approximately

$$\Delta m_i \approx \lambda(x_i)\Delta x.$$

The total mass of the rod is therefore approximated by the Riemann sum

$$M = \sum_{i=1}^{N} \lambda(x_i)\Delta x.$$

This, of course, is merely a Riemann sum for the integral of equation (8.14). Since this sum converges to the exact mass in the limit as $N \to \infty$, we recover the result of the theorem. ■

Mass of symmetric three-dimensional solids

Next, we consider the mass of a solid three-dimensional object whose density and cross-sectional area each vary as a function of the position along a single axis. This is similar to the volume problems discussed at the beginning of Section 8.3.

Theorem 8.4.2 Consider a solid object whose cross-sectional area A and volumetric density ρ both vary as a function of the position x along an axis. If the solid object is bounded by the planes $x = a$ and $x = b$, then its mass is given by

$$M = \int_a^b A(x)\rho(x)\, dx. \tag{8.15}$$

(Compare with equation (8.9) on page 496.)

Proof. We proceed by constructing a Riemann sum to approximate the total mass of the given solid. We begin by dividing the interval $[a, b]$ into N equal subintervals, each of width Δx. The ith slide of the solid is located on the subinterval $[x_{i-1}, x_i]$, and the volume of the ith slice is approximated by

$$\Delta V_i \approx A(x_i)\Delta x.$$

Since density is mass per unit volume, the mass of the ith slice is approximated by

$$\Delta m_i \approx A(x_i)\rho(x_i)\Delta x.$$

This is illustrated in Figure 8.33.

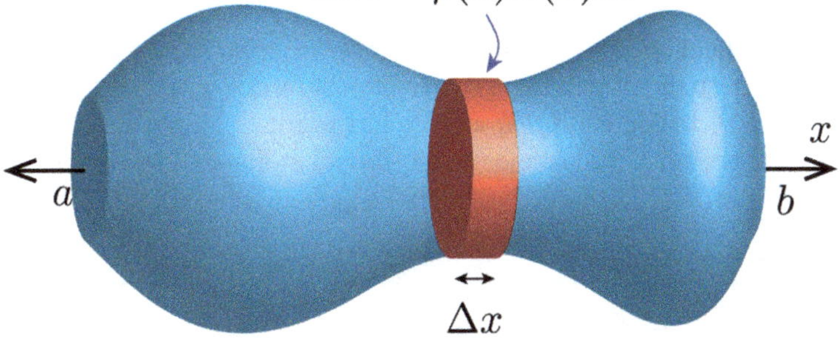

Figure 8.33: A three-dimensional solid with volumetric density $\rho(x)$ and perpendicular cross-sectional area $A(x)$.

The total mass of the solid is therefore approximated by the sum

$$M \approx \sum_{i=1}^{N} A(x_i)\rho(x_i)\Delta x.$$

In the limit as $N \to \infty$, this sum converges to the integral in equation (8.15), as it approaches the exact mass of the object. This completes the proof. ■

8.4.2 Center of Mass

The concept of *center of mass* is an important one in physics and engineering. The center of mass of a body (whether a one-dimensional rod, two-dimensional plate, or three-dimensional solid) is essentially the single balancing point of the body. In *mechanics*, a three-dimensional object can be described by the position of its center of mass coupled with information about the object's orientation in space to determine the dynamic evolution of the body's motion. A dramatic simplification results when formulating the equations of motion relative to the center of mass of the bodies comprising a given system. In this paragraph, we begin with a discussion of the center of mass of a distribution of collinear point masses, and then generalize the result to derive an equation for the center of mass of a one-dimensional rod.

Definition 8.4.2 Given a set of N point masses, with masses $m_1, \ldots, m_N$, arranged along an axis at positions $x_1, \ldots, x_N$, the *center of mass* for the system is the unique point $\overline{x}$ given by

$$\overline{x} = \frac{1}{M}\sum_{i=1}^{N} x_i m_i, \tag{8.16}$$

where M represents the *total mass*, as given by

$$M = \sum_{i=1}^{N} m_i. \tag{8.17}$$

To understand why the definition of center of mass is set the way it is, we must first understand something about rotational forces. A *torque* (or a *moment*) is a measure of the strength that a force has to rotate an object about an axis. Consider, for example, the bar and fulcrum shown in Figure 8.34. Two forces with

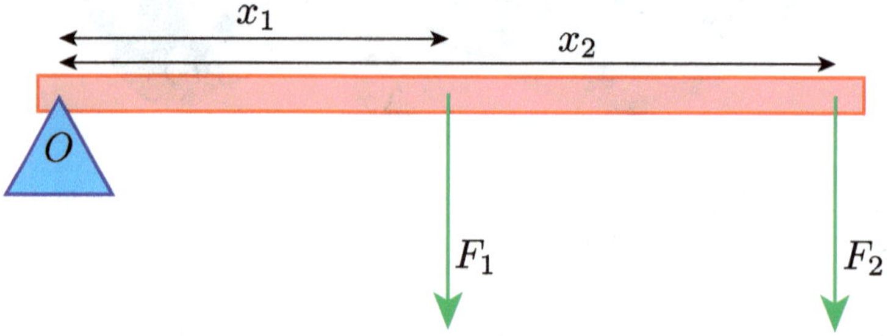

Figure 8.34: Two torques acting on a bar.

magnitudes F_1 and F_2 are acting on the bar, located at positions x_1 and x_2. The left end of the bar is fixed to the fulcrum, which can supply an opposite upward force to counteract the downward forces F_1 and F_2. The forces create a torque, however, causing the bar to rotate.

A door can be opened by pushing it at any point. The farther away from the hinges, however, the easier it is to open the door. This is why door handles are located on the side of the door opposite to the hinges. In Figure 8.34, it is easier to create a rotation in the bar when applying a force at the point x_2 than it is at the point x_1. In fact, F_1 would have to be twice as strong as F_2 in order to produce the same rotation, assuming that F_2 is twice as far to the pivot as F_1. In this way, we arrive at the following important observation: *the efficacy of a force to produce a rotation is linearly proportional to the distance between the force and the pivot point.* Thus, a force F acting at a distance d to a pivot point produces a *torque* τ equal to

$$\tau = Fd. \tag{8.18}$$

Torques are assigned opposite signs based on which side of the pivot they are applied. The sign of the force also comes into play, indicating the direction in which the force is acting.

The center of mass may now be understood as the unique point at which the gravitational torques acting on the mass distribution cancel. This is the concept that allows one to construct a hanging mobile.

Theorem 8.4.3 Given a set of N point masses, with masses $m_1, \ldots, m_N$, arranged along an axis at positions $x_1, \ldots, x_N$. If these masses were glued to a massless rod, then the rod would perfectly balance if supported at its center of mass $\overline{x}$, as given by equation (8.16).

Proof. First, let us imagine the masses in a uniform, downward gravitational field, with constant gravitational acceleration g. The gravitational field produces a downward force $F_i = m_i g$ on the ith point mass. To prove the theorem, consider a number of forces acting on a rigid, massless bar, as shown in Figure 8.35. Let us consider the "downward" forces to be positive, as shown in the figure.

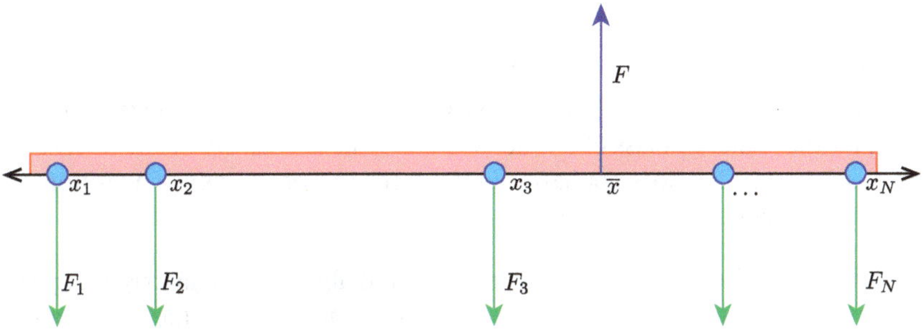

Figure 8.35: A set of point masses balance at their collective center of mass.

To proceed, let us define $\overline{x}$ to be the point at which the rigid bar balances. Thus, if an *upward* force

$$F = -\sum_{i=1}^{N} F_i = -(F_1 + F_2 + \cdots + F_N) = -(m_1 g + \cdots + m_N g).$$

were to act on the bar at the point $\overline{x}$—for instance, if the bar were to be hanged from a ceiling by a string—then the net torque would vanish. Defining the total mass M as in (8.17), the upward support force is given by

$$F = -Mg.$$

This is required so that there is zero net force acting on the system. (Otherwise the system would have a net acceleration, and it could not be statically suspended. The net torque (which results in rotation) acting on the bar is therefore given by

$$\tau = \sum_{i=1}^{N} (m_i g x_i) - M g \overline{x}.$$

Setting the net torque equal to zero (so that there is no resultant rotation) and solving for $\overline{x}$ yields equation (8.16). This proves the theorem. ■

▪ **Example 8.15** Three point masses are located at $x_1 = -1.5$, $x_2 = 0.5$, and $x_3 = 4$, with masses $m_1 = 2$, $m_2 = 5$, and $m_3 = 1$, respectively. Determine the location of the center of mass of the system.

First, we compute the total mass as

$$M = m_1 + m_2 + m_3 = 2 + 5 + 1 = 8.$$

Applying equation (8.16), we determine

$$\overline{x} = \frac{1}{M}\sum_{i=1}^{3} m_i x_i = \frac{1}{8}[(2)(-1.5) + (5)(0.5) + (1)(4)] = \frac{7}{16}.$$

This is the position of the center of mass. ▪

Next, we use the theory of integration and the previous results to determine a formula for the center of mass of a rigid nonhomogeneous bar. By *nonhomogeneous*, we mean that the bar does not have a uniform (or homogeneous) composition, i.e., its density is not constant.

Theorem 8.4.4 Consider a one-dimensional rod aligned with an axis with position variable x that lies between the points $x = a$ and $x = b$. If the nonnegative function $\lambda(x)$ represents the linear density of the rod as a function of the position x, then the *center of mass* of the rod is located at the point

$$\overline{x} = \frac{1}{M}\int_a^b x\lambda(x)\, dx, \tag{8.19}$$

where M represents the bar's *total mass*, which is given by equation (8.14).

Proof. We proceed as usual: first, we divide the interval $[a,b]$ into N equal subintervals, each of width Δx. If we treat each slice of the rod as a point mass, we may approximate the center of mass using equation (8.16). This yields the equation

$$\overline{x} \approx \frac{\sum_{i=1}^{N} x_i \lambda(x_i)\Delta x}{\sum_{i=1}^{N} \lambda(x_i)\Delta x}.$$

In the limit as $N \to \infty$, we have

$$\lim_{N\to\infty}\left(\sum_{i=1}^{N} x_i\lambda(x_i)\Delta x\right) = \int_a^b x\lambda(x)dx$$

and

$$\lim_{N\to\infty}\left(\sum_{i=1}^{N}\lambda(x_i)\Delta x\right) = \int_a^b \lambda(x)\,dx = M.$$

(This just follows from the definition of the definite integral.) This proves the result. ■

Naturally, we may extend this result to solids with a density that varies along an axis of symmetry.

Corollary 8.4.5 Consider a solid object whose cross-sectional area A and volumetric density ρ both vary as a function of the position x along an axis. If the solid object is bounded by the planes $x = a$ and $x = b$, then its center of mass has an x-coordinate of

$$\overline{x} = \frac{1}{M}\int_a^b xA(x)\rho(x)\,dx, \tag{8.20}$$

where M is the total mass, as given by equation (8.15).

Proof. This is analogous to the one-dimensional rod, except the amount of mass contained on the ith subinterval $[x_{i-1}, x_i]$ is given by

$$\rho(x)A(x)\Delta x,$$

instead of $\lambda(x)\Delta x$. Replacing λ with ρA in equation (8.19) produces the result. ■

8.4.3 Moments of Inertia

Another key concept in the physics of rotational mechanics is the *moment of inertia* of a body. For an object with constant mass, Newton's second law of motion may be stated (in its one-dimensional form) as

$$F = ma. \tag{8.21}$$

The force acting on an object equals the product of its mass times its acceleration. The equivalent equation in rotational dynamics is

$$\tau = I\alpha, \tag{8.22}$$

where τ is the net torque acting on the object (defined as force times the length of the moment arm), I is the object's moment of inertia, and α is the object's angular acceleration.

We will derive this latter equation as follows. Suppose that the force F acts on the body at a distance d to a fixed pivot point on the body. Multiplying equation (8.21) by d, we obtain

$$Fd = mad.$$

The left-hand side is equal to the torque τ. Recall that the object's velocity v is related to its angular velocity ω and the distance d to the pivot point via the equation

$$v = \omega d.$$

(In SI units, d has units of meters, ω has units of 1/s, and v has units of m/s.) Differentiating with respect to time, we obtain $a = \alpha d$. Substituting this in for a, we arrive at the equation

$$\tau = m\alpha d^2 = (md^2)\alpha.$$

We may therefore define the *moment of inertia* of a point mass with mass m located at a distance d from a pivot point as

$$I = md^2. \tag{8.23}$$

Given this definition, equation (8.22) follows.

Theorem 8.4.6 Consider a one-dimensional rod aligned with an axis with position variable x that lies between the points $x = a$ and $x = b$. If the nonnegative function $\lambda(x)$ represents the linear density of the rod as a function of the position x, then the *moment of inertia* of the rod about an axis passing through the point $x = p$ is given by

$$I_p = \int_a^b (x-p)^2 \lambda(x)\, dx. \tag{8.24}$$

Proof. We partition the rod into N subintervals, as usual. The distance between the ith subinterval and the pivot point $x = p$ is approximately $d_i = (x_i - p)$. The mass contained within the ith slice is $\Delta m_i = \lambda(x_i)\Delta x$. Therefore, the ith slice of the rod makes a contribution

$$\Delta I_i = (x_i - p)^2 \lambda(x_i)\Delta x.$$

Summing and taking the limit as $N \to \infty$, we recover our result. ∎

▪ **Example 8.16** Compute the moment of inertia of a rigid, uniform bar with length L about one of its endpoints, as shown in Figure 8.36.

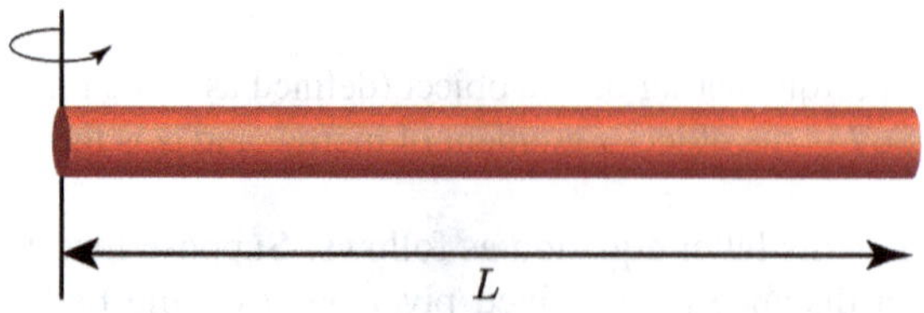

Figure 8.36: The moment of inertia of a uniform bar about its endpoint.

To begin, we need to define an axis for integration. This axis is parallel to the bar and set so that the left endpoint of the bar is located at the position $x = 0$. Following equation (8.24), the moment of inertia about the endpoint I_e is given by

$$I_e = \lambda \int_0^L x^2\, dx,$$

where λ is the (constant) linear density of the bar. Upon integrating, we obtain

$$I_e = \frac{\lambda L^3}{3}.$$

Since the total mass of the bar M is given by $M = \lambda L$, this may be rewritten as

$$I_e = \frac{ML^2}{3}. \tag{8.25}$$

■

■ **Example 8.17** Determine the moment of inertia of a uniform rigid bar of length L about an axis perpendicular to the bar and passing through the bar's center of mass.

The setup for this example is almost identical to the setup from Example 8.16, except we will set the x-axis so that the origin is coincident with the bar's center of mass. Since the bar is uniform, its center of mass is located at the halfway point. The moment of inertia about an axis passing through the center of mass I_c is therefore given by

$$I_c = \lambda \int_{-L/2}^{L/2} x^2\, dx = \frac{M}{L}\frac{L^3}{12} = \frac{ML^2}{12}. \tag{8.26}$$

We have again used the fact that the total mass is given by $M = \lambda L$. (This follows only since the bar is uniform, i.e., as it has a constant density.) ■

The points about which we computed the moment of inertia in Examples 8.16 and 8.17 lie a distance $d = L/2$ apart from each other. By comparing equations (8.25) and (8.26), we observe that

$$I_e = I_c + Md^2.$$

More than just a happy coincidence, this result holds in general. Our next theorem shows how a body's moment of inertial through an axis is related to the moment of inertia through a parallel axis passing through the body's center of mass. Though we restrict the theorem to one dimensional rods at this current juncture, it is a general theorem that holds even when applied to three-dimensional bodies.

Theorem 8.4.7 — Parallel-Axis Theorem. Consider a one-dimensional rod aligned with an axis with position variable x that lies between the points $x = a$ and $x = b$, where the nonnegative function $\lambda(x)$ represents the linear density of the rod as a function of the position x. Then the moment of inertia about the center of mass I_c and the moment of inertia about any other point I_p are related by the formula

$$I_p = I_c + Md^2, \tag{8.27}$$

where d represents the distance between the center of mass and the point p, and M represents the total mass of the bar, as given by equation (8.14).

Proof. The proof is simplified if we set the x-axis so that the center of mass is located at the origin $x = 0$. Following equation (8.24), the moments of inertia of the bar through the center of mass and through the point $x = p$ are given by

$$I_c = \int_a^b x^2 \lambda(x)\, dx, \tag{8.28}$$

$$I_p = \int_a^b (x-p)^2 \lambda(x)\, dx. \tag{8.29}$$

Expanding the quadratic function in the integrand of equation (8.29) and breaking apart the integral (using linearity), we see that

$$I_p = \int_a^b x^2 \lambda(x)\, dx - 2p \int_a^b x\lambda(x)\, dx + \int_a^b p^2 \lambda(x)\, dx.$$

The first integral in the preceding equation is simply I_c, as observed by comparing with equation (8.28). Recalling equation (8.19), we see that the second integral is equal to a constant times the location of the rod's center of mass. We choose our coordinates, however, so that the bar's center of mass is located at $\overline{x} = 0$. This implies that the second integral in the preceding equation vanishes. The third integral yields p^2 times the mass of the rod. Since the distance between the bar's center of mass $\overline{x} = 0$ and the point $x = p$ is $d = |p|$, the result follows. ■

The parallel-axis theorem is illustrated in Figure 8.37. Here, the moment of inertia about an axis passing through the body's center of mass (marked with the black-and-white checkerboard circle) is I_c, and the moment of inertia about a parallel axis passing through the point p is given by I_p. The distance between the center of mass and the point p is equal to d. The moments of inertia satisfy equation (8.27).

Exercises

For Exercises 1–10, determine (a) the total mass M, (b) the location of the center of mass $\overline{x}$, and (c) the moment of inertial with respect to the origin I_0 of the given one-dimensional rod with linear mass density $\lambda(x)$ g/cm.

1. $\lambda(x) = 2 + x$; $x \in [0,1]$.
2. $\lambda(x) = 0.5 + x^2$; $x \in [0,1]$.
3. $\lambda(x) = 1 + 0.1\sin(4\pi x)$; $x \in [0,1]$.
4. $\lambda(x) = e^{-x}$; $x \in [0,2]$.
5. $\lambda(x) = xe^{-x}$; $x \in [0,2]$.
6. $\lambda(x) = \ln(x+1)$; $x \in [1,e]$.
7. $\lambda(x) = \dfrac{1}{x}$; $x \in [1,2]$.
8. $\lambda(x) = \dfrac{1}{x^2}$; $x \in [1,2]$.

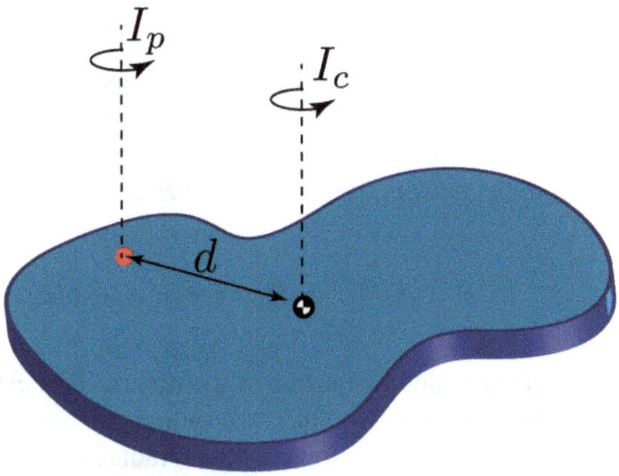

Figure 8.37: Illustration of the parallel-axis theorem, as applied to a planar lamina.

9. $\lambda(x) = 1 + x + 0.2\sin(4\pi x)$; $x \in [0,1]$.

10. $\lambda(x) = \cosh(x)$; $x \in [0,1]$.

For Exercises 11–20, determine (a) the mass of the indicated solid and (b) the location, along the axis of symmetry, of the solid's center of mass.

11. A hemisphere of radius R and constant density ρ.

12. A pyramid with constant density ρ, square base side length s, and height h.

13. A hemisphere of radius 2 and density $\rho(x) = 8e^{-x}$ at a vertical distance x from the base.

14. A pyramid with square base side length 50 cm, height 70 cm, and density $\rho(x) = 2 + 0.1x$ at a vertical distance x from the base.

15. A right-circular cone with base radius r and height h, and constant density ρ.

16. The solid obtained by revolving the region bounded by $y = 0$, $y = \sin(x)$, $x = 0$, and $x = \pi$ about the x-axis, with constant density ρ.

17. The solid obtained by revolving the region bounded by $y = 0$, $y = \sin(x)$, $x = 0$, and $x = \pi$ about the x-axis, with density $\rho(x) = x/8$.

18. The solid obtained by revolving the region bounded by $y = 0$ and $y = x(1-x)$ about the x-axis, with density $\rho(x) = x^3$.

19. The solid obtained by revolving the region bounded by $y = e^{-x}$, $y = 1$, $x = 0$, and $x = 1$ about the x-axis, with density $\rho(x) = x$.

20. The solid obtained by revolving the region bounded by $x = 0$, $y = x^2$, and $y = 1$ about the x-axis, with density $\rho(x) = x$.

Problems

21. Construct a definition for linear and planar density similar to the one presented for volumetric density in equation (8.13).
22. An oil spill emanates from a large oil tanker, and extends to a distance of 10 miles away from the tanker. The oil density (per unit area) of the oil is given by

$$\sigma(r) = 18e^{-r/10} \text{ gallons per square mile,}$$

as a function of the radial distance r to the tanker. Use circular shells to determine the total amount of oil that was spilled.
23. Consider a roll of toilet paper with inner (tube) radius r_0, outer radius r, such that each sheet of toilet paper has length L and thickness τ.
    (a) Use circular shells to construct a Riemann sum that approximates the total number of sheets in the roll.
    (b) Convert the Riemann sum from part (a) to a definite integral, and then evaluate this definite integral to determine the total number of sheets N in the roll, as a function of the outer radius r.
    (c) A particular roll has the parameter values $r_0 = 0.75$ in, $r = 2.25$ in, $L = 4$ in, and $\tau = 0.003$ in. Determine how many sheets of toilet paper are in this roll.
    (d) Invert the formula from part (b) to determine the radius r as a function of the number of sheets, i.e., $r = r(N)$. (This can be used, for example, to mark off the levels at which 100 sheets, 200 sheets, 300 sheets, etc., are remaining the roll.)
24. The city wall surrounding Squareville is a 10 km by 10 km square. The population density is constant along concentric squares centered at the origin. The population density anywhere along the square of side length L, centered at the city center, is $\sigma(L) = 1000e^{-L/2}$ people/km^2. How many people live in Squareville? (Round down to the nearest person.)
25. A spherical snowball has density $\rho(r) = e^{-r^3/64}$ g/cm^3 anywhere at a radial distance r to the center of the snowball. Its outer radius is 4cm. Find the mass of this snowball.
26. Let $\mathcal{R}$ be the interior of the oval $x^2 + 9y^2 = 9$ which lies in the first quadrant. A certain UFO has the shape obtained by revolving the region $\mathcal{R}$ about the y-axis. The UFO contains superheated ammonia vapor at 35°C with density $\rho = 5$ kg/m^3 and pressure 700 kPa.
    (a) Find the total mass of ammonia in the UFO.
    (b) Determe the location on the y-axis of the center of mass.
    (c) If the density of the gas anywhere on a constant y slice were instead to be given by the formula $\rho(y) = 5y$ kg/m^3, find the total mass of the gas that would be contained in the UFO.
27. The *Hernquist model* for spherical galaxies states that the matter density $\rho(s)$, measured in kg/m^3, where s is the distance to the center of the galaxy,

is given by:

$$\rho(s) = \frac{c}{2\pi}\frac{1}{s}\frac{1}{(1+s)^3}$$

where c is a positive constant.

(a) Set up the definite integral representing the total mass $M(R)$ contained in a sphere of radius R, as predicted by the Hernquist model.

(b) Evaluate this integral, obtaining an explicit expression for $M(R)$.

(c) Determine $\lim_{R\to\infty} M(R)$, and explain the meaning of this limit in practical terms.

28. Consider a cylindrical tank with height H, mass M, and cross-sectional area A, as shown in Figure 8.38. The tank is partially filled with a liquid with density ρ to height h.

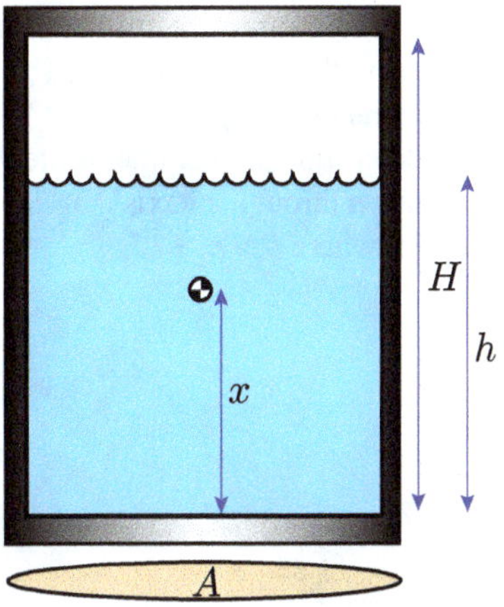

Figure 8.38: Center of mass for a cylindrical water tank.

(a) Show that the center of mass of the tank and liquid is located at height

$$x(h) = \frac{MH + \rho Ah^2}{2M + 2\rho Ah},$$

on the domain $h \in [0, H]$.

(b) Observe that $x(0) = H/2$ and $x(H) = H/2$. Explain what this means in practical terms. Does this make sense?

(c) Based on the result from part (b), the center of mass starts at $H/2$ when the tank is full, lowers as the tank empties, but at some point turns around and starts rising again. Determine a formula for the water level h^* at which the center of mass is the lowest.

(d) Determine the height of the center of mass when the water level is h^*, i.e., compute $x(h^*)$. Simplify this expression and compare $x(h^*)$ with h^*. What is the relationship between these two values?

29. Consider the region bounded by $x = a$, $x = b$, $y = f(x)$, and $y = g(x)$, where $g(x) < f(x)$, and let E represent the solid obtained by revolving this region about the y-axis.
    (a) Explain why the x-position of the center of mass of this planar lamina is located at
    $$\overline{x} = \frac{1}{A}\int x\left[f(x) - g(x)\right]\, dx,$$
    if it possess a constant area mass density σ.
    (b) Demonstrate, using the shell method, that the volume of E is given by
    $$V = 2\pi \int_a^b x\left[f(x) - g(x)\right]\, dx.$$
    (c) Combine the results from parts (a) and (b) to interpret the volume formula as the area of the lamina times the distance traveled by its center of mass during the revolution, i.e., $V = 2\pi\overline{x}A$. This result is an instance of the *theorem of Pappus*. .
    (d) Use the result to determine the volume of the torus obtained by revolving a circle of radius a through an axis that lies at a distance b from the center of the circle, where $b > a$.

8.5 Physics: Work and Pressure

In this section, we discuss two important applications of the theory of integration to physics: work and hydrostatic force.

8.5.1 Work

In physics, a *force* may be applied to an object to produce an acceleration, following Newton's second law, which, for bodies of constant mass, may be expressed as $F = ma$. A given force produces an acceleration that is inversely proportional to the object's mass. For example, a force of 1 N can accelerate a 1 kg object by 1 m/s^2, but it can only accelerate a 2 kg object by 0.5 kg/s^2. (One *Newton* (1 N), by definition, is the amount of force required to accelerate a 1 kg object by 1 m/s^2.) When a given force is used to accelerate an object through through a given distance, *work* is said to be done by the force on the object.

Definition 8.5.1 If a constant force acts on a body, thereby moving the body through a distance d of space, then the force is said to do *work* on the body in an amount equal to

$$W = Fd, \tag{8.30}$$

where F represents the component of the force in the direction of the body's motion.

The SI unit of work is the *Joule* (J); one Joule is the amount of work done by a one-Newton force pushing an object through a distance of one meter. (1 J = 1 N·m.)

▪ **Example 8.18** A man applies a 200 N force to pull a wooden crate through a distance of 12 m. Calculate the amount of work done on the crate.

Assuming that the force was applied in the direction of the crate's motion (and not at an angle to it), equation (8.30) allows us to calculate the total work done by

$$W = (200\text{ N})(12\text{ m}) = 2400\text{ N}\cdot\text{m} = 2400\ J.$$

If, on the other hand, the crate was being pulled by a force F that is directed at an angle θ from the horizontal, as shown in Figure 8.39, we must consider only the component of the force in the direction of motion. That is, if the crate is sliding over the horizontal plane, only the component $F\cos(\theta)$ of the force in the horizontal direction contributes to the work done by the force on the crate. Thus, the total work is given by

$$W = Fd\cos(\theta).$$

If there is no motion in the direction of the applied force, then the applied force does no work on the body. It is only the component of the force that lies in the direction of motion that matters. Here, the component of the force in the direction of motion is given as $F\cos(\theta)$, which follows from simple trigonometry. ▪

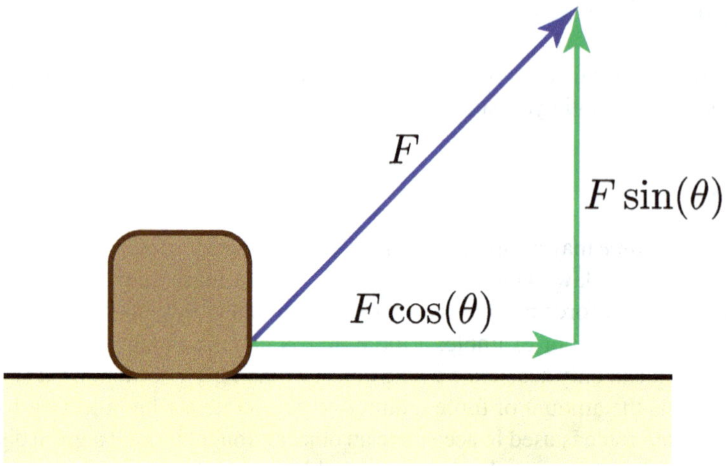

Figure 8.39: A wooden crate being pulled by a force F that makes an angle θ with the horizontal.

The theory of integration comes into play when the force required is not constant, but varies as a function of the particle's position. In general, a force that is a function of a particle's position is called a *force field*. The constant gravitational field, the inverse-square law gravitational field, the electric field, and the restoring force of a spring, as given by Hooke's law, each constitute a classical example of a force field. Our next theorem tells us how we may calculate the work done when moving a particle against a force field.

Theorem 8.5.1 Consider a particle restricted to one-dimensional motion along the x-axis. The particle is moved in the presence of a force field $F(x)$. Then the work done by the force field on the particle, as the particle moves from position $x = a$ to position $x = b$, is given by

$$W = \int_a^b F(x)\,dx. \tag{8.31}$$

(R) If the work done by the forcefield, as calculated using equation (8.31), is negative, this means that work must be done *against* the force field in order to achieve the motion. For example, if a brick is lifted upward to a certain height, the work done by the gravitational field will be negative, indicating that the motion was being resisted by the force field. On the other hand, if a brick falls, then the work done by the gravitational field is positive—gravity does work on the brick as it accelerates it from a given height to the ground.

■ **Example 8.19 — Hooke's Law.** Consider a block with mass m attached to a wall with a spring, as shown in Figure 8.40. In physics, the force that the spring exerts

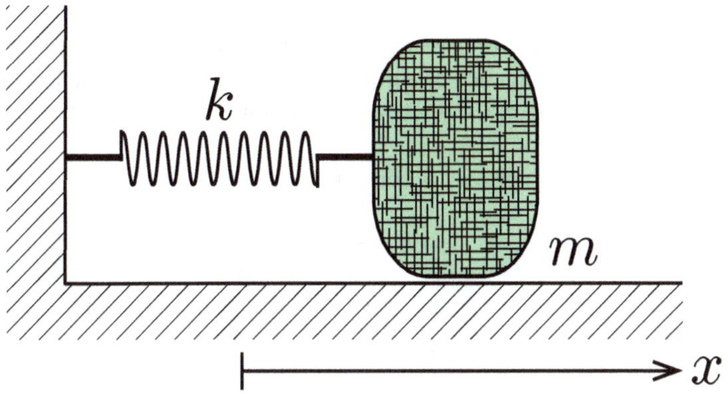

Figure 8.40: A block–spring system.

on the block is known as the *restoring force*, as its purpose is to restore the spring back to its resting length. We define the x-coordinate so that $x = 0$ corresponds to the block being positioned so that the spring has its equilibrium length, i.e., so that the spring is neither stretched nor compressed. *Hooke's law* states that the restoring force is proportional to the amount by which the spring is stretched or compressed, i.e.,

$$F(x) = -kx, \tag{8.32}$$

where k is a constant known as the *spring constant*. (This constant differs depending on physical properties of the given spring.)

(R) If the spring is stretched, resulting in a positive x-position, Hooke's law yields a negative force, indicating that spring is pulling the block in the negative x-direction, i.e., back toward its equilibrium position. Similarly, if the spring is compressed, resulting in a negative x-position, the restoring force, as given by Hooke's law, is positive, meaning that the spring pushes the block in the positive x-direction, again trying to restore the block to its equilibrium or resting position.

In this example, we compute the work done by the spring when moving the block from position x_1 to position x_2.

To proceed, we use the restoring force given by equation (8.32) in the work equation (8.31). We obtain

$$W = \int_{x_1}^{x_2} -kx\, dx = \frac{kx_1^2}{2} - \frac{kx_2^2}{2}. \tag{8.33}$$

We can perform a quick reality check to see whether the sign makes sense. Suppose the spring is initially stretched, and then the block is released. As the block moves from its stretched position to the equilibrium position, it picks up speed. The spring is doing work on the block in order to create this motion, and the work given from equation (8.33) is positive. ∎

∎ **Example 8.20** A block with mass m moves from a vertical height x_1 to a vertical height of x_2 under the influence of a gravitational field, with constant downward gravitational acceleration g. Compute the work done by the gravitational field on the block.

Here, the force due to gravity is constant:

$$F = -mg.$$

The minus sign is required since the gravitational acceleration is in the downward direction. Thus, the total work is given by

$$W = \int_{x_1}^{x_2} -mg\, dx = mgx_1 - mgx_2.$$

If, for example, the block is dropped from a height of $x_1 = H$, and lands on the ground, at height $x_2 = 0$, the total work done by the gravitational field is $W = mgH$. Thus, as the block falls, gravity does work on the block (thereby accelerating the block toward Earth). Conversely, if the block is launched upward from an initial height of $x_1 = 0$, then the total work done by gravity on the block is given by $W = -mgH$, in order for the block to reach a height $x_2 = H$. Here, the work is negative, as the gravitational field is opposing the block's upward motion. ∎

(R) As we have seen in the previous example, gravity does a negative amount of work on an object as it is lifted from one height to a greater height. If one were to use a machine to lift a heavy object, the amount of work done by the machine must make up for the negative amount of work done by the gravitational force. Thus, the negative amount of work done by gravity is opposing the upward motion, so the machine must produce an equal, positive amount of work to counteract this.

(R) *Work* may be thought of as a transfer of energy. When the gravitational field does work on an object, it is supply the object with energy. (In particular, as an object falls, its potential energy, stored in the gravitational field, is being converted into kinetic energy, which manifests as an increased velocity.) To lift an object to a certain height, the gravitational field does a negative amount of work, meaning that you must feed the gravitational field energy in order to perform the maneuver. This energy must be supplied either by the object's kinetic energy or by a machine, if the object is to be steadily lifted from one height to another.

∎ **Example 8.21** Compute the work required to pump the water out of a in-ground swimming pool with width w, length ℓ, and heigh h, as shown in Figure 8.41.

To proceed, let us assume that the water has constant (volumetric) mass density ρ. (In SI units, the density of water is $\rho = 1{,}000$ kg/m^3.) Each horizontal "slice"

Figure 8.41: A rectangular swimming pool.

of the water is to be pumped by a different height. Each horizontal slice has a cross-sectional area of

$$A = w\ell$$

and volume of

$$\Delta V = w\ell\,\Delta x,$$

where x represents the vertical distance from the given slice to ground level. The weight of each horizontal slice is therefore

$$\rho g w\ell\,\Delta x.$$

Since the work done against gravity equals the weight times the distance, and since the ith slice is a distance of x_i from the ground level, the work done to pump the ith slice of water out of the swimming pool is equal to

$$\Delta W_i = \rho g w\ell x_i\,\Delta x.$$

The total work done to pump the water out of the tank is therefore approximated by the Riemann sum

$$W \approx \sum_{i=1}^{N} \rho g w\ell x_i\,\Delta x.$$

Taking the limit as $N \to \infty$, this sum converges to the value of the definite integral

$$W = \int_0^h \rho g w\ell x_i\,dx = \frac{\rho g w\ell h^2}{2}.$$

∎

The preceding example can be generalized to a variety of situations. For example, if the horizontal cross-sectional area of each slice varies with the vertical direction, this area ($w\ell$ in the previous example) should be replaced by $A(x)$. Similarly, if the density of the liquid varies with vertical position, then ρ should similarly be replaced with a function $\rho(x)$.

■ **Example 8.22** Determine a formula for the total work required to build a stone pyramid of height h, side length s, and density ρ, where the acceleration due to gravity is g.

We proceed in a similar fashion to Example 8.13. Recall that the horizontal slices at a vertical distance x to the peak have cross-sectional area $A(x) = s^2x^2/h^2$, as shown in Figure 8.42. The weight of the ith horizontal slice, located at x_i, is

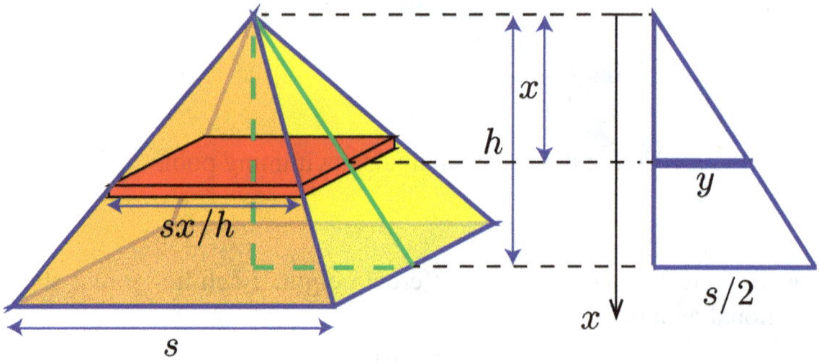

Figure 8.42: A pyramid with a horizontal square cross-sectional slice.

therefore given by

$$\frac{\rho s^2 x_i^2}{h^2}\,\Delta x.$$

Since the ith slice is a distance $(h - x_i)$ above the ground, the work done against gravity to lift the material for that slice is given by

$$\Delta W_i = \frac{\rho s^2 x_i^2 (h - x_i)}{h^2}\,\Delta x.$$

The total work required to build the pyramid is approximated by the Riemann sum

$$W \approx \sum_{i=1}^{N} \frac{\rho s^2 x_i^2 (h - x_i)}{h^2}\,\Delta x.$$

In the limit as $N \to \infty$, this sum converges to the value of the definite integral

$$\begin{aligned} W &= \int_0^h \frac{\rho s^2 x_i^2 (h - x_i)}{h^2}\, dx \\ &= \int_0^h \left(\frac{\rho s^2 x^2}{h} - \frac{\rho s^2 x^3}{h^2} \right) dx \\ &= \frac{1}{12} \rho s^2 h^2. \end{aligned}$$

The great pyramid of Giza, for example, was built with a height 146.5 m and side length 230.4 m. Assuming a constant density of $\rho = 2600$ kg/m^3, the total work required to build the pyramid was approximately 247 gigajoules. (This is equivalent to approximately 69 megawatt-hours.) ■

8.5.2 Work–Kinetic Energy Theorem

In the previous paragraph, we discussed how work may be viewed as a transfer for energy. We next take a closer look at this assertion, examining the relationship between work and kinetic energy.

Theorem 8.5.2 — Work-Kinetic Energy Theorem. The net work W done on a particle, by all the forces acting on it, is equal to the net change in its kinetic energy, i.e.,

$$W = \frac{mv_2^2}{2} - \frac{mv_1^2}{2}, \tag{8.34}$$

where m is the mass of the particle, and where v_1 and v_2 represent the particle's initial and final velocity, respectively.

Proof. By definition, the total work done on the particle is given by

$$W = \int_{x_1}^{x_2} F(x)\, dx.$$

Newton's second law of motion, however, relates the net force acting on the particle to the particle's acceleration, according to the formula

$$F = mv'(t).$$

Moreover, from the definition of velocity, we have the relationship

$$dx = v\, dt.$$

Thus, the total work is equivalent to

$$W = \int_{t_1}^{t_2} mv(t)v'(t)\, dt.$$

Recognizing that $v'(t)\,dt = dv$, we may play one additional change of variables to obtain

$$W = \int_{v_1}^{v_2} mv\,dv = \frac{mv^2}{2}\bigg|_{v_1}^{v_2} = \frac{mv_2^2}{2} - \frac{mv_1^2}{2}.$$

This proves the result. ■

8.5.3 Hydrostatic Pressure

Another useful concept in physics is the concept of *pressure*. In general, fluids exert a force on their surroundings. That force acts in every direction at every point within the fluid. *Pressure* is a measure of force per unit area. That is, if a fluid with constant pressure P acts on a flat plate with surface area A, then the total force of the fluid against the plate is given by $F = PA$. If the fluid is static (i.e., it is not moving), the pressure within the fluid is referred to as *hydrostatic pressure*.

Theorem 8.5.3 The (absolute, or total) hydrostatic pressure P of a liquid at depth h is given by

$$P(h) = P_0 + \rho g h, \tag{8.35}$$

where ρ is the density of the liquid, g is the acceleration due to gravity, and P_0 is the pressure at the surface of the liquid (usually given by atmospheric pressure). The term

$$p(h) = \rho g h$$

is called the *gauge pressure*.

(R) The gauge pressure $p(h)$ is given its name since pressure gauges only measure the difference between the pressurized fluid and the atmospheric pressure.

▪ **Example 8.23** Compute the net hydrostatic force acting on a vertical, rectangular, dam wall of length L and height H, as shown in Figure 8.43.

Since the pressure varies linearly with depth, we choose the x-coordinate to represent the depth below the water's surface. Proceeding as usual, we construct horizontal slices of length L and width Δx, as shown in Figure 8.44. Thus, the ith slice, located at position x_i, has area $L\Delta x$. To compute the net force on the wall, we use the gauge pressure. (Technically, the water exerts a pressure of $P = P_0 + \rho g x$ on the wall, and the air, on the other side of the wall, exerts an opposite pressure equal to P_0.) The net hydrostatic force acting on the ith slice of the wall is therefore given by

$$\Delta F_i = \rho g x L\,\Delta x.$$

The total hydrostatic force is thus approximated by the Riemann sum

$$F \approx \sum_{i=1}^{N} \rho g x L\,\Delta x,$$

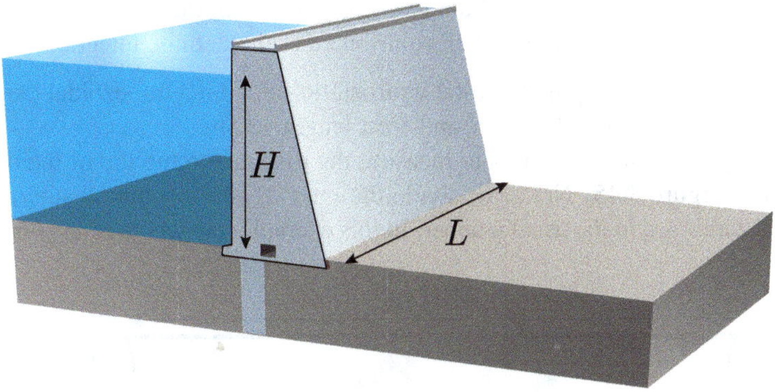

Figure 8.43: Hydrostatic force on a dam wall.

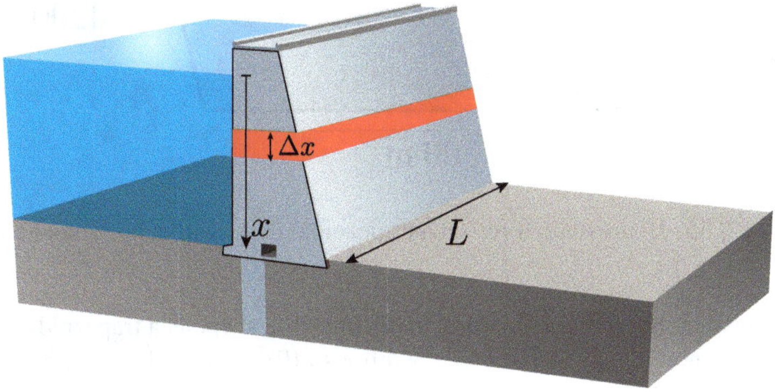

Figure 8.44: Hydrostatic force on a dam wall; horizontal slices.

which converges to the value of the definite integral

$$F = \int_0^H \rho g L x \, dx$$

in the limit as $N \to \infty$. The total hydrostatic force acting on this rectangular dam is therefore equal to

$$F = \frac{\rho g L H^2}{2}.$$

For a non-rectangular dam, one must determine the cross-sectional length L as a function of x. In this case, the hydrostatic force is found by computing the integral

$$F = \rho g \int_0^H L(x) x \, dx.$$

Similarly, if the density of the liquid varies with depth, then ρ would have to be replaced by $\rho(x)$, which would then contributed to the definite integral. ■

■ **Example 8.24** Compute the total hydrostatic force on a trapezoidal dam with height 200 m, base length 300 m, and crest length 400 m.

To begin, we set the x-axis to measure the distance to the top of the dam, as shown in Figure 8.45. Next, we divide the interval $[0,200]$ into N subintervals, each of width Δx, in the usual manner. In this example, the length of the ith slice is

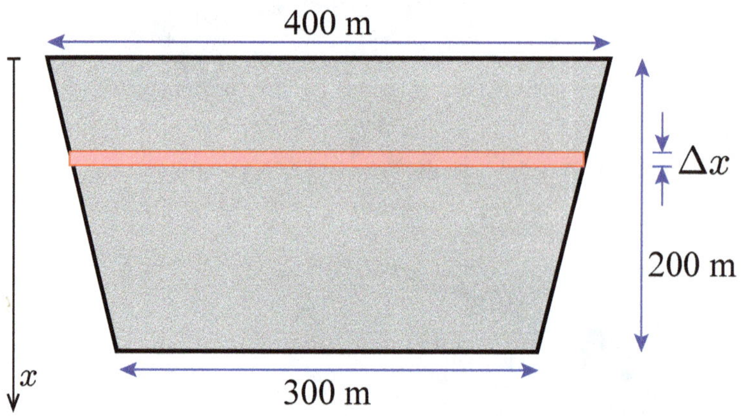

Figure 8.45: Hydrostatic force on a trapezoidal dam; with a horizontal slices.

not a constant, but rather a function of x. Since the shape is a trapezoid, we seek the linear function that satisfies the conditions $L(0) = 400$ and $L(200) = 300$. The linear function that satisfies these constraints is given by

$$L(x) = 400 - \frac{x}{2}.$$

The area of the ith slice, located at position x_i, is therefore approximated by

$$\Delta A_i \approx (400 - x/2)\ \Delta x.$$

(You may verify, as a quick check, that the value of the definite integral

$$\int_0^{200} (400 - x/2)\, dx$$

returns the area of the trapezoid.) Since the (gauge) pressure at depth x is given by

$$p(x) = \rho g x,$$

where $\rho = 1000$ kg/m^3 is the density of water and $g = 9.8$ m/s^2 is the acceleration due to gravity at sea level. Therefore, the hydrostatic force acting on the ith slice is approximated by

$$\Delta F_i \approx \rho g x (400 - x/2)\ \Delta x.$$

The total hydrostatic force acting on the dam wall is therefore approximated by the Riemann sum

$$F \approx \sum_{i=1}^{N} \rho g x(400 - x/2)\ \Delta x,$$

which converges to the value of the definite integral

$$F = 9800 \int_0^{200} \left(400x - x^2/2\right)\, dx$$

in the limit as $N \to \infty$. The total hydrostatic force on the dam therefore equals $F = 65.\overline{333}$ GN (i.e., $65.\overline{333}$ giganewtons). ∎

Exercises

For Exercises 1–5, determine the work done by a spring with spring constant k to move a block from position $x = a$ to position $x = b$; the spring is at equilibrium when $x = 0$.

1. $k = 0.2$; $a = 5$; $b = 2$.
2. $k = 3.1$; $a = -2$; $b = 3$.
3. $k = 1.7$; $a = -2$; $b = 5$.
4. $k = 2.67$; $a = -3.22$; $b = 3.22$.
5. $k = 0.4$; $a = 0.5$; $b = 2.5$.

For Exercises 6–10, a chain (or piece of rope) with length ℓ m and with linear mass density λ kg/m is given. Acceleration due to gravity is $g = 9.8$ m/s^2. If the chain is hanging off the ledge of a rooftop, determine the amount of work done against gravity that would be required to pull the chain up to the roof. (*Hint*: Each slice Δx travels a different distance to the rooftop.)

6. $\lambda = 4$; $\ell = 20$.
7. $\lambda = 6.2$; $\ell = 25$.
8. $\lambda = 0.58$; $\ell = 5.6$.
9. $\lambda = 0.88$; $\ell = 9.2$.
10. $\lambda = 1.55$; $\ell = 13$.

For Exercises 11–15, determine the work required to pump the water from a tank or swimming pool with the indicated shape. (Assume that the water is pumped to the level of the brim or top of the container.) Water has density $\rho = 1000$ kg/m^3 and acceleration due to gravity is $g = 9.8$ m/s^2.

11. The bottom half of a sphere with radius 10 m.
12. An inverted right-circular cone with base radius 5 m and height 5 m.
13. An inverted square pyramid with base side length 8 m and height 5 m.
14. An inverted square pyramid with base side length 5 m and height 7 m.
15. The side of the swimming pool can be described by the region bounded by $y = 0$, $y = -x/2$, $y = -3$, and $x = 12$. The width of the swimming pool (perpendicular to this side) has length 10 m.

For Exercises 16–20, determine the hydrostatic force acting on the given vertical plate, if the top of the plate is a depth h below the water's surface. (Use gauge pressure for determining the force.)

16. A circle with radius 0.5 m; $h =$ 3 m.
17. A circle with radius 1.2 m; $h =$ 7 m.
18. An inverted equilateral triangle with side length 1.5 m, so that the top edge is one of its sides; $h = 12$ m.
19. An equilateral triangle with side length 3 m, so that the bottom edge is one of its sides; $h = 6$ m.
20. The bottom half of a circle of radius 1.4 m; $h = 5$ m.

Problems

21. The top of a tank has the shape of an isosceles triangle with base width 10 m and height 20 m. The front of the tank (perpendicular to the base of the triangle) is a square with side length 10 m. The vertical cross sections parallel to the front are squares, as shown in Figure 8.46.

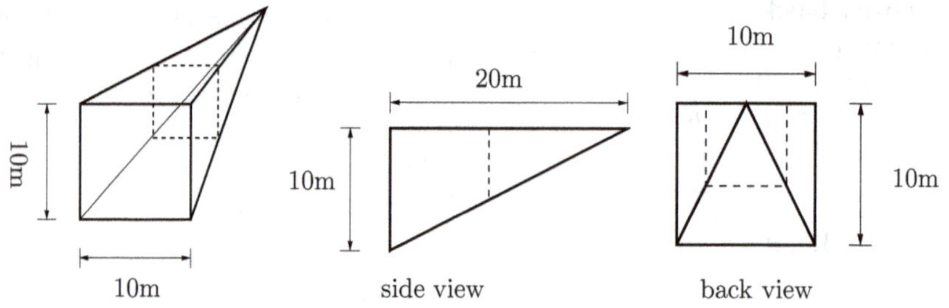

Figure 8.46: Tank from Problem 21.

If the tank is filled with hydrazine (a rocket fuel), which has density 795 kg/m^3, determine the work required to pump the rocket fuel to a height of 5 m above the tank's surface. (Take $g = 9.8$ m/s^2, as usual.)

22. A large semi-cylindrical trough of radius $R = 2$ m and length $L = 9$ m is filled with rain water (density $\rho = 1000$ kg/m^3; acceleration due to gravity $g = 9.8$ m/s^2), as shown in Figure 8.47. Determine the work required to

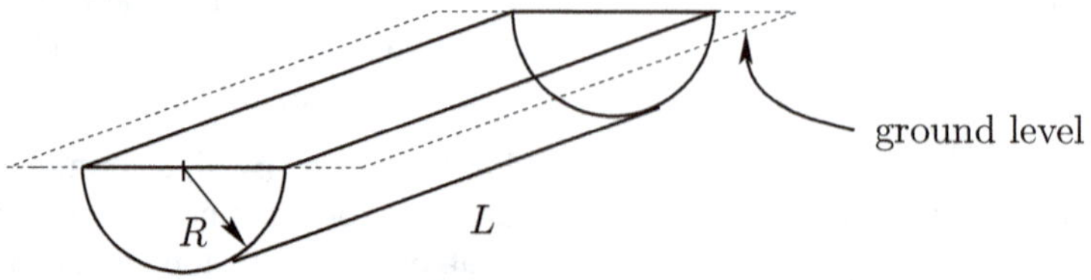

Figure 8.47: Semi-cylindrical trough; Problem 22.

pump all of the rain water out of the trough to ground level. Include the appropriate units in your answer.

23. Determine a general formula for the work required to pump the liquid from the tank in Problem 22 in terms of the parameters ρ, g, R, and L.
24. A cylindrical tank with base radius R and length L is laid on its side (so that its circular base forms a vertical wall). The tank is filled with a liquid with density ρ; acceleration due to gravity is g.
    (a) Determine the work required to pump the liquid from the tank to a height of h above the top of the tank.
    (b) Compare the formula from part (a) with the formula that would be obtained if all of the mass was initially located at the center of mass.
25. A certain dam stands 200 m high. Let x be the height in meters (m) from the base of the dam. Let $L(x)$ be the width (m) of the dam at height x and let $P(x)$ be the pressure (kPa) of the water behind the dam at height x. Some values are given in the table below.

Height x (m)	0	50	100	150	180	200
Pressure $P(x)$ (kPa)	1961	1471	981	490	196	0
Width $L(x)$ (m)	600	625	630	655	660	675

    (a) Write down a definite integral in terms of the functions $P(x)$ and $L(x)$ that represents the total hydrostatic force on the dam. Estimate the value of this integral using the trapezoid rule and the table provided.
    (b) Use the trapezoid rule with four equally spaced subintervals to approximate the integral

$$\int_0^{200} L(x)\,dx$$

    (c) What does the integral in part (b) represent physically? Include units in your answer.
26. A sphere of radius 2m is submerged under water, with its center at a depth of 10m below the water's surface. The density of water is $\rho = 1000\ \text{kg/m}^3$ and acceleration due to gravity is $9.8\ \text{m/s}^2$. Find the total hydrostatic force on the sphere. (*Hint*: First determine an integral that represents the surface area of the sphere.)

8.6 Applications to Probability

In this application, we take a look at how one may apply the concept of integration to the field of probability, in particular, to situations in which one has a continuous variable.

8.6.1 Probability Density Functions

We begin with the concept of a probability density function. A probability density function describes the probability distribution of a *continuous random variable*, which is any statistical quantity the varies continuously, such as height, weight, manufacturing time, etc., each of which can take on *any* numerical value along a continuum. This is to be contrasted with a *discrete random variable*, which can only take values within a discrete spectrum, such as the outcome of a die roll, which can only take on a value in the set $\{1,2,3,4,5,6\}$.

Definition 8.6.1 A *probability density function* p of a continuous random variable x is any function with the properties:

1. **[non-negativity]** The probability density must be everywhere non-negative, i.e.,

$$p(x) \geq 0, \qquad \text{for all } x \in \mathbb{R}; \tag{8.36}$$

2. **[normalization]** The total probability is *normalized* to unity, i.e.,

$$\int_{-\infty}^{\infty} p(x)\,dx = 1; \tag{8.37}$$

3. **[probability density]** The function p is a probability density, so that the integral

$$\int_a^b p(x)\,dx$$

represents the probability that the variable x takes a value between $x = a$ and $x = b$.

(R) The integral in property 3 of Definition 8.6.1 is sometimes alternatively interpreted as *the fraction of the population* whose x-value lies between $x = a$ and $x = b$. Thus, probability density functions can equally describe probability as well as statistical outcomes. This is simply a cosmetic distinction, as, in the latter interpretation, if one were to randomly poll a member of society, then the probability that his or her x-value lies between a and b would coincide with the fraction of the population whose x-value lies between a and b. We therefore stick with the probability interpretation of the definition, with the caveat that one might encounter this other, equivalent definition elsewhere.

A probability density function is literally a probability *density*—it represents a *change* in the probability per change in the continuous variable x. When one integrates over this function between two points, one therefore recovers the *total* probability that the continuous random variable x lies between those two points.

An example probability density function is shown in Figure 8.48. Notice

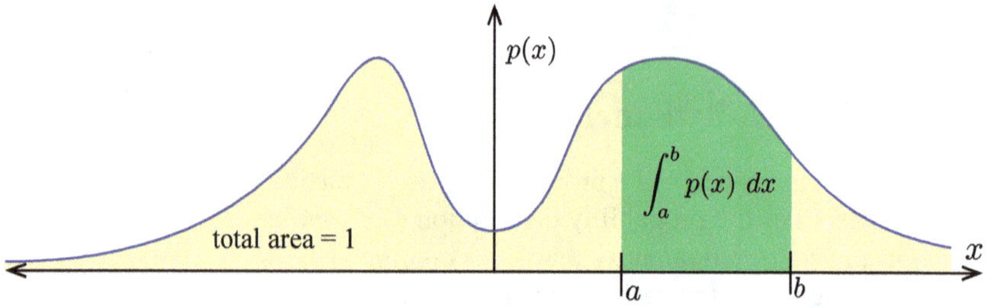

Figure 8.48: Schematic of a probability density function.

that each of the defining properties hold: the function p is non-negative, so that the probability of any region must be positive or zero. Second, the total area is normalized to equal one; thus, the probability that the variable x takes on *some* value—any value—is 100%. (As a consequence of this, it must be the case that $p(x)$ tends to zero as $x \to \pm\infty$; moreover, $p(x)$ must go to zero "fast enough" so that its integral converges.) Finally, if we compute the integral

$$\int_a^b p(x)\,dx,$$

we obtain the probability that the variable x takes on a value between $x = a$ and $x = b$.

In addition to this basic definition, there are a few characteristics that statisticians use to describe a probability distribution, as we outline in our next definition.

Definition 8.6.2 Given a probability density function p of a continuous random variable x, the *mean value* μ ("mu") of the distribution is the number

$$\mu = \int_{-\infty}^{\infty} xp(x)\,dx. \tag{8.38}$$

The *standard deviation* σ ("sigma") of the distribution is defined by the relation

$$\sigma^2 = \int_{-\infty}^{\infty} (x-\mu)^2 p(x)\,dx. \tag{8.39}$$

The quantity σ^2 is sometimes referred to as the *variance*.

R The mean is related to the center of mass of a one-dimensional rod with a mass-density function $p(x)$; it is literally the x-value at which one could "balance" the distribution on top one's fingertip.

The standard deviation can be thought of as the square root of the average of the distance to the mean squared. Similar root-mean-square statistics are used in electrical engineering and physics. Essentially, the standard deviation is a measure of how "spread out" the probability distribution is. By squaring the distance to the mean, one avoids consideration as to whether a particular x-value is below or above the mean—it is just the distance to the mean that counts. One averages the square of the distance to the mean, and then computes the square root. This final number is the standard deviation.

Definition 8.6.3 Given a probability density function p of a continuous random variable x, the *median* m is the number such that

$$\int_{-\infty}^{m} p(x)\,dx = \frac{1}{2} \tag{8.40}$$

R Given the normalization condition (8.37), it is also true that the median satisfies the relation

$$\int_{m}^{\infty} p(x)\,dx = \frac{1}{2}.$$

Intuitively, the median represents the x-value that divides the probability into halves: half of the total probability occurs for x-values below m, and half of the total probability occurs for x-values above m.

■ **Example 8.25 — Exponential Distribution.** In this example, we consider a commonly used probability density function known as the *exponential distribution*, which is defined by the equation

$$p(x) = \begin{cases} 0 & \text{for } x < 0 \\ ae^{-ax} & \text{for } x \geq 0 \end{cases}, \tag{8.41}$$

where a is a positive parameter, $a > 0$. The exponential distribution is plotted for the case $a = 1$ in Figure 8.49. Let us first show that this distribution is normalized;

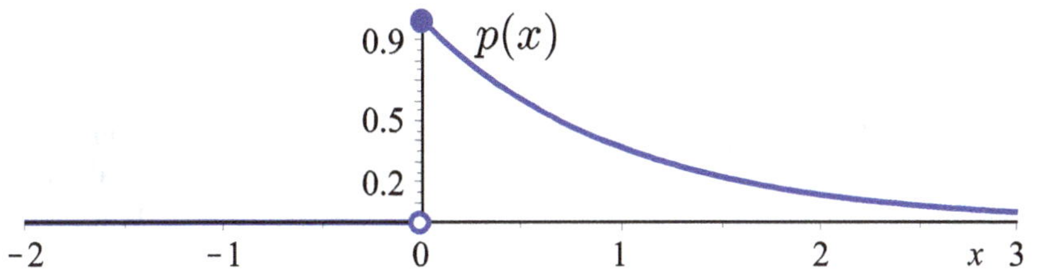

Figure 8.49: The exponential distribution.

then we will compute the mean, standard deviation, and median.

In order to show that the normalization condition (8.37) is satisfied, we compute the integral

$$\begin{aligned}
\int_{-\infty}^{\infty} p(x)\,dx &= \int_{-\infty}^{0} p(x)\,dx + \int_{0}^{\infty} p(x)\,dx \\
&= \int_{-\infty}^{0} 0\,dx + \int_{0}^{\infty} ae^{-ax}\,dx \\
&= \lim_{T\to\infty}\left(\int_{0}^{T} ae^{-ax}\,dx\right) \\
&= \lim_{T\to\infty}\left(-e^{-ax}\Big|_{x=0}^{x=T}\right) \\
&= \lim_{T\to\infty}\left(1-e^{-aT}\right) = 1.
\end{aligned}$$

Since this is a piecewise-defined function, notice that we must break up the definite integral as a separate integral over each piece of the domain.

Next, one may show that the mean and standard deviation are given by

$$\mu = \int_{0}^{\infty} axe^{-ax}\,dx = \frac{1}{a}, \tag{8.42}$$

$$\sigma^2 = \int_{0}^{\infty} \left(x-\frac{1}{a}\right)^2 ae^{-ax}\,dx = \frac{1}{a^2}. \tag{8.43}$$

These results will be computed in Problem 1.

To compute the median m, we require that

$$\int_{-\infty}^{m} p(x)\,dx = \int_{0}^{m} ae^{-ax}\,dx = \frac{1}{2}.$$

We compute the second integral as

$$\int_{0}^{m} ae^{-ax}\,dx = -e^{-ax}\Big|_{x=0}^{x=m} = 1-e^{-am}.$$

Setting this expression equal to 1/2, we may solve for m to obtain

$$m = \frac{\ln(2)}{a}. \tag{8.44}$$

We observer that, due to the antisymmetry of the distribution, the mean and the median do not occur at the same point. This is because the long "tail" of this distribution, which stretches out to infinity, skews the balancing point to the left of the halfway point. ∎

8.6.2 Cumulative Distribution Functions

Given a probability density function, one can always construct its associated *cumulative distribution function*, which, essentially, tells us the total probability that occurs up to a certain point.

Definition 8.6.4 Given a probability density function $p(x)$, its associated *cumulative distribution function* $C(T)$ is defined by

$$C(T) = \int_{-\infty}^{T} p(x)\, dx. \tag{8.45}$$

Thus, $C(T)$ represents the probability that the variable x takes on a value $x < T$.

The area representing integral (8.45) is depicted in Figure 8.50. We see that the cumulative distribution function captures the total probability *up to* the value $x = T$.

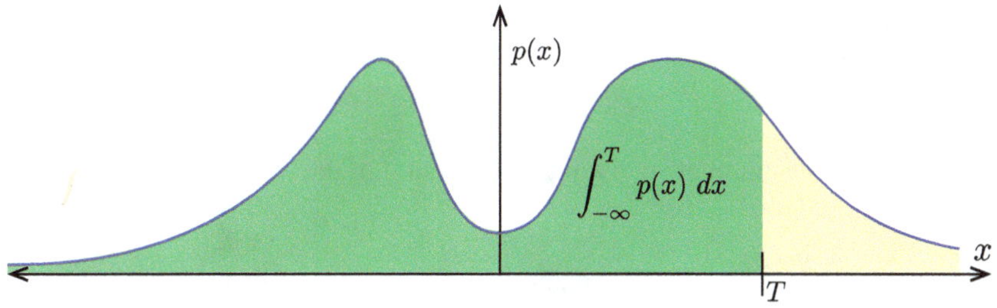

Figure 8.50: Schematic of the value of $C(T)$.

Some basic facts about cumulative distribution functions are highlighted in the following proposition.

Proposition 8.6.1 Suppose that $C(T)$ is the cumulative distribution function for the probability density function $p(x)$. Then the following hold:

1. C is non-negative, i.e., $C(T) \geq 0$ for all $T \in \mathbb{R}$;
2. C is an increasing function;
3. $\lim_{T\to-\infty} C(T) = 0$ and $\lim_{T\to\infty} C(T) = 1$.
4. C is an antiderivative of p, i.e., $C'(T) = p(T)$.

■ **Example 8.26** The cumulative distribution function associated with the exponential distribution, as defined by equation (8.41), is given by

$$C(T) = \int_{-\infty}^{T} p(x)\, dx.$$

Thus, if $T \leq 0$, we have $C(T) = 0$. If $T > 0$, we have

$$C(T) = \int_{0}^{T} ae^{-ax} = -e^{-ax}\Big|_{x=0}^{x=T} = 1 - e^{-aT}.$$

We therefore have

$$C(T) = \begin{cases} 0 & \text{for } T < 0 \\ 1 - e^{-aT} & \text{for } T \geq 0 \end{cases}. \tag{8.46}$$

This function is plotted in Figure 8.51. (Compare with its associated probability density function, as plotted in Figure 8.49.) Notice that $C(T)$ emanates from an

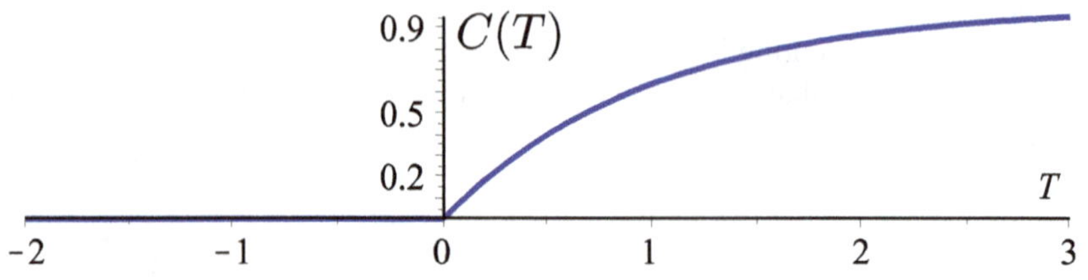

Figure 8.51: The cumulative distribution function of the exponential distribution.

output value of zero as $T \to -\infty$, and that $C(T)$ approaches the output value of $C = 1$ as $T \to \infty$. This represents the fact that no probability is accumulated "at negative infinity," but that 100% of the probability does accumulate "before positive infinity."

To show that the exponential distribution is properly normalized, we may therefore simply write

$$\lim_{T\to\infty} C(T) = 1.$$

Likewise, to determine the median, we simply set $C(m) = 1/2$ and solve for m. ∎

8.6.3 Gaussian Distributions

The most commonly used and well known probability density function is the Gaussian distribution, which is also known as the normal distribution or the bell-shaped curve. The Gaussian distribution is shown with its namesake, Karl Friedrich Gauss, on the ten-Mark bill in Figure 8.52.

Definition 8.6.5 The *Gaussian distribution* (or *normal distribution* or *bell-shaped curve*) is the probability density function defined by the equation

$$p(x) = \frac{1}{\sigma\sqrt{2\pi}} \exp\left(\frac{-(x-\mu)^2}{2\sigma^2}\right), \tag{8.47}$$

where μ is the mean value of the distribution, and σ is the standard deviation.

Three Gaussian distributions, each with a mean $\mu = 0$, and with standard deviations $\sigma = 0.5$, 1, and 2 are plotted in Figure 8.53. Notice that as σ increases, the

Figure 8.52: Karl Friedrich Gauss and the Gaussian distribution as shown on a ten-Mark bill.

probability becomes more spread out, whereas for small values of σ, the probability is more concentrated near the mean. Also, since the Gaussian distribution is symmetric about its mean, we can conclude that its mean and median coincide at that point.

The normalization coefficient $1/(\sigma\sqrt{2\pi})$ that appears in equation (8.47) follows from equation (7.23), i.e., it follows from the important identity

$$\int_{-\infty}^{\infty} e^{-x^2}\,dx = \sqrt{\pi}. \tag{8.48}$$

To see this, let us invoke the substitution

$$u = \frac{x-\mu}{\sigma\sqrt{2}} \qquad \text{and} \qquad du = \frac{dx}{\sigma\sqrt{2}}. \tag{8.49}$$

Computing the integral (8.37) for the Gaussian distribution (8.47) using this substitution and the identity (8.48), we obtain

$$\int_{-\infty}^{\infty} \frac{1}{\sigma\sqrt{2\pi}} \exp\left(\frac{-(x-\mu)^2}{2\sigma^2}\right) dx = \int_{-\infty}^{\infty} \frac{e^{-u^2}}{\sqrt{\pi}}\,du = 1.$$

This confirms that the Gaussian distribution is correctly normalized, so that the total probability equals 100%.

Similarly, one can show that the mean and standard deviation for the Gaussian distribution are μ and σ, respectively. See Problem 36 from Section 7.5 and Problem 2 from this section.

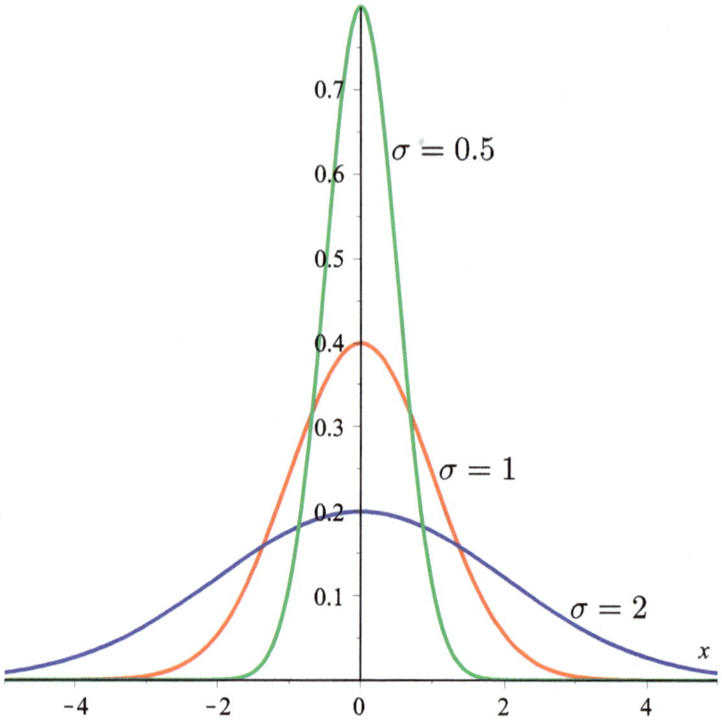

Figure 8.53: Gaussian distribution with $\mu = 0$ and $\sigma = 0.5$, 1, and 2.

The Gaussian distribution does *not* have an antiderivative that can be written in terms of elementary functions in a finite number of terms. Despite this shortcoming, its antiderivative can be numerically determined with a good deal of precision and tabulated. To facilitate this development, we first define a function that serves as an antiderivative of e^{-x^2}, and this function is known as the *error function.*

Definition 8.6.6 The *error function* $\text{erf}(x)$ is the function defined by

$$\text{erf}(x) = \frac{2}{\sqrt{\pi}} \int_0^x e^{-s^2}\, ds. \tag{8.50}$$

The error function is graphed in Figure 8.54.

Given the identity (8.48), one can show that

$$\lim_{x\to\infty} \text{erf}(x) = 1 \qquad \text{and} \qquad \lim_{x\to-\infty} \text{erf}(x) = -1.$$

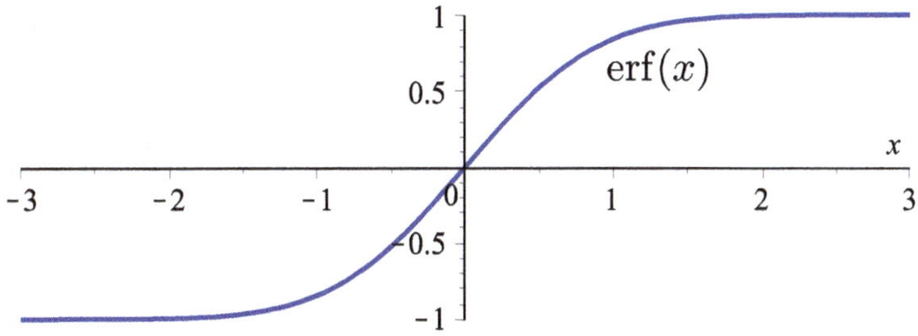

Figure 8.54: Graph of the *error function.*

Moreover, the derivative of the error function is given by

$$\text{erf}'(x) = \frac{2}{\sqrt{\pi}} e^{-x^2}.$$

The error function is useful in formalizing the cumulative distribution function for the Gaussian distribution, as we see in our next theorem.

Theorem 8.6.2 The cumulative distribution function $C(T)$ for the Gaussian distribution (8.47) may be expressed in terms of the error function as

$$C(T) = \frac{1}{2}\left[1 + \text{erf}\left(\frac{T-\mu}{\sigma\sqrt{2}}\right)\right]. \tag{8.51}$$

The cumulative distribution function (8.51) is plotted using $\mu = 0$ and $\sigma = 1$ in Figure 8.55. The proof of this theorem proceeds as follows.

Proof. Given the definition of the cumulative distribution function, as captured by equation (8.45), we have

$$C(T) = \int_{-\infty}^{T} \frac{1}{\sigma\sqrt{2\pi}} \exp\left(\frac{-(x-\mu)^2}{2\sigma^2}\right) dx.$$

Using the substitution (8.49), this becomes

$$\begin{aligned} C(T) &= \int_{-\infty}^{(T-\mu)/(\sigma\sqrt{2})} \frac{e^{-u^2}}{\sqrt{\pi}} du \\ &= \int_{-\infty}^{0} \frac{e^{-u^2}}{\sqrt{\pi}} du + \int_{0}^{(T-\mu)/(\sigma\sqrt{2})} \frac{e^{-u^2}}{\sqrt{\pi}} du \\ &= \frac{1}{2} + \frac{1}{2}\text{erf}\left(\frac{T-\mu}{\sigma\sqrt{2}}\right). \end{aligned}$$

The first term in the last line follows since e^{-u^2} is symmetric about $u = 0$. If the total area under this curve is $\sqrt{\pi}$, then half of the area must be $\sqrt{\pi}/2$. This completes the proof. ■

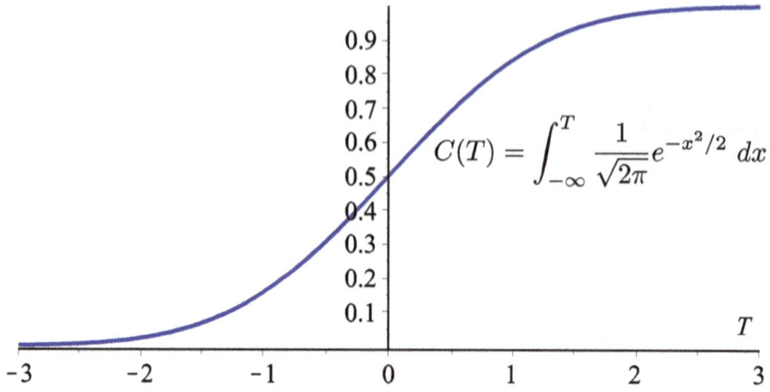

Figure 8.55: Cumulative distribution function for the Gaussian distribution, with $\mu = 0$ and $\sigma = 1$.

Our next paragraph will deal with how one may *practically* apply this theory to answer questions involving Gaussian distributions.

8.6.4 z-scores

Since a simple antiderivative for the Gaussian distribution is not available, one must rely on tabulated values in order to answer practical questions of probability and statistics involving bell-shaped curves. To facilitate this process, we must somehow "standardize" the notation so that one chart can be used for all purposes. This is achieved using an instrument known as the z-score.

Definition 8.6.7 Given a Gaussian distribution with mean μ and standard deviation σ, the *z-score* of a given x value is defined by

$$z = \frac{x-\mu}{\sigma}. \tag{8.52}$$

Thus, a z-score represents how many standard deviations above (if positive) or below (if negative) the mean a given x-value is.

Converting from an x-value, which has intrinsic meaning in terms of a real-world problem, to its associated z-value, which only has extrinsic meaning (i.e., it only has practical meaning by relating it back to its x-value), is an instance of a change of variables or a substitution. The standardize z-values tells one the number of standard deviations from the mean a given x-value is. For example, an x-value of 63.2 must be interpreted in terms of the problem: 63.2 has a direct physical meaning in terms of the problem variables, and must be related to the mean and standard deviation in order to understand where in the population this x-value lies.

Contrast this with a z-value of 0.78. Here, we do not directly know the physical measurement (if this is height, what height is $z = 0.78$?); it does tell us, on the other hand, that the measurement is above average, but only 78% of one standard deviation above average.

Given the substitution (8.52), the Gaussian distribution may be rewritten as

$$p(x)\,dx = p(z)\,dz = \frac{1}{\sqrt{2\pi}} e^{-z^2/2}\,dz. \tag{8.53}$$

Next, we introduce a convenient rule-of-thumb for figuring probabilities in terms of the first three standard deviations of the mean.

Theorem 8.6.3 — The 68–95–99.7 Rule. Given a Gaussian distribution, the approximate probabilities of the continuous random variable lying within one, two, and three standard deviations of the mean are:

1. Approximately a 68.27% probability of being one standard deviation (1-σ) from the mean;
2. Approximately a 95.45% probability of being two standard deviations (2-σ) from the mean; and
3. Approximately a 99.73% probability of being three standard deviations (3-σ) from the mean.

Theorem 8.6.3 is illustrated in Figure 8.56. This shows us approximately how

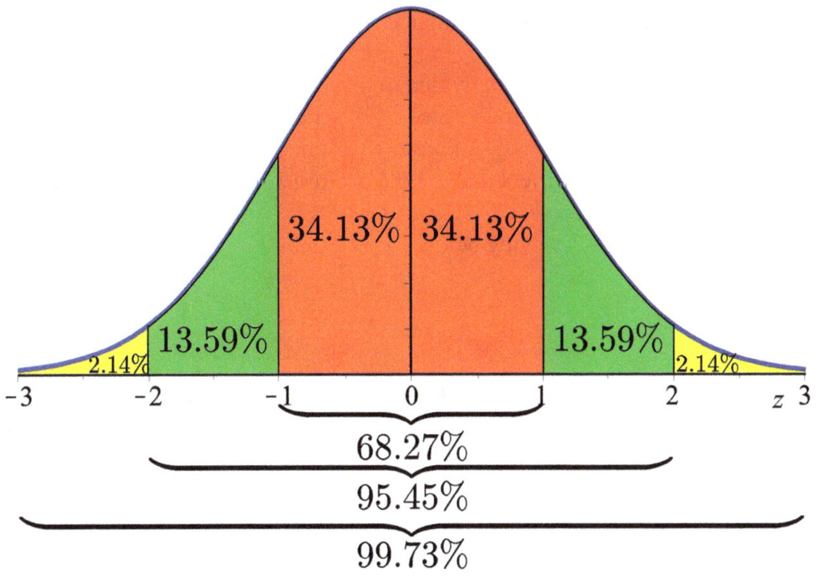

Figure 8.56: Illustration of the 68–95–99.7 rule.

probability is distributed within the first few standard deviations of the mean when using the Gaussian distribution. This is a good rule-of-thumb for memorizing, as it quickly gives one an idea of how much probability lies at these milestones.

One often requires greater accuracy, however. To achieve these ends, values for the cumulative distribution function (8.51) of the Gaussian distribution are tabulated in Appendix E. This table can be used in conjunction with the symmetry of the distribution to answer practical questions within the range of three standard deviations about the mean. For example, the probability of having a z-value less than 0.78 is given by 0.7823. These data are normalized using the z-values. Thus, the table literally provides the cumulative distribution function of $p(z)$, as given by equation (8.53), i.e.,

$$C(Z) = \frac{1}{\sqrt{2\pi}} \int_{-\infty}^{Z} e^{-z^2/2}\, dz. \tag{8.54}$$

Here, we use Z instead of T for the cumulative distribution function to remind us that this is the cumulative distribution function using the z-values, and not the intrinsic variables of the problem. In order to use Appendix E to compute the probabilities at *negative* z-values, we may utilize the following proposition.

Proposition 8.6.4 Suppose that $Z > 0$. Then

$$C(-Z) = 1 - C(Z), \tag{8.55}$$

where $C(Z)$ is the cumulative distribution function (8.54) for the Gaussian distribution, as expressed using z-values.

This proposition follows simply due to the symmetry. For example, the probability that $z < 1$ is identical to the probability that $z > 1$, and so forth. We may use equation (8.55) as follows: suppose, for example, we wish to compute the probability that a z-value is less than -1.32. We use Appendix E to determine that $C(1.32) = 0.9066$. Therefore, $C(-1.32) = 1 - C(1.32) = 0.0934$. We conclude that the probability of a z-value lying below -1.32 is 9.34%.

Problems

1. Compute the integrals in equations (8.42) and (8.43) in order to verify that the mean and standard deviation of the exponential distribution are both given by $\mu = \sigma = 1/a$.
2. In this problem, we consider the Gaussian distribution (8.47).
   (a) Using the substitution (8.49), compute the value of the integral

$$\int_{-\infty}^{\infty} \frac{x}{\sigma\sqrt{2\pi}} \exp\left(\frac{-(x-\mu)^2}{2\sigma^2}\right) dx$$

   and show that it is equal to μ.

(b) Compute the value of the integral

$$\int_{-\infty}^{\infty} \frac{(x-\mu)^2}{\sigma\sqrt{2\pi}} \exp\left(\frac{-(x-\mu)^2}{2\sigma^2}\right) dx$$

and show that it is equal to σ^2.

(c) Conclude that the parameters μ and σ, as they appear in the definition of the Gaussian distribution, really are the mean and standard deviation of the distribution.

3. The average height of men in the United States is 69.1 inches, with a standard deviation of 2.8 inches. Assume that heights are *normally distributed* (i.e., they follow a Gaussian distribution).
   (a) What fraction of men in the U.S. are taller than six feet?
   (b) What fraction of men are shorter than 5'3" feet?
4. The average height of women in the United States is 64.0 inches, with a standard deviation of 2.5 inches. Assume that heights are normally distributed.
   (a) What fraction of women in the U.S. are taller than 5'9"?
   (b) What fraction of women are shorter than five feet?
5. The *intelligence quotient* (i.e., "IQ") of a person is defined assuming that intelligence is normally distributed with a mean IQ of 100 and a standard deviation of 15.
   (a) What fraction of the population has an IQ greater than 140?
   (b) What fraction of the population has an IQ between 110 and 120?
6. The wait time at a call-in customer-service center is exponentially distributed with an average wait time of 12 minutes.
   (a) What fraction of callers should expect to wait for longer than 12 minutes?
   (b) What is the median wait time?
   (c) What is the probability that the total wait time for a given caller is between 10 and 15 minutes?
7. Suppose the average wait time at Hawaiian Hamburger is three minutes, and assume that wait time is exponentially distributed. Suppose the manager wants to advertise that, "If you wait longer than x minutes, you get a free hamburger." If the manager does not wish to give a free hamburger to more than 10% of the customers, what should x be?
8. Let x be the amount of time (in minutes) since the start of Professor X's 70-minute calculus class. Let $p(x)$ be a probability density function such that $\int_a^b p(x)\,dx$ is the probability that Professor X covers an interesting example between time $x = a$ and time $x = b$. The formula for $p(x)$ is as follows:

$$p(x) = \begin{cases} 0 & \text{for } x \le 0 \\ c & \text{for } 0 < x < 60 \\ 4c & \text{for } 60 \le x \le 70 \\ 0 & \text{for } 70 < x \end{cases}.$$

The graph of $p(x)$ is shown in Figure 8.57.

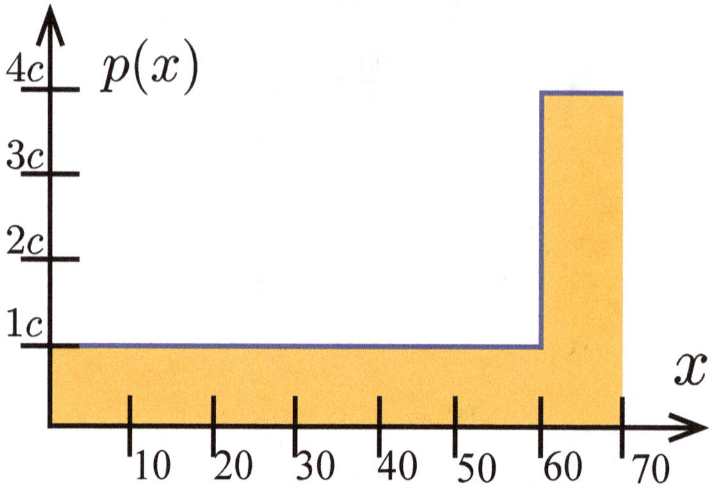

Figure 8.57: Graph of the probability density function for Problem 8.

(a) Find the constant c.
(b) Find the median value of x.
(c) Find the average value of x.
(d) Find the probability that Professor X includes an interesting example during the last forty minutes of class.

9. Consider the probability density function

$$p(x) = \begin{cases} 0 & \text{for } x < 0 \\ cxe^{-ax} & \text{for } x \geq 0 \end{cases}.$$

(a) Determine the value of c in terms of the parameter a, so that this distribution is guaranteed to satisfy the normalization condition (8.37).
(b) Determine the mean μ of this distribution in terms of the parameter a.
(c) Determine the standard deviation σ of this distribution in terms of the parameter a.
(d) Determine the cumulative distribution function $C(T)$.
(e) Show that the median m satisfies the transcendental equation

$$\frac{1}{2} = (1 + am)e^{-am}.$$

(f) Compute $C(\mu)$.
(g) What is the probability that x lies above the mean?

10. Consider the probability density function

$$p(x) = \begin{cases} 0 & \text{for } x < 0 \\ a\sin(bx) & \text{for } 0 \leq x \leq \pi/b \\ 0 & \text{for } x > \pi/b \end{cases}.$$

(a) Determine the value of the parameter b in terms of the parameter a so that the normalization condition (8.37) is satisfied.
(b) Determine the mean μ in terms of the parameter a.
(c) Determine the standard deviation σ in terms of the parameter a.
(d) Determine the cumulative distribution function $C(T)$.
(e) Determine the median m in terms of the parameter a.

11. The *logistic distribution* is the probability density function defined by

$$p(x) = \frac{\pi}{4\sigma\sqrt{3}}\operatorname{sech}^2\left[\frac{\pi}{2\sqrt{3}}\left(\frac{x-\mu}{\sigma}\right)\right].$$

(a) Graph the logistic distribution using $\mu = 0$ and $\sigma = 1$.
(b) Show that the associate cumulative distribution function is given by

$$C(T) = \frac{1}{2\sigma}\left(\tanh\left[\frac{\pi}{2\sqrt{3}}\left(\frac{x-\mu}{\sigma}\right)\right] + 1\right).$$

(c) Show that the logistic distribution is properly normalized, i.e., show that

$$\int_{-\infty}^{\infty} p(x)\,dx = 1.$$

(d) Show that the mean and standard deviation are μ and σ, respectively.

8.7 Applications to Economics

The value of a dollar has spatiotemporal variance: one dollar in Kansas is not worth one dollar in silicon valley; one dollar today is not equal to one dollar tomorrow. In this section, we seek to address the issue of the *temporal* variance of monetary value—i.e., inflation. In particular, we will examine the value of payment or income streams and utilize an analysis to help inform economic decisions.

8.7.1 Present and Future Values of Lump Sums

The question of valuing money earned or spent at a specific point in time relies on the underlying *inflation rate*, i.e., we assume that monetary value follows and exponential trend. A dollar in the future must be discounted or deflated to determine what it is worth today. Similarly, a dollar today must be inflated to determine what it will be worth in the future. In particular, we have the following definitions.

Definition 8.7.1 The *present value* V_o of an amount of money that exists in the future is equal to the value that it would be worth today.

The *future value* V_f of an amount of money is the value that the money will be worth at some specified future time.

For a single valuation, we can rely on the exponential growth formulas to perform simple computations. In this context, we are only dealing with *lump sum* payments, i.e., a single payment made today versus at some point in the future.

Proposition 8.7.1 Suppose that V_0 and V_f are the present and future values of a certain lump sum of capital, and that the future value is computed at a time T in the future. Assuming a *continuous inflation rate* r, the present and future values of the capital are related to each other by the equation

$$V_f = V_0 e^{rT}. \tag{8.56}$$

Alternatively, if the money appreciates with a growth factor b, (e.g., $b = 1+$ APR), then

$$V_f = V_0 b^T. \tag{8.57}$$

Example 8.27 Suppose that capital can be safely invested at a rate of 5% per year. If a payment of \$1,000 is to be received in one year's time, what is that payment worth today? In other words, if that future payment of \$1,000 can be traded for a payment of a lesser amount today, what is the minimum amount that one should accept?

To answer this, we need to compute the present value of that future payment, assuming a 5% continuous inflation rate. Here, the future value is just $V_f = 1000$. The present value is therefore

$$V_0 = V_f e^{-rT} = 1000e^{-0.05} = 951.23.$$

Therefore, a payment of \$1,000 is worth \$951.23 today, assuming a 5% inflation rate. This means that if we invest \$951.23 today, it is expected to be worth exactly \$1,000 in one year's time. Therefore, if we are offered an amount *greater* than \$951.23 in lieu of a payment of \$1,000 to be made one year from now, it is advantageous to accept the offer. ■

■ **Example 8.28** Suppose that \$1,000 is to be invested into an account with an APR of 2.7%. What is the future value of this lump sum payment in five years from now?

Since the annual percentage rate (APR) is 2.7%, the annual growth factor is $b = 1.027$. To compute the future value, we simply write

$$V_f = V_0 b^T = 1000(1.027)^5 = 1142.49.$$

Thus, the lump sum of \$1,000 will be worth \$1,142.49 in five years from now. ■

8.7.2 Present and Future Values of Annuities

In this paragraph, we will consider the present and future values of steady payments made at discrete intervals over time.

Definition 8.7.2 An *annuity* is a sequence of payments made at fixed intervals of time.

An *annuity-immediate* is an annuity in which payments are made at the end of each payment period. An *annuity-due* is an annuity in which payments are made at the beginning of each payment period.

The differente between an annuity-immediate and an annuity-due using N payment periods is illustrated in Figure 8.58. A mortgage payment is an example of an annuity-immediate, as interest is charged before each payment is made. Rent payments and insurance premiums are examples of annuities-due.

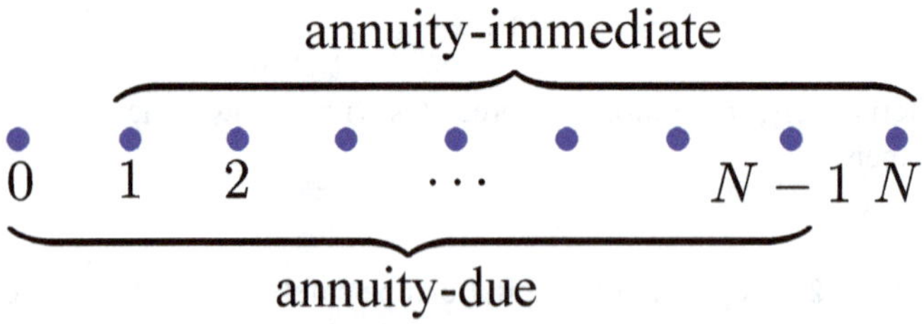

Figure 8.58: An annuity with N payment periods.

Since each payment of an annuity-immediate lags the corresponding payment of an annuity-due by exactly one payment period, the present value of a payment-immediate V_0^i is related to the present value of a payment-due V_0^d by the relation

$$V_0^d = bV_0^i, \tag{8.58}$$

where b is the *periodic growth factor*, i.e., the growth factor of inflation for each payment period. This formula holds because each payment in an annuity-immediate is worth one periodic growth factor more than its corresponding payment in an annuity-due.

Theorem 8.7.2 — Present and Future Values of an Annuity-Due. Consider an annuity-due with N periodic payments in an amount A, assuming a periodic growth factor b. Its present and future values are

$$V_0^d = A\frac{1-b^{-N}}{1-b^{-1}}, \tag{8.59}$$

$$V_f^d = A\frac{b^N-1}{1-b^{-1}}, \tag{8.60}$$

respectively.

Proof. For an annuity-due, N payments are made at the points $i = 0, 1, 2, \ldots, (N-1)$, i.e., at the beginning of each interval. (Compare this with the formulas for a left-hand Riemann sum.) When the ith payment is made, i periods of interest must be depreciated in order to compute the present value. Thus, we multiply the ith payment by b^{-i} in order to determine the present value of that single payment. The present value of the total annuity-due is therefore computed as the finite geometric series

$$V_0^d = \sum_{i=0}^{N-1} Ab^{-i} = A\frac{b^{-N}-1}{b^{-1}-1} = A\frac{1-b^{-N}}{1-b^{-1}}.$$

The last equality is done as a matter of style, since b^{-1} and b^{-N} are each less than one.

To compute the future value, we simply have to add N periods of interest to the present value; we thus obtain

$$V_f^d = b^N V_0^d = A\frac{b^N-1}{1-b^{-1}}.$$

This completes the proof. ■

Theorem 8.7.3 — Present and Future Values of an Annuity-Immediate. Similarly, consider an annuity-immediate with N periodic payments in an amount A, assuming a periodic growth factor b. Its present and future values are

$$V_0^i = A\frac{1-b^{-N}}{b-1}, \tag{8.61}$$

$$V_f^i = A\frac{b^N-1}{b-1}. \tag{8.62}$$

Proof. First, note that equations (8.61) and (8.62) are obtained by multiplying equations (8.59) and (8.60) by a factor of b^{-1}, respectively, in accordance with equation (8.58). To verify equation (8.58), let us consider the following. Since the payments of an annuity-immediate are made at the points $i = 1,2,\ldots,N$, the present value is given by

$$V_0^i = \sum_{i=1}^{N} Ab^{-i} = b^{-1} \sum_{i=0}^{N-1} b^{-i} = b^{-1} V_0^d.$$

The future value will similarly be related to the present value by factoring in N periods of growth; thus $V_f^i = b^N V_0^i$, and the desired result is obtained. ■

▪ **Example 8.29** Suppose that Jill wins the lottery, and is given the option of a lump sum payment of $1,000,000 today or a payment of $60,000 per year for 30 years, with the first payment being issued immediately. Assuming that Jill can invest money and earn a 5% interest rate, which option is worth more in today's dollars?

First, the present value of the lump sum payment is clearly $1,000,000, so this is the benchmark present value we wish to compare the annuity option to. Also, since the first annuity payment is issued immediately, the annuity type is an annuity-due. Even though the total payment amount in the annuity option is $1.8 million, this does not take into account inflation and interest. One must compute the present value of these payments. The present value of the annuity-due is calculated using equation (8.59), from which we obtain

$$V_0^d = (60000)\frac{1-(1.05)^{-30}}{1-(1.05)^{-1}} = \$968,464.41.$$

We conclude that, at 5% interest, the lump sum payment is presently worth more than the value of the annuity. Even though the annuity payments total $800,000 more than the lump sum payment, the lump sum payment is worth $31,535.58 more than the annuity in today's dollars. ▪

(R) One will notice the sensitivity in these calculators to the underlying inflation rate. The mathematics is nevertheless extremely useful when making financial decisions. First, future inflation rates, though not guaranteed, can be modeled and forecast. Second, these mathematical tools allow analysts to easily perform what-if analyses: if the inflation rate is r, what are the present values of our different options and what decision should we make?

8.7.3 Continuous Revenue Streams

In our next paragraph, we consider the case of an income or payment stream that is modeled as a continuous, instead of a discrete, function of time. This may be useful, for example, in a large company that might have a continuous stream of revenue that can be modeled over time. A continuous payment stream $P(t)$ over a period $[0,T]$ might appear as in Figure 8.59. Here, $P(t)$ represents the *rate* that payments are being made (or that revenue is coming in, etc.), measured in dollars

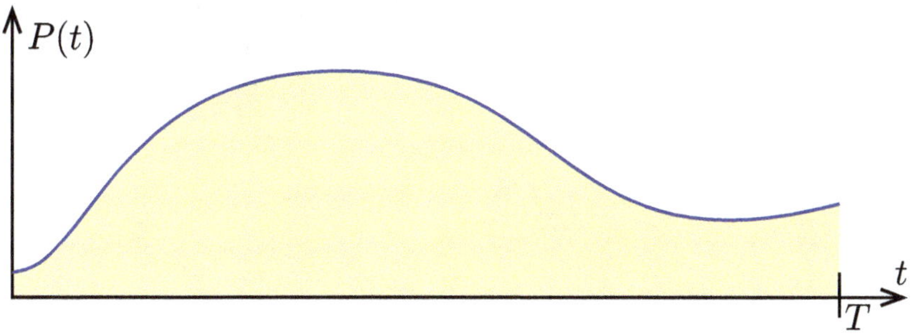

Figure 8.59: Example of a revenue stream over a period $[0, T]$.

per unit time. The present value of this revenue stream is its worth at time $t = 0$, whereas the future value of this revenue stream is its worth at time $t = T$. Of course, the present and future values are dependent upon the assumed continuous inflation rate r.

Theorem 8.7.4 — Present and Future Value of a Revenue Stream. Let $P(t)$ represent a revenue stream over a period $[0, T]$. For a given continuous inflation rate r, the present and future values of the revenue stream are

$$V_0 = \int_0^T P(t)e^{-rt}\,dt, \tag{8.63}$$

$$V_f = e^{rT}\int_0^T P(t)e^{-rt}\,dt, \tag{8.64}$$

respectively.

First, notice that the preceding theorem obeys equation (8.56): if we apply T years of interest (or T months, etc., depending on the units of time) to the present value, we obtain the future value.

Proof. To determine the present value of a revenue stream, we must first think about *approximating* the present value using a Riemann sum. Thus, let us break up the interval $[0, T]$ into N equal subintervals, each with width

$$\Delta t = \frac{T}{N}.$$

As usual, we label the points $t_i = i\Delta t$, for $i = 0, \ldots, N$, so that the ith subinterval is $[t_{i-1}, t_i]$.

Now, using the right end-points, we can approximate the revenue received during the ith subinterval as

$$P(t_i)\Delta t.$$

(This is the *rate* that revenue is being earned, approximated using the right endpoint, times the change in time.) This quantity is represented by the rectangle shown in Figure 8.60. A similar rectangle is constructed for *each* of the N subintervals.

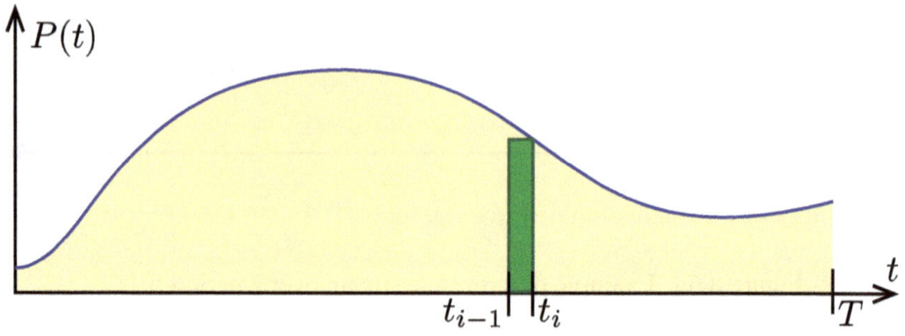

Figure 8.60: Approximate revenue earned during the ith subinterval $[t_{i-1}, t_i]$ is given by $P(t_i)\Delta t$.

Thus, we have approximated our *continuous* revenue stream as a variable annuity, with N payments $P(t_i)\Delta t$, for $i = 1, \ldots, N$. We next wish to depreciate these values in order to obtain their present values. The present value of the ith payment is given by

$$P(t_i)\Delta t e^{-rt_i}.$$

Therefore, the present value of the revenue stream is approximated by the sum

$$V_0 \approx \sum_{i=1}^{N} P(t_i)e^{-rt_i}\Delta t.$$

In the limit as $N \to \infty$, this approximation becomes exact, and we recover equation (8.63), since

$$\lim_{N\to\infty}\left(\sum_{i=1}^{N} P(t_i)e^{-rt_i}\Delta t\right) = \int_0^T P(t)e^{-rt}\,dt,$$

by the definition of the definite integral.

To compute the future value of this revenue stream, we must simply appreciate each payment by an amount $e^{r(T-t_i)}$, instead of depreciating it by e^{-rt_i}. (Note that $(T-t_i)$ is the amount of time between the payment and the future epoch at which we are appreciating the payment to.) Pulling out a common factor of e^{rT}, we recover equation (8.64). ■

▪ **Example 8.30** Suppose that a local small business can model its revenue stream as

$$P(t) = 155000 + 75000\cos(4\pi t)$$

dollars per year. This model captures seasonal fluctuations in the demand for their products. Suppose that there is an underlying continuous inflation rate of 3%. What is the present value of the revenue stream for the first five years?

To compute this, we apply equation (8.63), thereby obtaining

$$V_0 = \int_0^5 [155000 + 75000\cos(4\pi t)]\, e^{-0.03t}\, dt \approx 719677.44.$$

Thus, the revenue stream for the first five years is presently worth approximately \$719,677.44, assuming an underlying 3% continuous inflation rate. Compare this with the actual total revenue for the same period, ignoring inflation, which is \$755,000. ▪

8.7.4 Supply and Demand Curves

In a free market economy, the quantity and price of a certain item that is to be produced and sold are related to each other by the *supply and demand curves*. Supply and demand is a microeconomics concept that is used to determine the production and pricing equilibrium in a competitive market.

Definition 8.7.3 A *supply curve* is a relation that determines what quantity of goods manufacturers are willing and able to supply at a given price level.

A *demand curve* is a relation that determines what quantity of goods consumers are willing to buy at a given price level.

It is generally assumed that as the price level increases, manufacturers are willing to produce more and consumers are willing to buy less. In a competitive market, the price ultimately settles at an equilibrium value, at which the quantity manufacturers are willing to produce meets the consumer demand for the product.

Even though the quantity supplied and demanded is described as a function of the price, this relation is usually assumed to be invertible. Since it is customary to display the quantity q on the horizontal axis and the price p on the vertical axis, we will describe the supply and demand curves by the functions $p = S(q)$ and $p = D(q)$, respectively. Typical supply and demand curves for a competitive market are shown in Figure 8.61.

It is assumed that the market will settle at the equilibrium price p^*, at which a quantity q^* of goods will be produced and sold. In Figure 8.61, the price p_0 is the *lowest* price for which there is some manufacturer willing to produce a small quantity of the given commodity, whereas the price p_1 is the *highest* price for which there is some consumer willing to buy the commodity. Thus, the interval $[0, q^*]$ represents a range on which consumers are willing to pay *more* than the equilibrium price and manufacturers are willing to produce the item for *less* than the equilibrium price. Thus, both consumers and producers profit for having traded. This is quantified in the following definition.

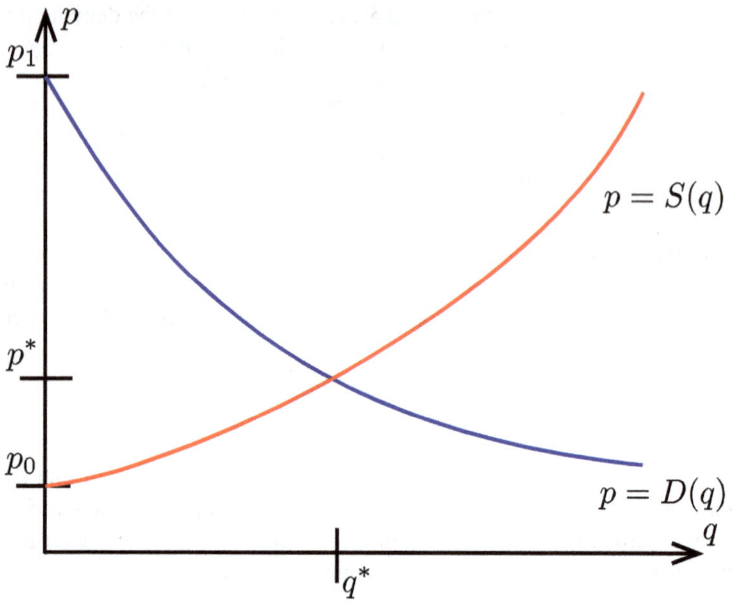

Figure 8.61: Supply and demand curves for a market.

Definition 8.7.4 The *consumers' surplus* is the total amount of monetary gain by the consumers, as quantified by the difference between the highest price each consumer was willing to pay and the equilibrium price at which they actually made the purchase.

The *producers' surplus* is the total amount of monetary gain by the producers, as quantified by the difference between the equilibrium price at which they actually sold the commodity and the lowest price at which they would have been willing to produce the commodity.

In order to quantify the consumers' and producers' surplus, let us consider Figure 8.62. In making the ith sale, there is one consumer who was willing to purchase one item of the commodity at the price $p = D(q_i)$. The marginal consumer surplus in making this sale is therefore $[D(q_i) - p^*]\Delta q$. (We can think of Δq as small enough to represent a single item.) In order to approximate the *total* consumers' surplus, we therefore sum

$$\sum_{i=1}^{N} \left[D(q_i) - p^*\right] \Delta q.$$

Similarly, the marginal producer surplus is given by $[p^* - S(q_i)]\Delta q$. In order to

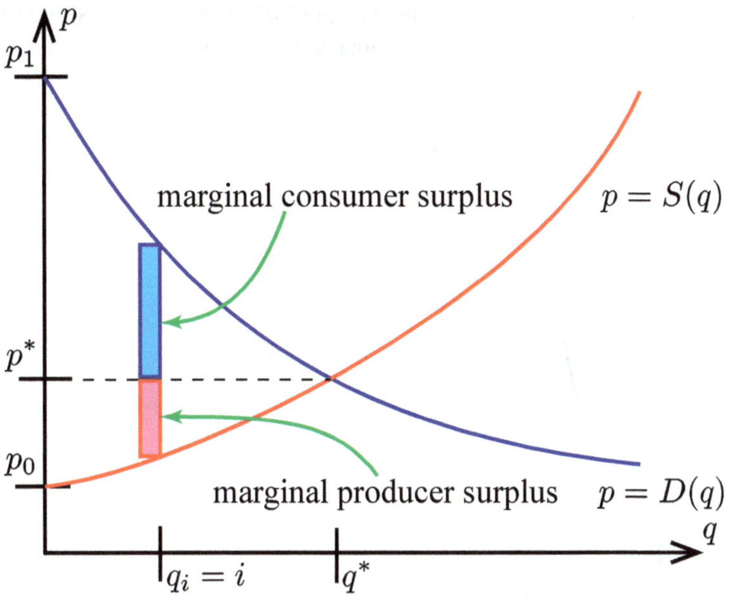

Figure 8.62: Marginal consumer and producer surplus.

approximate the total producers' surplus, we compute the sum

$$\sum_{i=1}^{N} \left[p^* - S(q_i)\right] \Delta q.$$

If we take the limit as $N \to \infty$, i.e., as $\Delta q \to 0$, we obtain an exact measure of the consumers' and producers' surplus, as these summations turn into definite integrals. We have therefore obtained the following theorem.

Theorem 8.7.5 Given a supply curve $p = S(q)$, a demand curve $p = D(q)$, and an equilibrium (q^*, p^*), the consumers' surplus is given by

$$\text{consumers' surplus} = \int_0^{q^*} \left(D(q) - p^*\right)\, dq. \tag{8.65}$$

Similarly, the producers' surplus is given by

$$\text{producers' surplus} = \int_0^{q^*} \left(D(q) - p^*\right)\, dq. \tag{8.66}$$

In accordance with the preceding theorem, the consumers' surplus and producers' surplus is illustrated in Figure 8.63. We observe that the consumers' surplus is represented by the total area bounded by the vertical axis, the demand curve, and the line $p = p^*$, which represents the equilibrium price level. Similarly, the producers' surplus is represented by the total area bounded by the vertical axis, the supply curve, and the line $p = p^*$. The *total economic surplus*, which is the sum of the consumers' and producers' surpluses, is the total area bounded by the vertical axis and the supply and demand curves.

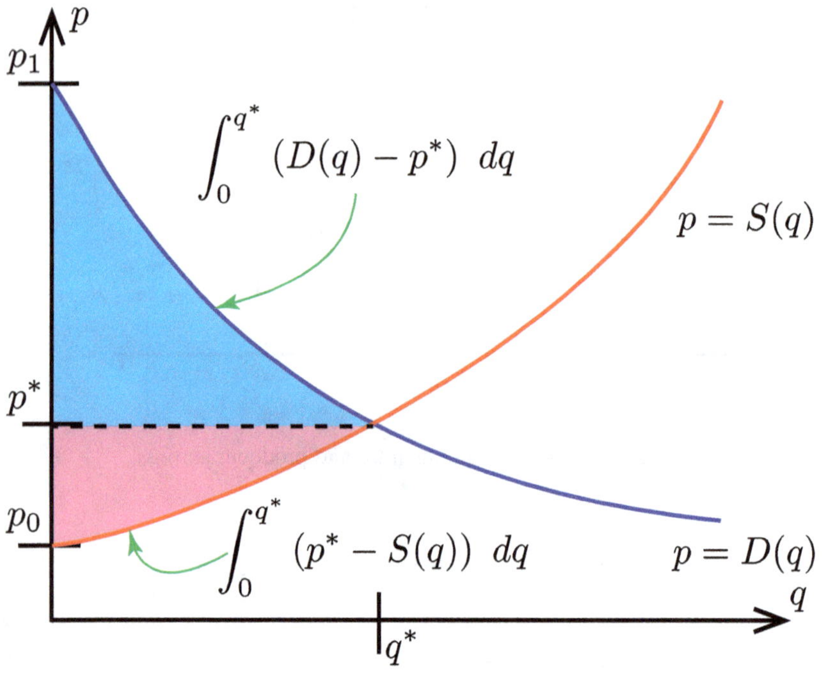

Figure 8.63: Consumers' surplus and producers' surplus.

Problems

1. Suppose that Manish and Padmaja take out a \$650,000 mortgage for a house. The 30-year mortgage bears an interest rate of 4.75%.
   (a) How much is their monthly mortgage payment?
   (b) If Manish and Padmaja always make their payments on time, and if they do not pay off the loan early, what is the present value of their 360 mortgage payments, assuming that the first payment is due one month *after* close of escrow, and assuming an underlying inflation rate of 3%?
   (c) If Manish and Padmaja pay an extra \$500 each month on top of their mortgage payment, towards the principal, how early will their mortgage be paid off? What is the present value of this new payment annuity?

2. A *perpetuity* is an annuity that continues indefinitely. Show that the present value of a perpetuity, in which payments are made at the end of each period, is equal to the amount of the periodic payment divided by the interest rate for each period.
3. A legal settlement reached between a company and a former employee consisted of a \$5.4 million lump sum payment plus a monthly payment of \$12,500 per month, made at the end of each month, for a period of 20 years. The press reports that the settlement is worth \$7.3 million. What assumptions did they make about the underlying APR interest rate to reach this conclusion? Use technology to solve your equation.
4. Suppose that a piece of art can be sold for a price $P(t) = a(1+0.2t^2)$ when it is t years old, where a is the original sale price of the piece. Assume that such a piece is purchased as an investment, and that the investor can alternatively invest in the stock market and receive a continuous growth rate of 10% at any time.
   (a) If the art piece is sold after t years, determine the present value of this transaction.
   (b) When should the investor sell her art piece in order to maximize the present value of the sale?
5. Suppose that a position at a company offers an initially salary of \$75,000 per year, and that the salary increases at a rate of 2% per year.
   (a) Assuming that the payment may be modeled as a continuous payment stream $P(t) = 75000e^{0.02t}$, and assuming a continuous inflation rate of 5%, determine the present cost of hiring someone for the position, assuming they will work for the company for 20 years.
   (b) If the payments are instead modeled as annual payments in the amount $A_i = 75000e^{0.02i}$, at the end of the ith year, for $i = 1, \ldots, 20$, how would this change the answer? (Still assuming a 5% continuous inflation rate.)
   (c) Which estimate of the present value is greater? Why?
6. Jake wins the lottery, and is offered the option of receiving a single lump-sum payment of \$23 million, or 30 annual payments of \$1.7 million each, the first payment being made immediately.
   (a) If the annual percentage rate is 5%, which option has the greater present value?
   (b) If the annual percentage rate is 10%, which option has the greater present value?
   (c) If Jake chooses the lump-sum option, what assumption is he making about the interest rate, assuming that his decision was made to maximize the present value of the payouts?
7. Suppose that the supply and demand curves for a given quantity are linear, so that

$$S(q) = p_0 + aq \qquad \text{and} \qquad D(q) = p_1 - bq.$$

   (a) Determine the equilibrium quantity q^* and the equilibrium price level p^* in terms of the parameters a, b, p_0, and p_1.

(b) Determine the consumers' surplus in terms of a, b, and q^*.
(c) Determine the producers' surplus in terms of a, b, and q^*.

8. Suppose that the price of a commodity is kept artificially low, at a price level of $a < p^*$. Draw a supply and demand curve and describe the effect of such a policy on (a) the consumers' surplus, (b) the producers' surplus, and (c) the total economic surplus.
9. Suppose that the price of a commodity is kept artificially high, at a price level of $b > p^*$. Draw a supply and demand curve and describe the effect of such a policy on (a) the consumers' surplus, (b) the producers' surplus, and (c) the total economic surplus.
10. Suppose that a tech startup has a patent on a new piece of technology, so that they are the only company that is able to supply it. Let $p = D(q)$ represent the demand curve, which relates the quantity q consumers are willing to buy at a price level p, and let $p = S(q)$ represent the supply curve, which relates the manufacturing price per item as a function of the quantity produced.
    (a) Express a formula for the revenue $R(q)$ and cost $C(q)$ to the company if q units are manufactured and sold.
    (b) Show that profit reaches an extremum (maximum or minimum) whenever the relation

$$q(D'(q) - S'(q)) + D(q) - S(q) = 0$$

    holds.
    (c) Since the startup has control of the market, via its patent, should it produce a quantity of goods less than or greater than the theoretical equilibrium q^*?
    (d)
11. Suppose that $D(q) = p_1 - aq$ and $S(q) = p_0 + bq^2$.
    (a) Determine the equilibrium quantity q^* in terms of the parameters (a, b, p_0, and p_1).
    (b) Assuming the company wishes to maximize profit, as described in Problem 10, determine the optimum quantity q_0 that will maximize profit.
    (c) Is the optimum quantity q_0 above or below the equilibrium quantity q^*?
12. Suppose that $a = 0.1$, $b = 0.01$, $p_0 = 1$, and $p_1 = 10$ in Problem 11.
    (a) Determine the values of q^* and q_0. Determine the equilibrium price level p^* and the optimal price level p_0.
    (b) Determine the consumers' and producers' surplus if the price level is set at the equilibrium value q^*.
    (c) Determine the consumers' and producers' surplus if the price level is set at the optimal value p_0

Chapter 9

Sequences and Series

9.1 Sequences and Series

In this section, we discuss the basic notions of sequences and series.

9.1.1 Sequences of Numbers

We begin with a brief introduction to sequences.

Definition 9.1.1 A *sequence* is an ordered list of numbers. The nth term of a sequence is denoted by s_n.

If a sequence has k terms, and if the enumeration begins with the first term, then the sequence may be denoted as

$$(s_n)_{n=1}^{k} = (s_1, s_2, s_3, \ldots, s_k). \tag{9.1}$$

If there are an infinite number of terms in a sequence, it may be denoted as

$$(s_n)_{n=1}^{\infty} = (s_1, s_2, s_3, \ldots), \tag{9.2}$$

or, equivalently, as $(s_n)_{n\in\mathbb{N}}$, or, more simply, as (s_n), if it is understood that the index n ranges over all natural numbers.

(R) We will almost exclusively consider infinite sequences. When we use the word *sequence*, it will therefore assume that it refers to an infinite sequence, unless the context requires otherwise.

Any sequence may alternatively be viewed as a function on the domain of natural numbers, i.e., a sequence if a function that sends each natural number into a real number.

Example 9.1 Write out the first several terms of the sequence

$$\left(\frac{1}{n}\right)_{n=1}^{\infty}.$$

Construct a graph of the sequence.

Here, the formula

$$s_n = \frac{1}{n}$$

is used to generate the nth term of the sequence. The first several terms of this sequence are therefore given by

$$\left(\frac{1}{n}\right)_{n=1}^{\infty} = \left(1, \frac{1}{2}, \frac{1}{3}, \frac{1}{4}, \frac{1}{5}, \ldots\right).$$

These first few terms are plotted in Figure 9.1. Notice that a sequence behaves and

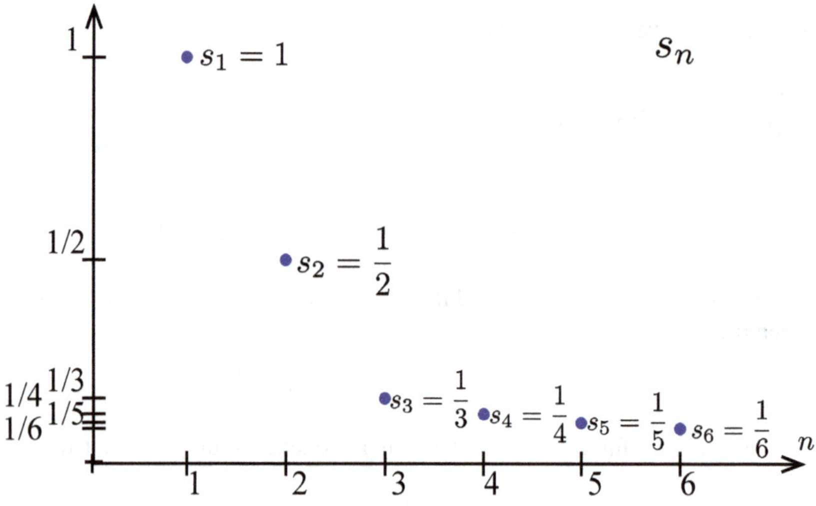

Figure 9.1: First few terms of the sequence $(1/n)$ from Example 9.1.

is graphed exactly like a function defined only on the domain $\mathbb{N} = \{1, 2, 3, 4, \ldots\}$, as we mentioned previously. ■

Not all sequences, however, are described with an explicit formula. In our next example, we study a sequence that is defined *recursively*. A *recursively defined sequence* is a sequence in which each term is defined based on the value of the previous term(s) in the sequence. The rule that generates the new term based on the previous term(s) is known as a *recursion relation*.

Example 9.2 Consider the sequence (s_n) defined by the initial value $s_1 = 41$ and the recursion relation

$$s_{n+1} = \frac{s_n + 1}{3},$$

for $n \in \mathbb{N}$. Determine the first several terms of this sequence.

Starting with the initial value $s_1 = 41$, we apply the recursion relation (using $n = 1$) to obtain

$$s_2 = \frac{s_1 + 1}{3} = \frac{42}{3} = 14.$$

Applying the recursion relation a second time, using $n = 2$, we obtain

$$s_3 = \frac{s_2 + 1}{3} = \frac{15}{3} = 5.$$

Continuing in a similar fashion, we uncover the first few terms of the sequence:

$$(s_n) = \left(41, 14, 5, 2, 1, \frac{2}{3}, \frac{5}{9}, \frac{14}{27}, \frac{41}{81}, \frac{122}{243}, \frac{365}{729}, \ldots\right).$$

■

Example 9.3 The *Fibonacci numbers* are the terms of the *Fibonacci sequence*, which is defined recursively using the initial values $s_0 = 1$, $s_1 = 1$, and the recursion relation

$$s_{n+2} = s_{n+1} + s_n.$$

This recursion relation generates the sequence

$$(s_n) = (1, 1, 2, 3, 5, 8, 13, 21, 34, 55, 89, 144, \ldots).$$

These numbers are known as the Fibonacci numbers. Interesting, the ratio of successive terms approaches

$$\lim_{n\to\infty} \frac{s_{n+1}}{s_n} = \frac{1 + \sqrt{5}}{2},$$

which is known as the *golden ratio*. ■

9.1.2 Convergence of Sequences

Next, we examine the concept of limits of (infinite) sequences. This is analogous to taking the limit of a function as the independent variable goes to infinity; in fact, both operations are governed by the same rules. Intuitively, we say that

$$\lim_{n\to\infty} s_n = L,$$

for a given sequence (s_n), if the output values s_n become closer and closer to the value L as the index n becomes arbitrarily large. The precise definition is as follows.

Definition 9.1.2 Given an infinite sequence (s_n), we say that *the limit of s_n as n goes to infinity is equal to L*, denoted

$$\lim_{n\to\infty} s_n = L,$$

if the following condition holds: for every $\varepsilon > 0$, there exists a number N, such that

$$|s_n - L| < \varepsilon \qquad \text{whenever} \qquad n > N. \tag{9.3}$$

Equation (9.3) means that no matter how close we wish to make the terms of the sequence s_n to the value L, if we wait long enough, our wish will be fulfilled. That is, for *any* value of $\varepsilon > 0$, no matter how small, there will always come a point in the sequence, after which the terms of the sequence are within an ε-distance to the value L.

This definition is illustrated in Figure 9.2. A positive number $\varepsilon_1 > 0$ defines a horizontal tube centered about the line $y = L$. Since $\lim_{n\to\infty} s_n = L$, there exists some point N, such that every term in the sequence satisfies the inequality $|s_n - L| < \varepsilon_1$, i.e., lies within the horizontal tube

$$L - \varepsilon_1 < s_n < L + \varepsilon_1,$$

for all $n > N_1$, i.e., for all points occurring later than N_1. But this statement can

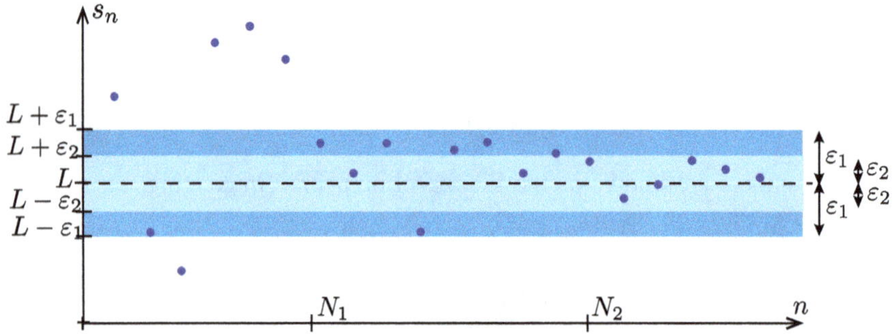

Figure 9.2: Illustration of the limit of a sequence (s_n).

also be made for a smaller positive number $\varepsilon_2 > 0$. Similarly, there exists some point N_2, such that the terms of the sequence are within the (smaller) horizontal tube

$$L - \varepsilon_2 < s_n < L + \varepsilon_2,$$

for all $n > N_2$, i.e., for all points occurring later than N_2.

Since the same trick works for *any* positive ε, regardless of how small, we may conclude that the farther we go out in the positive n-direction, the closer and closer the terms of the sequence (s_n) will be to the value L.

Definition 9.1.3 Consider an infinite sequence (s_n). If the limit

$$\lim_{n\to\infty} s_n$$

exists, we say that the sequence *converges*. Otherwise, we say the sequence *diverges*.

The notion of the convergence or divergence of a sequence is completely analogous to the notion of the convergence or divergence of the limit of a function as the independent variable goes to infinity.

Example 9.4 Determine whether the sequence (s_n), defined by the formula

$$s_n = \frac{n^3 + e^{-3n}}{7n^3 + 8 + e^{-n}}$$

converges or diverges.

In other words, we must determine the limit

$$\lim_{n\to\infty} \frac{n^3 + e^{-3n}}{7n^3 + 8 + e^{-n}}.$$

As the index n becomes large, the terms e^{-3n} and e^{-n} become smaller and smaller, i.e.,

$$\lim_{n\to\infty} e^{-3n} = 0 \qquad \text{and} \qquad \lim_{n\to\infty} e^{-n} = 0.$$

Aside from the terms that exhibit exponential decay, we have a cubic polynomial in both the numerator and denominator. Since the power of these polynomials is balanced, the limit exists. Moreover, the limit is equal to the ratio of the leading coefficients, or 1/7. We conclude by stating that the sequence (s_n) converges to the value $1/7$ (in the limit as $n \to \infty$). ■

The field of *analysis* is based on the precise definition of limits and their consequences. The road of analysis winds through a rigorous development of the theory of calculus, where everything is properly justified using precise definitions. The next two results will help broaden our intuitive feel of the behavior of certain sequences, and the proof will give us a glimpse into the flavor of elementary analysis. But first we introduce two ways in which sequences can be classified.

Definition 9.1.4 A sequence (s_n) is *bounded* if there exists a number $M > 0$ such that

$$|s_n| \leq M \tag{9.4}$$

for all $n \in \mathbb{N}$. Any number M satisfying the inequality (9.4) is said to be a *bound* for the sequence. If a sequence is not bounded, it is said to be *unbounded*.

A bounded sequence is thus simply a sequence whose terms cannot grow to be arbitrarily large, i.e., there is some maximum size that the terms cannot exceed.

Definition 9.1.5 A sequence (s_n) is *increasing* if its terms satisfy

$$s_n \leq s_{n+1},$$

for all $n \in \mathbb{N}$.

A sequence (s_n) is *decreasing* if its terms satisfy

$$s_n \geq s_{n+1},$$

for all $n \in \mathbb{N}$.

A sequence (s_n) is *monotone* if it is either increasing or decreasing.

The concept of an increasing or decreasing sequence parallels that of an increasing or decreasing function. Next, we explore the relation between convergence and whether or not a sequence is bounded.

Theorem 9.1.1 Every convergent sequence is bounded.

Proof. We begin with a convergent sequence (s_n). Since the sequence converges, the limit

$$\lim_{n\to\infty} s_n = L$$

must exist; let us define L to be that limit. By the definition of the limit: for every $\varepsilon > 0$, there exists some number N such that the logical implication (9.3) holds. In particular, let us consider $\varepsilon = 1$. For this value of ε, there must exist some natural number $N \in \mathbb{N}$ such that

$$|s_n - L| < 1 \qquad \text{whenever} \qquad n > N;$$

in other words,

$$L - 1 < s_n < L + 1, \tag{9.5}$$

for all $n > N$. Next, let us define

$$M_1 = \max\{|s_n| : n \leq N\}$$

and

$$M_2 = \max\{|L+1|, |L-1|\}.$$

By the definition of M_1, it follows that

$$|s_n| \leq M_1$$

for all $n \leq N$. Similarly, since the inequality (9.5) holds whenever $n > N$, it follows that

$$|s_n| \leq M_2$$

for all $n > N$. (Think about it!) Finally, by defining

$$M = \max\{M_1, M_2\},$$

we may conclude that $|s_n| \leq M$ for all $n \in \mathbb{N}$. This implies that the sequence is bounded, as claimed. ■

(R) Theorem 9.1.1 states that every convergent sequence must be bounded. The converse, however, need not be true. That is, a bounded sequence does not have to converge. Consider, for example, the sequence (s_n) defined by $s_n = (-1)^n$, i.e., the sequence $(-1, +1, -1, +1, -1, +1, \ldots)$. This sequence is bounded since every term satisfies $|s_n| \leq 1$, yet it fails to converge, as the terms of the sequence never "settle down" to a particular value.

Theorem 9.1.2 — Monotone Convergence Theorem. Every bounded, monotone sequence converges.

Proof. Let (s_n) be a bounded, monotone sequence. Since the sequence is monotone, it is either an increasing or decreasing sequence. Let us first consider the case in which (s_n) is an increasing sequence, so that $s_n \leq s_{n+1}$, for all $n \in \mathbb{N}$. Since the sequence is bounded, there exists a number M, such that $|s_n| \leq M$, for all $n \in \mathbb{N}$. In particular, it must also be true that $s_n \leq M$, i.e., that M is an *upper bound* for the sequence. Now, there are many upper bounds for the sequence, as any number greater than one upper bound must, itself, also be an upper bound. There is an axiom in mathematics—*the completeness axiom*—that guarantees that any set of real numbers that is bounded above must have a *least upper bound*, i.e., there is a unique number M that is an upper bound for the sequence, but is less than every other possible upper bound. To proceed, we will take M to be this least upper bound. We now claim that the sequence must converge to M, i.e., that

$$\lim_{n\to\infty} s_n = M.$$

Now let $\varepsilon > 0$ be arbitrary. (Since we take ε to be an *arbitrary* positive number, whatever we conclude from it must apply *to every* positive $\varepsilon > 0$.) The number $M - \varepsilon$ cannot be an upper bound for the sequence, since this number is *less than* the number M and since M is the *least* upper bound. Therefore, there must be some integer N for which we have

$$s_N > (M - \varepsilon).$$

(If this previous statement were false, then $s_n \leq (M - \varepsilon)$ *for all* n, meaning that $M - \varepsilon$ would be an upper bound that is smaller than the least upper bound—a logical contradiction.) But, since the sequence is increasing, the previous inequality further implies that

$$s_n > (M - \varepsilon), \qquad \text{for all } n > N.$$

But the s_ns are also bounded by M, so that

$$(M - \varepsilon) < s_n \leq M < (M + \varepsilon), \qquad \text{for all } n > N.$$

This previous statement is equivalent to

$$|s_n - M| < \varepsilon, \qquad \text{for all } n > N.$$

Since $\varepsilon > 0$ was arbitrary, there must be some number N, for which this inequality holds, for any positive $\varepsilon > 0$. We have just proved: for every $\varepsilon > 0$, there exists an N, such that

$$|s_n - M| < \varepsilon, \qquad \text{for all } n > N.$$

But this is the definition of the limit of the sequence converging to M.

If the sequence (s_n) is a decreasing bounded sequence, then the related sequence (t_n), defined by $t_n = -s_n$ would constitute an increasing bounded sequence, which, pursuant to our result thus far, must converge. Since (s_n) and (t_n) are a multiple of one another, this implies that the decreasing bounded sequence (s_n) must converge as well, which completes our proof. ■

(R) Do take some time to try to sift through the preceding proof. Draw your own pictures and try to gain a grasp on the arguments, one step at a time. You may find it surprisingly enjoyable!

Despite its complexity, the previous theorem is quite simple: an increasing (or decreasing) sequence that is bounded must level out at some limiting value.

■ **Example 9.5** Consider the sequence $(s_n)_{n=1}^{\infty}$, defined by

$$s_n = 1 - \frac{1}{n}.$$

This sequence is bounded above by 1, since $s_n < 1$, for all n. Moreover, this is an increasing sequence, since $1/n$ becomes smaller each time n is incremented by 1. Therefore, it must converge according to the monotone convergence theorem. (It, in fact, converges to 1, i.e., $\lim s_n = 1$.) ■

9.1.3 Infinite Series

In Chapter 5, we gave a brief introduction to infinite series and, in particular, discussed infinite geometric series. We proceed now in greater detail.

Definition 9.1.6 Given an infinite sequence $(a_n)_{n=1}^{\infty}$ of real numbers, one may formally define the *associated infinite series* as

$$\sum_{n=1}^{\infty} a_n = a_1 + a_2 + a_3 + \cdots. \tag{9.6}$$

The *kth partial sum* S_k of the series (9.6) is defined as the sum of its first k terms:

$$S_k = \sum_{n=1}^{k} a_n = a_1 + a_2 + \cdots + a_k. \tag{9.7}$$

R If the index n of a series begins at $n=0$, then the kth partial sum will instead be given by the formula

$$S_k = \sum_{n=0}^{k-1} a_n = a_0 + a_1 + \cdots + a_{k-1}. \tag{9.8}$$

A given series may be indexed one way or another; the important thing is that to include the first k terms, regardless of how they are enumerated, when writing the kth partial sum.

■ **Example 9.6** Consider the series

$$\sum_{n=1}^{\infty} \frac{1}{n}.$$

The first partial sum is given by

$$S_1 = \sum_{n=1}^{1} \frac{1}{n} = 1.$$

The second partial sum is given by

$$S_2 = \sum_{n=1}^{2} \frac{1}{n} = 1 + \frac{1}{2} = \frac{3}{2}.$$

The third partial sum is given by

$$S_3 = \sum_{n=1}^{3} \frac{1}{n} = 1 + \frac{1}{2} + \frac{1}{3} = \frac{11}{6}.$$

The fourth partial sum is given by

$$S_4 = \sum_{n=1}^{4} \frac{1}{n} = 1 + \frac{1}{2} + \frac{1}{3} + \frac{1}{4} = \frac{25}{12}.$$

■

Given an infinite series, a different partial sum may be formed for each natural number $k \in \mathbb{N}$. In this way, the set of all partial sums may, itself, be considered an infinite sequence. That is, given a series in the form of equation (9.6), the partial sums S_k, for $k = 1, 2, 3, \ldots$, form a sequence known as *the sequence of partial sums*: $(S_k)_{k=1}^{\infty} = (S_1, S_2, S_3, \ldots)$.

■ **Example 9.7** Continuing Example 9.6, the corresponding sequence of partial sums is given by the sequence

$$(S_k)_{k=1}^{\infty} = \left(1, \frac{3}{2}, \frac{11}{6}, \frac{25}{12}, \ldots\right).$$

The kth term in this sequence is simply the kth partial sum of the series. ■

The concept of the sequence of partial sums is important in defining convergence of a series.

Definition 9.1.7 An infinite series is said to *converge* if its sequence of partial sums converges. Otherwise, the series is said to *diverge*.

In particular, if (S_k) represents the sequence of partial sums and if

$$\lim_{k\to\infty} S_k = S,$$

then we ascribe the value S to the infinite series.

Example 9.8 Recall the infinite geometric series

$$\sum_{n=0}^{\infty} \frac{1}{2^n} = 1 + \frac{1}{2} + \frac{1}{4} + \frac{1}{8} + \cdots .$$

The kth partial sum is given by

$$S_k = \sum_{n=0}^{k-1} \frac{1}{2^n} = 1 + \frac{1}{2} + \frac{1}{4} + \cdots + \frac{1}{2^{k-1}} = \frac{1-(1/2)^k}{1-(1/2)}.$$

(The final equality follows since this is a finite geometric series; see Theorem 5.2.2.) This formula gives us an explicit formula for the kth partial sum. Taking the limit as $k \to \infty$, we obtain

$$\lim_{k\to\infty} \frac{1-(1/2)^k}{1-(1/2)} = 2.$$

We conclude that the series converges; in particular, it has the value

$$\sum_{n=0}^{\infty} \frac{1}{2^n} = 2.$$

■

We conclude with two fundamental properties of the convergence of series.

Theorem 9.1.3 If two series are identical, except for a finite number of terms, then either both series converge or both series diverge.

The preceding theorem is straightforward enough: changing a finite number of terms of any convergent series results in a convergent series. This is because the notion of *convergence* only deals with whether or not the series, in the long run, sums to a finite value or not. By tampering with only a finite number of terms, one may only change the value of a series by a finite amount. Though this amount may be astronomically large, it is finite nonetheless; the convergence (or divergence) of the series therefore is unchanged.

Theorem 9.1.4 Consider a sequence (a_n), from which an infinite series $\sum_{n=1}^{\infty} a_n$ is constructed. If the series converges, then the terms a_n must approach zero as *n* goes to infinity, i.e., convergence of the series implies that

$$\lim_{n\to\infty} a_n = 0.$$

R The converse of Theorem 9.1.4 is not necessarily true! That is, it is possible to construct a divergent series with terms that go to zero. (The catch is that they do not go to zero *fast enough* for the sum to be limited to a finite value.) We will see an example of this in our next section.

Theorem 9.1.4 can, in some cases, be used to prove that a series does *not* converge. Namely, since every convergent series must have the property that its terms go to zero, whenever the terms do not approach zero, the series cannot be convergent.

Example 9.9 Determine whether or not the series

$$\sum_{n=0}^{\infty} \frac{3n^4 + 9x^2 + \pi}{7n^4 + 8n^3 + e}$$

converges.

First, we note that

$$\lim_{n\to\infty} \frac{3n^4 + 9x^2 + \pi}{7n^4 + 8n^3 + e} = \frac{3}{7}.$$

Thus, the terms of the series become closer and closer to the ratio $3/7$ as $n \to \infty$. This, in particular, prevents the series from converging. (How could the sum of an infinite number of $3/7$s converge?) If the series converged, then, by Theorem 9.1.4, this limit would have to equal zero. Since this limit is not zero, the series cannot converge. ∎

Example 9.10 What does Theorem 9.1.4 have to say about the convergence of the series

$$\sum_{n=1}^{\infty} \frac{1}{n} = 1 + \frac{1}{2} + \frac{1}{3} + \frac{1}{4} + \cdots?$$

The terms of this series go to zero, i.e., the condition

$$\lim_{n\to\infty} \frac{1}{n} = 0$$

is clearly satisfied. Theorem 9.1.4 therefore gives us no information about the convergence or divergence of this series! The series may converge or it may diverge. (We will see in our next section that this series, in fact, diverges.) Theorem 9.1.4 only states that every convergent series satisfies the condition that the terms go to zero as the index goes to infinity; it does not preclude divergent series from also satisfying the very same condition. ∎

Exercises

For Exercises 1–5, write out the first several terms of the indicated sequence.

1. $\left(\frac{n}{n+1}\right)_{n=1}^{\infty}$.
2. $\left(\frac{n+1}{n!}\right)_{n=1}^{\infty}$.
3. $\left((-1)^n\right)_{n=0}^{\infty}$.
4. $\left(\frac{(-1)^n}{n}\right)_{n=1}^{\infty}$.
5. $\left(\frac{1}{n}-\frac{1}{n+1}\right)_{n=1}^{\infty}$.

For Exercises 6–10, determine whether the indicated sequence converges. If it converges, determine its limit as $n \to \infty$.

6. $\left(\frac{1+(1/2)^n}{3+1/n}\right)_{n=1}^{\infty}$.
7. $\left(e^{1/n}\right)_{n=1}^{\infty}$.
8. $(\sin(n))_{n=1}^{\infty}$.
9. $(\sin(\pi n))_{n=1}^{\infty}$.
10. $\left(\frac{e^n+e^{-n}}{n^2}\right)_{n=1}^{\infty}$.

For Exercises 11–15, determine the partial sums S_1, S_2, S_3, and S_4 for the indicated series.

11. $\sum_{n=1}^{\infty} n$.
12. $\sum_{n=1}^{\infty} \frac{1}{n}$.
13. $\sum_{n=1}^{\infty} 0.5^n$.
14. $\sum_{n=0}^{\infty} (-1)^n$.
15. $\sum_{n=0}^{\infty} (-1)^n n^2$.

Problems

16. Provide an example of a bounded sequence that does not converge.
17. Provide an example of a monotone sequence that does not converge.
18. True or false:
$$1+\frac{2013}{2014}+\frac{2013^2}{2014^2}+\frac{2013^3}{2014^3}+\cdots=2014.$$
Justify your answer.
19. Determine whether the series
$$\sum_{n=1}^{\infty}\frac{4n^9+8n^3+15n}{16n^9+23n^7+42}$$
converges or diverges. Justify your answer.
20. Suppose that the series $\sum_{n=1}^{\infty} a_n$ converges, and that the sequence (b_n) is defined by the relations
$$b_n=\begin{cases} n^n & \text{for } n=1,\ldots,1{,}000{,}000{,}000 \\ a_n & \text{for } n>1{,}000{,}000{,}000 \end{cases}.$$
Consider the series
$$S=\sum_{n=1}^{\infty} b_n.$$

(a) Determine the first ten partial sums of the series S.
(b) Determine whether or not the series S converges. Cite any theorem or result that you use.

21. Suppose that a certain sequence of numbers $\{a_n\}_{n=0}^{\infty}$ has the property

$$\lim_{s\to\infty} a_n = \frac{\sqrt{\pi}}{2},$$

and that the sequence (b_n) is defined by the formula

$$b_n = \begin{cases} n! & \text{for } 0 \le n \le 1{,}500{,}000 \\ a_n & \text{for } 1{,}500{,}001 \le n \end{cases}.$$

Does the series $\sum_{n=0}^{\infty} b_n$ converge, diverge, or is there not enough information given? Explain your answer.

22. Suppose that a certain sequence of numbers $\{a_n\}_{n=0}^{\infty}$ has the property

$$\lim_{s\to\infty}\left(\sum_{n=0}^{s} a_n\right) = \frac{\sqrt{\pi}}{2},$$

and that the sequence (b_n) is defined by the formula

$$b_n = \begin{cases} n! & \text{for } 0 \le n \le 1{,}500{,}000 \\ a_n & \text{for } 1{,}500{,}001 \le n \end{cases}.$$

Does the series $\sum_{n=0}^{\infty} b_n$ converge, diverge, or is there not enough information given? Explain your answer.

23. The fundamental theorem of calculus allows us to express a differentiable function $f(x)$ in the form

$$f(x) = f(a) + \int_a^x f'(t)\,dt.$$

(a) Use integration by parts to show that

$$f(x) = f(a) + f'(a)(x-a) + \int_a^x (x-t)f''(t)\,dt. \qquad (9.9)$$

(b) Given a function $f(x)$ and a point $x = a$, we define the *first-order Taylor polynomial* $T_1(x)$ by the relation

$$T_1(x) = f(a) + f'(a)(x-a).$$

Show that $f(a) = T_1(a)$ and $f'(a) = T_1'(a)$.

(c) For each of the following functions, derive its associated first-order Taylor polynomial; then plot f and T_1 on the same graph:

i. $f(x) = e^x, \quad a = 0;$
ii. $f(x) = \sin x, \quad a = 0;$
iii. $f(x) = \sin(x), \quad a = \frac{\pi}{2}.$

(d) The *error* of the first degree Taylor polynomial at x, $E_1(x)$, is defined by

$$E_1(x) = |f(x) - T_1(x)|.$$

Let I be a closed interval containing the point $x = a$ and let

$$M_2 = \max_{x \in I} |f''(x)|$$

be the maximum value of the function $|f''(x)|$ on the interval I. Show that, for all $x \in I$,

$$E_1(x) \leq \frac{M_2}{2}|x - a|^2.$$

Conclude that the error goes to zero faster than $|x - a|^2$, as x goes to a.

9.2 Tests for Convergence

In this section, we discuss a variety of tests that can be used to determine whether a given series converges.

9.2.1 Integral Comparison Test

Our first test involves comparing a series to a related improper integral. It unfolds as follows.

Theorem 9.2.1 — Integral Comparison Test. Consider a series $\sum_{n=1}^{\infty} a_n$, and suppose a function f can be found that satisfies the conditions:

1. f is nonnegative, i.e., $f(x) > 0$ for all $x \in [1, \infty)$;
2. f is a decreasing function, i.e., $f(y) \le f(x)$ whenever $x < y$; and
3. $f(n) = a_n$, for all $n \in \mathbb{N}$.

In such a case, the following logical statements are valid:

1. If the integral $\int_1^{\infty} f(x)\,dx$ converges, then the series $\sum_{n=1}^{\infty} a_n$ also converges.
2. If the integral $\int_1^{\infty} f(x)\,dx$ diverges, then the series $\sum_{n=1}^{\infty} a_n$ also diverges.

In other words, if a nonnegative, decreasing function f can be found such that $f(n) = a_n$, for all $n \in \mathbb{N}$, then:

$$\int_1^{\infty} f(x)\,dx \text{ converges} \qquad \text{if and only if} \qquad \sum_{n=1}^{\infty} a_n \text{ converges.}$$

This is an astonishing result! Suppose, for instance, that the convergence of the integral implies convergence of the series. In some sense, that is to say that the series is "smaller" than the integral, so that if the integral is finite, so, too, must be the series. Following this logic, if the integral is, instead, infinite, one would expect that no information could be obtained about the series (i.e., one would expect that the series could still be either finite or infinite). But this is not correct: if the integral is infinite, then, so too must be the series. In other words, the integral and the series must be of similar "size," in the sense that either both are finite or both are infinite.

Proof. Consider an infinite series $\sum_{n=1}^{\infty} a_n$ and suppose that a nonnegative, decreasing function f can be found such that $f(n) = a_n$, for all natural numbers n. An illustration of such a function is shown in Figure 9.3.

Case 1. Suppose that the integral

$$\int_1^{\infty} f(x)\,dx$$

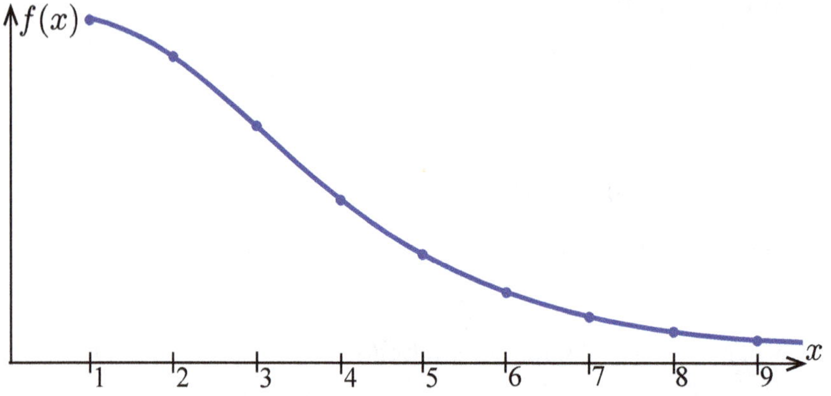

Figure 9.3: A nonnegative, decreasing function f that satisfies $f(n) = a_n$.

diverges. The left-hand Riemann sum for this integral, using a $\Delta x = 1$, may be expressed as

$$\sum_{n=1}^{\infty} f(n) = \sum_{n=1}^{\infty} a_n,$$

as shown in Figure 9.4. Notice that the Riemann sum is an overestimate of the of

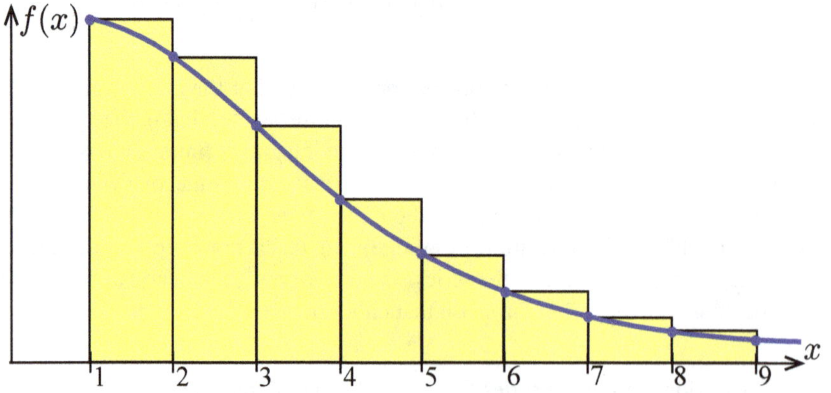

Figure 9.4: Left-hand Riemann sum for $\int_1^\infty f(x)\,dx$.

the value of the integral. Since the value of the Riemann sum is equal to the value of our infinite series, we may conclude that

$$\int_1^\infty f(x)\,dx \leq \sum_{n=1}^{\infty} a_n. \tag{9.10}$$

Since this integral diverges, by definition, so too must the series.

Case 2. Suppose that the integral

$$\int_1^\infty f(x)\,dx$$

converges.

Since the function is decreasing, we know that its right-hand Riemann sum is an underestimate for the integral, i.e.,

$$\sum_{n=2}^{\infty} f(n) \leq \int_1^\infty f(x)\,dx.$$

Since $f(n) = a_n$, we may add the number a_1 to both sides of this inequality to obtain

$$\sum_{n=1}^{\infty} a_n \leq a_1 + \int_1^\infty f(x)\,dx. \tag{9.11}$$

Since the integral is finite, by supposition, adding the number a_1 must result in a finite value. Therefore, the series is less than or equal to some finite number. Since each term in the series is nonnegative, the value of the series is bounded below by zero. Thus, the series must converge. This completes the proof. ■

■ **Example 9.11** Use the integral comparison test to determine whether or not the series

$$\sum_{n=1}^{\infty} e^{-n}$$

converges.

To proceed, we define the function $f(x) = e^{-x}$. Notice that this function is a nonnegative, decreasing function, such that $f(n) = e^{-n}$, for every natural number n. The conditions of the integral comparison test are therefore satisfied. Recall that

$$\int_1^\infty e^{-x}\,dx$$

converges. (In fact, this integral has a value of e^{-1}.) Therefore, the series $\sum_{n=1}^{\infty} e^{-n}$ must also converge, according to the integral comparison test.

Moreover, by combining inequalities (9.10) and (9.11), we may conclude that

$$e^{-1} \leq \sum_{n=1}^{\infty} e^{-n} \leq 2e^{-1}.$$

■

As an additional example of the integral comparison test, we next consider the p-test for series.

Theorem 9.2.2 — p-test for Series. The p-series

$$\sum_{n=1}^{\infty} \frac{1}{n^p}$$

1. converges if $p > 1$; and
2. diverges if $p \leq 1$.

Proof. First, note that the terms of the sequence

$$\left(\frac{1}{n^p}\right)_{n=1}^{\infty}$$

are nonnegative and decreasing. Consider now the nonnegative, decreasing function

$$f(x) = \frac{1}{x^p},$$

which possesses the property that $f(n) = 1/n^p$. The integral $\int_1^{\infty} 1/x^p \, dx$ converges if $p > 1$ and diverges if $p \leq 1$, pursuant to Theorem 7.5.1 on page 461. The result therefore follows by the integral comparison test (Theorem 9.2.1). ■

▪ **Example 9.12 — Harmonic Series.** Determine whether the series

$$\sum_{n=1}^{\infty} \frac{1}{n} = 1 + \frac{1}{2} + \frac{1}{3} + \frac{1}{4} + \cdots$$

converges or diverges. This series is known as the *harmonic series.*

By the p-test, the series diverges. Alternatively, we can use the integral comparison test: since the improper integral

$$\int_1^{\infty} \frac{1}{x} \, dx$$

diverges, so, too, must the corresponding series. (Note that the function $f(x) = 1/x$ is a nonnegative, decreasing function with the property that $f(n) = 1/n$.) ▪

Recall that Theorem 9.1.4 states that if a series converges, then its terms must go to zero. The converse of this statement—"if the terms go to zero, then the series must converge"—is false. In particular, the harmonic series' terms go to zero, and yet the series diverges.

9.2.2 Series and Limit Comparison Tests

Next, we visit two similar and rather straightforward results.

Theorem 9.2.3 — Series Comparison Test. Consider two series

$$\sum_{n=1}^{\infty} a_n \quad \text{and} \quad \sum_{n=1}^{\infty} b_n,$$

the terms of which satisfy the inequality

$$0 \leq a_n \leq b_n,$$

for all n sufficiently large. Then:

1. If the series $\sum_{n=1}^{\infty} b_n$ converges, then the series $\sum_{n=1}^{\infty} a_n$ also converges.
2. If the series $\sum_{n=1}^{\infty} b_n$ diverges, then the series $\sum_{n=1}^{\infty} a_n$ also diverges.

This theorem says something rather intuitive: if the larger series is finite, then the smaller series must also be finite. Similarly, if the smaller series is infinite, then the larger series must also be infinite. The limit comparison test is useful whenever we may massage a series—by judicious application of inequalities—to look like a series with known convergence properties.

■ **Example 9.13** Use the series comparison test to determine whether the series

$$\sum_{n=1}^{\infty} \frac{n^2+17n}{5n^4+3n+2}$$

converges or diverges.

The terms of this series approach $1/(5n^2)$ as n become large; we therefore suspect that the series converges according to the p-test. To prove it, however, we must set up an inequality. In particular, we want to show that the terms are *less* than a multiple of $1/n^2$.

To begin, we would like to show that the numerator is less than some multiple of n^2. Notice that

$$17n \leq 17n^2,$$

for all $n \in \mathbb{N}$. Therefore

$$n^2 + 17n \leq 18n^2$$

for all $n \in \mathbb{N}$.

Next, we consider the denominator. Since inequalities are reversed upon reciprocation, we seek to show that the denominator is *greater* than a multiple of n^4. This is readily obtained:

$$5n^4 + 3n + 2 \geq n^4,$$

for $n \in \mathbb{N}$. Taking the reciprocal, we obtain

$$\frac{1}{5n^4+3n+2} \leq \frac{1}{n^4}.$$

Combining with our previous result, we have

$$\frac{n^2+17n}{5n^4+3n+2} \le \frac{18n^2}{n^4} = \frac{18}{n^2}.$$

The series $\sum_{n=1}^{\infty} \frac{18}{n^2}$ converges according to the p-test. Therefore, the series $\sum_{n=1}^{\infty} \frac{n^2+17n}{5n^4+3n+2}$ converges by comparison with $\sum 1/n^2$. ■

■ **Example 9.14** Use the comparison test to determine whether or not the series

$$\sum_{n=1}^{\infty} \frac{10n^3-8}{n^4+8n^2+12}$$

converges or diverges.

Since the terms of this series approach $1/n$ as n becomes large, we conjecture that the series diverges. To see this, we must compare the series with a similar series with smaller terms. First, we observe that $8 \le 8n^3$, for all $n \in \mathbb{N}$. Therefore, $-8 \ge -8n^3$, which allows us to write

$$10n^3 - 8 \ge 10n^3 - 8n^3 = 2n^3.$$

In the denominator, we have the inequality

$$n^4 + 8n^2 + 12 \le n^4 + 8n^4 + 12n^4 = 21n^4,$$

from which we obtain

$$\frac{1}{n^4+8n^2+12} \ge \frac{1}{21n^4}.$$

Combining these results, we have

$$\frac{10n^3-8}{n^4+8n^2+12} \ge \frac{2n^3}{21n^4} = \frac{2}{21n}.$$

The series in question therefore diverges by comparison with the series $\sum(1/n)$. ■

Theorem 9.2.4 — Limit Comparison Test. Suppose that

$$\lim_{n\to\infty} \frac{a_n}{b_n} = c, \tag{9.12}$$

where $a_n > 0$ and $b_n > 0$, for all $n \in \mathbb{N}$. Then the series $\sum_{n=1}^{\infty} a_n$ and $\sum_{n=1}^{\infty} b_n$ either both converge or both diverge.

Essentially, equation (9.12) tells us that $a_n \to cb_n$, as $n \to \infty$. Whenever this condition is satisfied, the conclusion is that both series must behave the same: either they both converge or they both diverge.

■ **Example 9.15** Determine whether or not the series in Example 9.14 converges by using the limit comparison test.

As mentioned previously, the terms of series in question $\left(\sum_{n=1}^{\infty} \frac{10n^3 - 8}{n^4 + 8n^2 + 12}\right)$ closely resemble $1/n$ as n becomes large. Motivated by this observation, we set $b_n = 1/n$, where our a_ns are given by

$$a_n = \frac{10n^3 - 8}{n^4 + 8n^2 + 12}.$$

The ratio a_n/b_n is therefore

$$\frac{a_n}{b_n} = \frac{10n^4 - 8n}{n^4 + 8n^2 + 12}.$$

This ratio has limit

$$\lim_{n\to\infty} \frac{a_n}{b_n} = \lim_{n\to\infty} \frac{10n^4 - 8n}{n^4 + 8n^2 + 12} = 10.$$

Since the series $\sum(1/n)$ diverges, we conclude by the limit comparison test that the series $\sum(10n^3 - 8)(n^4 + 8n^2 + 12)$ must diverge as well. ■

9.2.3 Ratio Test

Our next convergence test is perhaps the most important, as it will be our go-to test when studying convergence of power series.

Theorem 9.2.5 — Ratio Test. Consider a series $\sum_{n=1}^{\infty} a_n$, where $a_n \neq 0$ for all $n \in \mathbb{N}$, and suppose that

$$\lim_{n\to\infty} \left| \frac{a_{n+1}}{a_n} \right| - L. \tag{9.13}$$

The following logical implications hold:

1. if $L < 1$, then the series converges;
2. if $L > 1$, then the series diverges.

If $L = 1$, the ratio rest cannot determine whether or not the series converges.

The ratio test is especially potent in its use to determine the convergence of series involving factorials. Before we proceed to example, we provide a brief review of factorial arithmetic.

Recall that, given a natural number $n \in \mathbb{N}$, its factorial is defined as the product of all natural numbers less than or equal to itself, i.e.,

$$n! = n(n-1)(n-2)\cdots 3\cdot 2\cdot 1.$$

The factorial of the number zero is defined separately, so that

$$0! = 1.$$

The reason for this is twofold: first, the gamma function (see Problem 38 in Section 7.5 on page 470), which extends the factorial to non-integers, yields a value of $0! = 1$. Second, whenever we have a summation formula involving factorials, the formulas work correctly for the $n = 0$ term only if we set $0! = 1$. (That is, there is also a convenience factor.) We will see this explicitly when we derive the formulas for Taylor series.

There is a certain "factorial arithmetic" that is useful when applying the ratio test. For example, note that

$$\frac{(n+1)!}{n!} = n+1.$$

Why? Because $(n+1)! = (n+1)n!$, and the $n!$s therefore cancel. Case in point:

$$\frac{5!}{4!} = \frac{5\cdot 4\cdot 3\cdot 2\cdot 1}{4\cdot 3\cdot 2\cdot 1} = 5.$$

In a similar vein, we have

$$\frac{(n+2)!}{n!} = (n+2)(n+1),$$

and so forth.

▪ **Example 9.16 — Geometric Series.** Use the ratio test to determine the values of a for which the geometric series

$$\sum_{n=0}^{\infty} a^n$$

converges.

The ratio of subsequent terms is given by

$$\frac{a^{n+1}}{a^n} = a.$$

Therefore, the limit in equation (9.13) yields

$$\lim_{n\to\infty} |a| = |a|.$$

We conclude, by the ratio test, that the geometric series converges if $|a| < 1$ and diverges if $|a| > 1$.

If $|a| = 1$, the ratio test yields no information, and we must fall on other techniques. First, let us consider the case $a = 1$. Here, the geometric series degenerates to the simple sum

$$\sum_{n=0}^{\infty} 1 = 1+1+1+\cdots.$$

This series clearly diverges, as the sum of an infinite number of 1s must be infinite. (To be precise, we can use Theorem 9.1.4: since $1^n \not\to 0$, the series cannot converge.) Next, consider the case $a = -1$. Here, the series becomes

$$\sum_{n=0}^{\infty} (-1)^n = 1-1+1-1+1-1+\cdots.$$

The sequence of partial sums is $(1,0,1,0,1,0,\ldots)$. This sequence does not converge (as the numbers never "settle down" to any particular value); therefore, the geometric series, with $a=-1$, diverges. We conclude that the geometric series converges if and only if $|a|<1$. ■

■ **Example 9.17** Use the ratio test to determine whether or not the series

$$\sum_{n=0}^{\infty}\frac{1}{n!}=1+1+\frac{1}{2}+\frac{1}{6}+\frac{1}{24}+\cdots$$

converges or diverges.

Defining $a_n=1/n!$, we may write the ratio a_{n+1}/a_n as

$$\frac{a_{n+1}}{a_n}=\frac{1}{(n+1)!}\cdot\frac{n!}{1}=\frac{n!}{(n+1)!}=\frac{1}{n+1}.$$

Upon taking the limit, we obtain

$$\lim_{n\to\infty}\left|\frac{a_{n+1}}{a_n}\right|=\lim_{n\to\infty}\frac{1}{n+1}=0.$$

Since $0<1$, the ratio test tells us that this series converges. ■

■ **Example 9.18** Use the ratio test to determine whether or not the series

$$\sum_{n=0}^{\infty}\frac{(n!)^2 3^n}{(2n)!}$$

converges or diverges.

Defining $a_n=(n!)^2 3^n/(2n)!$, we may write the $(n+1)$-term as

$$a_{n+1}=\frac{[(n+1)!]^2 3^{n+1}}{[2(n+1)]!}=\frac{[(n+1)!]^2 3^{n+1}}{(2n+2)!}.$$

Dividing by a_n (equivalent to multiplying by the reciprocal of a_n) we obtain

$$\frac{a_{n+1}}{a_n}=\frac{[(n+1)!]^2 3^{n+1}}{(2n+2)!}\cdot\frac{(2n)!}{(n!)^2 3^n}=\frac{(2n)!(n+1)!(n+1)!3^{n+1}}{(2n+2)!(n!)(n!)3^n}.$$

This expression may be simplified as

$$\frac{a_{n+1}}{a_n}=\frac{3(n+1)(n+1)}{(2n+2)(2n+1)}=\frac{3n^2+6n+3}{4n^2+6n+2}.$$

Taking the limit we obtain

$$\lim_{n\to\infty}\left|\frac{a_{n+1}}{a_n}\right|=\lim_{n\to\infty}\frac{3n^2+6n+3}{4n^2+6n+2}=\frac{3}{4}.$$

Since $3/4<1$, we conclude, by the ratio test, that this series converges.

Notice that it is unlikely that one would be able to intuit such a result. In particular, notice that if we replace the 3^n with a $(3.9)^n$, the series still converges, though it does not converge if we were to instead replace it with a $(4.1)^n$. In these cases, the only way to determine the convergence is by actually constructing the ratio a_{n+1}/a_n and taking the limit as $n\to\infty$. ■

9.2.4 Absolute and Conditional Convergence

In this paragraph, we examine how the sign of the terms of the series may affect its convergence.

Definition 9.2.1 The series $\sum_{n=1}^{\infty} a_n$ is said to *absolutely convergent* if the related series $\sum_{n=1}^{\infty} |a_n|$ converges.

A series that is convergent, but not absolutely convergent, is said to be *conditionally convergent.*

Absolute convergence is a stronger condition than convergence alone. If a series converges, it might be aided by cancellation between terms of opposing sign. Absolute convergence, on the other hand, says that the series converges *even if you sum the absolute values of the terms.* We may formalize this as follows.

Proposition 9.2.6 Absolute convergence implies convergence.

Proof. Consider a series

$$S = \sum_{n=1}^{\infty} a_n$$

that converges absolutely, that is, such that the related series

$$R = \sum_{n=1}^{\infty} |a_n|$$

converges. We wish to use the limit comparison test (Theorem 9.2.4) to show that the series S must also converge. The problem, however, is that the terms of the series S may be positive or negative. This problem is alleviated if we instead consider the series

$$T = \sum_{n=1}^{\infty} a_n + |a_n|.$$

(Notice that

$$a_n + |a_n| = \begin{cases} 2a_n & \text{if } a_n > 0 \\ 0 & \text{if } a_n \le 0 \end{cases},$$

from which we may infer that the terms of the series T are nonnegative.) Since $a_n \le |a_n|$ (clearly), we have the inequality

$$0 \le a_n + |a_n| \le 2|a_n|.$$

The convergence of the series R therefore implies the convergence of the series T, by the aforementioned limit comparison test. The difference $T - R$ is given by

$$T - R = \sum_{n=1}^{\infty} (a_n + |a_n|) - \sum_{n=1}^{\infty} |a_n|.$$

Since each of the series on the right-hand side converges, we are allowed to combine them in a single sum. Thus, the series

$$T - R = \sum_{n=1}^{\infty} (a_n + |a_n| - |a_n|) = \sum_{n=1}^{\infty} a_n = S$$

converges, competing the proof. ■

Though, as we just saw, an absolutely convergent series must converges. (If the sum of the absolute values of the terms of a series too, then so too must the sum without the absolute values, which may only aid in its convergence due to cancellation between terms.) The opposite statement—that convergent series must be absolutely convergent—is false. We will see a counterexample momentarily. But first, we must brave one last result on series convergence properties, following our next definition.

Definition 9.2.2 An *alternating series* is one in which the terms alternate sign from one term to the next.

Thus, any series with a pattern of plus–minus–plus–minus, etc., is called an alternating series. Alternating series possess a special property, following our next theorem.

Theorem 9.2.7 — Alternating-Series Test. If the alternating series $\sum_{n=1}^{\infty} a_n$ has the property that

$$\lim_{n \to \infty} a_n = 0, \tag{9.14}$$

then the series must converge.

We previously saw in Theorem 9.1.4 that a convergent series must possess the property that its terms tend to zero in the limit as $n \to \infty$. There are examples of series, however, whose terms go to zero, yet they do not converge. (For example, the harmonic series.) The alternating-series test tells us that for the special case of alternating series, as long as their terms go to zero, the series will converge.

■ **Example 9.19** Determine the convergence of the series

$$\sum_{n=1}^{\infty} \frac{(-1)^{n+1}}{n} = 1 - \frac{1}{2} + \frac{1}{3} - \frac{1}{4} + \frac{1}{5} - \frac{1}{6} + \cdots .$$

This series is similar to the harmonic series (which diverges), except for the fact that the terms now alternate in sign. Since this is an alternating series, whose terms approach zero as $n \to \infty$, we conclude, by the alternating-series test, that it must converge.

Moreover, this is an example of a conditionally convergent series, as it converges, but not absolutely. (We know it is not absolutely convergent as the harmonic series diverges.) ■

Exercises

For Exercises 1–20, determine whether the indicated series converges or diverges. Cite any theorem used.

1. $\sum_{n=1}^{\infty} \frac{1}{\sqrt{n}}$.
2. $\sum_{n=1}^{\infty} \frac{1}{n^3}$.
3. $\sum_{n=1}^{\infty} \frac{1}{n^2+n^3}$.
4. $\sum_{n=1}^{\infty} \frac{(-1)^n n^5}{8n^7+19n^5+\pi}$.
5. $\sum_{n=1}^{\infty} \frac{\cos(n\pi)}{n}$.
6. $\sum_{n=1}^{\infty} \sin(n)e^{-n}$.
7. $\sum_{n=1}^{\infty} \frac{3+2\sin(n)}{n^3+2}$.
8. $\sum_{n=1}^{\infty} \frac{(-1)^n 5^n}{n!}$.
9. $\sum_{n=1}^{\infty} \frac{3^n}{(2n)!}$.
10. $\sum_{n=1}^{\infty} \frac{8n^4+16n^2+12}{9n^5+11n}$.
11. $\sum_{n=1}^{\infty} \frac{(-1)^n}{2+(1/2)^n}$.
12. $\sum_{n=1}^{\infty} \frac{1}{n+\sqrt{n}}$.
13. $\sum_{n=1}^{\infty} \frac{n+e^{-n}}{n^3+\pi}$.
14. $\sum_{n=1}^{\infty} \frac{n+2}{100n+e^{-n}}$.
15. $\sum_{n=1}^{\infty} \frac{(-1)^n 3^n (n!)^2}{(2n)!}$.
16. $\sum_{n=1}^{\infty} \frac{(-1)^n n!}{(2n)!}$.
17. $\sum_{n=1}^{\infty} \frac{7^n}{8^n}$.
18. $\sum_{n=1}^{\infty} \frac{15^n}{16^n+n}$.
19. $\sum_{n=1}^{\infty} \frac{8^n (n!)^3}{(3n)!}$.
20. $\sum_{n=1}^{\infty} \frac{10^n (n!)^3}{(3n)!}$.

Problems

21. True or false: The series

$$\sum_{n=1}^{\infty} -n^{-0.1}$$

converges by the comparison test, since $-n^{-0.1} < 1/n^2$ and since the series

$$\sum_{n=1}^{\infty} \frac{1}{n^2}$$

converges (by the p-test).

If the above statement is false, find the mistake in the above "proof."

22. True or false: If $a_n < 1/n^2$, then $\sum a_n$ must converge. Explain your reasoning.
23. True or false: If $\sum |a_n|$ converges, then $\sum (-1)^n a_n$ must also converge. Explain your reasoning.

24. Construct an example of a series that is absolutely convergent, but not convergent.
25. Use the ratio test to determine all possible values of a for which each of the following series converges:
    (a) $\sum_{n=1}^{\infty} \frac{a^n (n!)^2}{(2n)!}$.
    (b) $\sum_{n=1}^{\infty} \frac{(3n)!}{a^n (n!)^3}$.
26. Consider a sequence defined by an initial value $a_0 = a$ and the recursion relation

$$a_{n+1} = ra_n,$$

for $n \in \{0, 1, 2, \ldots\}$.
    (a) Determine a formula for the nth term a_n.
    (b) What type of series is $\sum_{n=0}^{\infty} a_n$?
    (c) Explain (heuristically) why a series with a convergent limit given by equation (9.13) approaches a geometric series as $n \to \infty$, and use this observation to explain why the result of the ratio test makes sense in this context.
27. This is a continuation of Problem 23 in Section 9.1.
    (a) Use equation (9.9) and integration by parts to show that

$$f(x) = f(a) + (x-a)f'(a) + \frac{(x-a)^2}{2} f''(a) + \int_a^x f'''(t) \frac{(x-t)^2}{2}\, dt. \tag{9.15}$$

    (b) Given a function $f(x)$ and a point $x = a$, we define the *second-order Taylor polynomial* $T_2(x)$ by the relation

$$T_2(x) = f(a) + f'(a)(x-a) + \frac{f''(a)}{2!}(x-a)^2.$$

    Show that $f(a) = T_2(a)$; $f'(a) = T_2'(a)$; and $f''(a) = T_2''(a)$.
    (c) For each of the following functions, derive its associated second-order Taylor polynomial; then plot f and T_2 on the same graph:
        i. $f(x) = e^x, \quad a = 0$;
        ii. $f(x) = \sin x, \quad a = 0$;
        iii. $f(x) = \sin(x), \quad a = \frac{\pi}{2}$.
    (d) The *error* of the second-degree Taylor polynomial at x, $E_2(x)$, is defined by

$$E_2(x) = |f(x) - T_2(x)|.$$

    Let I be a closed interval containing the point $x = a$ and let

$$M_3 = \max_{x \in I} |f'''(x)|$$

    be the maximum value of the function $|f'''(x)|$ on the interval I. Show that, for all $x \in I$,

$$E_2(x) \le \frac{M_3}{3!} |x-a|^3.$$

Conclude that the error goes to zero faster than $|x-a|^3$ as x approaches a.

9.3 Power Series

In this section, we introduce the basic notions of *power series*. Power series are infinite series with terms that depend on an independent variable. As such, they converge to a function, not a number. Instead of simply asking the question of whether the series converges, we must ask on what intervals does the series converge. In this section, we outline the basic notation and results that will be used later during our examination of Taylor series.

9.3.1 Radius of Convergence

To begin, let us define what we mean by power series.

Definition 9.3.1 A *power series, centered about* $x = a$, is an infinite series of the form

$$P(x) = \sum_{n=0}^{\infty} c_n (x-a)^n, \tag{9.16}$$

for some selection of constant coefficients $(c_n)_{n=0}^{\infty}$.

When analyzing power series, we ask the question: for *which* x-values does the series converge?

Definition 9.3.2 A power series (9.16) converges at the point $x = b$ if the series

$$\sum_{n=0}^{\infty} c_n (b-a)^n$$

converges.

In other words, to determine whether a given power series converges at a given x-value, we test the convergence of the series while regarding x as a fixed constant.

Example 9.20 Determine the domain on which the power series

$$P(x) = \sum_{n=0}^{\infty} ax^n$$

converges, where a is a fixed constant.

For this power series, each of the coefficients take on the same constant value. (In general, the coefficient of the nth term will depend on n.) We also notice this is a power series that is "centered about the origin $x = 0$."

Now, as it turns out, we already know the answer to this question. This series is simply the geometric series, which converges for $|x| < 1$ and diverges otherwise. In particular, we know from Theorem 5.2.3 (on page 358) that this series converges to the function

$$P(x) = \sum_{n=0}^{\infty} ax^n = \frac{a}{1-x},$$

for $|x| < 1$ (i.e., for $-1 < x < 1$). This function is graphed using the value $a = 1$ in Figure 9.5.

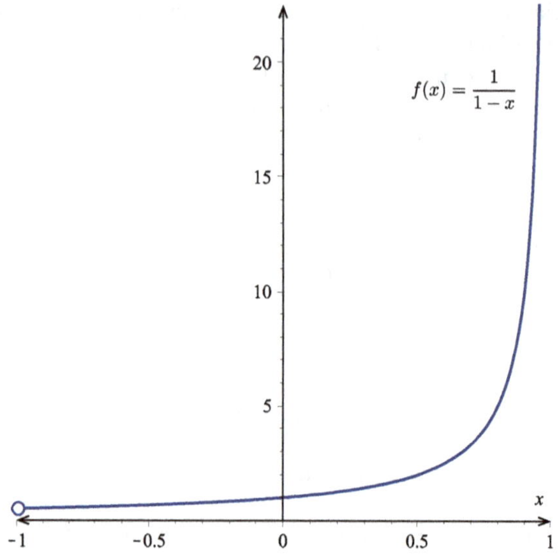

Figure 9.5: Graph of the power function $P(x) = \sum_{n=0}^{\infty} x^n$.

Moreover, we can visualize the first several partial sums,

$$\begin{aligned} P_0(x) &= 1, \\ P_1(x) &= 1 + x, \\ P_2(x) &= 1 + x + x^2, \\ P_3(x) &= 1 + x + x^2 + x^3, \\ P_4(x) &= 1 + x + x^2 + x^3 + x^4, \end{aligned}$$

as graphed in Figure 9.6. In general, the partial sum $P_k(x)$ will equal $P_k(x) = 1 + x + \cdots + x^k$. As $k \to \infty$, these functions become closer and closer to our function $P(x) = 1/(1-x)$, as is already apparent in Figure 9.6.

Recall that we proved Theorem 5.2.3 by studying the associated finite geometric series and then taking their limit, that is, by taking the limit of the partial sums. We also regarded x as a constant; in fact, we called it r. For most power series, however, the approach of calculating actual formulas for the partial sums is untenable. We do have, though, a number of tools at our disposal that were developed in the preceding sections. The main tool is the ratio test. For point of instruction, we will proceed to analyze the infinite geometric series using the ratio test. This will be our main approach for analyzing the Taylor series to come.

To begin, we must freeze x—that is, we must view x as a constant. (If x were a constant, after all, then we can apply all of our tools from Section 5.2.) The ratio

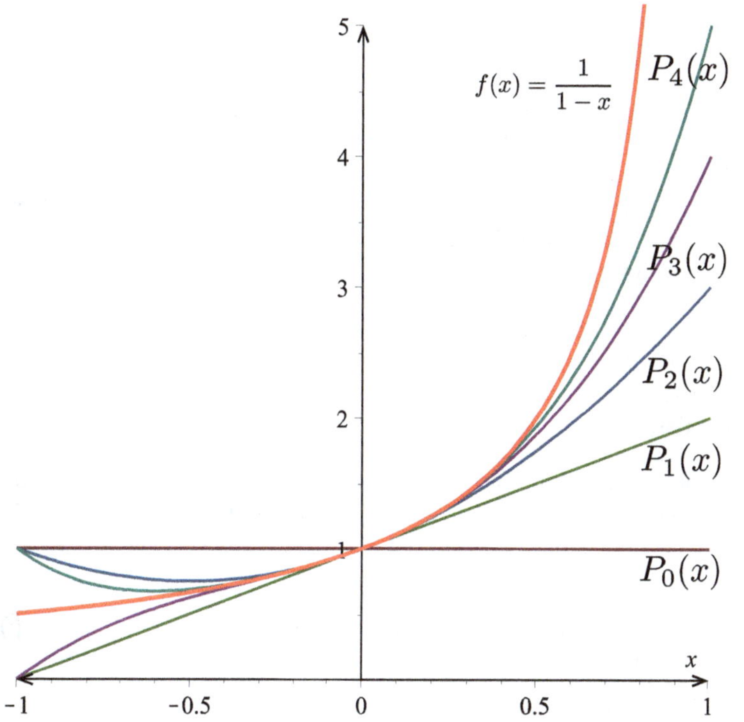

Figure 9.6: First several partial sums of the infinite geometric series.

of subsequent terms of our series is now given by

$$\frac{a_{n+1}}{a_n} = \frac{ax^{n+1}}{ax^n} = x$$

(where $a_n = ax^n$ represents the nth term). Taking the limit of equation (9.13), we therefore obtain

$$\lim_{n\to\infty} \left| \frac{ax^{n+1}}{ax^n} \right| = \lim_{n\to\infty} |x| = |x|.$$

(Remember, for the purpose of applying the ratio test, the variable x is regarded as a fixed constant.) By the ratio test (Theorem 9.2.5), we may therefore conclude that the infinite geometric series converges whenever $|x| < 1$, and diverges for $|x| > 1$. (The ratio test, by itself, does not tell us what happens at the two boundary points $x = -1$ and $x = +1$.) In this way, the ratio test has confirmed our prior result from Section 5.2. ∎

In our preceding example, we saw that the power series in question converged on a symmetric interval that was centered at—well—the point at which we said the series was "centered at" in Definition 9.3.1. As it turns out, this is a feature of power series in general.

Theorem 9.3.1 Let $P(x)$ be a power series, centered about the point $x = a$, with nonzero coefficients c_n. Then one of the following statements must be true:

1. The power series $P(x)$ converges *only* at the point $x = a$.
2. There exists a positive number R, such that the power series $P(x)$ converges for $|x-a| < R$ and diverges for $|x-a| > R$.
3. The power series $P(x)$ converges for all values of x.

Thus, a power series may (1) converge only at the single point at which it is centered, (2) converge for all points on an interval that is centered at the point $x = a$, or (3) converge for all values of x. To aid in our ability to describe these possibilities, we invoke the following terminology.

Definition 9.3.3 The number R, in Theorem 9.3.1, is called the *radius of convergence* for the given power series. In Case 1, we say that $R = 0$. In Case 3, we say that the radius of convergence is infinity, i.e., we may informally write $R = \infty$.

The radius of convergence R may therefore be any value on the set $[0,\infty) \cup \{\text{"}\infty\text{"}\}$, i.e., any nonnegative number of the symbol "∞". Additionally, we say the following.

Definition 9.3.4 The domain on which the power series converges is called the *interval of convergence.*

Next, we offer a sketch of a proof for Theorem 9.3.1.

Proof. Consider a power series of the form (9.16). Define the nth term, which is a function of the variable x, as

$$a_n = c_n(x-a)^n.$$

We imagine the variable x to be held fixed and ask, for a fixed value of x, does the power series converge? To begin, we observe that the power series *must* converge at least at the single point $x = a$, since setting x equal to a results in every term except the first term vanishing to zero: $P(a) = c_0$. So our question reduces to: other than the point $x = a$ itself, at what other x-values does a given power series converge? To answer this, we apply the ratio test. The ratio of subsequent terms is given by

$$\frac{a_{n+1}}{a_n} = \frac{c_{n+1}(x-a)^{n+1}}{c_n(x-a)^n}.$$

We again stress that for the purpose of the ratio test, we are regarding x as a fixed constant. The limit (9.13) may therefore be written as

$$\lim_{n\to\infty}\left|\frac{a_{n+1}}{a_n}\right| = \lim_{n\to\infty}\left|\frac{c_{n+1}}{c_n}\right| \cdot |x-a| = |x-a| \lim_{n\to\infty}\left|\frac{c_{n+1}}{c_n}\right|.$$

This final equality holds due to the fact that the factor $|x-a|$ is constant, relative to the index n, and therefore can be pulled out of the limit. Defining

$$\ell = \lim_{n\to\infty} \left| \frac{c_{n+1}}{c_n} \right|, \tag{9.17}$$

the limit used in the ratio test becomes $L|x-a|$. Now, there are three possibilities: either (1) $\ell = \infty$, (2) $0 < \ell < \infty$, or (3) $\ell = 0$. In the first case, the quantity $\ell|x-a|$ will also be infinity, for all values of x, except the value $x = a$, at which we already know the series converges. By the ratio test, the series diverges everywhere except at the single point $x = a$, and we have recovered Case 1 of the theorem. In the second case, ℓ takes a positive, finite value. The ratio test guarantees convergence of the series whenever

$$\ell|x-a| < 1,$$

and divergence of the series whenever

$$\ell|x-a| > 1.$$

The radius of convergence can therefore be identified with $R = 1/\ell$, and we have recovered Case 2 of the theorem. In the final case, when $\ell = 0$, the quantity $\ell|x-a|$ will equal zero, for every value of x. It follows that the inequality

$$\ell|x-a| < 1$$

will always be satisfied (since $\ell = 0$), indicating that the series will converge for every value of x. This completes the proof. ■

9.3.2 Differentiation and Integration of Power Series

Our next result concerns what happens when one computes the term-by-term derivative of a power series, i.e., what happens if one formally differentiates a power series one term at a time.

Theorem 9.3.2 Consider a power series centered about $x = a$:

$$P(x) = \sum_{n=0}^{\infty} c_n (x-a)^n.$$

Let

$$D(x) = \sum_{n=1}^{\infty} n c_n (x-a)^{n-1}$$

be the power series obtained by formally differentiating $P(x)$ term-by-term. Then the power series P and D have the same radius of convergence.

Proof. Recall from the proof of Theorem 9.3.1, the radius of convergence of a power series is uniquely determined by the value of the limit ℓ, as defined by equation (9.17). For the power series $P(x)$, this limit is

$$\ell = \lim_{n\to\infty} \left| \frac{c_{n+1}}{c_n} \right|,$$

whereas, for the derived power series $D(x)$, this limit is given by

$$\ell_D = \lim_{n\to\infty} \left| \frac{(n+1)c_{n+1}}{nc_n} \right|.$$

But the two limits are identical, i.e., $\ell_D = \ell$. This conclusion follows since

$$\lim_{n\to\infty} \frac{n+1}{n} = 1.$$

This proves that the two power series must have the same radius of convergence. ∎

Corollary 9.3.3 Consider a power series centered about $x = a$:

$$P(x) = \sum_{n=0}^{\infty} c_n(x-a)^n.$$

Let

$$I(x) = \sum_{n=0}^{\infty} \frac{c_n}{n+1}(x-a)^{n+1}$$

be the power series obtained by formally integrating $P(x)$ term-by-term. Then the power series P and I have the same radius of convergence.

Proof. The relationship between the power series I and P is completely analogous to the relationship between the power series P and D from Theorem 9.3.2. This result is therefore simply a restatement of that theorem. ∎

We have thus proven that if one obtains a new power series by either integrating or differentiating an existing power series term-by-term, the result will have the same radius of convergence as the original power series. This is useful, as it will allow us to form new Taylor series, which are power-series representations of specific functions, by using the tools of differentiation and integration.

Exercises

For Exercises 1–20, use the ratio test to determine the radius of convergence of the given power series.

1. $\sum_{n=0}^{\infty} x^n$.
2. $\sum_{n=0}^{\infty} \frac{x^n}{n}$.
3. $\sum_{n=0}^{\infty} n^4 x^n$.
4. $\sum_{n=0}^{\infty} \frac{(-1)^n(x-1)^n}{n}$.

5. $\sum_{n=0}^{\infty} \frac{(-1)^n x^n}{n!}$.

6. $\sum_{n=0}^{\infty} \frac{2^n (x-3)^n}{n^2}$.

7. $\sum_{n=0}^{\infty} \frac{(n!)^2 x^n}{(2n)!}$.

8. $\sum_{n=0}^{\infty} \frac{(-1)^n (n!)^3 x^n}{(3n)!}$.

9. $\sum_{n=0}^{\infty} n! x^n$.

10. $\sum_{n=0}^{\infty} (4x)^n$.

11. $\sum_{n=0}^{\infty} \frac{3^n (x-7)^n}{n^3}$.

12. $\sum_{n=0}^{\infty} \frac{88^n n^{88} x^n}{n!}$.

13. $\sum_{n=0}^{\infty} \frac{(-1)^n (2n)!(x-5)^n}{(n!)^2}$.

14. $\sum_{n=0}^{\infty} e^{-n} x^n$.

15. $\sum_{n=0}^{\infty} \frac{n!(x-2)^n}{n!+n}$.

16. $\sum_{n=0}^{\infty} \frac{(-1)^n x^{2n}}{(2n)!}$.

17. $\sum_{n=0}^{\infty} \frac{(n!)^2 x^{2n}}{(2n)!}$.

18. $\sum_{n=0}^{\infty} \frac{x^{3n}}{8^n}$.

19. $\sum_{n=0}^{\infty} \frac{(-1)^n x^{2n+1}}{(2n+1)!}$.

20. $\sum_{n=0}^{\infty} \frac{n^3 x^{2n}}{2^{2n}}$.

Problems

21. In applied mathematics, there are a variety of situations in which elementary functions are inadequate to model given physical situations, such as vibrating drumheads, heat conduction, quantum mechanics, or describing planetary motion. To remedy this, scientists have turned to a host of non-elementary functions, based on power series with special properties. One such class of special functions is the set of *Bessel functions*. The zeroth-order Bessel function J_0 is defined by the series

$$J(x) = \sum_{n=0}^{\infty} \frac{(-1)^n x^{2n}}{2^{2n}(n!)^2}. \tag{9.18}$$

   (a) Determine the radius of convergence for $J(x)$.
   (b) Write down an expression for the first partial four partial sums of this series: $S_0(x)$, $S_1(x)$, $S_2(x)$, and $S_3(x)$. Plot these partial sums on a graph.
   (c) By differentiating term-by-term, derive a power-series expression for the derivative functions $J'(x)$ and $J''(x)$.
   (d) The zeroth-order Bessel functions satisfies the differential equation

$$x^2 J'' + xJ' + x^2 J = 0.$$

Substitute the results from (c) into this equation to show that it is satisfied. (Hint: reindex the original series J so that the summation ranges from $n = 1$ to infinity.)

22. This is a continuation of Problem 27 in Section 9.2.
    (a) Use equation (9.15) and integration by parts to show that

$$\begin{aligned} f(x) &= f(a)+(x-a)f'(a)+\frac{(x-a)^2}{2}f''(a) \\ &+\frac{(x-a)^3}{3!}f'''(a)+\int_a^x f^{(iv)}(t)\frac{(x-t)^3}{3!}\,dt. \end{aligned} \tag{9.19}$$

    (b) Given a function $f(x)$ and a point $x = a$, we define the *third-order Taylor polynomial* $T_3(x)$ by the relation

$$T_3(x) = f(a)+f'(a)(x-a)+\frac{f''(a)}{2!}(x-a)^2+\frac{f'''(a)}{3!}(x-a)^3.$$

    Show that $f(a) = T_3(a)$; $f'(a) = T_3'(a)$; $f''(a) = T_3''(a)$; and $f'''(a) = T_3'''(a)$.

    (c) For each of the following functions, derive its associated third-order Taylor polynomial; then plot f and T_3 on the same graph:
        i. $f(x) = e^x, \quad a = 0$;
        ii. $f(x) = \sin x, \quad a = 0$;
        iii. $f(x) = \sin(x), \quad a = \frac{\pi}{2}$.

    (d) The *error* of the third-degree Taylor polynomial at x, $E_3(x)$, is defined by

$$E_3(x) = |f(x)-T_3(x)|.$$

    Let I be a closed interval containing the point $x = a$ and let

$$M_4 = \max_{x\in I}|f^{(iv)}(x)|$$

    be the maximum value of the function $|f^{(iv)}(x)|$ on the interval I. Show that, for all $x \in I$,

$$E_3(x) \le \frac{M_4}{4!}|x-a|^4.$$

    Conclude that the error goes to zero faster than $|x-a|^4$ as x approaches a.

9.4 Taylor Series

Next, we construct a specific method for generating power series from basic functions. These resulting *Taylor series* constitute a powerful tool in applied mathematics, as they can be used to generate a sequence of successively better approximations to a function near a given point of interest.

9.4.1 Taylor Polynomials

We begin by reviewing the basics of Taylor Polynomials. First, recall that the tangent line of a function f at a point $x = a$ may be expressed by the linear function

$$T_1(x) = f(a) + f'(a)(x-a). \tag{9.20}$$

This function is called the *first order Taylor polynomial* for f centered about $x = a$. Essentially, it is the unique linear function that resembles the original function f as accurately as possible near the point $x = a$, in the sense that both its value and the value of its derivative match the values of the function f and its derivative f' at the point $x = a$:

$$T_1(a) = f(a) \qquad \text{and} \qquad T_1'(a) = f'(a).$$

(Do verify this by evaluating $T_1(a)$ and $T_1'(a)$.)

Motivated by this observation, we next define the *second order Taylor polynomial* of f about $x = a$ by the relation

$$T_2(x) = f(a) + f'(a)(x-a) + \frac{f''(a)}{2}(x-a)^2. \tag{9.21}$$

This is a quadratic function in x defined so that

$$T_2(a) = f(a), \qquad T_2'(a) = f'(a), \qquad \text{and} \qquad T_2''(a) = f''(a),$$

as may be verified by direct computation. Thus, T_2 is the single quadratic function that most closely resembles the function f near the point $x = a$.

Equations (9.20) and (9.21) may, of course, be generalized, as we state in our following definition.

Definition 9.4.1 Given a function f with at least k continuous derivatives, its *Taylor polynomial of degree k*, centered about a given point $x = a$, is defined by the relation

$$T_k(x) = \sum_{n=0}^{k} \frac{f^{(n)}(a)(x-a)^n}{n!}. \tag{9.22}$$

Equation (9.22) may alternatively be expressed without summation notation as

$$T_k(x) = f(a) + f'(a)(x-a) + \frac{f''(a)}{2}(x-a)^2 + \frac{f'''(a)}{3!}(x-a)^3 + \cdots + \frac{f^{(k)}(a)}{k!}(x-a)^k.$$

As we should expect by now, this is just a kth degree polynomial with coefficients selected in a way such that the values of its first k derivatives match the values of the function f's first k derivatives at the point $x = a$, as we state in the following proposition.

Proposition 9.4.1 The kth order Taylor polynomial of a function f about a point $x = a$ matches the values of the function f and its derivatives at the given point, up to the kth derivative, i.e.,

$$T_k(a) = f(a), \quad T_k'(a) = f'(a), \quad \cdots, \quad T_k^{(k)}(a) = f^{(k)}(a).$$

The proof can be dizzying due to the proliferation of notation. Do not let this deter your understanding, as the process is no different than the simpler, non-generic case of first and second order Taylor polynomials we began the section discussing. To see what is going on, try to tailor your own proof of this theorem for the specific case of a fourth order Taylor polynomial.

Proof. When evaluating $T_k(x)$ at $x = a$, each term containing a factor of $(x - a)$ will vanish, as $a - a = 0$:

$$T_k(a) = f(a) + f'(a)(a-a) + \frac{f''(a)}{2}(a-a)^2 + \cdots + \frac{f^{(k)}(a)}{k!}(a-a)^k = f(a).$$

Differentiating $T_k(x)$, we obtain:

$$T_k'(x) = f'(a) + f''(a)(x-a) + \frac{3f'''(a)}{3!}(x-a)^2 + \cdots + \frac{kf^{(k)}(a)}{k!}(x-a)^{k-1},$$

which of course reduces to

$$T_k'(x) = f'(a) + f''(a)(x-a) + \frac{f'''(a)}{2!}(x-a)^2 + \cdots + \frac{f^{(k)}(a)}{(k-1)!}(x-a)^{k-1},$$

since $n/n! = 1/(n-1)!$. Evaluating at $x = a$, we obtain $T_k'(a) = f'(a)$. Again, the other terms in $T_k'(x)$ each have a factor of $(x - a)$, and therefore vanish when we evaluate at $x = a$.

Continuing in this manner, the second derivative is given by

$$T_k''(x) = f''(a) + f'''(a)(x-a) + \frac{f^{(iv)}(a)}{2!}(x-a)^2 + \cdots + \frac{f^{(k)}(a)}{(k-2)!}(x-a)^{k-2}.$$

Evaluating, we obtain $T_k''(a) = f''(a)$. Continuing in a like manner, we eventually reach the kth derivative

$$T_k^{(k)}(x) = f^{(k)}(a).$$

Since this function is a constant, evaluating at $x = a$ will return the same: $T_k^{(k)}(a) = f^{(k)}(a)$. This proves the result.

■

■ **Example 9.21** Consider the function $f(x) = \sqrt{x}$. Determine the first, second, and third order Taylor polynomials for f centered about $x = 1$.

To proceed, we will need at our disposal the derivatives of the function f up to third order. Differentiating, we find

$$\begin{aligned} f(x) &= \sqrt{x}, \\ f'(x) &= \frac{1}{2\sqrt{x}}, \\ f''(x) &= \frac{-1}{4x^{3/2}}, \\ f'''(x) &= \frac{3}{8x^{5/2}}. \end{aligned}$$

The formula for the third order Taylor polynomial about $x = 1$ is

$$T_3(x) = f(1) + f'(1)(x-1) + \frac{f''(1)}{2}(x-1)^2 + \frac{f'''(1)}{3!}(x-1)^3.$$

Evaluating the function f and its derivatives at the point $x = 1$ and inputing these values into the formula, we obtain the third order Taylor polynomial for $\sqrt{x}$ about $x = 1$:

$$T_3(x) = 1 + \frac{x-1}{2} - \frac{(x-1)^2}{8} + \frac{(x-1)^3}{16}.$$

Of course, this automatically gives us the first and second order Taylor polynomials as part of the spoils of war:

$$\begin{aligned} T_1(x) &= 1 + \frac{x-1}{2}, \\ T_2(x) &= 1 + \frac{x-1}{2} - \frac{(x-1)^2}{8}. \end{aligned}$$

The function f is plotted with its first three Taylor polynomials about $x = 1$ in Figure 9.7. Notice that the second order Taylor polynomial does a better job of approximating the function f than the first order, and the third order better than the second order. The higher the order of the Taylor polynomial, the more accurate the approximation is near the point $x = a$. ■

Error in Taylor Polynomials

We previously observed how higher degree Taylor polynomials more accurately represent a function than its lower degree Taylor polynomials. In this section, we provide a more precise analysis of the error when using Taylor polynomials to approximate a function.

Definition 9.4.2 Given a function f that possesses a kth degree Taylor polynomial T_k about the point $x = a$. Then the *error* in using T_k to approximate f at a

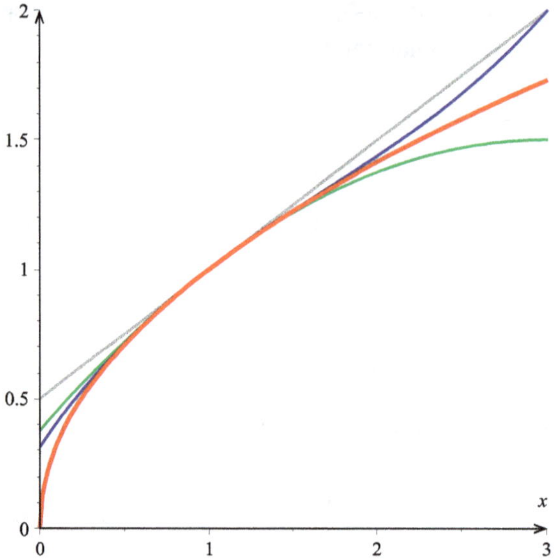

Figure 9.7: The function $f(x) = \sqrt{x}$ (red), plotted with T_1 (gray), T_2 (green), and T_3 (blue).

point x is defined by

$$E_k(x) = |f(x) - T_k(x)|. \tag{9.23}$$

In other words, the error is simply the absolute value of the difference between the exact value $f(x)$ and the approximation $T_k(x)$. The following theorem allows us to quantify how quickly the error goes to zero as $x \to a$.

Theorem 9.4.2 Let f be a function with derivatives of all orders. Given a closed interval I containing the point $x = a$, we define the quantity

$$M_k = \max_{x \in I} \left| f^{(k)}(x) \right|$$

as the maximum value of the kth derivative of f on the interval I. (This is well defined due to the extreme value theorem.) Then the error E_k in approximating the function f with the kth Taylor polynomial T_k is bounded by

$$E_k(x) = \frac{M_{k+1}}{(k+1)!} |x-a|^{k+1}, \tag{9.24}$$

for all $x \in I$.

We outline a proof below.

Proof. The fundamental theorem of calculus allows us to express the function f in the form

$$f(x) = f(a) + \int_a^x f'(t)\,dt.$$

The trick is to integrate by parts, with the selection

$$\begin{aligned} u &= f'(t) & du &= f''(t)\,dt \\ dv &= dt & v &= (t-x). \end{aligned}$$

From this, we can write

$$\begin{aligned} \int_a^x f'(t)\,dt &= f'(t)(t-x)\Big|_{t=a}^{t=x} - \int_a^x (t-x)f''(t)\,dt \\ &= f'(a)(x-a) + \int_a^x (x-t)f''(t)\,dt. \end{aligned}$$

Combining this with our original equation, we uncover the formula

$$f(x) = f(a) + f'(a)(x-a) + \int_a^x (x-t)f''(t)\,dt. \tag{9.25}$$

The error $E_1(x)$ of using the first order Taylor polynomial is defined by

$$E_1(x) = |f(x) - T_1(x)| = |f(x) - f(a) - f'(a)(x-a)|.$$

From equation (9.27), we may express this error as

$$\begin{aligned} E_1(x) &= \left|\int_a^x (x-t)f''(t)\,dt\right| \\ &\le \int_u^x \left|(x-t)f''(t)\right|\,dt \\ &\le M_2 \int_a^x |x-t|\,dt = \frac{M_2}{2}(x-a)^2, \end{aligned}$$

whenever $x \in I$.

This process can be repeated to uncover the general case. Integrating equation (9.27) by parts, for instance, reveals the equation

$$f(x) = f(a) + f'(a)(x-a) + \frac{f''(a)}{2}(x-a)^2 + \int_a^x (x-t)^2 f'''(t)\,dt. \tag{9.26}$$

From this equation, it can be deduced

$$E_3(x) \le \frac{M_3}{3!}(x-a)^3,$$

for $x \in I$. This process may be continued by induction, thus proving the theorem. ■

9.4.2 Taylor Series

How far can we go? For an infinitely differentiable function—that is, a function with derivatives of *all* orders—we can write down the corresponding *Taylor series*. The Taylor series is just the infinite series that one obtains by following the pattern we uncovered in the Taylor polynomials. But does the resultant series actually converge to the function we started with? The following theorem confirms that it does. At least, it does for all points within its interval of convergence.

Theorem 9.4.3 — Taylor's Theorem. Given an infinitely differentiable function f, its *Taylor series*

$$T(x) = \sum_{n=0}^{\infty} \frac{f^{(n)}(a)}{n!}(x-a)^n \tag{9.27}$$

converges to the function f for all points in its interval of convergence.

The general Taylor series in equation (9.27) may alternatively be expressed by the equation

$$T(x) = f(a) + f'(a)(x-a) + \frac{f''(a)}{2!}(x-a)^2 + \frac{f'''(a)}{3!}(x-a)^3 + \cdots. \tag{9.28}$$

For the special case of $a = 0$, we obtain the so-called *Maclaurin series*:

$$T(x) = \sum_{n=0}^{\infty} \frac{f^{(n)}(0)}{n!}x^n = f(0) + f'(0)x + \frac{f''(0)}{2!}x^2 + \frac{f'''(0)}{3!}x^3 + \cdots. \tag{9.29}$$

Taylor Series for Elementary Functions

In the next several examples, we will derive the Taylor-series expansions for a number of elementary functions. We will summarize these results at the end of the paragraph.

■ **Example 9.22** Determine the Taylor-series expansion for the function $f(x) = e^x$ about the point $x = 0$ and its corresponding radius of convergence.

Since the derivative (of any order) of e^x is again e^x, so that $f^{(k)}(0) = e^0 = 1$, we obtain from equation (9.29) the Taylor series

$$e^x = 1 + x + \frac{x^2}{2!} + \frac{x^3}{3!} + \cdots = \sum_{n=0}^{\infty} \frac{x^n}{n!}.$$

In particular, this (amazingly enough) gives us a series expansion for the number e:

$$e = 1 + 1 + \frac{1}{2!} + \frac{1}{3!} + \cdots = \sum_{n=0}^{\infty} \frac{1}{n!}.$$

In fact, the partial sum

$$1 + 1 + \frac{1}{2!} + \frac{1}{3!} + \cdots + \frac{1}{10!}$$

yields an approximate value $e \approx 2.718281801146385$, which has an error of approximately 10^{-8}. Thus with a few steps of simple arithmetic, we were able to approximate the value of e accurately to eight significant figures.

Next, we apply the ratio test to determine the radius of convergence. Regarding x as an arbitrary fixed constant (for the purpose of the ratio test), the ratio of subsequent terms yields

$$\frac{a_{n+1}}{a_n} = \frac{x^{n+1}/(n+1)!}{x^n/n!} = \frac{x^{n+1}n!}{x^n(n+1)!} = \frac{x}{n+1}.$$

The limit (9.13) becomes

$$L = \lim_{n\to\infty}\left|\frac{a_{n+1}}{a_n}\right| = \lim_{n\to\infty}\frac{|x|}{n+1} = 0,$$

for any fixed value of x. Since $L = 0$, regardless of the x-value in question, the ratio test will guarantee convergence of the series independent of the value for x. We conclude that the radius of convergence $R = \infty$, and that the Taylor series for e^x converges for all real numbers. ■

■ **Example 9.23** Determine the Taylor-series expansion for the function $f(x) = \sin(x)$ about the point $x = 0$ and its corresponding radius of convergence.

The first several derivatives of the function $f(x) = \sin(x)$ are given by

$$\begin{aligned} f(x) &= \sin(x), \\ f'(x) &= \cos(x), \\ f''(x) &= -\sin(x), \\ f'''(x) &= -\cos(x), \end{aligned}$$

and so forth. Evaluating these functions at $x = 0$, we obtain

$$f(0) = 0, \quad f'(0) = 1, \quad f''(0) = 0, \quad f'''(0) = -1, \quad f^{(iv)}(0) = 0, \ldots.$$

The general form of a Maclaurin series (i.e., a Taylor series centered about $x = 0$), as given by equation (9.29), therefore yields the expansion

$$\sin(x) = x - \frac{x^3}{3!} + \frac{x^5}{5!} + \cdots.$$

Representing this series in summation notation requires certain care, as only the odd powers of x appear in the series. The key to achieving this is the observation that the mapping $(n) \mapsto 2n+1$ sends the nonnegative integers into the positive, odd integers, i.e., it sends $0 \mapsto 1$, $1 \mapsto 3$, $2 \mapsto 5$, and so forth. We may leverage this fact to express

$$\sin(x) = \sum_{n=0}^{\infty} \frac{(-1)^n x^{2n+1}}{(2n+1)!}.$$

(Do replace n with 0, 1, and 2 to see that this formula indeed recovers the Taylor series expansion for the sine function.)

We may write the ratio of subsequent (nonzero) terms as

$$\frac{a_{n+1}}{a_n} = \frac{(-1)^{n+1}x^{2n+3}/(2n+3)!}{(-1)^n x^{2n+1}/(2n+1)!} = \frac{-x^2}{(2n+3)(2n+2)}.$$

Regarding x as a fixed constant, we may now compute the limit (9.13), thereby obtaining

$$L = \lim_{n\to\infty} \frac{x^2}{(2n+3)(2n+2)} = 0.$$

Again, this limit is independent of the actual x-value in question. Since $0 < 1$, we conclude that this Taylor series converges for every value of x, and therefore its radius of convergence is given by $R = \infty$. ■

We will leave the derivation of the Taylor series for $\cos(x)$ as an exercise.

■ **Example 9.24** Determine the Taylor-series expansion for the function $f(x) = \ln(x)$ about the point $x = 1$ and its corresponding radius of convergence.

The first several derivatives are given by

$$\begin{aligned} f(x) &= \ln(x), \\ f'(x) &= \frac{1}{x}, \\ f''(x) &= \frac{-1}{x^2}, \\ f'''(x) &= \frac{2}{x^3}, \\ f^{(iv)}(x) &= \frac{-3!}{x^4}, \end{aligned}$$

and so forth. Evaluating these functions at $x = 1$ yields

$$f(1) = 0, \quad f'(1) = 1, \quad f''(1) = -1, \quad f'''(1) = 2!, \quad f^{(iv)}(1) = -3!, \cdots.$$

Substituting these numbers into the general form (9.30), we obtain the Taylor series

$$\begin{aligned} T(x) &= (x-1) - \frac{(x-1)^2}{2!} + \frac{2!(x-1)^3}{3!} - \frac{3!(x-1)^4}{4!} + \cdots, \\ &= (x-1) - \frac{(x-1)^2}{2} + \frac{(x-1)^3}{3} - \frac{(x-1)^4}{4} + \cdots. \end{aligned}$$

This series may be represented in summation notation as

$$T(x) = \sum_{n=1}^{\infty} \frac{(-1)^{n+1}(x-1)^n}{n}.$$

The ratio of subsequent terms is given by

$$\frac{a_{n+1}}{a_n} = \frac{(-1)^{n+2}(x-1)^{n+1}/(n+1)}{(-1)^{n+1}(x-1)^n/n} = \frac{-n(x-1)}{n+1}.$$

Taking the limit, we obtain

$$L = \lim_{n\to\infty}\left|\frac{a_{n+1}}{a_n}\right| = |x-1|\lim_{n\to\infty}\frac{n}{n+1} = |x-1|.$$

We conclude, by the ratio test, that this Taylor series converges for $|x-1| < 1$ and diverges for $|x-1| > 1$. Thus, it converges on the interval $(0,2)$. We really could not have expected better than this, as the function $ln(x)$ is not even defined for $x = 0$. ∎

Summarizing our results thus far, we have the following.

Proposition 9.4.4 The elementary functions e^x, $\sin(x)$, $\cos(x)$ and $\ln(x)$ have Taylor series representations

$$e^x = \sum_{n=0}^{\infty}\frac{x^n}{n!} = 1 + x + \frac{x^2}{2!} + \frac{x^3}{3!} + \cdots, \tag{9.30}$$

$$\sin(x) = \sum_{n=0}^{\infty}\frac{(-1)^n x^{2n+1}}{(2n+1)!} = x - \frac{x^3}{3!} + \frac{x^5}{5!} + \cdots, \tag{9.31}$$

$$\cos(x) = \sum_{n=0}^{\infty}\frac{(-1)^n x^{2n}}{(2n)!} = 1 - \frac{x^2}{2!} + \frac{x^4}{4!} + \cdots \tag{9.32}$$

$$\ln(x) = \sum_{n=1}^{\infty}\frac{(-1)^{n+1}(x-1)^n}{n} = (x-1) - \frac{(x-1)^2}{2} + \frac{(x-1)^3}{3} - \cdots. \tag{9.33}$$

The series (9.30)–(9.32) converge for all values of x, whereas the series (9.33) converges on the interval $(0,2)$.

Exercises

For Exercises 1–10, determine the first four nonzero terms for the Taylor series of the indicated function.

1. $f(x) = \sqrt{x}$; about $x = 4$.
2. $f(x) = \ln(1+x)$; about $x = 0$.
3. $f(x) = \sqrt{1+x}$; about $x = 0$.
4. $f(x) = (1+x)^{3/2}$; about $x = 0$.
5. $f(x) = \frac{1}{x}$; about $x = 1$.
6. $f(x) = \tan(x)$; about $x = 0$.
7. $f(x) = e^{-2x}$; about $x = 0$.
8. $f(x) = \sqrt{1-x^2}$; about $x = 0.5$.
9. $f(x) = \tan^{-1}(x)$; about $x = 0$.
10. $f(x) = \sin(x)$; about $x = \frac{\pi}{6}$.

For Exercises 11–15, determine a value for the indicated sum. (*Hint*: First recognize the series as a Taylor series evaluated at a particular point.)

11. $1 + 1 + \frac{1}{2!} + \frac{1}{3!} + \frac{1}{4!} + \cdots$.
12. $\frac{\pi}{6} - \frac{\pi^3}{1296} + \frac{\pi^5}{933120} - \cdots$.
13. $1 - \frac{\pi^2}{72} + \frac{\pi^4}{31104} - \cdots$.
14. $\pi - \frac{\pi^3}{3!} + \frac{\pi^5}{5!} - \cdots$.
15. $1 + 2 + 2 + \frac{4}{3} + \frac{2}{3} + \cdots$.

Problems

16. Derive the formula for the Taylor series of the cosine function (equation (9.32)). Confirm that its radius of convergence is infinite.
17. Differentiate the Taylor series for $\sin(x)$ term-by-term. What series is obtained?
18. Differentiate the Taylor series for $\cos(x)$ term-by-term. What series is obtained?
19. (a) Derive the Taylor series for e^{2x}.
    (b) Substitute $(2x)$ for x in the Taylor series for e^x (equation (9.30)). Show that the result is identical to your result from part (a).
20. (a) Derive the Taylor series for e^{x^2}.
    (b) Substitute (x^2) for x in the Taylor series for e^x (equation (9.30)). Show that the result is identical to your result from part (a).
21. Approximate $\ln(1.1)$ accurately to six decimal places, using only addition, subtraction, multiplication, and division.
22. Approximate $\sqrt{10}$ using only addition, subtraction, multiplication, and division.
23. In physics and engineering, two commonly used approximations are $\sin(\theta) \approx \theta$ and $\cos(\theta) \approx 1$, valid for small values of the angle θ (which must be measured in radians). Justify these approximations using Taylor series.
24. Determine the Taylor series for the hyperbolic sine function $\sinh(x)$ about $x = 0$. Express the result in summation notation. Determine its radius of convergence.
25. Determine the Taylor series for the hyperbolic cosine function $\cosh(x)$ about $x = 0$. Express the result in summation notation. Determine its radius of convergence.
26. The Taylor series for a particular function f is given by

$$f(x) = x^2 - \frac{x^6}{3!} + \frac{x^{10}}{5!} - \frac{x^{14}}{7!} + \cdots.$$

Determine $f^{(10)}(0)$ (i.e, the tenth derivative of f at $x = 0$).

9.5 Further Topics in Taylor Series

9.5.1 Creating New Taylor Series

We begin by examining a variety of approaches for creating new Taylor series from old ones.

Substitution

The first technique is *substitution*. We show how one may create a Taylor series for a composite function by substituting the inner function for the independent variable in the Taylor series for the outer function. We proceed by example.

■ **Example 9.25** Determine the Taylor series for the function

$$f(x) = \sin(3\pi x).$$

Instead of deriving this Taylor expansion from scratch—by differentiating until a pattern emerges and then inputing the resultant values into the general form of a Taylor series—we can leverage our existing Taylor series for $\sin(x)$, as given by equation (9.31). Let us rewrite that equation with u substituted in for x:

$$\sin(u) = \sum_{n=0}^{\infty} \frac{(-1)^n u^{2n+1}}{(2n+1)!} = u - \frac{u^3}{3!} + \frac{u^5}{5!} + \cdots .$$

(We use u, in place of x, to avoid confusion in our next step.) Next, we replace u with $3\pi x$:

$$\sin(3\pi x) = \sum_{n=0}^{\infty} \frac{(-1)^n (3\pi x)^{2n+1}}{(2n+1)!} = 3\pi x - \frac{(3\pi x)^3}{3!} + \frac{(3\pi x)^5}{5!} + \cdots .$$

By replacing the independent variable in equation (9.31) with $3\pi x$, we have thus obtained a new Taylor series for $\sin(3\pi x)$. ■

■ **Example 9.26** Determine the Taylor series for the function

$$f(x) = e^{-x^2},$$

centered about $x = 0$.

We begin by recalling the Taylor series for e^x, as given by equation (9.30). In order to obtain the Taylor series for e^{-x^2}, we simply replace x with $-x^2$ in equation (9.30), thereby obtaining

$$e^{-x^2} = \sum_{n=0}^{\infty} \frac{(-x^2)^n}{n!} = \sum_{n=0}^{\infty} \frac{(-1)^n x^{2n}}{n!}. \tag{9.34}$$

This may alternatively be expressed as

$$e^{-x^2} = 1 - x^2 + \frac{x^4}{2!} - \frac{x^6}{3!} + \cdots .$$

By using substitution, we obtained a new Taylor series from an existing one with relative ease. ■

Differentiation and Integration of Taylor Series

In Theorem 9.3.2 and Corollary 9.3.3, we saw that the term-by-term differentiation or integration of a power series invariably results in a new power series with the same radius of convergence. Since Taylor series are a type of power series, this result is equally applicable here. Moreover, the term-by-term derivative (or integral) of a Taylor series converges to the derivative (or integral) of the originating function, as we state precisely in the following.

Theorem 9.5.1 Consider an infinitely differentiable function f and its corresponding Taylor series

$$T(x) = \sum_{n=0}^{\infty} \frac{f^{(n)}(a)}{n!}(x-a)^n, \tag{9.35}$$

which converges to f for all $x \in (a-R, a+R)$, where R is the radius of convergence. Then the associated series

$$D(x) = \sum_{n=1}^{\infty} \frac{f^{(n)}(a)}{(n-1)!}(x-a)^{n-1}, \tag{9.36}$$

obtained by term-by-term differentiation of the series (9.35), converges to the derivative f' on the interval $(a-R, a+R)$, and the associated series

$$I(x) = \sum_{n=0}^{\infty} \frac{f^{(n)}(a)}{(n+1)!}(x-a)^{n+1}, \tag{9.37}$$

defined by term-by-term integration of the series (9.35), converges to the particular antiderivative

$$F(x) = \int_a^x f(t)\, dt \tag{9.38}$$

on the interval $(a-R, a+R)$.

(R) Recall that the particular antiderivative given by equation (9.38) is the one that possesses the property $F(a) = 0$. Thus, by integrating a Taylor series term-by-term, without adding a constant of integration, yields the particular antiderivative that has an x-intercept at the point $x = a$ at which the series is centered. Other antiderivatives may be obtained by adding a constant of integration to this particular antiderivative.

■ **Example 9.27** Determine the Taylor series for the error function, as defined by equation (8.50), repeated here for convenience:

$$\operatorname{erf}(x) = \frac{2}{\sqrt{\pi}} \int_0^x e^{-s^2}\, ds. \tag{9.39}$$

We would normally begin by deriving the Taylor series for e^{-x^2}, but this task has already been accomplished; the result was recorded in equation (9.34).

Theorem 9.5.1 tells us that the Taylor series for the error function, as expressed in equation (9.39), may be obtained using term-by-term integration, from which we obtain

$$\operatorname{erf}(x) = \frac{2}{\sqrt{\pi}} \sum_{n=0}^{\infty} \frac{(-1)^n x^{2n+1}}{(2n+1)n!}.$$

Expanded, this series may alternatively be expressed

$$\operatorname{erf}(x) = x - \frac{x^3}{3} + \frac{x^5}{10} - \frac{x^7}{42} + \frac{x^9}{216} + \cdots.$$

(Compare with the expanded form of equation (9.34), which follows thereafter.) ■

The previous example illustrates how one might go about using these technologies to develop a Taylor series for even *non-elementary* functions, such as the error function. Given the widespread use of non-elementary functions in applied mathematics and engineering, the practical applications of Taylor series should not be underestimated. After all, what is sine or cosine or the natural logarithm other than an infinite degree polynomial expressed as a Taylor series! Taylor series allow your handheld calculator to compute arbitrary values of these basic functions for a virtually endless supply of possible function inputs you might conjure. Special functions—not expressible in terms of the elementary functions—are, in a sense, no more difficult to handle computationally than a sine or cosine function, as computers can leverage their series expansions to produce all the precision one might need. In particular, when studying differential equations—a class of differential instructions for nature that permeates physics and engineering applications—only the simplest examples have solutions that may be described using elementary functions. Series representations of functions therefore give us added tools for coping with the mathematical abyss of nature herself.

9.5.2 Binomial Series

Another quintessential example in Taylor series is that of the binomial series.

Theorem 9.5.2 — Binomial-series Expansion. The power function

$$f(x) = (1+x)^p,$$

for arbitrary real power $p \in \mathbb{R}$, has a Taylor-series expansion

$$T(x) = 1 + px + \frac{p(p-1)x^2}{2!} + \frac{p(p-1)(p-2)x^3}{3!} + \cdots. \tag{9.40}$$

Moreover, the Taylor series (9.40) converges to the function $(1+x)^p$ for $|x| < 1$.

■ **Example 9.28** Use the binomial-series expansion to determine the Taylor series for

$$f(x) = \sqrt{1+x}.$$

Here, we identify the power p with $p = 1/2$. Writing out the first several terms of the binomial series (9.40), we have

$$\sqrt{1+x} = 1 + \frac{x}{2} - \frac{x^2}{8} + \frac{x^3}{16} - \frac{x^4}{640} + \cdots.$$

Just for fun, we can use the fourth-order Taylor polynomial (i.e., just the terms shown in the previous equation) to approximate $\sqrt{1.1}$—

$$\sqrt{1.1} \approx 1.04881234375.$$

We thus obtained five significant figures of accuracy in computing $\sqrt{1.1}$ using only elementary arithmetic. ■

9.5.3 Applications of Taylor Series

Next, we consider several particular examples of the application of Taylor series to science and engineering.

■ **Example 9.29** The *relativistic kinetic energy* K of a particle with rest mass m and velocity v is given by

$$K(v) = \frac{mc^2}{\sqrt{1-(v/c)^2}} - mc^2, \tag{9.41}$$

where the constant c represents the speed of light in vacuum. Use the binomial series to approximate $K(v)$ for small values of v.

To begin, let us first write down the binomial series for the case of $p = -1/2$. From equation (9.40), we have

$$\frac{1}{\sqrt{1+x}} = 1 - \frac{x}{2} + \frac{3x^2}{8} - \frac{5x^3}{16} + \frac{35x^4}{128} + \cdots,$$

for $|x| < 1$. Next, let us substitute $x = (v/c)^2$, and insert the result into equation (9.41):

$$K(v) = mc^2\left(1 - \frac{v^2}{2c^2} + \frac{3v^4}{8c^4} - \frac{5v^6}{16c^6} + \frac{35v^8}{128c^8} + \cdots\right) - mc^2.$$

Of course, the leading term of this expansion cancels with the mc^2 following it, so that we obtain the simplification

$$K(v) = \frac{mv^2}{2} + \frac{3mv^4}{8c^2} - \frac{5mv^6}{16c^4} + \frac{35mv^8}{128c^6} + \cdots.$$

This result is valid for $(v/c)^2 < 1$, i.e., for $0 \le v < c$. Notice, however, that the first nonzero term of the Taylor expansion for the relativistic kinetic energy of a particle is precisely the classical, Newtonian kinetic energy. Special relativity therefore unravels that higher-order relativistic effects of velocity on a particle's energy as the speed of the particle starts to become appreciable as compared with the speed of light. ■

■ **Example 9.30** The *electric potential* created by a point charge, with charge q, at a distance r, is given by the formula

$$V(r) = \frac{1}{4\pi\varepsilon_0}\frac{q}{r},$$

where ε_0 is a constant known as the *permitivity of free space*. (In SI units, this constant takes the value $\varepsilon_0 = 8.854187817 \times 10^{-12}$ F/m.)

Consider two electric point charges, with opposite charge q and $-q$, situated with a distance a between them, as shown in Figure 9.8. Determine the electric

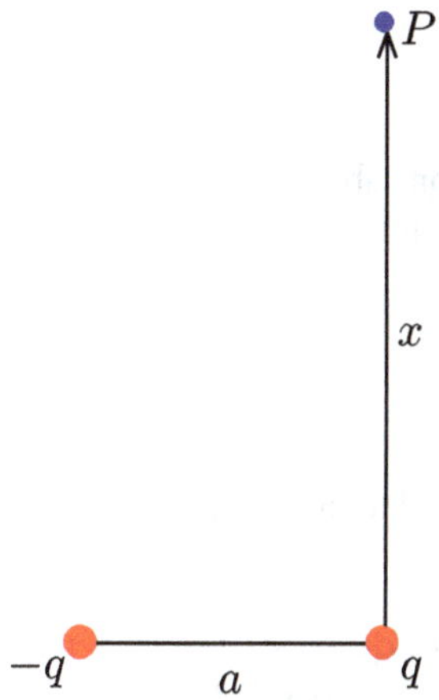

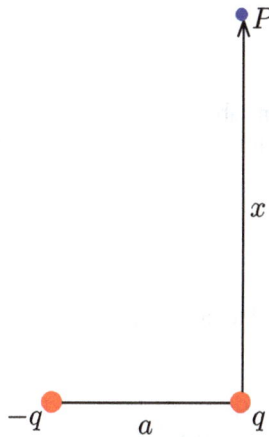

Figure 9.8: An electric dipole; Example 9.30.

potential at the point P, which is located at a distance x from the positive charge, and use Taylor series to approximate the potential function for the case $x >> a$. This is a reasonable assumption as the opposing charges in an electric dipole are in very close proximity to one another, for example, in a water molecule.

To begin, the electric potential of the dipole is obtained by simply adding the resulting electric potentials of each individual point charge:

$$V(x) = \frac{q}{4\pi\varepsilon_0}\left(\frac{1}{x} - \frac{1}{\sqrt{x^2+a^2}}\right).$$

This follows from the *principle of superposition* in physics. Since we are interested in studying the regime in which x is much larger than a, we factor an x from the second term, so that we are left with the ratio a/x:

$$V(x) = \frac{q}{4\pi\varepsilon_0 x}\left[1 - \frac{1}{\sqrt{1+(a/x)^2}}\right].$$

The second term within the parentheses can be approximated using the binomial series with $p = -1/2$ in the variable $(a/x)^2$, that is, we may replace p with $-1/2$

and x with a^2/x^2 in equation (9.40) to obtain the approximation

$$\begin{aligned} V(x) &= \frac{q}{4\pi\varepsilon_0 x}\left[1-\left(1-\frac{a^2}{2x^2}+\frac{3a^4}{8x^4}-\frac{5a^6}{16x^6}+\cdots\right)\right] \\ &= \frac{q}{4\pi\varepsilon_0}\left(\frac{a^2}{2x^3}-\frac{3a^4}{8x^5}+\frac{5a^6}{16x^7}+\cdots\right). \end{aligned}$$

Since the distance a is typically several orders of magnitude smaller than the distance x, the electric potential can be accurately approximating with the leading term:

$$V(x) \approx \frac{qa^2}{8\pi\varepsilon_0 x^3}.$$

These sorts of approximations abound in physics and engineering, as they afford scientists the ability to accurately represent phenomena with simplified expressions.

■

9.5.4 Euler's Formula

Taylor series may also be used to prove a powerful theorem in the field of complex variables, as we will see in our next theorem.

Theorem 9.5.3 — Euler's Equation. Given a real number θ, the imaginary exponential $e^{i\theta}$ is a complex number that may be expressed by *Euler's equation*:

$$e^{i\theta} = \cos(\theta) + i\sin(\theta). \tag{9.42}$$

(The imaginary number i is defined as $i = \sqrt{-1}$.)

If we identify an imaginary number $z = x + iy$ with a point (x, y) in the Euclidean plane, we observe that the complex exponential $e^{i\theta}$ returns a point on the unit circle, since the identification forces the relations $x = \cos(\theta)$ and $y = \sin(\theta)$, which, as the reader is sure to recall, constitutes a parameterization of the unit circle. In fact, this observation is arguably the basis for the field of complex analysis. It is also used extensively in the fields of electrical engineering and physics (especially quantum mechanics).

Proof. Before we proceed to proof, let us casually observe a pattern when computing powers of the imaginary unit i:

$$i = i, \qquad i^2 = -1, \qquad i^3 = -i, \qquad i^4 = 1,$$

$$i^5 = i, \qquad i^6 = -1, \qquad i^7 = -i, \qquad i^8 = 1,$$

and so on. The powers of i are therefore cyclical: they follow the repeating pattern $i, -1, -i, +1$.

To prove Euler's equation, we simply substitute $x = i\theta$ in equation (9.30), from which we obtain:

$$\begin{aligned} e^{i\theta} &= 1 + (i\theta) + \frac{(i\theta)^2}{2!} + \frac{(i\theta)^3}{3!} + \frac{(i\theta)^4}{4!} + \frac{(i\theta)^5}{5!} + \frac{(i\theta)^6}{6!} + \frac{(i\theta)^7}{7!} + \frac{(i\theta)^8}{8!} \cdots \\ &= 1 + i\theta - \frac{\theta^2}{2!} - i\frac{\theta^3}{3!} + \frac{\theta^4}{4!} + i\frac{\theta^5}{5!} - \frac{\theta^6}{6!} - i\frac{\theta^7}{7!} + \frac{\theta^8}{8!} + \cdots . \end{aligned}$$

Now comes the clever part—let us group the real and imaginary terms separately, factoring out the number i from the latter group. We obtain

$$e^{i\theta} = \left(1 - \frac{\theta^2}{2!} + \frac{\theta^4}{4!} - \frac{\theta^6}{6!} + \frac{\theta^8}{8!} + \cdots\right) + i\left(\theta - \frac{\theta^3}{3!} + \frac{\theta^5}{5!} - \frac{\theta^7}{7!} + \cdots\right).$$

But, behold what we hath wrought! The real and imaginary components of this number are just the Taylor series for the functions $\cos(\theta)$ and $\sin(\theta)$, respectively! We may therefore conclude that

$$e^{i\theta} = \cos(\theta) + i\sin(\theta).$$

We have thus recovered the formula that was first published by Leonhard Euler in 1748, and that would later be called by Nobel-prize winner Richard Feynman (physics) both "our jewel" and "the most remarkable equation in all of mathematics." ■

As a direct consequence of Euler's equation, we have the following astonishing result.

Corollary 9.5.4 Applying Euler's equation with $\theta = \pi$ yields the famous relation

$$e^{i\pi} + 1 = 0. \tag{9.43}$$

This simple equation combines the five most important number in all of mathematics—0, 1, e, π, and i.

Exercises

For Exercises 1–10, determine the first few nonzero terms of the Taylor series of the indicated function using a previously derived Taylor series and substitution. State the radius of convergence for each.

1. $f(x) = \sin(2x)$; about $x = 0$.
2. $f(x) = \cos(2x)$; about $x = 0$.
3. $f(x) = \ln(2x)$; about $x = 1/2$.
4. $f(x) = e^{2x}$; about $x = 0$.
5. $f(x) = \sqrt{1+4x}$; about $x = 0$.
6. $f(x) = \sqrt{1-2x}$; about $x = 0$.
7. $f(x) = \sqrt{1-4x^2}$; about $x = 0$.
8. $f(x) = e^{-x}$; about $x = 0$.
9. $f(x) = \dfrac{1}{\sqrt{4-x}}$; about $x = 0$.
10. $f(x) = e^{-x^4}$; about $x = 0$.

Problems

11. Determine the Taylor series for $f(x) = 1/x$ about $x = 1$ by differentiating term-by-term the Taylor series for $\ln(x)$ about $x = 1$.
12. (a) By making the substituting $x \mapsto (1+x)$, determine the Taylor series for $\ln(1+x)$ about $x = 0$.
    (b) By differentiating term-by-term, determine the Taylor series for $1/(1+x)$ about $x = 0$. What is its radius of convergence?
13. (a) Use substitution and the result from Problem 12 to determine the Taylor series for $\frac{1}{1+x^2}$.
    (b) Use the result from part (a) to determine the Taylor series for $\tan^{-1}(x)$ about $x = 0$.
14. The Fresnel S function is defined by:

$$\text{FresnelS}(x) = \int_0^x \sin\left(\frac{\pi u^2}{2}\right)\, du$$

    (a) Find the first three *nonzero* terms of the Taylor Series of the FresnelS function about $x = 0$.
    (b) Use your result to estimate FresnelS(0.5) accurately to 6 decimals.
15. Consider a spherical planet with mass M, surrounded by an orbiting planetary ring, with linear mass density λ and radius a. A satellite with mass m is located along the planet's polar axis (perpendicular to the plane of the rings) at a distance z to the planet's center. The gravitational force acting on the satellite is given by

$$F(z) = -Gm\left[\frac{M}{z^2} + \frac{2\pi a\lambda z}{(a^2+z^2)^{3/2}}\right].$$

    (a) Show that this expression is equivalent to

$$F(z) = -Gm\left[\frac{M}{z^2} + \frac{2\pi a\lambda}{z^2[1+(a/z)^2]^{3/2}}\right].$$

    (b) Determine a series expansion for $F(z)$, valid for large $z >> a$. (In particular, it should become more accurate as $z \to \infty$.)
16. A termite's position is given by

$$\begin{aligned} x(t) &= t \\ y(t) &= t^2 \end{aligned}$$

where x and y are in meters and t is in seconds.
    (a) Find $v(t)$, the termite's speed as a function of t.
    (b) Let $L(t)$ be the arc length of the path traced out by the termite between the times $[0,t]$. Write a definte integral that represents $L(t)$.
    (c) Find the Taylor Series for $v(t)$ about $t = 0$. What is its interval of convergence?

(d) Find the Taylor Series for $L(t)$ about $t = 0$. What is its interval of convergence?

(e) Estimate $L(0.1)$ to 6 decimals. (Write down and use the fifth-order Taylor polynomial.)

17. The speed of a particle (m/s) moving along the x-axis is given by the following power series:

$$v(t) = \frac{(t-1)^2}{2\cdot(2\cdot 1)} + \frac{(t-1)^3}{2^2\cdot(3\cdot 2)} + \frac{(t-1)^4}{2^3\cdot(4\cdot 3)} + \frac{(t-1)^5}{2^4\cdot(5\cdot 4)} + \cdots = \sum_{n=1}^{\infty} \frac{(t-1)^{n+1}}{2^n(n+1)n}$$

where t is the time measured in seconds.

(a) At what value of t is this power series centered?

(b) Find the radius of convergence of this power series.

(c) Find the power series of $v'(t)$, the particle's acceleration in units of m/s.

(d) At what moment in time is the particle's acceleration equal to zero? *Hint:* The particle's acceleration passes through zero only once.

(e) Use the second derivative test to determine whether the particle's velocity is a local maximum or a local minimum at that moment in time.

(f) Find a power series for the particle's position $x(t)$ as a function of time, given that the particle is initially located at $x(0) = 1$.

18. Determine the limit

$$\lim_{x\to\infty} \sum_{n=0}^{\infty} \frac{(-1)^n x^{2n+1}}{(2n+1)n!}.$$

Chapter 10

Ordinary Differential Equations

We begin the text with a treatment of first-order differential equations, for which the rate of change of a given quantity depends on the current value of the quantity and, possibly, time. We pay particular attention to *separable* and *linear* first-order differential equations, and discuss techniques for solving the differential equation in each case. We also introduce several modeling problems, including exponential growth and decay, Newton's law of heating and cooling, and logistic population growth. We conclude the chapter with a section on numerical solutions to differential equations.

10.1 Separable Equations and Exponential Growth

In this section, we introduce differential equations, and then we discuss first order separable equations and their application to exponential growth.

10.1.1 What are Differential Equations?

In algebra, one is given a relation that describes an unknown variable x, and one must solve this algebraic equation to determine the possible values of x that make the equation hold.

In differential equations, one is given a relation that describes an unknown function y, of an independent variable t, and its various derivatives. One must solve this differential equation to determine the possible functions $y(t)$ that makes the differential equation hold.

More precisely, we have the following.

Definition 10.1.1 An nth order *ordinary differential equation* (ODE) is an equation that relates an unknown function y and its first n derivatives, i.e.,

$$F(y^{(n)}, y^{(n-1)}, \ldots, y'', y', y; t) = 0.$$

Example 10.1 The following equations are each an example of a differential equation:

$$\begin{aligned} y'' + 3y' + 4y &= 0. \\ t^2 y''' + \sin(t) y'' - 4ty &= 0. \\ y^{iv} + t^3 y'' + y &= \cos(\sqrt{t}). \\ y'' + yy' &= 0. \end{aligned}$$

The first three differential equations are examples of *linear* differential equations, whereas the fourth is nonlinear. We'll come back to this point later. (Can you spot the difference between the first three examples and the fourth?) ■

Definition 10.1.2 The *solution* to a differential equation is the set of functions which, when substituted into the differential equation for y, satisfy the equation for all x.

Example 10.2 Consider the second-order differential equation

$$y'' + 9y = 0.$$

Its solution is given by

$$y = A\cos(3t) + B\sin(3t).$$

To see this, we compute the derivatives of the function y:

$$\begin{aligned} y' &= -3A\sin(3t) + 3B\cos(3t), \\ y'' &= -9A\cos(3t) - 9B\sin(3t). \end{aligned}$$

Substituting y, y', and y'' back into the differential equation, we obtain:

$$y'' + 9y = (-9A\cos(3t) - 9B\sin(3t)) + 9(A\cos(3t) + B\sin(3t)) = 0.$$

The differential equation is therefore satisfied for all t. ■

Definition 10.1.3 An *initial value problem* (IVP) is a differential equations with prescribed initial conditions (IC's), i.e., it is an equation and a set of initial conditions of the form

$$\begin{aligned} F(y^{(n)}, y^{(n-1)}, \ldots, y'', y', y; t) &= 0, \\ y(t_0) &= y_0, \\ y'(t_0) &= y'_0, \\ &\vdots \\ y^{(n-1)}(t_0) &= y_0^{(n-1)}, \end{aligned}$$

where $y_0, y'_0, \ldots, y_0^{(n-1)}$ are prescribed constants.

Typically $t_0 = 0$ (hence *initial* conditions), though this is not a requirement. Notice that there are as many initial conditions as the order of the differential equation.

When solving an initial value problem, one *first* solves the differential equation, obtaining the general solution. (In this step, the initial condition(s) do not come into play.) Once the general solution is obtained, one can go back and use the initial conditions to solve for the constants of integration.

Example 10.3 Consider the initial value problem

$$\begin{aligned} y'' + 9y &= 0, \\ y(0) &= 4, \\ y'(0) &= 6. \end{aligned}$$

As we previous saw (though we didn't derive the result), the general solution to the differential equation is given by

$$y(t) = A\cos(3t) + B\sin(3t).$$

Computing $y(0)$ and $y'(0)$, using the general solution, we have

$$\begin{aligned} y(0) &= A, \\ y'(0) &= 3B. \end{aligned}$$

Comparing this to the initial conditions ($y(0) = 4, y'(0) = 6$), we therefore deduce that $A = 4$ and $B = 2$. The solution to this initial value problem is

$$y(t) = 4\cos(3t) + 2\sin(3t).$$

One can then go back and check that it satisfies both the differential equation and the initial conditions. ■

10.1.2 Separable Equations

In this paragraph, we will discuss a special class of first order differential equations known as *separable equations*.

Definition 10.1.4 A *separable equation* is a first order differential equation that can be rewritten in the form

$$y' = f(y)g(t). \tag{10.1}$$

■ **Example 10.4** The differential equations

$$\begin{aligned} y' &= y, \\ y' &= y^2t^3, \\ y' &= ty+y \end{aligned}$$

are separable, whereas the differential equations

$$\begin{aligned} y' &= y^2+x^2, \\ y' &= y+t, \\ y' &= y^2-t^2 \end{aligned}$$

are not separable. ■

To solve a separable equation, you first *separate variables*, and then you integrate. In particular, starting with equation (10.1), we obtain:

$$\begin{aligned} \frac{1}{f(y)}y' &= g(t), \\ \int \frac{1}{f(y)}y'dt &= \int g(t)dt, \\ \int \frac{1}{f(y)}dy &= \int g(t)dt. \end{aligned}$$

When carrying out this computation, it is important to *only add "+C" to one side of the equation*! Technically, each indefinite integral carries its own "+C." However, the left side and right side will have *different* constants of integration, so therefore one should technically add "+C_1" to the left side and "+C_2" to the right side. But one can always subtract C_1, so that there is no constant of integration on the left side, whereas the right side has the constant $C_2 - C_1$. But since C_1 and C_2 are each arbitrary constants, then so is their difference $C_2 - C_1$. We can therefore rename this difference C.

Once the integrals are preformed, one must solve for y, thereby obtaining $y = y(t)$. We discuss our first example of separation of variables in the next paragraph, on exponential growth and decay.

■ **Example 10.5 [Exponential Growth and Decay]** Exponential functions arise as solutions to the differential equation

$$y' = ky. \tag{10.2}$$

This is a separable equation. To solve it, we separate variables and integrate:

$$\begin{aligned}\frac{1}{y}\frac{dy}{dt} &= k,\\ \int \frac{dy}{y} &= \int k dt,\\ \ln|y| &= kt + \tilde{C}.\end{aligned}$$

Taking the exponential of both sides yields:

$$|y(t)| = e^{kt+\tilde{C}} = e^{kt}e^{\tilde{C}}.$$

Since $\tilde{C}$ is an arbitrary constant, it follows that $e^{\tilde{C}}$ must be an arbitrary *positive* constant. (The function $f(t) = e^t$ always returns a positive number, i.e., the range of the exponential function is $(0, \infty)$.) We therefore obtain

$$y(t) = Ce^{kt}, \tag{10.3}$$

where C is an arbitrary constant. (The constant C is given in terms of our original content $\tilde{C}$ by the relation $C = \pm e^{\tilde{C}}$.)

Technically, C is a *nonzero* constant. However, notice that when we divided both sides of equation (10.2) by the variable y, we made an implicit assumption that $y \neq 0$. Considering the case $y = 0$ as a separate case, one easily sees that $y(t) = 0$ is itself a solution to the differential equation (10.2). (It is called the *trivial solution.*) By allowing the constant C that appears in the solution (10.3) to be any arbitrary number (including zero–no longer restricted to *nonzero* values), we see that equation (10.3) describes *all* of the possible solutions to the differential equation (10.2). ■

■ **Example 10.6** Solve the initial value problem

$$y' = (y-3)\cos t, \qquad y(0) = 7.$$

First, we separate variables, integrate, and determine the general solution. Then we go back and use the initial condition $y(0) = 7$ to solve for the constant of integration.

Separating variables, we obtain

$$\frac{dy}{y-3} = \cos(t)\, dt.$$

Integrating and solving for y yields:

$$\begin{aligned}\int \frac{dy}{y-3} &= \int \cos(t)\, dt\\ \ln|y-3| &= \sin(t) + C\\ y-3 &= Ae^{\sin(t)},\end{aligned}$$

where $A = \pm e^C$, similar to what we saw in the previous example. Solving for y yields

$$y(t) = 3 + Ae^{\sin(t)}.$$

Now, to impose the initial condition $y(0) = 7$, we must use our formula for the general solution to first compute $y(0)$, obtaining

$$y(0) = 3 + A.$$

Since we wish $y(0) = 7$, we therefore set $A = 4$. The solution to the initial value problem is thus given by

$$y(t) = 3 + 4e^{\sin(t)}.$$

■

10.1.3 Existence and Uniqueness Theorem

Though its proof lies outside the scope of this text, the following theorem plays a fundamental role in the theory of differential equations.

Theorem 10.1.1 Suppose that a function $f : \mathbb{R}^2 \to \mathbb{R}$ and its partial derivative $\partial f / \partial y$ are continuous in a rectangle $R = [t_0 - a, t_0 + a] \times [y_0 - b, y_0 + b]$, then there exists a positive constant $h < a$, such that the initial value problem

$$\begin{aligned} \frac{dy}{dt} &= f(t,y) \\ y(t_0) &+ y_0 \end{aligned}$$

has a unique solution on the interval $[t_0 - h, t_0 + h]$.

■ **Example 10.7** Consider the initial value problem

$$\frac{dy}{dt} = \frac{2y}{t}, \qquad y(0) = 0.$$

By separating variables, we obtain the solution

$$y(t) = ct^2,$$

for any constant $c \in \mathbb{R}$. Thus, there does not exist a unique solution to the initial value problem. Notice, however, that $f(t,y) = 2y/t$ is not continuous at $(0,0)$, so the conditions of the uniqueness and existence theorem are not satisfied. ■

10.1.4 Modeling with Differential Equations

A main focus early on in this text is to develop the skills of mathematical modeling and problem solving. These are the skills that give engineers and scientists value to their employers. In this book, you will spend time studying and solving problems that have been understood for several hundred years (that is not to say that they are "easy;" the people who invented them devoted their careers to doing just that), but in your profession—whether you become an engineer or a scientist or an academic—you will be faced with problems that *nobody* has solutions for. You'll have to work in teams for months or even years to come up with new solutions to new problems. This paragraph is aimed at providing a brief overview of some common modeling assumptions to help you get started.

Conservation Principles

Conservation principles abound in engineering and the mathematical sciences. The two most important conservation principles are *the conservation of mass* and *the conservation of energy*. Basically, they state that mass can be neither created nor destroyed (conservation of mass) and that energy can be neither created or destroyed (conservation of energy). Absent nuclear reactions, in which matter is converted into energy *a la* $E = mc^2$, these principle always hold throughout the universe.

Incompressible Fluid Assumptions. To a very good approximation, liquid water is incompressible, meaning that its density is constant. (It requires a tremendous amount of pressure to affect a very small change in the density of water.) Thus, conservation of mass taken together with an incompressibility assumption implies that the *volume* of water is conserved.

Control-Volume Analysis. How can one use a conservation principle to assist in modeling a physical system? How does one express a conservation principle mathematically? We can achieve these things with a construct known as a *control volume*. A control volume is an imaginary region of space defined by a boundary. One can place a control volume around an entire system, or any subsystem. In Figure 10.1, a container contains a certain quantity $Q(t)$, which depends on time. (Q may represent the volume of water, the mass of a substance, or energy, depending on what is being conserved.) The quantity is flowing into the control volume at a rate R_{in}, and is flowing out of the control volume at a rate R_{out}. (The units of R_{in} and R_{out} must equal the units of Q divided by the units of time t.) The conservation of Q therefore yields the following differential equation:

$$\frac{dQ}{dt} = R_{\text{in}} - R_{\text{out}}, \tag{10.4}$$

that is, the rate of change of quantity stored in the tank is equal to *the rate in minus the rate out*.

If the quantity is flowing in at multiple locations, R_{in} represents the *sum* of all the in-flow rates. Similarly, R_{out} must be interpreted as the *sum* of all of the outflow rates.

Steady–State Assumption

The *steady–state assumption* is that the system is not changing in time. In particular, the left-hand side of equation (10.4) must vanish, i.e., $Q'(t) = 0$. In this scenario,

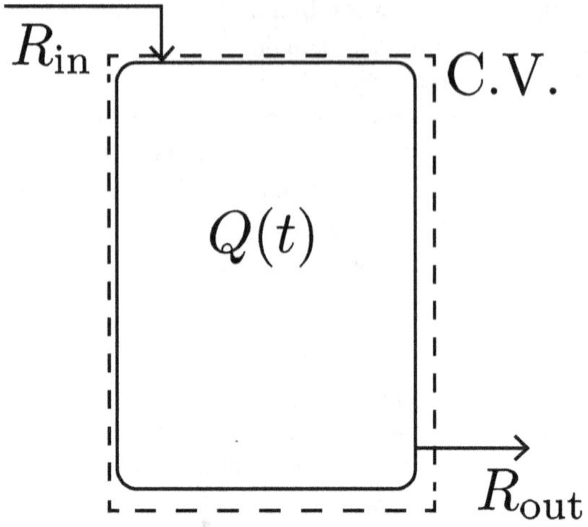

Figure 10.1: A typical control-volume analysis.

$R_{in} = R_{out}$: the net rate of the quantity entering any control volume must equal the net rate of the quantity leaving that control volume, or *rate in equals rate out.*

In general, when one has a differential equation that models some physical process, the *steady–state* or *equilibrium solution* is any solution that is constant in time. Thus, if one is studying the first-order differential equation

$$y' = f(y,t),$$

and if y_{eq} is a fixed constant such that

$$f(y_{eq},t) = 0$$

for all t, then $y(t) = y_{eq}$ is an equilibrium solution of the differential equation. y_{eq} is also called a *fixed point.*

It is common to observe generic solutions to a differential equation, $y = y(t)$, approach a steady–state solution as $t \to \infty$, especially if the differential equation models a physical process and if there is only one steady–state solution.

Proportionality

Another common assumption encountered in word problems is *proportionality*. Consider the following examples:

1. The rate of change of a population y is proportional to the size of the population:

$$\frac{dy}{dt} = ky.$$

2. The gravitational force F is inversely proportional to the square of the distance r from the attracting body:

$$F = \frac{k}{r^2}.$$

3. The gravitational potential energy U is inversely proportional to the distance r from the attracting body:

$$U = \frac{k}{r}.$$

In the preceding examples, k is taken to be the (positive) constant of proportionality. Sometimes you must rely on your intuition in order to determine whether or not the given term should come with a plus or minus sign. (Is it increasing or decreasing?)

Sometime only *one* of the rates of change is assumed to be proportional to something. That will model a single rate of change (e.g., R_{in}, or one of the components constituting R_{in}), to be summed with the other rates of change of the quantity coming in or out.

Dimensional Analysis

Dimensional Analysis is a method for comparing the dimensions (units) of physical quantities in the problem. Since, in mathematics, one can only add terms *with like units*, the overall units of each term in a differential equation must be consistent. Thus, R_{in}, R_{out} and $Q'(t)$ must all have the same units in equation (10.4). Thinking about the units is a common way to gather clues as to what the correct expression should be. (Though, sometimes the constant of proportionality will carry its own units, which work out to be whatever is necessary for the units to be consistent across the differential equation.)

Multiple-Component Analysis

Often an engineering or physical system is comprised of multiple components, each of which is conducive to individual analysis. The number of components of a given system can range from a few to hundreds, or even thousands. A separate control volume can be defined for any single component or group of components. The inflows and outflows of that particular control volume are then balanced with the net rate of change of quantity stored; if the system is operating in steady state, then the net inflow is exactly balanced with the net outflow.

For a simple example, consider the jet engine depicted in Figure 10.2. A jet engine consists of four main components: a compressor, a combustor, a turbine, and a nozzle. Thus, one can place a control volume around any of these components and do a complete analysis on that component, invoking the laws and relationships of thermodynamics to derive equations relating the inflow and outflow properties of the gas. (In a jet-engine analysis, one needs to use thermodynamic relationships to compute the pressure, temperature, density, enthalpy, and entropy at each stage of the engine, as well as the ambient (atmospheric) conditions and the stagnation

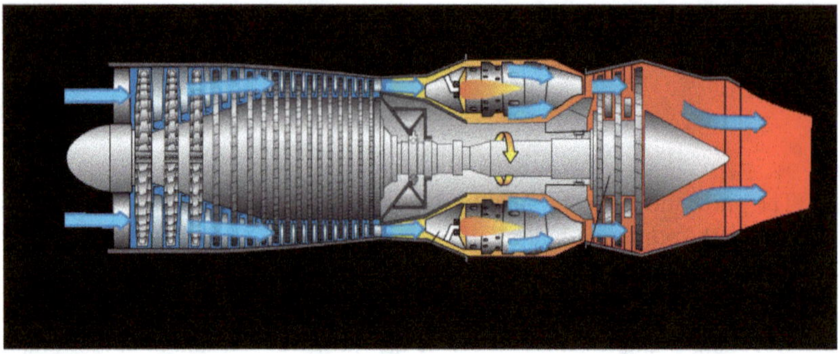

Figure 10.2: Cross section of a jet engine (courtesy Federal Aviation Administration).

conditions at the air inlet.) A multiple-component analysis may also be necessary, for example, when analyzing electrical circuits, power pants, or automobile engines: individual components of the system are each amenable to an individual analysis that works together with the results obtained from the other components.

Exercises

For Exercises 1–6, first find the general solution to the given differential equation. Then use the initial condition to solve for C.

1. $y' = -y^2 \sec^2(t), \qquad y(0) = 3.$
2. $y' = ty - 3t - 2y + 6, \qquad y(0) = 5.$
3. $y' = 2t\sqrt{9 - y^2}, \qquad y(0) = 0.$
4. $y' = -2ty, \qquad y(0) = 7.$
5. $y' = \sin y, \qquad y(0) = \pi/2.$
6. $y' = 16 - y^2, \qquad y(1) = 8.$

Problems

7. The mass of a single bacterium is approximately 10^{-15}kg, and the doubling time of a certain bacterial population is 15 minutes. Assuming exponential growth, how long will it take for a single bacterium to spawn a bacterial population that equals the mass of Earth (approximately 6×10^{24}kg)?
8. Consider a jet engine, similar to the one shown in Figure 10.2, operating under a steady state condition. The intake air mass flow rate is 77kg/s, and fuel is being injected into the combustion chamber at a rate of 1.3kg/s. (The fuel mixes with the compressed intake air, causing combustion.) Suppose,

further, that the exhaust nozzle has an exit area of 0.15m^2, and that the density of the exhaust gas is 0.75kg/m^3. Compute the velocity of the exhaust gas (relative to the engine).

9. Leaves in a forest fall and accumulate on the ground at a rate of r kg per day. At the same time, the fallen leaves decompose at a rate proportional to the leaf mass on the ground, with proportionality factor α. Assume that the constants $r, \alpha > 0$.
   (a) Let $m(t)$ be the leaf mass on the ground. Give the differential equation for $m(t)$.
   (b) Over a long period of time, the leaf mass approaches an equilibrium value M. Find M in terms of the annual accumulation rate r and the decomposition factor α.
   (c) Assuming $m(0) = m_0$, where $m_0 > 0$ is a constant that represents the initial leaf mass, solve the differential equation to find $m(t)$.
   (d) A flood washes away all the leaves. How long does it take for the leaf mass to recover to 50% of its equilibrium value?

10. *Newton's law of heating and cooling* states that the time rate of change of a body's temperature is proportional to the difference between its temperature and the ambient temperature. Let $u(t)$ be the temperature of the object, u_0 its initial temperature at time $t = 0$, R the ambient temperature, and $k > 0$ the constant of proportionality.
    (a) Set up a differential equation for $u(t)$.
    (b) Solve the differential equation to determine the general solution $u(t)$.
    (c) A frozen turkey with temperature 30°F is placed into a hot oven with temperature 350°F. One hour later, the turkey's temperature was 75°F. How long does it take for the turkey's temperature to reach 165°F?

11. A woman jumps out of an airplane at an altitude of 10,000 ft, falls freely for 20s, then opens her parachute, so that, at any moment, her acceleration is given by

$$\frac{dv}{dt} = -g - kv,$$

where $v = dx/dt$. $x(t)$ is the woman's altitude t seconds after bailing out of the plane. Assume linear air resistance kv ft/s^2, taking $k = 0.2\text{s}^{-1}$ without the parachute and $k = 2.0\text{s}^{-1}$ with the parachute. Take $g = 32\text{ft/s}^2$.
    (a) At what altitude (to the nearest foot) was the woman when her parachute opened?
    (b) At what time did she land on the ground? (Use technology.)

12. The vertical velocity of a particle moving under the influence of a uniform gravitational field with *quadratic* air resistance is modeled by the differential equation

$$\frac{dv}{dt} = -g - kv|v| = \begin{cases} -g - kv^2, & \text{for } v > 0 \\ -g + kv^2, & \text{for } v \leq 0 \end{cases}. \tag{10.5}$$

Define the parameter $v_c = \sqrt{g/k}$, which corresponds to a critical velocity.
    (a) Find the general solution of the differential equation (10.5) for the case in which $v > 0$, in terms of an arbitrary initial velocity $v(0) = v_0 > 0$.

(b) Find the general solution of the differential equation (10.5) for the case in which $v < 0$, in terms of an arbitrary initial velocity $v(0) = v_0 < 0$.
(c) Find the general solution of the differential equation (10.5) for the case in which the velocity is zero at time t_0: $v(t_0) = 0$.
(d) Consider a particle launched straight up from the ground. How long does it take (t_0) before the particle reaches the top of its trajectory and starts falling back to Earth?
(e) Consider again a particle launched straight up from the ground. Determine its height as a function of time.
(f) Let $k = 0.01\text{m}^{-1}$ and $g = 9.8\text{ms}^{-2}$. If a projectile is launched straight up with initial speed 200m/s, determine the total time before it lands.

13. A tank of water has a spout at the bottom that drains water from the tank at a rate (volume per time) proportional to the square root of the water depth. Take $k > 0$ to be the constant of proportionality. (Explicitly write down the minus sign in your equation if needed.) Assume the water level is initially equal to H.
    (a) If the tank has the shape of a right circular cylinder, with cross-sectional area A, set up and solve a differential equation for the water level $x(t)$ as a function of time t, and use the solution to determine the length of time before the tank is emptied.
    (b) Repeat part (a) if the shape of the tank is a right circular cone, and A is the cross sectional area at the height of the initial water level H.
14. A large tank initially contains 200L of pure water. Brine with a salt concentration of 3g/L is pumped into the tank at a constant rate of 20L/min. The solution in the tank is well mixed at all times, and is drained from the tank at a rate of 20L/min. Determine the mass of salt in the tank as a function of time.
15. **[Conservation of Energy–1d Case]** Consider a particle moving in a straight line, subject to a conservative force field

$$\mathbf{F} = -\frac{dU}{dt},$$

for a potential function U. Newton's second law (for constant mass) states that

$$m\frac{dv}{dt} = -\frac{dU}{dx}.$$

Use the chain rule, in the form

$$\frac{dv}{dt} = \frac{dv}{dx}\frac{dx}{dt} = v\frac{dv}{dx},$$

to prove the law of conservation of energy, i.e., that

$$E = \frac{1}{2}v^2 + U(x)$$

is constant along particle trajectories.

16. **[Conservation of Energy]** Consider a particle moving in a conservative force field

$$\mathbf{F} = -\nabla U,$$

for a potential function U. Newton's second law (for constant mass) states that

$$m\frac{d\mathbf{v}}{dt} = -\nabla U.$$

Use the chain rule, in the form

$$\frac{d\mathbf{v}}{dt} = \mathbf{v} \cdot \nabla \mathbf{v},$$

to prove the law of conservation of energy, i.e., that

$$E = \frac{1}{2}\mathbf{v} \cdot \mathbf{v} + U(\mathbf{x})$$

is constant along particle trajectories.

10.2 Linear Equations and Integrating Factors

In this section, we introduce *linear* differential equations, and discuss a technique that can be used to solve general linear first order differential equations, whether they are separable or not.

10.2.1 Linear Equations and Homogeneous Equations

We next introduce several more important pieces of nomenclature.

Definition 10.2.1 A *linear* differential equation is an equation that is linear in y and its derivatives; i.e., the general nth order linear ordinary differential equation is an equation of the form

$$a_n(t)y^{(n)} + \cdots + a_1(t)y' + a_0(t)y = f(t), \tag{10.6}$$

where $a_0, \ldots, a_n$, and f are given functions (possibly nonlinear) of the independent variable t.

A linear differential equation is *homogeneous* if each term contains a single factor of y or one of y's derivatives, i.e., if $f(t)$ in equation (10.6) is identically zero.

A linear differential equation that is not homogeneous is said to be *nonhomogeneous*. The term $f(t)$ in a nonhomogeneous differential equation (10.6) is called the *nonhomogeneous term.*

Thus, in order for a differential equation to be *linear* each term either be a function of t alone, or contain exactly one single factor of y or one of y's derivatives. The term that is only a function of the independent variable t is called the *nonhomogeneous* term.

We shall see later in this text that homogeneous differential equations have special properties, such as the principle of superposition, that make them quite useful. (In fact, to solve a nonhomogeneous differential equation, the first step, as we shall shortly see, is to solve a related homogeneous equation.)

Example 10.8 The following are examples of linear differential equations:

$$\begin{aligned}
y''' + t^2y'' + e^{\sin(t^2)}y' + \tan(t^3)y &= 0, \\
y'' + \sqrt{t}\sin(t)y' + e^{-t^2}y &= 0, \\
y''' + \ln(t)y'' + t^{3/2}y' + ty &= \cosh(t^3 + t^2), \\
y'' + 3y' + ty &= \sqrt{t}.
\end{aligned}$$

The first two of the preceding equations are homogeneous, whereas the last two are nonhomogeneous.

The following are examples of nonlinear differential equations:

$$
\begin{aligned}
y'' + yy' &= 0, \\
y'' + 3y' + y^2 &= 0, \\
y''' + t^2yy' + y &= 3t, \\
y'' + ty' + \sqrt{y} &= 7.
\end{aligned}
$$

Note that the term *homogeneous* is a term used to describe *linear* differential equations. So it does not make sense to classify a nonlinear differential equation as being homogeneous or nonhomogeneous. It's just nonlinear. ■

10.2.2 The Integrating Factor Technique

In this paragraph, we consider linear first-order differential equations of the form

$$y' + p(t)y = g(t). \tag{10.7}$$

We next introduce a technique that can be used to solve general first-order differential equations of this form. The idea is to introduce an *integrating factor*: a function of the independent variable t that, when you multiply it by the left-hand side of equation (10.7), you obtain an exact differential. In other words, we wish to determine a function $\mu(t)$ such that

$$\frac{d}{dt}\left[\mu y\right] = \mu y' + \mu p(t)y. \tag{10.8}$$

Differentiating the left-hand side (using the product rule), we obtain

$$\frac{d}{dt}\left[\mu y\right] = \mu y' + \mu' y.$$

Comparing this with the right-hand side of equation (10.8), we require that

$$\mu' = \mu p(t).$$

Applying separation of variables, we can integrate this to show that $\mu(t) = Ae^{P(t)}$, where $P(t)$ is an antiderivative of $p(t)$, i.e., $P'(t) = p(t)$. This motivates the following definition.

Definition 10.2.2 The *integrating factor* of the first-order linear equation (10.7) is the function

$$\mu = e^{P(t)}, \tag{10.9}$$

where, $P(t) = \int p(t)dt$ is *any one* antiderivative for $p(t)$, i.e., $P'(t) = p(t)$.

(R) In order for the equation (10.9) to work, the function $p(t)$ must be the y-coefficient, such that the y'-coefficient equals 1. If the coefficient of y' is not 1, then one must first divide the differential equation (10.7) by the y'-coefficient, in order to obtain the standard form given by equation (10.7).

The solution to the differential equation (10.7) then proceeds as follows: First, multiply equation (10.7) by the integrating factor (10.9), thereby obtaining

$$e^{P(t)}y' + e^{P(t)}p(t)y = e^{P(t)}g(t).$$

Next, observe that the left-hand side is a perfect derivative, so that the left-hand side can be replaced as follows:

$$\frac{d}{dt}\left[e^{P(t)}y\right] = e^{P(t)}g(t).$$

Integrating the left-hand and right-hand sides of the preceding equation with respect to t then yields

$$e^{P(t)}y = \int e^{P(t)}g(t)dt + C.$$

The constant of integration $+C$ is written out explicitly here for convenience. (Even though, technically, it is a part of the indefinite integral.) Solving for y, we thus obtain the following solution to the differential equation (10.7):

$$y = e^{-P(t)}\int e^{P(t)}g(t)dt + Ce^{-P(t)}.$$

Do not memorize the formula, but the process by which the solution is obtained.

■ **Example 10.9** Consider the differential equation

$$y' + 2ty = 3t.$$

The integrating factor is given by

$$\mu = e^{\int 2t\,dt} = e^{t^2}.$$

Multiplying the differential equation by the integrating factor, we obtain

$$e^{t^2}y' + 2te^{t^2}y = 3te^{t^2}.$$

The left-hand side can be rewritten as a perfect derivative, yielding

$$\frac{d}{dt}\left[e^{t^2}y\right] = 3te^{t^2}.$$

Integrating, we obtain

$$e^{t^2}y = \frac{3e^{t^2}}{2} + C.$$

Dividing both sides by the integrating factor, we obtain the general solution

$$y(t) = \frac{3}{2} + Ce^{-t^2}.$$

■

Exercises

For Exercises 1–5, first find the general solution to the given differential equation. Then use the initial condition to solve for C.

1. $y' + 3y = 5, \qquad y(0) = 7.$
2. $ty' + y = 2t, \qquad y(1) = 3.$
3. $y' + \cos(t)y = 2\cos(t), \qquad y(0) = 3.$
4. $\cos^2(t)y' + 3y = 9, \qquad y(0) = 10.$
5. $y' - \tan(t)y = 4, \qquad y(0) = 7.$

Problems

6. A ball is shot vertically upward from a canon with initial velocity v_0. Attached to the ball is a small rocket that provides a sinusoidal forcing. During its flight, the ball is subjected to linear air resistance. The velocity v of the ball is modeled by the differential equation

$$\frac{dv}{dt} = -g - \alpha v + a\cos(\omega t),$$

where g is the acceleration due to gravity, a is the maximum thrust, and ω is the thrusting frequency.
   (a) Solve for the ball's velocity as a function of time.
   (b) Integrate your result to find the ball's height as a function of time.

7. A large tank with a capacity of 500L initially contains 200L of pure water. Brine with a salt concentration of 3g/L is pumped into the tank at a constant rate of 20L/min. The solution in the tank is well mixed at all times, and is drained from the tank at a rate of 10L/min.
   (a) Determine the mass of salt in the tank as a function of time.
   (b) How much salt is in the tank a the moment it overflows?

8. A rocket is launched vertically upward with initial velocity v_0. Burning fuel at a constant rate of r, its mass is given by

$$m(t) = M_0 - rt.$$

Newton's second law states that

$$\frac{d}{dt}[mv] = -mg,$$

where g is the acceleration due to gravity.
   (a) Determine the rocket's velocity as a function of time.
   (b) Determine the height of the rocket as a function of time.

9. Two interconnected tanks are setup as shown in Figure 10.3. Each initially contains 1,000L of pure water. Brine with a salt concentration of 4g/L flows into tank 1 at a constant rate of 50L/min. Brine flows out of tank 1 and into tank 2 at a constant rate of 50L/min, and brine flows out of tank 2 at a constant rate of 50L/min. Let $Q_1(t)$ and $Q_2(t)$ represent the total mass (g) of salt in tanks 1 and 2, respectively, after t minutes. Assume each tank is well mixed.
   (a) Set up and solve a differential equation for $Q_1(t)$.
   (b) Use your result from part (a) to set up and solve a differential equation for $Q_2(t)$.

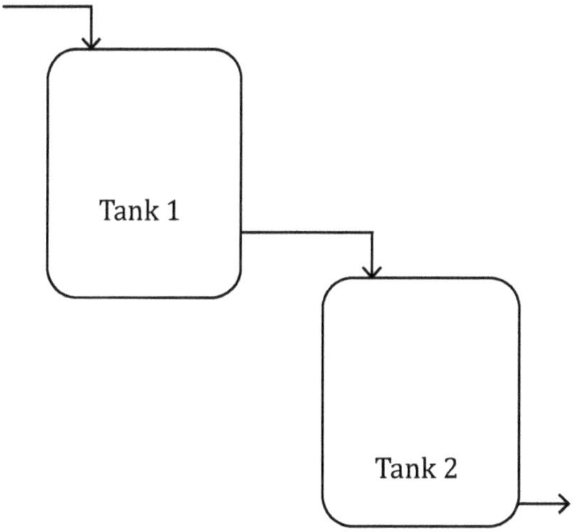

Figure 10.3: Tank-mixing problem, as given in Problem 9

10. There is a freezer in a house in the arctic. Initially, the temperature of the arctic is 0°F, the temperature of the house is 80°F, and the temperature of the freezer is 0°F. At time $t = 0$, the power goes out. Let $h(t)$ represent the average temperature inside the house, t hours after the power goes out, and let $u(t)$ represent the temperature inside the freezer, t hours after the power goes out. Assume that the house temperature $h(t)$ satisfies Newton's law of heating and cooling, with proportionality constant $k > 0$, and with the arctic temperature used as the ambient temperature. Further assume that the freezer satisfies Newton's law of heating and cooling, with proportionality constant $r > 0$, and with the house temperature used as the ambient temperature.
    (a) Set up and solve an initial value problem for $h(t)$.
    (b) Set up and solve an initial value problem for $u(t)$.
    (c) Explain why $u(t)$ can never be negative, in terms of your solution.

Consider two cases: (i) $r < k$ and (ii) $r > k$.

(d) Determine the time at which the temperature inside the freezer is maximum.

10.3 Autonomous Equations and Population Dynamics

In this section, we discuss *autonomous* differential equations, as well as our first *qualitative* method for understanding solutions to differential equations. As an application, we then discuss *logistic growth*, a model for population growth.

10.3.1 Autonomous Equations and Equilibrium Solutions

Definition 10.3.1 An *autonomous* differential equation is a first order differential equation that does not explicitly depend on the independent variable, i.e., it is a first order differential equation of the form

$$\frac{dy}{dt} = f(y). \tag{10.10}$$

Autonomous equations conveniently loan themselves to a qualitative technique: phase-line analysis. *Qualitative* techniques allow one to gain an understanding of the solutions to a differential equation, without actually solving it. Many times, even if one is obtained, an exact analytic solution will bear no additional insight into the problem.

To implement a *phase-line analysis* for an autonomous differential equation, proceed as follows:

1. Draw a horizontal axis and label it with the name of the *dependent* variable (e.g., y). This is called the *phase line.*
2. Draw a sketch of the function $f(y)$ that determines the differential equation (10.10). Pay particular attention to the domain on which $f(y)$ is positive and on which it is negative.
3. Draw arrows along the y-axis in the direction of the solution flow: on the domain for which $f(y) > 0$, the arrows should point towards the right (i.e., towards increasing y-values); on the domain for which $f(y) < 0$, the arrows should point towards the left (i.e., towards decreasing y-values).

Once the phase line is drawn, one can draw a slope field on the ty-plane. Additionally, the zeros of the function f correspond to *steady state* or *equilibrium* solutions.

Definition 10.3.2 An *equilibrium solution* to a differential equation is any constant solution, i.e., any solution of the form

$$y(t) = y_e,$$

for some constant value y_e (known as the *equilibrium value*).

As we saw previously, a differential equation can be presented by itself or as part of an initial-value problem. Alternatively, solutions can be described as the *flow* of the differential equation.

Definition 10.3.3 Given a general first-order differential equation

$$\frac{dy}{dt} = f(t,y), \tag{10.11}$$

the *flow* of the differential equation (10.11) is the function $\varphi : \mathbb{R}\times\mathbb{R} \to \mathbb{R}$ defined by the conditions

$$\begin{aligned} \frac{d\varphi(t;y_0)}{dt} &= f(\varphi(t,y_0),t), && (10.12) \\ \varphi(0;y_0) &= y_0. && (10.13) \end{aligned}$$

In other words, the flow $\varphi(t;y_0)$ of the differential equation (10.11) can be thought of as the solution to the initial value problem consisting of the differential equation (10.11) and the initial condition (10.13). The concept of the solution flow is a technical one, and we shall not rely too heavily on it throughout these pages. It will, however, be used in a moment when we define stability.

The equilibrium solutions of the autonomous equation (10.10) are further classified in terms of their stability.

Definition 10.3.4 An equilibrium solution $y(t) = y_e$ is *stable* if there exists a positive number ε, such that the condition

$$|y_0 - y_e| < \varepsilon$$

implies that

$$|\varphi(t;y_0) - y_e| \leq |y_0 - y_e|,$$

for all $t \in [0,\infty)$. If no such $\varepsilon > 0$ exists, the equilibrium solution $y(t) = y_e$ is said to be *unstable*.

In other words, a *stable* equilibrium solution is one with the property that if you start off *close enough* to it, you will remain nearby for all future time.

■ **Example 10.10** Consider the differential equation

$$y' = \sin y. \tag{10.14}$$

In Exercise 5 of § 10.1, you showed that the general solution to this differential equation is given by

$$y(t) = 2\tan^{-1}(Ae^t).$$

This exact analytic solution, however, likely gives you no real insight into how the solution curves to the differential equation (10.14) actually look, or how they depend on the initial condition—i.e., it likely gives you no insight into what the *flow* of the differential equation (10.14) looks like. In this example, we will apply phase-line analysis to better understand the solution flow of the differential equation (10.14).

To proceed, we begin by plotting the function $f(y)$, as shown in Figure 10.4. We

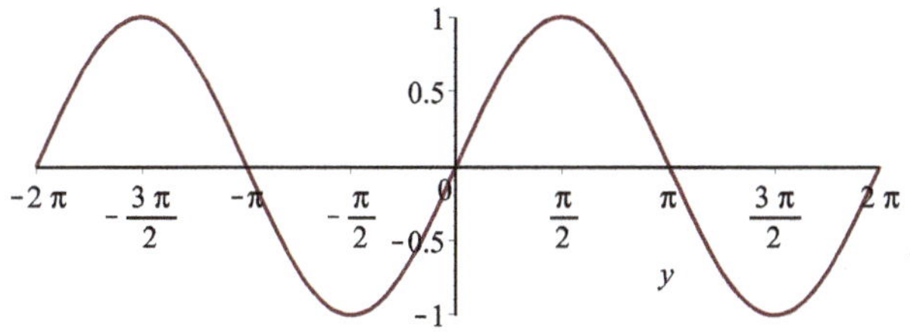

Figure 10.4: Graph of the function $f(y) = \sin y$, as given in Example 10.10

observe the following. First, the equilibrium solutions, on the interval $y \in [-2\pi, 2\pi]$, correspond to the values $y = 0, \pm\pi, \pm 2\pi$. Second, the solution to the differential equation (10.14) is increasing whenever $y(t) \in (-2\pi, -\pi)$ or $y(t) \in (0, \pi)$, and is decreasing whenever $y(t) \in (-\pi, 0)$ or $y(t) \in (\pi, 2\pi)$.

We can use this information to plot a direction field on the phase line, as shown in Figure 10.5. Right-pointing arrows are added on the intervals for which the function y is increasing; left-pointing arrows are added where y is decreasing. Additionally, the equilibrium points are marked with open or closed circles. Open

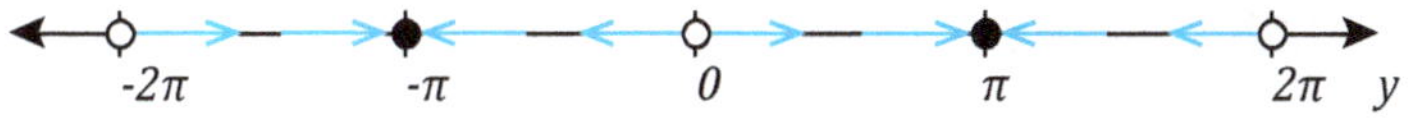

Figure 10.5: Phase line for the differential equation given in Example 10.10, with direction field.

circles represent unstable equilibria, whereas closed (filled) circles represent stable equilibria.

Solutions of the differential equation (10.14) are plotted for various initial conditions in Figure 10.6. Observe that whenever the initial conditions are within the interval $y_0 \in (-2\pi, -\pi)$, the resulting solution increases for all time, and approaches the stable equilibrium $y_e = -\pi$ asymptotically, as $t \to \infty$. Similarly, solutions with initial conditions in the interval $(-\pi, 0)$ remain in that interval, decrease for all time, and approach the stable equilibrium solution $y_e = -\pi$ asymptotically, as $t \to \infty$. Similar statements can be made for initial conditions in the intervals $(0, \pi)$ and $(\pi, 2\pi)$.

Also observe the following fundamental result of differential equations: solution paths never cross. How could they? Each point in the (t, y) plane has a specified slope. If two solution curves crossed at any given point, the slope assigned to that particular point (t, y) would have to take two values, which is a contradiction. ∎

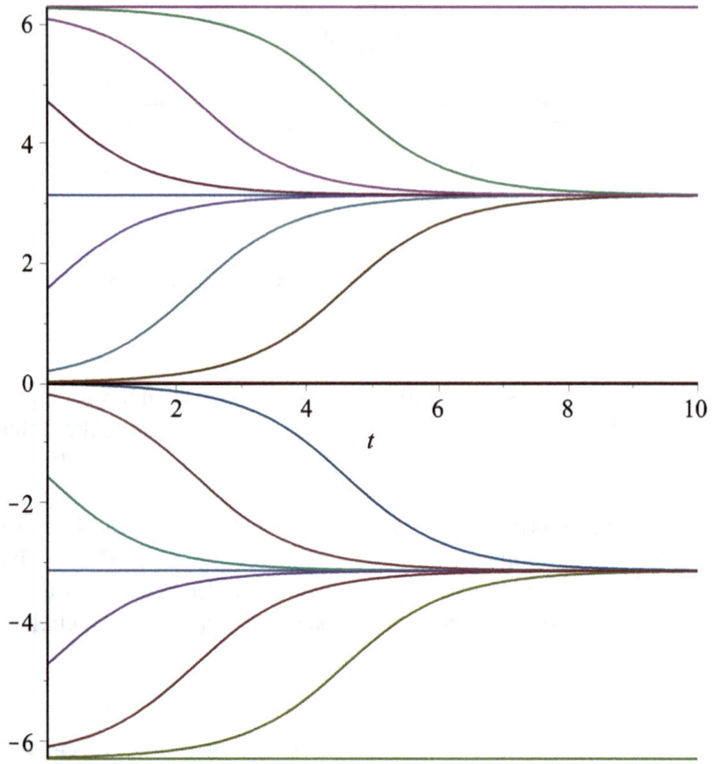

Figure 10.6: Various solutions for the differential equation (10.14), as given in Example 10.10.

10.3.2 Logistic Growth

Exponential growth has its limitations. As we saw in Exercise 7 from §10.1, if a bacterial population was allowed to double every fifteen minutes, then a single bacterium would be able to populate a colony with a mass equal to the mass of planet Earth in a little longer than a single day. The exponential growth model, therefore, breaks down at a certain point: exponential growth cannot be continued in perpetuity.

To resolve these difficulties, mathematicians came up with a more accurate model for things that grow exponentially—things like populations. This model is called the *logistic model*, and its solutions are described as possessing *logistic growth*. The logistic model is the autonomous differential equation

$$y' = ky\left(1 - \frac{y}{K}\right). \tag{10.15}$$

Equation 10.15 is called the *logistic equation.*

If the population size $y(t)$ is small compared to the parameter K, i.e., if $0 < y << K$, then the logistic equation can be approximated as

$$y' \approx ky.$$

Thus, for small population sizes, the growth will appear exponential. When y becomes appreciable to K, however, the nonlinear terms become relevant, and the behavior changes. The parameter K is known as the *carrying capacity* of the population, and it is a theoretical maximum population size that can be supported by the given environment.

Logistic Growth: Analytic Solution

To solve the logistic equation, we separate variables and integrate. Separating variables yields

$$\int \frac{dy}{y(1-y/K)} = \int k dt.$$

Integrating, we obtain

$$\ln\left|\frac{y}{1-y/K}\right| = kt + \tilde{C},$$

which may be rewritten as

$$y = (1-y/K)Ce^{kt},$$

or

$$\left(1+\frac{Ce^{kt}}{K}\right)y = Ce^{kt}.$$

Thus

$$y(t) = \frac{Ce^{kt}}{1+Ce^{kt}/K}.$$

This is equivalent to

$$y(t) = \frac{K}{Ae^{-kt}+1}, \tag{10.16}$$

where A is obtained from the initial condition, yielding

$$A = \frac{K}{y_0} - 1. \tag{10.17}$$

Equations (10.16) and (10.17) represent the full solution to the logistic equation (10.15).

Notice that

$$\lim_{t\to\infty} y(t) = K,$$

for all initial conditions. The constant K is known as the *carrying capacity*. Physically, it represents the population size that can be supported by the given environment.

Logistic Growth: Phase-Line Analysis

In this paragraph, we analyze the logistic equation (10.15) qualitatively using a phase-line analysis. The function $f(y)$ on the right-hand side of equation (10.15) is an up-side-down parabola with y-intercepts $y = 0$ and $y = K$. It follows that $f(y)$ is positive on the interval $(0, K)$, and negative on the intervals $(-\infty, 0)$ and (K, ∞). Solution curves are therefore increasing for $0 < y < K$, and decreasing otherwise. The equilibrium solutions are $y(t) = 0$ and $y(t) = K$. The phase line is depicted in Figure 10.7.

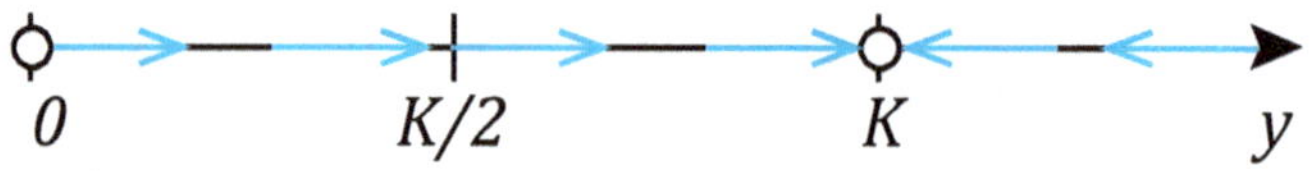

Figure 10.7: Phase line for logistic equation (10.15).

Also notice that $y' = f(y)$ is greatest when $y = K/2$. This represents an inflection point in the solution curves, since, when $y(t) < K/2$, the solution curve y will be increasing at a greater and greater rate, and when $K/2 < y(t) < K$, the solution curve y will be increasing at a smaller and smaller rate.

Various solutions to the logistic equation are plotted in Figure 10.8, for the case $K = 10$, $k = 0.1$. Notice that each solution possesses a change of concavity when it passes through the line $y = K/2$. The equilibrium solutions $y = 0$ and $y = K$ are also shown. For initial conditions above the carrying capacity, i.e., for $y_0 > K$, the solution decays exponentially to the carrying capacity. This case corresponds to an overpopulated region, with a population incapable of being supported by the resources. The population shrinks down until it levels off at the carrying capacity.

Exercises

For Exercises 1–5, draw the phase line for the given differential equation and then determine the equilibrium solutions and classify their stability.

1. $y' = y^2 - 9$.
2. $y' = 9 - y^2$.
3. $y' = -y \ln |y|$.
4. $y' = y^3 - y$.
5. $y' = 0.1\left(1 - \frac{y}{10}\right)\left(1 - \frac{y}{100}\right)\left(1 - \frac{y}{1000}\right)$.

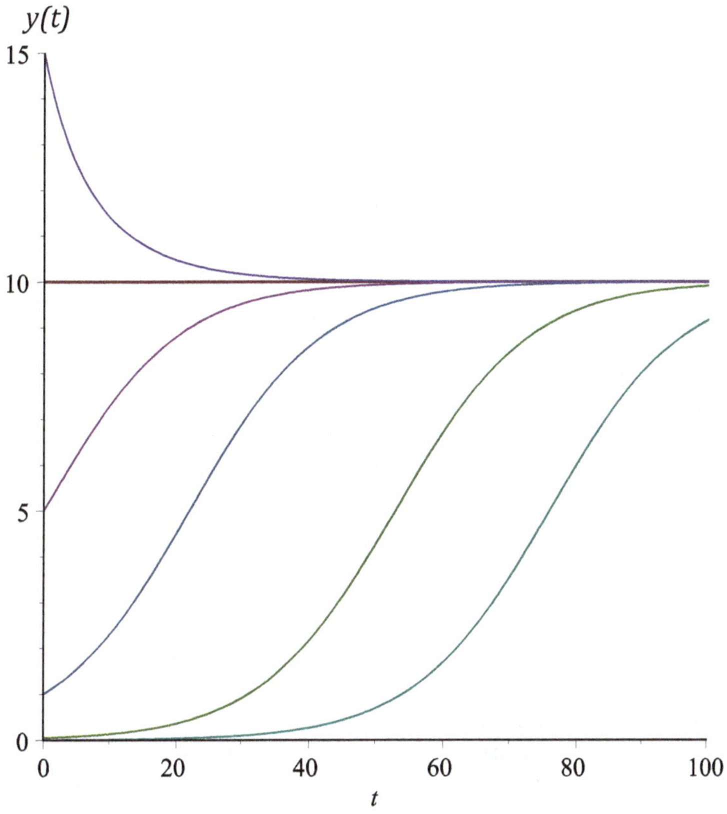

Figure 10.8: Various solutions for the for logistic equation (10.15).

Problems

6. Draw the phase line for Newton's law of heating and cooling, described in Problem 10 from §10.1. Identify the equilibrium solution(s). Sketch a graph of several solutions.
7. A rumor spreads at a rate proportional to the product of the fraction of the population who have heard the rumor and the fraction of the population who have not heard the rumor. A certain rumor starts with ten people at 8:00AM in a small town of 1,000 inhabitants. By noon, half the town has heard the rumor. By what time will 90% of the townspeople have heard the rumor?
8. Differentiate the logistic equation (10.15) implicitly with respect to time to obtain a relation for y'' in terms of y and y'. Use the result to prove that if $y(t)$ is a solution to the logistic equation, it has an inflection point at $t = t_0$ if and only if $y(t_0) = K/2$.
9. **[A population with a threshold]** The following differential equation models

a population with a carrying capacity and a *threshold population*:

$$\frac{dy}{dt} = ky\left(1 - \frac{T}{y}\right)\left(1 - \frac{y}{K}\right), \tag{10.18}$$

where $k > 0$ is a positive parameter, and where the parameters T and K satisfy $0 < T < K$.

(a) Draw a phase line for the differential equation (10.18). Identify all equilibrium points and state their stability.
(b) How do the solution curves depend on the initial conditions? How is the limiting behavior of the solution flow affected by the initial conditions?
(c) Explain the meaning of the parameter T in physical terms.
(d) Solve the differential equation (10.18).
(e) What does the analytical solution predict if $0 < y_0 < T$? How does this relate to your understanding of the phase line? Explain any discrepancy.
(f) Consider a population with $K = 100$, $T = 10$, and $k = 0.2$. If $y_0 = 9$, exactly how long does it take for the population to go extinct?

10. **[A semi-stable critical point]** Draw the phase line for the differential equation

$$\frac{dy}{dt} = y^2.$$

The critical point $y_c = 0$ is called *semi-stable*, as it is stable from one direction, but not the other. It is drawn with a circle that is half filled on the stable side.

11. **[Bifurcation diagram]** A *bifurcation diagram* is a plot that can be used to understand how the qualitative behavior of an autonomous differential equation depends on a parameter. Consider the differential equation

$$\frac{dy}{dt} = y(a - y),$$

for some parameter $a \in \mathbb{R}$.

(a) Draw the phase line, identifying each critical point and its stability, for three cases: (i) $a < 0$, (ii) $a = 0$, and (iii) $a > 0$.
(b) Draw the *bifurcation diagram* for the system: The horizontal axis will describe the parameter a, and the vertical axis will describe the critical points y. Draw stable critical points with a solid line, and unstable critical points with a dashed line. *Solution*: See Figure 10.9.

12. **[A pitchfork bifurcation]** Consider the differential equation

$$\frac{dy}{dt} = rx - x^3,$$

which depends on a parameter $r \in \mathbb{R}$.

(a) Draw the phase line, identifying each critical point and its stability, for three cases: (i) $r < 0$, (ii) $r = 0$, and (iii) $r > 0$.
(b) Draw the bifurcation diagram for the system. (See Problem 11.) The resultant graph should look like a pitchfork.

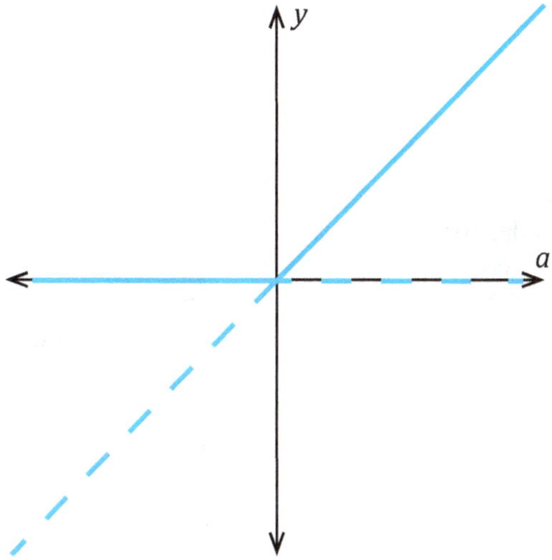

Figure 10.9: The bifurcation diagram for Problem 11.

(c) A *bifurcation* is a sudden, qualitative change in the behavior of a system as a parameter passes through a certain value known as the *bifurcation point*. What is the bifurcation point of this system? How does the behavior of the system change as the parameter value passes through the bifurcation point?

13. Consider the following chemical reaction:

$$CO + 3H_2 \leftrightarrows CH_4 + H_2O$$

Suppose that initially there are exactly 4 moles of CO present and 4 moles of H_2 present, and furthermore that these are the only species initially present. The *law of mass action* states that a reaction occurs at a rate proportional to the product of the number of moles of each reactant. Assume the constant of proportionality is 1 for the forward reaction and 3 for the backward reaction. Let $x(t)$ be the number of moles of CH_4 molecules present at time t.

(a) In terms of x, determine how many moles of CO, H_2, and H_2O are present at time t.

(b) Write an initial value problem for $x(t)$.

(c) Solve the initial value problem to determine $x(t)$.

(d) As $t \to \infty$, the reaction reaches its equilibrium mixture. How many moles of CO are present in its equilibrium mixture?

14. The *Gompertz* equation

$$\frac{dy}{dt} = -ky \ln\left(\frac{y}{K}\right) \tag{10.19}$$

is a modified version of the logistic equation that has been successfully used to model tumor growth.

(a) Determine the critical points of the differential equation (10.19). Draw the phase line. How is the phase line similar or different to the phase line for logistic growth?

(b) Solve the differential equation (10.19). Express the integration constant in terms of the initial condition $y(0) = y_0$.

15. The SIR-model for epidemics consists of a nonlinear first-order *system* with interrelated dependent variables $S(t)$, $I(t)$, and $R(t)$, which describe the number of susceptible, infected, and recovered people in a population, respectively. The differential equations modeling the dynamics of disease spread through the population are given by

$$\frac{dS}{dt} = -\alpha S I \tag{10.20}$$

$$\frac{dI}{dt} = \alpha S I - \beta I \tag{10.21}$$

$$\frac{dR}{dt} = \beta I, \tag{10.22}$$

where α and β are positive constants determined by the particular outbreak or disease and characteristics of the population. The model is based on two underlying assumptions:

- The rate at which susceptible people (healthy individuals who are not immune) become infected is proportional to the product of the susceptible and infected populations. (If you double the size of either I or S, one would expect that this doubles the number of encounters between susceptible and infected individual, and therefore doubles the rate that the disease is spreading.
- The rate at which infected people recover is proportional to the number of infected people. (I.e., if there are no susceptible individuals, the number of infected individuals will decay exponentially.)

(a) Define $N = S + I + R$. Explain what this variable represents physically. Use the differential equations (10.20)–(10.22) to show that $dN/dt = 0$, which implies that $N(t)$ is a constant function of time.

(b) The *peak outbreak* is defined as the moment in time at which the number of infected individuals is the greatest. Show that the peak outbreak occurs when $S = \beta/\alpha$. Use the second derivative test to prove that $I(t)$ is actually a maximum (and not a minimum) when this condition is satisfied. (*Hint*: Differentiate equation (10.21) implicitly with respect to time.)

(c) The number of infected people can be also considered as a function of the number of susceptible. Determine a formula for dI/dS in terms of I and S.

(d) Solve the differential equation you determined in part (c) to obtain $I(t)$ as a function of $S(t)$ and the initial conditions S_0 and I_0.

(e) An epidemic begins in a creation population with 1% of the population infected and 99% of the population susceptible. One week later, 5% of the population is infected and 92% of the population is susceptible. What fraction of the total population will be infected at peak outbreak?

10.4 Numerical Solutions

In this section, we discuss a basic numerical algorithm known as Euler's method, numerical error, and conclude with an example MATLAB program. We also briefly discuss a common fourth-order method known as the Runge–Kutta method and compare the accuracy of the two methods.

10.4.1 Euler's Method

Euler's method is a numerical method for approximating solutions to first order differential equations of the form given in equation (10.11). It is based on the *tangent-line approximation* to a function $y = f(x)$ at x_0:

$$T(x) = f(x_0) + f'(x_0)(x - x_0).$$

(This is equivalent to the first order Taylor polynomial of f centered about $x = x_0$.)

Given an initial value problem of the form

$$\frac{dy}{dt} = f(t,y) \qquad (10.23)$$

$$y(t_0) = y_0, \qquad (10.24)$$

for some specified initial condition (t_0, y_0), the *Euler's method with step size h* is the approximate solution obtained by the following algorithm:

1. Set y_0, t_0, and h (given).
2. For $i = 0, \ldots, (n-1)$:
   (a) Set $t_{i+1} = t_i + h$.
   (b) Set $y_{i+1} = y_i + f(t_i, y_i)h$.
   - End
3. Closing commands; e.g., plot the y_i's versus t_i's, compute the error, etc.

The values y_i, for $i = 1, \ldots, n$, are the approximate solutions to the differential equation (10.23) with initial condition (10.24), i.e.,

$$y_i \approx y(t_i),$$

if $y(t)$ is the solution to the initial value problem given by equations (10.23) and (10.24).

The "for–loop" is cycled through for various values of i, starting with $i = 0$[1], and stopping when $i = (n-1)$. The code between the start of the for–loop and the "End" is repeated for various values of i, as i runs from 0 to $(n-1)$ in increments of 1.

For example, consider $n = 4$. A computer program (or human) would follow the Euler's method algorithm by performing: step 1; set $i = 0$, step 2a, step 2b; set $i = 1$, step 2a, step 2b; set $i = 2$, step 2a, step 2b; set $i = 3$, step 2a, step 2b; step 3. Notice that the for–loop commands are repeated, once for each value of i

---

[1] though, in MATLAB, the initial conditions are index by 1 instead of 0

starting with $i = 0$, until i reaches its terminal value. In this way, the for–loop tells the computer how to compute t_{i+1} and y_{i+1} if it has already computed t_i and y_i.

Thus, Euler's method yields an approximate solution to a first-order initial value problem at a discrete sequence of points by using tangent-line approximations. The method is illustrated diagrammatically in Figure 10.10. The slope at each point is obtained using the differential equation (10.23).

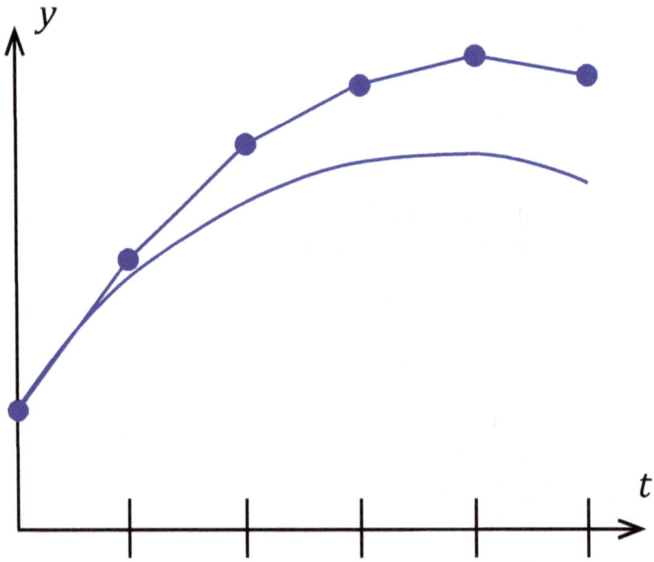

Figure 10.10: A sequence of points (t_i, y_i) generated using Euler's method, plotted along with the exact solution.

▪ **Example 10.11** Consider the initial value problem

$$y' = \frac{t+1}{y}, \qquad y(0) = 2.$$

Let's approximate $y(1)$ with a step size of $h = 0.2$.

First, we identify the step size $h = 0.2$ and the initial condition $t_0 = 0$, $y_0 = 2$. Also, the function $f(t, y)$ is given by the right-hand side of the differential equation, i.e., $f(t, y) = \frac{t+1}{y}$. We then execute the equations 2a and 2b of the Euler method, using $i = 0$. For $i = 0$, those two lines, specifically, state that

$$\begin{aligned} t_1 &= t_0 + h, \\ y_1 &= y_0 + hf(t_0, y_0). \end{aligned}$$

Incrementing the t-variable, we therefore obtain $t_1 = 0+0.2 = 0.2$. Incrementing the y-variable, we obtain

$$y_1 = 2+0.2f(0,2) = 2+0.2\left(\frac{1}{2}\right) = 2.1.$$

We now have our approximate solution at the next step, i.e., $t_1 = 0.2$ and $y_1 = 2.1$. Because we have reached the "End" of the for–loop, we cycle back to the beginning, incrementing i by 1.

Next, we use our values for t_1 and y_1 to compute the values for t_2 and y_2. Equations 2a and 2b, for the case $i = 1$, now read

$$\begin{aligned} t_2 &= t_1 + h, \\ y_2 &= y_1 + hf(t_1, y_1). \end{aligned}$$

Thus, the new value for the t-variable is $t_2 = 0.4$, and the new value for the y-variable is

$$y_2 = 2.1+0.2\left(\frac{1.2}{2.1}\right) \approx 2.21428571.$$

We repeat in a similar fashion: next we take $i = 2$. We compute $t_3 = t_2 + h = 0.6$, and

$$y_3 = y_2 + hf(t_2, y_2) \approx 2.21428571 + 0.2\left(\frac{1.4}{2.21428571}\right) \approx 2.34073732.$$

Continuing on, we have

$$\begin{aligned} t_4 &= 0.8, \\ y_4 &= y_3 + hf(t_3, y_3) \approx 2.34073732 + 0.2\left(\frac{1.6}{2.34073732}\right) \approx 2.47744638, \\ t_5 &= 1.0, \\ y_5 &= y_4 + hf(t_4, y_4) \approx 2.47744638 + 0.2\left(\frac{1.8}{2.47744638}\right) \approx 2.62275729. \end{aligned}$$

Thus, our approximation for $y(1)$ is given by $y(1) \approx y_5 = 2.62275729$. ■

Once you get the rhythm down, the Euler's method may be easily implemented with the aid of a table. Once can setup a table as shown in Table 10.1, i.e., for $i = 0$, we start by writing down $t_0 = 0$ and $y_0 = 2$. The table is initialized using the initial

i	t_i	y_i	$f(t_i, y_i)$
0	0	2	

Table 10.1: Table Initialization for Euler's method.

conditions from Example 10.11.

i	t_i	y_i		$f(t_i, y_i)$
0	0	2	$\rightarrow$	0.5

Table 10.2: Computing $f(t_0, y_0)$.

Next, we use t_0 and y_0 to compute $f(t_0, y_0)$, thereby completing the first row of the table, as shown in Table 10.2.

To begin the next row, we add h to the value for t_0, and then we compute $y_1 = y_0 + hf(t_0, y_0)$, as shown in Table 10.3.

i	t_i	y_i	$f(t_i, y_i)$
0	0	2	0.5
	$\downarrow$	$\downarrow$	$\swarrow$
1	0.2	2.1	

Table 10.3: Computing t_1 and y_1.

Continuing further, we use t_1 and y_1 to compute $f(t_1, y_1)$, as shown in Table 10.4.

i	t_i	y_i		$f(t_i, y_i)$
0	0	2		0.5
1	0.2	2.1	$\rightarrow$	0.57142857

Table 10.4: Computing $f(t_1, y_1)$.

We continue in a like manner until we reach our desired endpoint; see Table 10.5. Notice that the "new" t_i value is always the old value plus the step size h. Similarly, the "new" y_i value is always the old y_i value plus h times the old $f(t_i, y_i)$ value. In this way, the table can be filled out left to right, top to bottom. The final value of $f(t_5, y_5)$ does not need to be computed, as it would only be useful in finding y_6, which lies outside the interval on which we are interested.

10.4.2 Numerical Error

During the implementation of a numerical method, two types of error are relevant: local truncation error and global truncation error. *Local truncation error* e_n is the error that is incurred with each step of the numerical method. *Global truncation error* E_n is the cumulative error in performing n steps to approximate the solution at a specific final time $t_f > t_0$.

The local truncation error is given by the formula

$$e_i = y(t_{i+1}) - y_{i+1},$$

i	t_i	y_i	$f(t_i, y_i)$
0	0	2	0.5
1	0.2	2.1	0.57142857
2	0.4	2.21428571	0.63225807
3	0.6	2.34073732	0.68354530
4	0.8	2.47744638	0.72655457
5	1.0	2.62275729	

Table 10.5: Completed table for Example 10.11.

assuming that $y_i = y(t_i)$. The approximate solution y_{i+1} is obtained using Euler's method, thereby yielding

$$y_{i+1} = y_i + hf(t_i, y_i).$$

Note that $f(t_i, y_i)$ is the exact value of $y'(t_i)$.

To approximate the exact solution $y(t_{i+1})$, consider the Taylor series about t_n

$$y(t_i + h) = y(t_i) + hy'(t_i) + \frac{h^2 y''(t_i)}{2} + O(t^3).$$

For the purpose of computing the local truncation error, we may assume that $y(t_i) = y_i$. Comparing the preceding equations, we see that

$$|e_i| \leq \frac{Mh^2}{2},$$

where

$$M = \max_{t \in [t_0, t_f]} |y''(t)|.$$

Thus, the local truncation error is proportional to the square of the step size, and further depends on a factor related to the maximum value of the second derivative of the solution.

Suppose, now, that we approximate $y(t_f)$ using n steps. The step size is therefore given by

$$h = \frac{t_f - t_0}{n}.$$

We previously saw that the local truncation error for each step is proportional to h^2: if one cuts the step size in half, the error should be cut in quarter; if one reduces the step size by 90%, the error should be reduced by 99%, etc. The total number of steps n, however, is inversely proportional to h. The global truncation error is

$$E_n = y(t_n) - y_n,$$

where $t_n = t_f$, and where y_n is computed using n steps, from the starting value $y_0 = y(t_0)$. Since each step results in an error e_i proportional to h^2, and since there are $h^{-1}(t_f - t_0)$ steps, the global truncation error must be proportional to h:

$$|E_n| \leq kh,$$

where $k = M(t_f - t_0)/2$.

The local and global truncation error are each examples of *absolute error*: an absolute measurement of the difference between the exact and the approximate solution.

Another useful concept is that of *relative error*, obtained by dividing the absolute error by the magnitude of the exact solution.

To see how the concept of relative error is useful, suppose the exact value of a quantity we are attempting to approximate is 0.01. Suppose our approximate value is 0.015. The absolute error is only 0.005, which looks pretty good. However, dividing by the exact value reveals that the relative error is 50%, quite a difference!

10.4.3 A MATLAB Example: Euler's Method

Below is code for a MATLAB program, `examp01.m`, which approximates the solution to the initial value problem

$$y' = 0.5y \qquad (10.25)$$

$$y(0) = 1, \qquad (10.26)$$

on the interval $[0,2]$. It then outputs the absolute and relative error for the approximation of $y(2) = e$.

A word of caution: in MATLAB, one must use the index $i = 1$ for the initial conditions, instead of $i = 0$. This is because the index in MATLAB refers to the component of the vector. In other words, the approximate solution $\{(t_i, y_i)\}$ is approximated by computing two vectors, **t** and **y**, one component at a time. The first component of the two vectors, t_1 and y_1, correspond to the initial condition. The for–loop will therefore run between $i = 1$ and $i = n$, and y_{n+1} will be the approximation for the final y-value.

```
% examp01.m

% This program approximates the solution to the
% initial value problem
% y' = 0.5y,
% y(0) = 1,
% on the interval [0, 2], using Euler's method.
% The exact solution is y(t) = exp(0.5t), and y(2)= e.

%clear the memory
clear

n = input('Enter number of steps:  ');

%Define stepsize
h = (2-0)/n;

%initiate approximate solution
yeuler(1) = 1;
t(1) = 0;

%for-loop to implement Euler's method
for(i=1:n)
    %increment time by h
    t(i+1) = t(i) + h;
    %increment solution
    yeuler(i+1) = yeuler(i)
           + h*examp01funct(t(i), yeuler(i));
end

disp('absolute error'),
abs(yeuler(n+1)-exp(1)),
disp('relative error'),
abs(yeuler(n+1)-exp(1))/exp(1),

% defines texact as a 1x100 vector with equally
%spaced values between 0 and 2
texact = linspace(0, 2);
%defines a solution vector for the exact answer.
yexact = exp(texact/2);

plot(texact,yexact,'k','LineWidth',2)
```

```
hold on
plot(t,yeuler,'--','LineWidth,2)
plot(t,yeuler,'o','LineWidth,2)
legend('exact','approximate')
hold off
```

The program examp01.m calls on a *function m-file* named examp01funct.m, the code for which is displayed below.

```
% examp01funct.m

function F = examp01funct(t,y)
F = 0.5*y;
end
```

The program examp01.m was run using 5 steps as the input. (The step size is therefore $h = 0.4$.) The output plots are shown in Figure 10.11.

The program was subsequently run using a variety of different values for the number of steps. The absolute and relative errors for these runs are tabulated in Table 10.6. Notice that the error goes to zero as fast as h: reducing h by a factor of 10 results in a reduction in the error by a factor of 10. This is why Euler's method is called an *first-order method*: the global truncation error (absolute or relative) goes like $O(h)$. The error for a second-order method would go like $O(h^2)$, and so forth.

n	h	absolute error	relative error
10	0.2	0.1245	0.0458
100	0.02	0.0135	0.0050
1,000	0.002	0.0014	4.9954e-4
10,000	0.0002	1.3590e-4	4.9995e-5

Table 10.6: Absolute and relative error using different step sizes for the MATLAB program examp01.m

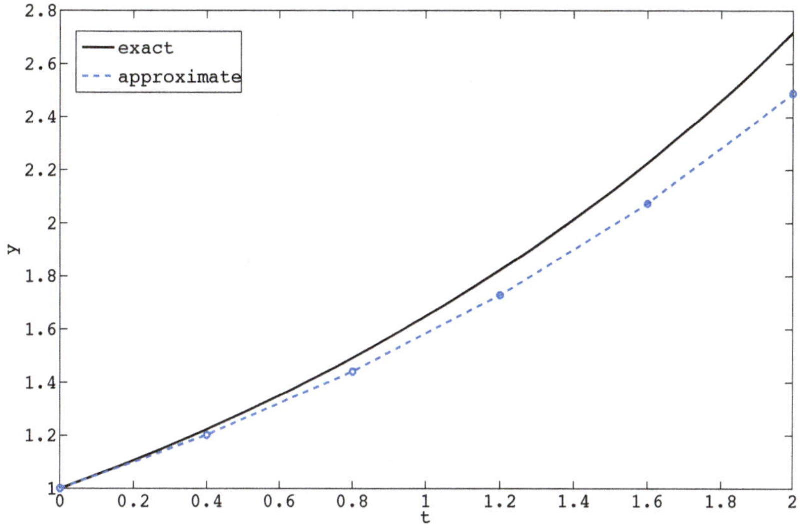

Figure 10.11: MATLAB output of `examp01.m` using 5 steps.

10.4.4 Runge–Kutta Method

In this paragraph, we discuss a higher-order algorithm known as the *Runge–Kutta method*. Recall that Euler's method was a first-order method: numerical error decreases linearly with step size. The Runge–Kutta method, on the other hand, is a fourth-order method, so that the numerical error is bounded by $O(h^4)$, the fourth power of the step size. In other words, if one were to decrease the step size by a factor of ten, the Euler's method would result in an error that is also decreased by a factor of ten, whereas the Runge–Kutta method would result in an error that is decreased by a factor of 10,000.

The Runge–Kutta method is a multi-stage method, meaning that to compute the next successive approximation (y_{i+1} from t_i and y_i), one performs several algebraic computations, each known as a stage, instead of a single one. In particular, to perform a single step of the Runge–Kutta method, one must compute the values of four parameters, k_1, k_2, k_3, and k_4, based on the current approximate values of t_i and y_i, before implementing the formula to compute y_{i+1}. Each of the parameters k_1, k_2, k_3, and k_4 is a slope at a particular point in the (t, y)-plane.

Given an initial value problem

$$\frac{dy}{dt} = f(t, y), \qquad y(t_0) = y_0,$$

the *Runge–Kutta* method with step size h is an approximate solution obtained by the following algorithm:

1. Initialize h, t_0, and y_0.
2. For $i = 0,\ldots,(n-1)$:
   (a) Define the parameters:

$$\begin{aligned} k_1 &= f(t_i, y_i), \\ k_2 &= f(t_i + \frac{h}{2}, y_i + \frac{h}{2}k_1), \\ k_3 &= f(t_i + \frac{h}{2}, y_i + \frac{h}{2}k_2), \\ k_4 &= f(t_i + h, y_i + hk_3). \end{aligned}$$

   (b) Set:

$$\begin{aligned} t_{i+1} &= t_i + h, \\ y_{i+1} &= y_i + \frac{h}{6}(k_1 + 2k_2 + 2k_3 + k_4). \end{aligned}$$

   ◦ End.
3. Closing commands, e.g., plot solution, compute error, etc.

The first observation is that the Runge–Kutta uses a weighted average of the slope at four distinct points on the (t, y) plane.

The parameter k_1 is defined by evaluating the slope function $f(t, y)$ at the previous approximation point (t_i, y_i). The parameter k_2 is obtained by taking a step of size $\Delta t = h/2$, using the slope k_1, and then evaluating the slope function at this new point. The evaluation of k_1 and k_2 is depicted in Figure 10.12.

Once k_2 is computed, one takes a new step, starting back at (t_i, y_i), with step size $h/2$, with slope k_2 (instead of k_1). The parameter k_3 is then obtained by evaluating the slope function at this new point, as shown in Figure 10.13.

Finally, a step with slope k_3 is taken, again starting at the point of the previous approximation (t_i, y_i), of step size h. The parameter k_4 is then obtained by evaluating the slope function at this final point, as shown in Figure 10.14.

■ **Example 10.12** Consider the initial value problem

$$y' = \frac{1+t}{y}, \qquad y(0) = 2.$$

We will approximate $y(0.2)$ using a step size of $h = 0.2$, i.e., we will only compute one step of the Runge–Kutta method.

Computing k_1 is identical to computing the slope in the Euler's method:

$$k_1 = f(t_0, y_0) = \frac{1}{2} = 0.5.$$

To compute k_2, we take a half step using the slope k_1, and approximate the slope at this virtual point:

$$k_2 = f(t_0 + \frac{h}{2}, y_0 + \frac{h}{2}k_1) = f(0.1, 2.05) = \frac{1.1}{2.05} = 0.536585366.$$

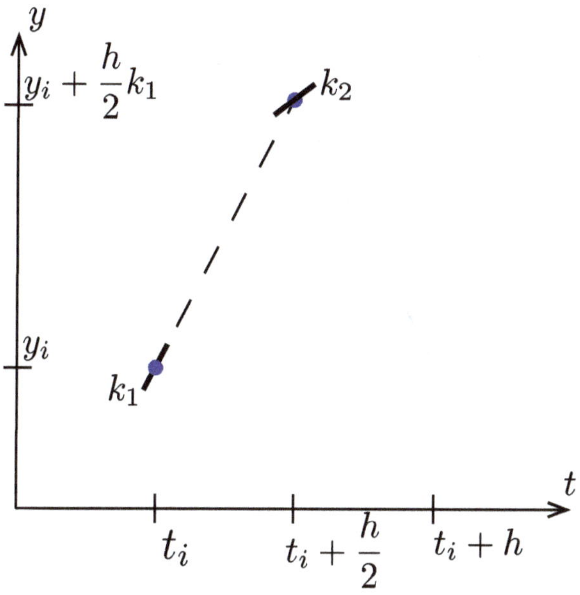

Figure 10.12: Runge–Kutta method: an illustration (steps k_1 and k_2).

To compute k_3, we go back to the initial point (t_0, y_0), and take a half-step using the slope k_2:

$$k_3 = f(t_0 + \frac{h}{2}, y_0 + \frac{h}{2}k_2) = f(0.1, 2.053658537) = \frac{1.1}{2.053658537} = 0.535629454.$$

Finally, to compute k_4, we go back to the initial point (t_0, y_0), and take a full step using the slope k_3:

$$k_4 = f(t_0 + h, y_0 + hk_3) = f(0.2, 2.107125891) = \frac{1.2}{2.107125891} = 0.569496111.$$

Next, we increment t_i by the step size, obtaining $t_1 = 0.2$, and y_i using a weighted average of the parameters k_1, k_2, k_3, k_4, obtaining

$$y_1 = y_0 + \frac{h}{6}(k_1 + 2k_2, +2k_3 + k_4) = 2.107130858.$$

If we were to continue further, we would go back to the beginning and redo these computations with the new values t_1 and y_1 in place of the old t_0 and y_0. At the end of our second loop, we would then have an approximation for t_2 and y_2. (We would have obtained $t_2 = 0.4$ and $y_2 = 2.2271059257$.) ■

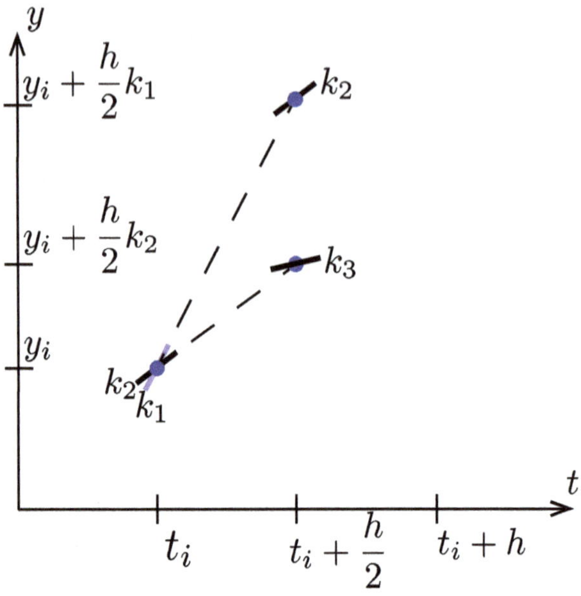

Figure 10.13: Runge–Kutta method: an illustration (step k_3).

10.4.5 A MATLAB Example: the Runge–Kutta Method

Below is code for a MATLAB program, examp02.m, which approximates the solution to the initial value problem given by equations (10.25) and (10.26) on the interval $[0,2]$. It then outputs the absolute and relative error for the approximation of $y(2) = e$.

```
%examp02.m

% This program approximates the solution to the
% initial value problem
% y' = 0.5y,
% y(0) = 1,
% on the interval [0, 2], using the Runge-Kutta method.
% The exact solution is y(t) = exp(0.5t), and y(2)= e.

%clears the memory
clear
```

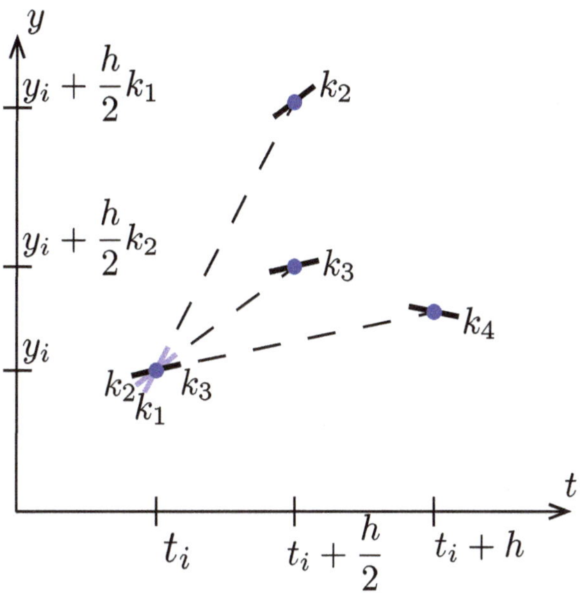

Figure 10.14: Runge–Kutta method: an illustration (step k_4).

```
n = input('Enter number of steps:  ');

%Define stepsize
h = (2-0)/n;

%initiate approximate solution
yRK(1) = 1;
t(1) = 0;

%for loop to implement the Runge-Kutta method
for(i=1:n)
    %increment time by h
    t(i+1) = t(i) + h;
    %Define parameters k1, k2, k3, k4
    k1 = examp01funct(t(i),yRK(i));
    k2 = examp01funct(t(i) + h/2, yRK(i) + h/2*k1);
    k3 = examp01funct(t(i) + h/2, yRK(i) + h/2*k2);
    k4 = examp01funct(t(i) + h, yRK(i) + h*k3);

    %increment solution
    yRK(i+1) = yRK(i) + h*(k1+2*k2+2*k3+k4)/6;
```

```
end

disp('absolute error'),
abs(yRK(n+1)-exp(1)),
disp('relative error'),
abs(yRK(n+1)-exp(1))/exp(1),

% defines texact as a 1x100 vector with equally
%spaced values between 0 and 2
texact = linspace(0, 2);
%defines a solution vector for the exact answer.
yexact = exp(texact/2);

plot(texact,yexact,'k','LineWidth',2)
hold on
plot(t,yRK,'--','LineWidth',2)
plot(t,yRK,'o','LineWidth',2)
hold off
```

The program `examp02.m` was run using 5 steps as the input. (The step size is therefore $h = 0.4$.) The output plots are shown in Figure 10.15.

By comparing Figure 10.11, which displays the result of Euler's method, and Figure 10.15, which displays the result of the Runge–Kutta method, using the same step size, it is immediately clear the advantage in accuracy that the Runge–Kutte method offers: greater accuracy obtained from a fewer number of steps.

The program was subsequently run using a variety of different values for the number of steps. The absolute and relative errors for these runs are tabulated in Table 10.7. Notice that the error goes to zero as fast as h^4: reducing h by a factor of 10 results in a reduction in the error by a factor of 10,000. This is why the Runge–Kutta method is called an *fourth-order method*: the global truncation error (absolute or relative) goes like $O(h^4)$.

n	h	absolute error	relative error
10	0.2	2.0843e-6	7.6678e-7
100	0.02	2.2464e-10	8.2641e-11
1,000	0.002	2.0872e-14	7.6785e-15
10,000	0.0002	1.1102e-14	4.0843e-15

Table 10.7: Absolute and relative error using different step sizes for the MATLAB program `examp02.m`

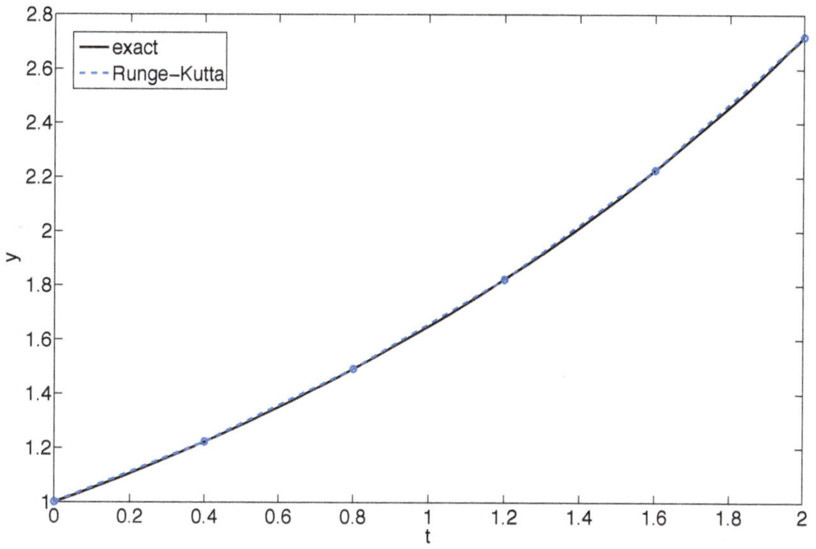

Figure 10.15: MATLAB output of `examp02.m` using 5 steps.

Notice, however, that there is no reduction in numerical error from going from 1,000 steps to 10,000 steps. (Or, though not shown in the table, from going from 10,000 steps to a million, or a billion, etc.) This is due to *rounding error*: an intrinsic error due to the fact that the computer only holds so many decimal places with each computation. Standard computer precision results in 16 decimal places. Hence, the numerical error that results from the algorithm eventually becomes less than the error due to the fact that MATLAB is only carrying around 16 decimal places at a time. (No matter how accurate the numerical method, an approximate solution will never be more accurate than the computer precision of the machine you implement it on.)

Also note that the Runge–Kutta method reaches computer precision using only 1,000 steps, whereas it would take Euler's method approximately 100,000,000,000,000 steps to obtain the same accuracy. Moreover, assuming that MATLAB can perform approximately 500,000 steps of the Euler's method each second, it would take MATLAB approximately 6.34 years, using Euler's method, to achieve the same accuracy that the Runge–Kutta achieves in about 5 milliseconds.

Exercises

For Exercises 1–5, use Euler's method to approximate $y(2)$ using a step size of $h = 0.5$.

1. $y' = t + y^2, \qquad y(0) = 0.$

2. $y' = \sin(yt)$, $y(0) = 3$.
3. $y' = \sqrt{y+t}$, $y(0) = 2$.
4. $y' = \dfrac{1}{t^2+y^2}$, $y(0) = 0.5$.
5. $y' = e^{yt}$, $y(0) = 0$.

For Exercises 6–10, use the Runge–Kutta method to approximate $y(1)$ using a step size of $h = 0.5$.

6. $y' = t + y^2$, $y(0) = 0$.
7. $y' = \sin(yt)$, $y(0) = 3$.
8. $y' = \sqrt{y+t}$, $y(0) = 2$.
9. $y' = \dfrac{1}{t^2+y^2}$, $y(0) = 0.5$.
10. $y' = e^{yt}$, $y(0) = 0$.

Problems

11. (a) Determine that the exact solution of the initial value problem

$$\frac{dy}{dt} = 60\sin(10t) - 1.1y, \qquad y(0) = 0. \tag{10.27}$$

Physically, $y(t)$ represents the current at time t in the following circuit shown in Figure 10.16, with $R = 1.1\ \Omega$, $L = 1$ h, and $E(t) = 60\sin(10t)$ V (the switch is closed for $t \geq 0$).

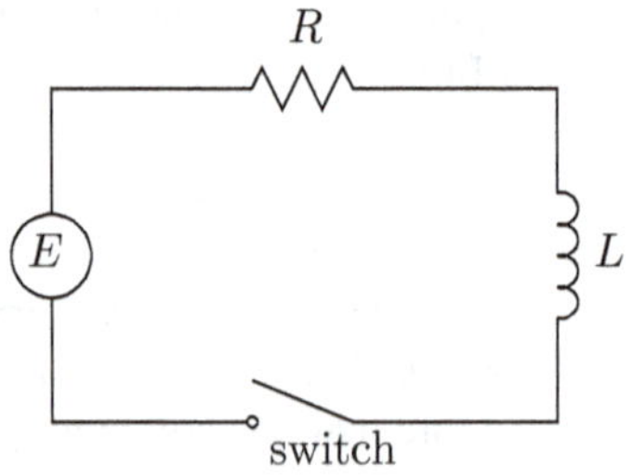

Figure 10.16: The circuit diagram for Problem 11.

(b) Write a program that does the following:

- **inputs:** Write an m-file first asks the user to input the stepsize h.
- **internal computation:** The program should then solve the initial value problem (10.27) numerically on the interval $t \in [0,3]$ using Euler's method and the Runge-Kutta method. The program should also compute the exact solution directly you obtained in part (a), using the grid `texact = linspace(0,3,1000)`.
  At this point you should have 5 vectors:

variable name	number of components
`texact`	1000 components
`yexact`	1000 components
`t`	$3/h+1$ components
`yeuler`	$3/h+1$ components
`yimp`	$3/h+1$ components
`yRK`	$3/h+1$ components

The program should also compute the error in the approximation of $y(3)$ obtained by each methods.

- **plot output:** On the interval $x \in [0,3]$, your program should plot the exact solution in black, the Euler's method solution in blue and the Runge-Kutta method solution in red, all on the same plot (include a legend).
- **display outputs:** Your program should also display the absolute Error in the approximation for $y(3)$ using both methods, (Error = $|y(3)_{\text{exact}} - y(3)_{\text{estimated}}|$).
- **Turn in:** Turn in a copy of your mfile. Turn in a copy of your plots for $h = 0.5,\ 0.1,\ 0.01$, and a filled out version of Table 10.8. Report errors using 4 decimals and correct scientific notation, (eg. 2.4561e-5).

stepsize	Euler Method Error	Runge-Kutta Method Error
0.5		
0.1		
0.01		
0.001		

Table 10.8: Error table for Problem 11.

Appendix A

Set Notation

In this paragraph, we review some basic notation that is often used throughout the text. This notation deals with *sets* and *set membership*. Often, we will desire to refer to certain sets of numbers: natural numbers, integers, open and closed intervals, etc. This is efficiently accomplished using set notation.

Definition A.0.1 A *set* is a collection of objects. These objects are called the *elements* of the set.

In this text, the objects that constitute a set are always numbers. A set can be defined by explicitly listing its elements; the *curly brackets* are used for this purpose. For example, the set

$$S = \{1, 3, 5, 7\}$$

is the set that consists of the numbers 1, 3, 5, and 7.

There are a few famous sets we all should know.

Definition A.0.2 The following symbols are reserved for the specified sets:

$$\begin{aligned} \emptyset &= \text{the empty set,} \\ \mathbb{N} &= \text{the set of all natural numbers,} \\ \mathbb{Z} &= \text{the set of all integers,} \\ \mathbb{Q} &= \text{the set of all rational numbers,} \\ \mathbb{R} &= \text{the set of all real numbers,} \\ \mathbb{C} &= \text{the set of all complex numbers.} \end{aligned}$$

(R) The sets $\mathbb{N}$, $\mathbb{Z}$, $\mathbb{Q}$, $\mathbb{R}$, and $\mathbb{C}$ are written using a typeface known as *black-board bold*, which is to be reserved for these special sets. Customized sets, which we will define along our way, are typically denoted with capital letters.

Sets basically define membership classes: a number is either in a given set, or it is not. It is therefore convenient to have notation to describe such membership relations.

Definition A.0.3 — Set Membership. Let x be an object (for us, a number) and S be a set. We say that *x is an element of S*, written

$$x \in S,$$

if the set S contains the object (number) x as one of its elements. The symbol "$\in$" is used for this relationship. Conversely, we say that *x is not an element of S*, written

$$x \notin S,$$

if x is *not* one of the elements that comprise the set S.

Example A.1 Since π is a real number, we may write $\pi \in \mathbb{R}$. Since it is an irrational number, however, we find that $\pi \notin \mathbb{Q}$, i.e., π is not a rational number. Similarly, $\sqrt{2} \in \mathbb{R}$, but $\sqrt{2} \notin \mathbb{Q}$. ∎

Definition A.0.4 Given two sets A and B. We say that *A is a subset of B*, written

$$A \subset B,$$

if every element of A is also an element of B, i.e., if the logical implication

$$x \in A \Longrightarrow x \in B$$

holds for every $x \in A$.

Similarly, we say that *A is a superset of B*, written

$$A \supset B,$$

if every element of B is also an element of A, i.e., if B is a subset of A.

Example A.2 We have the following set inclusions:

$$\mathbb{N} \subset \mathbb{Z} \subset \mathbb{Q} \subset \mathbb{R}.$$

That is, every natural number is an integer; every integer a rational; every rational a real. ∎

It is further useful to define sets of numbers based on given conditions. For this, we require the following notation.

Definition A.0.5 A set may be defined using *set-builder notation*, which consists of three parts: a variable definition, a colon or vertical bar, and a logical predicate. These three parts are contained within the customary curly brackets:

$$S = \{x \mid \text{condition on } x\}, \text{ or}$$
$$S = \{x : \text{condition on } x\}.$$

This is read: the set of all x such that the stated condition holds. (The vertical bar or colon is read as "such that.")

For example, set builder notation may be used to define various types of intervals.

Definition A.0.6 Given $a, b \in \mathbb{R}$, such that $a < b$, we may define the following *intervals*:

$$[a,b] = \{x \in \mathbb{R} : a \leq x \leq b\},$$
$$(a,b) = \{x \in \mathbb{R} : a < x < b\},$$
$$(a,b] = \{x \in \mathbb{R} : a < x \leq b\},$$
$$[a,b) = \{x \in \mathbb{R} : a \leq x < b\}.$$

These intervals are referred to as *closed*, *open*, *left-open*, and *right-open*, respectively.

Definition A.0.7 *Infinity* ∞ is a symbol used to denote that which is greater than every real number. *Negative infinity* $-\infty$ is a symbol used to denote that which is less than every real number.

Since ∞ and $-\infty$ are not numbers, they are never included in an interval. The infinities can, however, be used to describe certain semi-infinite intervals as follows.

Definition A.0.8 Given a number $a \in \mathbb{R}$, we may construct the following *semi-infinite intervals*:

$$[a,\infty) = \{x \in \mathbb{R} : a \leq x\},$$
$$(a,\infty) = \{x \in \mathbb{R} : a < x\},$$
$$(-\infty,a] = \{x \in \mathbb{R} : x \leq a\},$$
$$(-\infty,a) = \{x \in \mathbb{R} : x < a\}.$$

The semi-infinite intervals $[a,\infty)$ and $(-\infty,a]$ are closed intervals, whereas the semi-infinite intervals (a,∞) and $(-\infty,a)$ are *open intervals*.

(R) The semi-infinite intervals $[a,\infty)$ and $(-\infty,a]$ really are *closed* intervals, not half-open intervals. This follows because ∞ and $-\infty$ are not really numbers at all, but theoretical limits. The only actual boundary occurs at the single point a; as such, if the set includes a it is closed and if it does not it is open.

Definition A.0.9 Given two sets A and B, we say that the *union of A and B*, denoted $A \cup B$, is the set consisting of the elements from A plus all the elements from B, i.e.,

$$A \cup B = \{x : x \in A \text{ or } x \in B\}.$$

The *intersection of A and B*, denoted $A \cap B$, is the set that consists only of elements that are contained in *both* A and B, i.e.,

$$A \cap B = \{x : x \in A \text{ and } x \in B\}.$$

■ **Example A.3** We have the following examples:

$$\begin{aligned} (-\infty, 0] \cap [0, \infty) &= \{0\}, \\ (-\infty, 0] \cup [0, \infty) &= \mathbb{R}. \end{aligned}$$

■

Appendix B

Trigonometric Identities

Pythagorean Theorem

$$\sin^2(x) + \cos^2(x) = 1 \tag{B.1}$$
$$\tan^2(x) + 1 = \sec^2(x) \tag{B.2}$$

Even and Odd Properties

$$\sin(-x) = -\sin(x) \tag{B.3}$$
$$\cos(-x) = \cos(x) \tag{B.4}$$

Addition and Subtraction Formulas

$$\sin(x \pm y) = \sin(x)\cos(y) \pm \cos(x)\sin(y) \tag{B.5}$$
$$\cos(x \pm y) = \cos(x)\cos(y) \mp \sin(x)\sin(y) \tag{B.6}$$

Double-angle Formulas

$$\sin(2x) = 2\sin(x)\cos(x) \tag{B.7}$$
$$\cos(2x) = \cos^2(x) - \sin^2(x) \tag{B.8}$$

Half-angle Formulas

$$\cos^2(x) = \frac{1 + \cos(2x)}{2} \tag{B.9}$$
$$\sin^2(x) = \frac{1 - \cos(2x)}{2} \tag{B.10}$$

Appendix C

Selected Answers

Section 1.1

7. 237.6 million.
9. 303.24 million.
13. x is a function of y, but y is not a function of x.

Section 1.2

1. $f(x) = 4 - x/2$.
3. $f(x) = \dfrac{17 - x}{3}$.
5. $f(x) = \dfrac{13x - 91}{4}$.
8. $T(x) =$

$$\begin{cases} 0.1x & x \in [0, 8925] \\ 0.15x - 446.25 & x \in (8925, 36250] \\ \vdots & \vdots \end{cases}$$

10. (a) $C(x) = 18300 + 0.072x$; (c) $\Pi(x) = 0.488x - 18300$; (d) 37500.

Section 1.3

1. $f(x) = \dfrac{1.5^x}{3}$.
3. $f(x) = 3\left(\dfrac{1}{2}\right)^{x/16}$.
5. $f(x) = 3.5(2)^{x/7}$.
11. (a) $T(x) = 20000 + 2000x$.
12. approximately 62.77 kPa, or 0.612 atm.

Section 1.4

1. $\log(325) \approx 2.51188$.
7. $T \approx 6.116$.
7. $T \approx 6.116$.
12. $\dfrac{3\ln(3)}{\ln(2)}$.
15. (a) *Hint*: What happens as $t \to \infty$? (d) approximately 7 hours and 59 minutes.
16. 27,982.5 years old.

Section 1.5

1. (a) $\sqrt{x^2+3}$.
3. (b) $e^{\sin(x)}$.
7. $f^{-1}(x) = \frac{e^x}{3} - 3$.
21. $f(x+1) = 3x^2 + 8x + 12$.
28. (b) $R(x) = 0.36x - 54.4$; (c) \$206.67.

Section 1.6

1. $f(t) = 3 + 2\cos(4t)$.
5. $f(t) = 7 + 8\cos\left(\frac{\pi t}{4} - 9\right)$.
17. $\left\{\frac{7\pi}{6}, \frac{11\pi}{6}\right\}$.
23. $T(t) = 45 + 55\cos\left(\frac{3\pi t - 17\pi}{18}\right)$.

Section 1.7

1. 1.
4. 5.
2. $k = 0$.
6. $k = e^2 - 2$.
21. 6.
26. 12.
39. (b)(iv) \$11,066.61.

Section 1.8

1. 0.
4. 0.
7. $\frac{1}{9}$.
11. 40.
15. 5.

Section 2.1

1. (a) $\frac{3}{\pi}$.
5. (a) ln(2).
7. (a) 1; (b) e; (c) e^2.
9. (a) 1; (b) 0.5; (c) 0.25.
18. AROC(6,12) $\approx \frac{11}{12}$ ft/hr.
19. (c) AROC(15,20) = 0.7928.

Section 2.2

15. $IROC(0) = 1$; $IROC(\pi/4) = \frac{1}{\sqrt{2}}$; $IROC(\pi/2) = 0$.
17. $IROC(0) = 1$; $IROC(1) = e$; $IROC(2) = e^2$.
25. $h'(1) \approx -7.06$ ft/s.
27. $P'(4000) \approx 10.19$ kPa/m.
29. $\lim_{t\to 4^-} v'(t) \approx 7$ m/s^2.

Section 2.3

8. (b) $W'(1) \approx 44$ kg/min; (e) $f'(1) \approx 1.3693$ min^{-1}.
12. (a) Initially, $u'(0) \approx 3$ degrees/s; (c) $f'(100) \approx 0.9$ s/degree.
13. $\rho'(100) \approx -0.0297905$ kg/(m^{3}K).

Section 2.4

1. $f'(x) = 3x^2 + 7x^6$.

11. $\frac{2}{x}$.

13. $\frac{3}{x} - 3\cos(3x)$.

35. (a) $\lim_{M\to\infty} f(M) = \frac{\gamma+1}{\gamma-1}$; (c) $g'(2) = \frac{7}{3}$.

35. (a) $c \approx 146.18$ m/s.

Section 2.5

1. $(2.1)^3 \approx 9.2$.

3. $e^{1.2} \approx 1.2e$.

5. $\frac{1}{\sqrt{4.05}} \approx \frac{159}{320}$.

7. $\sin(5\pi/12) \approx \frac{12\sqrt{3}+\pi}{24}$.

9. $\tan(34\pi/99) \approx \sqrt{3} + \frac{4\pi}{99}$.

Section 3.1

1. $f'(x) = 16x + 9$.

3. $f'(x) = 30x^2 + 3$.

5. $f'(x) = e^x + \frac{1}{x}$.

7. $f'(x) = \frac{-1}{x^2}$.

9. $f'(x) = 6xe^{3x^2+7}$.

11. $f'(x) = 1$.

13. $f'(x) = \frac{-3}{2\sqrt{x^5}}$.

15. $f'(x) = \frac{32x^3 + 39x^2}{8x^4 + 13x^3 + 19}$.

17. $f'(x) = \tan(x)$.

19. $f'(x) = 2\pi\cos(2\pi x + 7)$.

21. $f'(x) = 3(e^x + \ln|x|)^2\left(e^x + \frac{1}{x}\right)$.

23. $f'(x) = 35e^{5\sin(7x+2)}\cos(7x+2)$.

25. $f'(x) = \frac{e^{\sqrt{x}}}{2\sqrt{x}}$.

27. $g'(t) = 2te^{a^2+t^2}$.

29. $g'(w) = \frac{-(\cos(w) + 1/w)}{(\sin(w) + \ln(w))^2}$.

31. $f'(\theta) = 0$.

33. $v'(u) = \frac{3\cos(3u+\pi) + e^u}{\sin(3u+\pi) + e^u}$.

35. $g'(t) = \frac{1}{2t}$.

Section 3.2

1. $f'(x) = 88x^{87} + 704x^7$.

3. $f'(x) = 16x^3 + 9x^2 + 4x + 1$.

5. $f'(x) = 3x^2 + 3$.

7. $f'(x) = 72x^8 + 72x^7$.

9. $f'(x) = \frac{-12(4x^3 + 27x^2)}{\sqrt[3]{(x^4 + 9x^3 + 7)^7}}$.

19. (a) $f(x) = 60/x$.

Section 3.3

1. $f'(x) = 8\pi e^{8\pi x+7}$.

3. $f'(x) = e^{e^x}e^x$.

5. $f'(x) = 8^{e^x+\pi}e^x\ln(8)$.

7. $f'(x) = \frac{5^x\ln(5) + 5x^4}{5^x + x^5}$.

9. $f'(x) = \frac{6xe^{3x^2} + x^{-1}}{e^{3x^2} + \ln|5x|}$.

Section 3.4

1. $f'(x) = 100x^{99}\cos(x^{100})$.
3. $f'(x) = \cos(\sin(x))\cos(x)$.
5. $f'(x) = \dfrac{1}{2x^{3/2}}\sin\left(\dfrac{1}{\sqrt{x}}\right)$.
7. $f'(x) = -120\pi\sin\left(6\pi x + \dfrac{\pi}{4}\right)$.
9. $f'(x) = 272\pi\cos(16\pi t)$.
11. $f' = e^{\sin(x)+\cos(x)}(\cos(x) - \sin(x))$.
13. $f' = 3\sin^2(x)\cos(x) - 3\cos^2(x)\sin(x)$.
15. $f'(x) = -\tan(x)$.
17. $f'(x) = \dfrac{-3\cos(x)}{2\sqrt{\sin^3 x}}$.
19. $f'(x) = 3e^{\sin(3x+7)}\cos(3x+7)$.
21. $f'(x) = 8\pi\sec(8\pi x + 9)\tan(8\pi x + 9)$.
23. $f'(x) = -e^{\csc(x)}\csc(x)\cot(x)$.
25. $f' = -\sec(\cos(x))\tan(\cos(x))\sin(x)$.
29. (a) $S(d) = \dfrac{51}{4} + \dfrac{136}{15}\cos\left[\dfrac{\pi(t-172)}{183}\right]$.

    (b) $S'(d) = \dfrac{-136\pi}{2745}\sin\left[\dfrac{\pi(t-172)}{183}\right]$.

    (c) $d_{ae} \approx 268.324$.

Section 3.5

1. $f'(x) = \dfrac{\cos(\sqrt{x})}{2\sqrt{x}}$.
3. $f'(x) = 1$.
5. $f'(x) = 2\sin(x)\cos(x)$.
7. $f'(x) = \cos(\sin(x))\cos(x)$.
9. $f'(x) = 24\pi\cos(3\pi x + 2)$.
13. $f'(x) = \dfrac{3\cos(6x+7)}{\sqrt{\sin(6x+7)}}$.
15. $f'(x) = \dfrac{-8\cos(8x+9)}{\sin^2(8x+9)}$.
17. $f'(x) = \dfrac{-1}{2x}$.
19. $f'(x) = e^{e^x}e^x$.

Section 3.6

1. $f'(x) = \cos^2(x) - \sin^2(x)$.
3. $f'(x) = 1 + \ln|x|$.
5. $f'(x) = \dfrac{x\cos(x) - \sin(x)}{x^2}$.
7. $f'(x) = 8x^7\sin(\sqrt{x}) + \dfrac{x^8\cos(\sqrt{x})}{2\sqrt{x}}$.
9. $f'(x) = 6x^5\sin(x) + x^6\cos(x)$.
11. $f'(x) =$ $4\sin^3(x)\cos^9(x) - 8\sin^5(x)\cos^7(x)$.
13. $f'(x) = e^{e^x}(xe^x + 1)$.
15. $f'(x) = \dfrac{32\pi\ln|x|\cos(4\pi x+12) - 8\sin(4\pi x+12)/x}{64(\ln|x|)^2}$.
17. $f'(x) = \sin(\sin(x))\cos(x)$ $+\cos(\sin(x))\sin(x)\cos(x)$.

Section 3.7

1. $\dfrac{dy}{dx} = \dfrac{-x^2}{y^2}$; $T(x) = 2.25 - 0.25x$.
3. $\dfrac{dy}{dx} = \tan(x)\tan(y)$; $T(x) = x - \frac{\pi}{6}$.
5. $\dfrac{dy}{dx} = \dfrac{-y\ln|y|}{x}$; $T(x) = 2 + (2 - x)\ln(2)$.
7. $\dfrac{dy}{dx} = \dfrac{-3y}{2x}$; $T(x) = 5 - 1.5x$.
9. $\dfrac{dy}{dx} = -\sqrt{\dfrac{y}{x}}$; $T(x) = 4 - x$.
11. $3z^2\,dz = 3x^2\,dx + 3y^2\,dy$.
13. $(3x^2y^5 + 5x^4y^3)dx + (5x^3y^4 + 3x^5y^2)dy = 0$.
15. $(2x + 8y)dx + (2y + 8x)dy = 0$.

17. $3x^2y^5dx + 5x^3y^4dy = 0.$

19. $P\,dv + v\,dP = R\,dT.$

27. $\Delta V \approx \frac{22\pi}{15}$ cm^3.

Section 4.1

1. critical point $x \approx 3.2078$ (global min); global max at $x = 5$.

5. critical point $x \approx 0.678097$ (global min); global max at $x = 0$.

Section 4.2

1. 0.5.

2. $r = \sqrt[3]{\frac{V}{2\pi}}.$

4. $x = 25.$

5. (c) Approximately 18.56 miles from campus.

6. $x^* = \frac{H}{2}.$

7. $\sqrt[3]{50} \times \sqrt[3]{50} \times \frac{25}{\sqrt[3]{50^2}}.$

9. $V - \frac{2R^3\pi}{9\sqrt{3}}$

10. $D^* = \frac{v_0^2}{g}.$

12. $x = 64$ miles.

14. $\mu = \tan(\theta).$

Section 4.3

1. 588 W.

4. $\frac{dD}{dt} = \frac{-6}{13}$ ft/s.

6. $\frac{dx}{dt} = \frac{25}{9\pi}.$

12. $\frac{dV}{dt} = -432\pi$ inch^3/hr.

14. $\frac{dh}{dt} \approx -51.2$ ft/s.

Section 4.6

1. $P(q) = \begin{cases} 3.925q & \text{if } q \in [0,24] \\ 3.425q & \text{if } q \in (24,80] \\ 3.05q & \text{if } q \in (80,\infty) \end{cases}.$

3. 216.

4. $v^* = 12$ mph.

8. $115.

10. (b) $q^* = 775.$

Section 4.7

1. 1/3.

3. 5/9 .

5. 1.

7. α/β .

9. 12.

Section 5.1

1. $\frac{x^5}{5} + \frac{5x^3}{3} + 17x + C.$

3. $\frac{2x^{3/2}}{3} + 2\sqrt{x} + C.$

5. $\frac{x^9}{9} - \frac{\cos(3x)}{3} + \frac{e^{8x}}{8} + C.$

7. $\ln|\sin(x)| + C.$

9. $0.01(1+x)^{100} + C.$

11. $\dfrac{\sin^3(x)}{3}+C.$

13. $-\dfrac{\cos(3x+5)}{3}+C.$

15. $\dfrac{-\cos(x^2)}{2}+C.$

17. $\dfrac{-e^{-x^2}}{2}+C.$

19. $e^{\sin(3x)}+C.$

21. $2\sqrt{\sin(x)}+C.$

23. $3\ln|x|+C.$

25. $-3\cos(3x)+4\sin(2x)+C.$

27. $\ln|\sin(x)+\pi|+C.$

29. $\dfrac{\sin^{10}(x)}{10}+C.$

31. 1.2755 m; 3.3784 m; 7.8125 m.

33. $x(t)=t-\ln(t).$

Section 5.2

1. 1705.

3. 25,333.

5. $\dfrac{2N^3+9N^2+13N}{6}.$

7. 1.

9. $6\left[1-\left(\dfrac{2}{3}\right)^{100}\right].$

11. $\displaystyle\sum_{i=1}^{100} i.$

13. $\displaystyle\sum_{i=1}^{20} \frac{1}{i^2}.$

15. $\displaystyle\sum_{i=1}^{10} \frac{1}{2i}.$

17. $\displaystyle\sum_{i=0}^{5} \frac{1}{2^i}.$

19. $\displaystyle\sum_{i=1}^{5} \frac{(-1)^{i+1}}{i!}.$

24. (b) $A_n=\dfrac{1600}{7}\left(1-\dfrac{1}{8^n}\right).$

Section 6.1

1. 0.05 miles.

3. 34.5 miles; 35.25 miles; or 34.875 miles (best estimate).

5. 100π m.

6. (d) $D(v)=\dfrac{11v}{3}+\dfrac{11v^2}{225}.$

(e) 9.625 car lengths at 30 mph; 24.75 car lengths at 60 mph.

Section 6.2

1. $L_6=7.5$; $R_6=10.5$.

3. $L_{10}=0.285$; $R_{10}=0.385$.

5. $L_4\approx 4.9243$; $R_4\approx 8.1189$.

7. $L_3=\dfrac{\pi}{12}\left(1+\sqrt{3}\right);$

$R_3=\dfrac{\pi}{12}\left(3+\sqrt{3}\right).$

9. $L_4=\dfrac{25}{12}$; $R_4=\dfrac{77}{60}$.

11. 21,600.

13. 158.

15. 172.

17. Between 437.34 and 454.34 million users.

Section 6.3

7. 4.

9. -7.

13. 25π m.

15. $L_6 \approx 0.6016$, $R_6 \approx 0.7116$.

Section 6.4

1. $\frac{a^2}{2}$.

3. $e - 1$.

5. $\frac{2-\sqrt{3}}{2}$.

7. $\frac{8}{3}$.

9. 1.

11. $e - 1$.

13. 3.

15. $\ln(2)$.

17. 2.

19. $\frac{1}{\ln(2)}$.

27. Approximately 121 m.

29. Approximately 445.73 million users.

31. (b) $q^* = \frac{1}{r} \ln \frac{a}{c}$.

(c) $\Pi(q) = \frac{a}{r}\left(1 - e^{-rq}\right) - (c_0 + cq)$.

Section 6.5

1. $x(t) = 2t + \frac{t^3}{2}$.

3. $x(t) = 11t + 12(e^{-0.5t} - 1)$.

5. $x(t) = 1 - t - \frac{t^4}{12} + \frac{t^3}{6}$.

7. $f'(x) = \ln|x|$.

9. $f'(x) = 2x\csc(x^2) - \csc(x)$.

11. Approximately 1067 meters.

Section 7.1

1. $\frac{-\cos(x^2)}{2} + C$.

3. $\frac{1}{3}\ln|1 + x^3| + C$.

5. $\frac{2}{3}\left(1 + e^x\right)^{3/2} + C$.

7. $\frac{1}{3}\sin^3(x) + C$.

9. $\frac{1}{2}(\ln|x|)^2 + C$.

11. $\frac{1}{2}\left(1 - \cos(\pi^2)\right)$.

13. $\frac{1}{2}$.

15. $2\ln(2)$.

17. $\frac{\ln(1.5)}{2}$.

19. $\frac{1}{2}$.

21. Approximately 44.729 g.

Section 7.2

1. $\sin(x) - x\cos(x) + C$.

3. $\frac{x + \sin(x)\cos(x)}{2} + C$.

5. $\frac{e^{5x}}{25}(5x - 1) + C$.

7. $2x\sin(x) + 2\cos(x) - x^2\cos(x) + C$.

9. $\frac{e^x}{2}(\sin(x) - \cos(x)) + C$.

11. $\frac{\pi}{4}$.

13. $\frac{\pi}{2}$.

15. 1.

17. $\frac{2}{5}\left(1 + e^{\pi/2}\right)$.

19. $\frac{1}{2}-e^{-1}$.

Section 7.3

1. $\frac{\ln|x-2|}{2}-\frac{\ln|x|}{2}$.

3. $\frac{1}{2}\ln\left|\frac{2x-2}{2x+2}\right|$.

5. $\ln\left|\frac{x-8}{x-7}\right|$.

7. $4\ln|x-3|-3\ln|x-2|$.

9. $\frac{17}{27}\ln|x-3|+\frac{10}{27}\ln|x+3|+\frac{7}{9x}$.

11. $\frac{-3}{x-3}+2\ln\left|\frac{x-2}{x-3}\right|$.

13. $\frac{7}{x}+\frac{7}{2x^2}-7\ln|x|+8\ln|x-1|$.

15. $x-3\tan^{-1}(x/3)$.

17. $\ln\left|x-8+\frac{32}{x}\right|$.

19. $\frac{1}{8(x^2+4)}-\frac{\ln|x^2+4|}{32}+\frac{\ln|x|}{16}$.

Section 7.4

1. $\frac{-1}{3}\left(a^2-x^2\right)^{3/2}$.

3. $\sqrt{x^2-a^2}$.

5. $\ln\left|x+\sqrt{x^2+a^2}\right|$.

7. $-\sqrt{a^2-x^2}$.

9. $(a^2-x^2)^{-1/2}$.

Section 7.5

1. 1.

3. 2.

5. 1.

7. $\frac{1}{3}$.

9. $\frac{1}{2}$.

11. $\frac{3}{2}$.

13. diverges.

15. 2.

19. 1.

Section 8.1

1. $\sqrt{17}-\frac{1}{4}\ln(\sqrt{17}-4)$.

2. $\frac{22\sqrt{22}-8}{27}$.

3. $\ln(1+\sqrt{2})$.

4. $\pi/2$.

5. $\sqrt{2}+\ln(1+\sqrt{2})$.
Hint: $\sqrt{1+\frac{1}{x}}=\frac{1+x}{\sqrt{x}}\to u=\sqrt{x}$.

6. $\sqrt{b^2+d^2}$.

7. $\frac{13\sqrt{13}-8}{27}$.

8. $5\sqrt{5}-8$.

9. $\sqrt{2}+\ln(1+\sqrt{2})$.

10. $\sqrt{10}+\frac{1}{\sqrt{2}}\ln(2+\sqrt{5})$.

11. $4\pi-6\sqrt{3}$.

13. $\frac{3}{16}$.

15. 4.

17. $1-\frac{\pi}{2}+\frac{\pi^3}{48}$.

19. $\frac{\pi^2}{8}-1$.

Section 8.2

1. $V = \frac{\pi}{5}$; $S = \dfrac{9\pi\sqrt{5}}{16} - \dfrac{\pi}{32}\ln(\sqrt{5}+2)$.

3. $V = \dfrac{\pi}{7}$; $S = \dfrac{(10\sqrt{10}-1)\pi}{27}$.

5. $V = \dfrac{\pi^2}{2}$; $S = 2\pi\left[\sqrt{2} + \ln\left(1+\sqrt{2}\right)\right]$.

7. $V = \pi(1+\sinh(1)\cosh(1))$;
$S = 2\pi(1+\sinh(1)\cosh(1))$.

9. $V = \dfrac{68\pi}{315}$;
$S = 2\pi\left[\dfrac{5\sqrt{2}-1}{18} - \dfrac{\ln(\sqrt{2}-1)}{6}\right]$.

11. $V = \dfrac{3\pi}{10}$.

13. $V = 8\pi + \dfrac{\pi^2}{2}$.

15. $V = \pi\left(\dfrac{1}{2} - \dfrac{e^{-1.4}}{2} - \dfrac{(0.7)^5}{5}\right)$.

17. $V = \dfrac{23\pi}{210}$.

18. $V = 90\pi^2$.

19. $V = \dfrac{\pi}{10}$.

Section 8.3

1. $V = \dfrac{32}{5}$.

3. $V = \dfrac{\pi^2}{2}$.

5. $V = \dfrac{h^3}{2\sqrt{3}}$.

7. $V = \dfrac{1-e^{-2}}{2}$.

9. $V = \dfrac{\pi}{4}$.

11. $V = 8\pi$.

13. $V = 4\pi^2(2\pi - 1)$.

15. $V = \dfrac{207\pi}{10}$.

17. $V = \dfrac{128\pi}{5}$.

19. $V = 8\pi$.

Section 8.4

1. $M = \dfrac{5}{2}$; $\overline{x} = \dfrac{8}{15}$; $I_p = \dfrac{11}{12}$.

3. $M = 1$; $\overline{x} \approx 0.49204$; $I_p \approx 0.6587$.

5. $M = 1 - 3e^{-2}$; $\overline{x} = \dfrac{2-10e^{-2}}{1-3e^{-2}}$; $I_p = 6 - 38e^{-2}$.

7. $M = \ln(2)$; $\overline{x} = \dfrac{1}{\ln(2)}$; $I_p = \dfrac{3}{2}$.

9. $M = 1.5$; $\overline{x} \approx 0.544945$; $I_p \approx 0.90075$.

11. $M = \dfrac{2\pi\rho R^3}{3}$; $\overline{x} = \dfrac{3R}{8}$; x is measured vertically from base.

13. $M = 16\pi + 48\pi e^{-2}$;
$\overline{x} = \dfrac{-16\pi + 208\pi e^{-2}}{16\pi + 48\pi e^{-2}}$.

15. $M = \dfrac{\pi\rho r^2 h}{3}$; $\overline{x} = \dfrac{h}{4}$; x is measured vertically from base.

17. $M = \dfrac{\pi^3}{32}$; $\overline{x} = \dfrac{2\pi^2 - 3}{3\pi}$.

19. $M = \dfrac{\pi + 3\pi e^{-2}}{4}$; $\overline{x} = \dfrac{e^2+15}{3e^2+9}$.

28. (c) $h^* = \dfrac{-M + \sqrt{M^2 + \rho AMH}}{\rho A}$.

Section 8.5

1. $W = 2.1$.

3. $W = -17.85$.

5. $W = -1.2$.

7. $W = 18987.5$ J.

9. $W = 364.96768$ J.

11. $W = 24.5\pi$ MJ (megajoules).

13. $W \approx 1.307$ MJ.

15. $W = 3.528$ MJ.

17. $F = 115718.4\pi$ N.

19. $F \approx 295300$ N.

21. $W = 64.925$ MJ.

23. $W = \dfrac{2\rho g L R^3}{3}$.

24. $W = \rho g L(h+R)\pi R^2$.

26. $F = 160\pi\rho g$.

Section 8.6

3. (a) Approximately 15%.

5. (a) Approximately 0.4%.

   (b) Approximately 16%.

7. 7 minutes.

Section 8.7

1. (b) \$804,240.41. (c) after approximately 274 payments; \$771,055.

3. The monthly growth factor is approximately 1.00413555, resulting in an APR of approximately 5%.

4. (b) After 19.75 years.

5. (a) \$1,127,970.91.

   (b) \$1,111,135.94.

7. (a) $q^* = \dfrac{p_1 - p_0}{a+b}$; $p^* = \dfrac{ap_1 + bp_0}{a+b}$.

   (b) consumers' surplus $= \dfrac{b(q^*)^2}{2}$;

   (c) producers' surplus $= \dfrac{a(q^*)^2}{2}$.

Section 9.1

1. $\left(\frac{1}{2}, \frac{2}{3}, \frac{3}{4}, \frac{4}{5}, \ldots\right)$.

3. $(1, -1, 1, -1, \ldots)$.

5. $\left(\frac{1}{2}, \frac{1}{6}, \frac{1}{12}, \frac{1}{20}, \ldots\right)$.

7. converges to 1.

9. converges to 0.

11. $S_1 = 1$, $S_2 = 3$, $S_3 = 6$, $S_4 = 10$.

13. $S_1 = 0.5$, $S_2 = 0.75$, $S_3 = 0.875$, $S_4 = 0.9375$.

Section 9.2

1. diverges; p-test or integral comparison test.

3. converges; comparison test.

5. converges; alternating series test.

7. converges; comparison test.

9. converges; ratio test.

11. diverges; terms fail to go to zero.

13. converges; comparison test.

15. converges; ratio test.

17. converges; geometric series.

19. diverges; ratio test.

Section 9.3

1. $R = 1$.

3. $R = 1$.

5. $R = \infty$.

7. $R = 4$.

9. $R = 0$.

11. $R = 1/3$.

13. $R = 1/4$.

15. $R = 1$.

17. $R = 2$.

19. $R = \infty$.

21. (c) $J'(x) = \sum\limits_{n=1}^{\infty} \dfrac{(-1)^n 2n x^{2n-1}}{2^{2n}(n!)^2}$;

$J''(x) = \sum\limits_{n=1}^{\infty} \dfrac{(-1)^n 2n(2n-1)x^{2n-2}}{2^{2n}(n!)^2}$.

(d) reindex J so that

$$J(x) = \sum_{n=1}^{\infty} \frac{(-1)^{n-1} x^{2n-2}}{2^{2n-2}(n-1)!}.$$

Then show that $x^2 J'' + x J' + x^2 J$ is equal to

$$\sum_{n=1}^{\infty} \frac{(-1)^n x^{2n}}{2^{2n}(n!)^2}\left[-4n^2 + 2n + 2n(2n-1)\right].$$

Section 9.4

1. $2 + \dfrac{(x-4)}{4} - \dfrac{(x-4)^2}{64} + \dfrac{(x-4)^3}{512} + \cdots$.

3. $1 + \dfrac{x}{2} - \dfrac{x^2}{8} + \dfrac{x^3}{16} + \cdots$.

5. $1 - (x-1) + (x-1)^2 - (x-1)^3 + \cdots$.

7. $1 - 2x + 2x^2 - \dfrac{4x^3}{3} + \cdots$.

9. $x - \dfrac{x^3}{3} + \dfrac{x^5}{5} - \dfrac{x^7}{7} + \cdots$.

11. e.

13. $\dfrac{\sqrt{3}}{2}$.

15. e^2.

24. $\sinh(x) = \sum\limits_{n=0}^{\infty} \dfrac{x^{2n+1}}{(2n+1)!}$.

$\qquad = x + \dfrac{x^3}{3!} + \dfrac{x^5}{5!} + \cdots$

25. $\cosh(x) = \sum\limits_{n=0}^{\infty} \dfrac{x^{2n}}{(2n)!}$.

$\qquad = 1 + \dfrac{x^2}{2!} + \dfrac{x^4}{4!} + \cdots$

Section 9.5

1. $2x - \dfrac{4x^3}{3} + \dfrac{4x^5}{15} + \cdots$; $R = \infty$.

3. $(2x-1) - \dfrac{(2x-1)^2}{2} + \dfrac{(2x-1)^3}{3} - \dfrac{(2x-1)^4}{4}$;

$R = 1/2$.

5. $1 + 2x - 2x^2 + 4x^3 - 10x^4$; $R = 0.25$.

7. $1 - 2x^2 - 2x^4 - 4x^6 - 10x^8 + \cdots$; $R = 1/2$.

9. $\dfrac{1}{2} + \dfrac{x}{16} + \dfrac{3x^2}{256} + \dfrac{5x^3}{2048} + \cdots$; $R = 4$.

11. $1 - (x-1) + (x-1)^2 - (x-1)^3 + \cdots$;

$R = 1$.

13. (a) $1 - x^2 + x^4 - x^6 + \cdots$; $R = 1$.

(b) $x - \dfrac{x^3}{3} + \dfrac{x^5}{5} - \dfrac{x^7}{7} + \cdots$; $R = 1$.

15. (b)

$$F(z) = Gm\left[\frac{M + 2\pi a\lambda}{z^2}\right.$$

$$\left. + 2\pi a\lambda\left(-\frac{3a^2}{2z^4} + \frac{15a^4}{8z^6} - \frac{35a^6}{16z^8} + \cdots\right)\right]$$

Section 10.1

1. $y(t) = \dfrac{1}{\tan(t) + C}$, $C = 1/3$.

2. $y(t) = 3 + Ce^{t^2/2 - 2t}$, $C = 2$.

3. $y(t) = 3\sin(t^2 + C)$, $C = 0$, solution valid for $t \in \left(-\sqrt{\tfrac{\pi}{2}}, \sqrt{\tfrac{\pi}{2}}\right)$.

4. $y(t) = Ce^{-t^2}$, $C = 7$.

5. $y(t) = 2\tan^{-1}(Ae^t)$, $C = 1$.

6. $y(t) = \dfrac{4Ce^{8t}-4}{1+Ce^{8t}}$, $C = -3e^{-8}$.

7. Approximately 33 hours, 2 minutes, 6 seconds.

8. 696m/s.

9. $$m(t) = \frac{r}{\alpha} + \left(m_0 - \frac{r}{\alpha}\right)e^{-\alpha t},$$
$T = \frac{\ln(2)}{\alpha}$.

10. (b) $u(t) = R + (u_0 - R)e^{-kt}$.
(c) Approximately 3 hours, 36 minutes, 57 seconds.

11. (a) 7585 ft.
(b) 524 s. after the parachute opens.

12. (a) $v(t) = v_c \tan(C - v_c kt)$, $C = \tan^{-1}(v_0/v_c)$.

(b) $v(t) = v_c \dfrac{Ae^{-2v_c kt}-1}{Ae^{-2v_c kt}+1}$, $A = \dfrac{v_c + v_0}{v_c - v_0}$.

(c) $v(t) = v_c \dfrac{e^{-2v_c k(t-t_0)}-1}{e^{-2v_c k(t-t_0)}+1}$, $t \geq t_0$.

(d) $t_0 = \dfrac{C}{v_c k}$.

(e) $x(t) = \dfrac{1}{k}\ln\left|\dfrac{\cos(C - v_c kt)}{\cos C}\right|$, for $t \leq t_0$; and x(t) =
$$\frac{1}{k}\left[\ln(\sec C) + \ln\left(\frac{1+e^{2C-2v_c kt_0}}{1+e^{2C-2v_c kt}}\right)\right]$$
$-v_c(t - t_0)$, for $t > t_0$.

(f) Approximately 12.7 seconds.

13. (a) $$x(t) = \left(\sqrt{H} - \frac{kt}{2A}\right)^2,$$
$T = \dfrac{2A\sqrt{H}}{k}$.

(b) $$x(t) = \left(\sqrt{H} - \frac{5kH^2 t}{2A}\right)^{2/5},$$
$T = \dfrac{2A\sqrt{H}}{5kH^2}$.

14. $Q(t) = 600\left(1 - e^{-t/10}\right)$.

Section 10.2

1. $y(t) = \dfrac{5}{3} + Ce^{-3t}$, $C = 16/3$.

2. $y(t) = t + \dfrac{C}{t}$, $C = 2$.

3. $y(t) = 2 + Ce^{-\sin(t)}$, $C = 1$.

4. $y(t) = 3 + Ce^{-3\tan(t)}$, $C = 7$.

5. $y(t) = 4\tan(t) + C\sec(t)$, $C = 7$.

6. (a)
$$\begin{aligned} v(t) &= \frac{-g}{\alpha} \\ &\quad + \frac{a}{\alpha^2+\omega^2}\left[\alpha\cos(\omega t) + \omega\sin(\omega t)\right] \\ &\quad + Ce^{-\alpha t}, \end{aligned}$$
$$C = v_0 + \frac{g}{\alpha} - \frac{a\alpha}{\alpha^2+\omega^2}.$$
(b)
$$\begin{aligned} x(t) &= -\frac{gt}{\alpha} \\ &\quad + \frac{a}{\alpha^2+\omega^2}\left[\frac{\alpha}{\omega}\sin(\omega t) - \cos(\omega t)\right] \\ &\quad - \frac{C}{\alpha}e^{-\alpha t} + D, \end{aligned}$$
$$D = \frac{C}{\alpha} + \frac{a}{\alpha^2+\omega^2}.$$

7. (a) $Q(t) = \dfrac{1200t + 30t^2}{20+t}$. (b) 1260g.

8. (a) $v(t) = \dfrac{rgt^2 - 2M_0 gt}{2(M_0 - rt)} + \dfrac{v_0 M_0}{M_0 - rt}$.

(b)
$$\begin{aligned} x(t) &= \frac{-gt^2}{4} + \frac{gM_0 t}{2r} + \frac{gM_0^2}{2r^2}\ln\left(1 - \frac{rt}{M_0}\right) \\ &\quad - \frac{M_0 v_0}{r}\ln\left(1 - \frac{rt}{M_0}\right). \end{aligned}$$

9. (a) $Q_1(t) = 4000(1 - e^{-t/20})$.

(b) $Q_2(t) = 4000(1 - e^{-t/20}) - 200te^{-t/20}$.

10. (a) $h(t) = 80e^{-kt}$.

(b) $u(t) = \dfrac{80r}{r-k}\left(e^{-kt} - e^{-rt}\right)$.

(d) $T = \dfrac{1}{k-r}\ln\left(\dfrac{k}{r}\right)$.

14. (b) $y(t) = Ke^{Ae^{-kt}}$, $A = \ln(y_0/K)$.

15. (c) $\dfrac{dI}{dS} = \dfrac{\beta - \alpha S}{\alpha S}$.

(d) $I(t) = I_0 + \dfrac{\beta}{\alpha}\ln\left(\dfrac{S(t)}{S_0}\right) - [S(t) - S_0]$.

(e) approximately 40.91% of the population.

Section 10.3

7. $8 + 4\log_{99}(891)$, or approximately 1:54PM.

9. (d) $y(t) = \dfrac{K + ATe^{-\alpha t}}{1 + Ae^{-\alpha t}}$, $\alpha = k(K-T)/K$; $A = \frac{K-y_0}{y_0-T}$.

(f) $T = \dfrac{50\ln(9.1)}{9}$.

13. (a) CO: $4 - x(t)$; H_2: $4 - 3x(t)$; H_2O: $x(t)$.

(c) $x(t) = 1 - e^{-16t}$.

Section 10.4

1. $y(2) \approx 1.83642578125$.
2. $y(2) \approx 2.8425300989$.
3. $y(2) \approx 5.91769927$.
4. $y(2) \approx 2.6965224878$.
5. $y(2) \approx 31.77756266$.
6. $y(1) \approx 0.55833959$.
7. $y(1) \approx 3.52767637$.
8. $y(1) \approx 3.81430786$.
9. $y(1) \approx 1.51401221$.
10. $y(1) \approx 1.76537522$.

Appendix D

Table of Integrals

Basic Forms

$$\int u^n \, du = \frac{u^{n+1}}{n+1} + C, \; n \neq 1 \tag{D.1}$$

$$\int \sin u \, du = -\cos u + C \tag{D.2}$$

$$\int \cos u \, du = \sin u + C \tag{D.3}$$

$$\int e^u \, du = e^u + C \tag{D.4}$$

$$\int \frac{du}{u} = \ln|u| + C \tag{D.5}$$

$$\int u dv = uv - \int v du \tag{D.6}$$

Rational Functions

$$\int \frac{du}{a+bu} = \frac{1}{b} \ln|a+bu| + C \tag{D.7}$$

$$\int \frac{u}{a+bu} du = \frac{u}{b} - \frac{a \ln|a+bu|}{b^2} + C \tag{D.8}$$

$$\int \frac{du}{a^2+u^2} = \frac{1}{a} \tan^{-1} \frac{u}{a} + C \tag{D.9}$$

$$\int \frac{du}{(a^2+u^2)^2} = \frac{u}{2a^2(u^2+a^2)} + \frac{1}{2a^3} \tan^{-1}\left(\frac{u}{a}\right) + C \tag{D.10}$$

$$\int \frac{du}{a^2-u^2} = \frac{1}{2a} \ln\left|\frac{a+u}{a-u}\right| + C \tag{D.11}$$

$$\int \frac{du}{(a+bu)(c+du)} = \frac{1}{ad-bc} \ln\left|\frac{c+du}{a+bu}\right| + C \tag{D.12}$$

Roots

$$\int \frac{du}{\sqrt{a^2-u^2}} = \sin^{-1}\frac{u}{a}+C \tag{D.13}$$

$$\int \sqrt{a^2-u^2}du = \frac{u}{2}\sqrt{a^2-u^2}+\frac{a^2}{2}\sin^{-1}\frac{u}{a}+C \tag{D.14}$$

$$\int \frac{du}{\sqrt{a^2+u^2}} = \ln\left|u+\sqrt{a^2+u^2}\right|+C \tag{D.15}$$

$$\int \sqrt{a^2+u^2}du = \frac{u}{2}\sqrt{a^2+u^2}+\frac{a^2}{2}\ln\left|u+\sqrt{a^2+u^2}\right|+C \tag{D.16}$$

Logarithmic Functions

$$\int \ln u\, du = u\ln u-u+C \tag{D.17}$$

$$\int u\ln u\, du = \frac{1}{2}u^2\ln u-\frac{u^2}{4}+C \tag{D.18}$$

Trigonometric Functions

$$\int \tan u \, du = -\ln|\cos u| + C \qquad \text{(D.19)}$$

$$\int \cot u \, du = \ln|\sin u| + C \qquad \text{(D.20)}$$

$$\int u \sin u \, du = \sin u - u \cos u + C \qquad \text{(D.21)}$$

$$\int u \cos u \, du = \cos u + u \sin u + C \qquad \text{(D.22)}$$

$$\int \sin^2 u \, du = \frac{u}{2} - \frac{\sin(2u)}{4} + C \qquad \text{(D.23)}$$

$$\int \cos^2 u \, du = \frac{u}{2} + \frac{\sin(2u)}{4} + C \qquad \text{(D.24)}$$

$$\int \tan^2 u \, du = \tan u - u + C \qquad \text{(D.25)}$$

$$\int \cos^2 u \, du = -\cot u - u + C \qquad \text{(D.26)}$$

$$\int \sin^3 u \, du = -\frac{1}{3}\left(2 + \sin^2 u\right) \cos u + C \qquad \text{(D.27)}$$

$$\int \cos^3 u \, du = \frac{1}{3}\left(2 + \cos^2 u\right) \sin u + C \qquad \text{(D.28)}$$

$$\int \tan^3 u \, du = \frac{1}{2} \tan^2 u + \ln|\cos u| + C \qquad \text{(D.29)}$$

$$\int \sin(au)\sin(bu) \, du = \frac{\sin[(a-b)u]}{2(a-b)} - \frac{\sin[(a+b)u]}{2(a+b)} + C \qquad \text{(D.30)}$$

$$\int \cos(au)\cos(bu) \, du = \frac{\sin[(a-b)u]}{2(a-b)} + \frac{\sin[(a+b)u]}{2(a+b)} + C \qquad \text{(D.31)}$$

$$\int \sin(au)\cos(bu) \, du = -\frac{\cos[(a+b)u]}{2(a+b)} - \frac{\cos[(a-b)u]}{2(a-b)} + C \qquad \text{(D.32)}$$

Inverse Trigonometric Functions

$$\int \sec u\, du = \ln|\sec u + \tan u| + C \tag{D.33}$$

$$\int \csc u\, du = \ln\left|\tan\frac{u}{2}\right| + C = -\ln|\csc u + \cot u| + C \tag{D.34}$$

$$\int \cot u\, du = \ln|\sin u| + C \tag{D.35}$$

$$\int \sec^2 u\, du = \tan u + C \tag{D.36}$$

$$\int \csc^2 u\, du = -\cot u + C \tag{D.37}$$

$$\int \sec u \tan u\, du = \sec u + C \tag{D.38}$$

$$\int \csc u \cot u\, du = -\csc u + C \tag{D.39}$$

Exponentials and Trigonometric Functions

$$\int e^{au} \sin bu\, du = \frac{e^{au}}{a^2+b^2}(a\sin bu - b\cos bu) + C \tag{D.40}$$

$$\int e^{au} \cos bu\, du = \frac{e^{au}}{a^2+b^2}(a\cos bu + b\sin bu) + C \tag{D.41}$$

$$\int \frac{ae^{-bu}-1}{ae^{-bu}+1}\, du = -\frac{2}{b}\ln(1+ae^{-bu}) - t + C \tag{D.42}$$

Hyperbolic Trigonometric Functions

$$\int \sinh u\, du = \cosh u + C \tag{D.43}$$

$$\int \cosh u\, du = \sinh u + C \tag{D.44}$$

$$\int \operatorname{sech}^2 u\, du = \tanh u + C \tag{D.45}$$

$$\int \operatorname{csch}^2 u\, du = -\coth u + C \tag{D.46}$$

$$\int \tanh u\, du = \ln(\cosh u) + C \tag{D.47}$$

$$\int \coth u\, du = \ln|\sinh u) + C \tag{D.48}$$

Appendix E

Gaussian Distribution

Table E.1 contains data values for the cumulative distribution function for the normal distribution,

$$C(Z) = \frac{1}{\sqrt{2\pi}} \int_{-\infty}^{Z} e^{-z^2/2} dz.$$

	0	0.01	0.02	0.03	0.04	0.05	0.06	0.07	0.08	0.09
0	0.5000	0.5040	0.5080	0.5120	0.5160	0.5199	0.5239	0.5279	0.5319	0.5359
0.1	0.5398	0.5438	0.5478	0.5517	0.5557	0.5596	0.5636	0.5675	0.5714	0.5753
0.2	0.5793	0.5832	0.5871	0.5910	0.5948	0.5987	0.6026	0.6064	0.6103	0.6141
0.3	0.6179	0.6217	0.6255	0.6293	0.6331	0.6368	0.6406	0.6443	0.6480	0.6517
0.4	0.6554	0.6591	0.6628	0.6664	0.6700	0.6736	0.6772	0.6808	0.6844	0.6879
0.5	0.6915	0.6950	0.6985	0.7019	0.7054	0.7088	0.7123	0.7157	0.7190	0.7224
0.6	0.7257	0.7291	0.7324	0.7357	0.7389	0.7422	0.7454	0.7486	0.7517	0.7549
0.7	0.7580	0.7611	0.7642	0.7673	0.7704	0.7734	0.7764	0.7794	0.7823	0.7852
0.8	0.7881	0.7910	0.7939	0.7967	0.7995	0.8023	0.8051	0.8078	0.8106	0.8133
0.9	0.8159	0.8186	0.8212	0.8238	0.8264	0.8289	0.8315	0.8340	0.8365	0.8389
1	0.8413	0.8438	0.8461	0.8485	0.8508	0.8531	0.8554	0.8577	0.8599	0.8621
1.1	0.8643	0.8665	0.8686	0.8708	0.8729	0.8749	0.8770	0.8790	0.8810	0.8830
1.2	0.8849	0.8869	0.8888	0.8907	0.8925	0.8944	0.8962	0.8980	0.8997	0.9015
1.3	0.9032	0.9049	0.9066	0.9082	0.9099	0.9115	0.9131	0.9147	0.9162	0.9177
1.4	0.9192	0.9207	0.9222	0.9236	0.9251	0.9265	0.9279	0.9292	0.9306	0.9319
1.5	0.9332	0.9345	0.9357	0.9370	0.9382	0.9394	0.9406	0.9418	0.9429	0.9441
1.6	0.9452	0.9463	0.9474	0.9484	0.9495	0.9505	0.9515	0.9525	0.9535	0.9545
1.7	0.9554	0.9564	0.9573	0.9582	0.9591	0.9599	0.9608	0.9616	0.9625	0.9633
1.8	0.9641	0.9649	0.9656	0.9664	0.9671	0.9678	0.9686	0.9693	0.9699	0.9706
1.9	0.9713	0.9719	0.9726	0.9732	0.9738	0.9744	0.9750	0.9756	0.9761	0.9767
2	0.9772	0.9778	0.9783	0.9788	0.9793	0.9798	0.9803	0.9808	0.9812	0.9817
2.1	0.9821	0.9826	0.9830	0.9834	0.9838	0.9842	0.9846	0.9850	0.9854	0.9857
2.2	0.9861	0.9864	0.9868	0.9871	0.9875	0.9878	0.9881	0.9884	0.9887	0.9890
2.3	0.9893	0.9896	0.9898	0.9901	0.9904	0.9906	0.9909	0.9911	0.9913	0.9916
2.4	0.9918	0.9920	0.9922	0.9925	0.9927	0.9929	0.9931	0.9932	0.9934	0.9936
2.5	0.9938	0.9940	0.9941	0.9943	0.9945	0.9946	0.9948	0.9949	0.9951	0.9952
2.6	0.9953	0.9955	0.9956	0.9957	0.9959	0.9960	0.9961	0.9962	0.9963	0.9964
2.7	0.9965	0.9966	0.9967	0.9968	0.9969	0.9970	0.9971	0.9972	0.9973	0.9974
2.8	0.9974	0.9975	0.9976	0.9977	0.9977	0.9978	0.9979	0.9979	0.9980	0.9981
2.9	0.9981	0.9982	0.9982	0.9983	0.9984	0.9984	0.9985	0.9985	0.9986	0.9986
3	0.9987	0.9987	0.9987	0.9988	0.9988	0.9989	0.9989	0.9989	0.9990	0.9990

Table E.1: Cumulative distribution function data for the normal distribution.

Index

www.ingramcontent.com/pod-product-compliance
Lightning Source LLC
Chambersburg PA
CBHW071312210326
41597CB00015B/1207
* 9 7 8 1 9 4 1 0 4 3 1 4 1 *